Direct Alcohol Fuel Cells

Horacio R. Corti • Ernesto R. Gonzalez
Editors

Direct Alcohol Fuel Cells

Materials, Performance, Durability and Applications

Springer

Editors
Horacio R. Corti
Departamento de Física de la Materia Condensada
Centro Atómico Constituyentes, CNEA, and INQUIMAE (Universidad de Buenos Aires – CONICET)
Buenos Aires, Argentina

Ernesto R. Gonzalez
Instituto de Química de São Carlos-USP
São Carlos, São Paulo, Brazil

ISBN 978-94-007-7707-1 ISBN 978-94-007-7708-8 (eBook)
DOI 10.1007/978-94-007-7708-8
Springer Dordrecht Heidelberg New York London

Library of Congress Control Number: 2013955722

Printed on acid-free paper

Springer is part of Springer Science+Business Media (www.springer.com)

Preface

Fuel cell energy generation is one of the fundamental electrochemical discoveries of the nineteenth century (William Grove 1839), which became a technology reality at the middle of the twentieth century mainly due to the development of alkaline fuel cells for the aerospace industry. All the fuel cells developed during the last century, from the low-temperature proton exchange membrane (PEM) fuel cells to the high-temperature solid oxide fuel cells, are fueled with hydrogen, although the high-temperature fuel cells can electrooxidize hydrocarbons, such as methane.

Low-temperature PEM fuel cells have reached a certain degree of maturity, and they are nowadays commercialized for some specific applications and are thought to compete with advanced batteries for powering future electric vehicles. One of the major drawbacks in the massive utilization of PEM fuel cells is associated with the storage and distribution of hydrogen, both as compressed gas and as a liquid at very low temperatures. The use of hydrogen-containing room-temperature liquids is an interesting alternative that has been proposed since long ago, alcohols being the best candidates to replace hydrogen due to their availability and low cost. However, the electrooxidation of the simplest alcohol, methanol, in a PEM fuel cell is still far from the electrochemical efficiency obtained with hydrogen. Thus, the direct methanol fuel cells technology is still in its early development stage, and the use of higher alcohols, such as ethanol or glycerol, is restricted to laboratory prototypes.

Several books and book chapters dealing with components of direct methanol fuel cells, mainly electrocatalysts and membranes, have been written during the last decade, some of them providing excellent revisions of the state of the art of the research and technology in the area. This book intends to cover a gap existing in the fuel cell literature, including not only the new electrocatalyst and ionomeric membranes developed in the last 5 years for direct methanol fuel cells, but extending the survey to materials for fuel cells fed with higher alcohols. We include recently developed anodic electrocatalysts for alcohol oxidation, alcohol- and CO-tolerant cathodes, and low alcohol permeability membranes, along with other relevant components of direct alcohol fuel cells, such as catalyst supports, gas diffusion layers, and bipolar plates with flow fields adapted to the two-phase fluid dynamics present in a direct alcohol fuel cell.

In the first chapter, we introduce the concept of methanol economy, as an alternative to the most popular but still elusive hydrogen economy, and we also provide a brief historical description of fundamental research on electrochemical oxidation of methanol and the development of the first alkaline direct methanol fuel cells more than 60 years ago. The operating principles of PEM and alkaline direct alcohol fuel cells are analyzed, as well as their components, configuration, and operation modes, with a final remark on the state of the art of the technology.

The second chapter is dedicated to analyzing the mechanism of methanol electroxidation on platinum, bi- and tri-metallic platinum materials, with a special emphasis on the effect of CO and formic acid oxidation on these electrocatalysts.

The electroxidation mechanism of ethanol and ethylene glycol is discussed in Chap. 3, along with an analysis of the use of these alcohols in fuel cells, and the search for tolerant cathodes. Chapter 4 is devoted to the electroxidation of 3-carbon alcohols (mainly glycerol), including fundamental studies and a comparison of direct glycerol fuel cells with hydrogen and direct ethanol fuel cells. The recent developments of platinum and non-platinum-based catalysts as methanol- and ethanol-tolerant oxygen reduction catalysts for direct alcohol fuel cells are described in Chap. 5.

Chapter 6 deals with the description of different membranes used in direct alcohol fuel cells. Firstly, the properties of Nafion and its inorganic and organic composites are analyzed, focused on the proton conductivity and alcohol permeability, which determine the alcohol selectivity of the modified Nafion membranes. Then, a number of alternative non-fluorinated proton conducting membranes, including sulfonated polyimides, poly(arylene ether)s, polysulfones, poly(vinyl alcohol), polystyrenes, and acid-doped polybenzimidazoles, are described in relation to their selectivity in comparison to Nafion. The chapter includes a comprehensive summary of the relative selectivity of these membranes and their performance in direct alcohol fuel cells. Anion exchange membranes for alkaline direct alcohol fuel cells are also reviewed.

The analysis of carbon materials used as catalyst support, gas diffusion layer, and current collector and bipolar plates is performed in Chap. 7. A number of carbon materials including carbon blacks, nanotubes, nanofibers, and structured porous carbon materials are analyzed and compared as catalyst support in direct methanol fuel cells. Commercial and non-commercial gas diffusion layers are described along with the role of the mesoporous layer on the fuel cell performance. Finally, synthetic graphite and carbon composites used as current collector and bipolar plates are discussed, focusing on their mechanical and electrical properties and production costs.

The physical modeling for fundamental understanding, diagnostics, and design of new materials and operation conditions of direct alcohol fuel cells is addressed in Chap. 8. The modeling of mechanisms and processes at atomistic and single-cell levels is critically reviewed, as well as the theoretical challenges. A multiscale model developed by the author is described, and a possible extension to direct alcohol fuel cells is discussed.

Last but not least, fundamental issues for the deployment of direct alcohol fuel cells—application niches, costs, and durability—are introduced in Chap. 9. The major drawbacks for commercialization, such as miniaturization, product balance, cost reduction, and lifetime extension, are addressed.

It is our expectation that this book will provide updated information on the direct alcohol fuel cell principles, materials, and performance for students, electrochemists, chemical engineers, and material scientists, as well as public interested in cleaner energy sources.

A major contribution of this book has been to collect the results of many experimental studies cited in more than 1,400 references, and partially reflected in 140 figures selected. We salute all the researchers who have contributed to the advancement of our knowledge in this field of fuel cell science and technology.

We would like to express our gratitude to Springer for inviting us to lead this book project, particularly to Sonia Ojo and Karin de Bie for their support and guidance in the preparation and editing phases. Finally, it is a great pleasure to thank all the young coauthors that have enthusiastically contributed in seven chapters of the book, providing their reliable expertise and critical views. H.R.C dedicates this book to Catalina and Martin, his grandchildren born this year, who hopefully will grow up in a more sustainable world.

Buenos Aires, Argentina
São Carlos, São Paulo, Brazil
August 2013

Horacio R. Corti
Ernesto R. Gonzalez

Contents

Chapter 1
Introduction to Direct Alcohol Fuel Cells

Horacio R. Corti and Ernesto R. Gonzalez

Abstract Fuel cells are strongly linked to renewable energies, particularly to the so-called "Hydrogen Economy". For decades the development of fuel cells able to convert hydrogen and oxygen in electrical energy with water as unique byproduct has motivated huge activity in fundamental and applied electrochemistry.

In this chapter we introduce the concept of methanol economy and discuss its status and perspectives. To be a reality the methanol and other alcohol economies depend on the development of alcohol feed fuel cells, whose components, operation modes and general performance are analyzed.

1.1 World Energy Consumption: Current Status and Tendencies

> The Stone Age did not end for lack of stone, and the oil age will end long before the world runs out of oil.
>
> Sheikh Ahmed Yamani (former Saudi Arabia's Oil Minister)

The world energy consumption at the beginning of this decade was 12 billion tonnes of oil equivalent (toe),[1] and 87 % is generated by burning fossil fuels (oil 33.7 %, natural gas 23.6 % and coal 29.7 %) [1], which are non-renewable and increase the CO_2 content of the atmosphere, considered as the major man-made

[1] Toe is defined as the amount of energy released when one tonne of crude oil is burned. It is equivalent to 41.87 GJ or 11.63 MWh.

H.R. Corti (✉)
Departamento de Física de la Materia Condensada, Centro Atómico Constituyentes, CNEA, and INQUIMAE (Universidad de Buenos Aires – CONICET), Buenos Aires, Argentina
e-mail: hrcorti@cnea.gov.ar

E.R. Gonzalez
Instituto de Química de São Carlos-USP, São Carlos, SP, Brazil
e-mail: ernesto@iqsc.usp.br

H.R. Corti and E.R. Gonzalez (eds.), *Direct Alcohol Fuel Cells: Materials, Performance, Durability and Applications*, DOI 10.1007/978-94-007-7708-8_1,

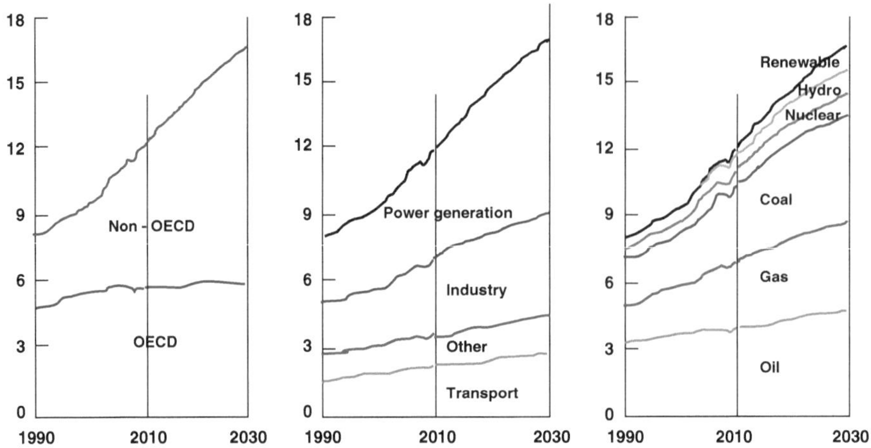

Fig. 1.1 Power demand in billions toe by region (*left*), primary use (*middle*), and fuel (*right*) (Adapted from Ref. [2])

cause of global warming. Nuclear and hydro contributed with almost 12 %, while renewable accounted for only 1.3 % of the energy global demand. Certainly, there are countries with an energy matrix having a significant lower dependence of fossil fuel, like France with more than 80 % nuclear or Brazil with 35 % hydro.

Around a half of the world energy demand corresponds to developed countries that originally signed the Convention on the Organisation for Economic Cooperation and Development (OECD), including United States, Canada, Japan and European Community. An important fraction of the other half is consumed by Russia and the fast-developing countries; particularly China and India, being coal, the worst fossil fuel in terms of CO_2 emissions, and the most vastly employed in these highly populated regions.

An important fraction, close to 40 % of the fuel resources is used for power generation, while industry and transport demand around 30 % and 20 %, respectively [1, 2]. Projections of energy demands for the next decades is a subject that concerns oil-related industries [2–4], governments and energy planners. British Petroleum projections till 2030 [2] are summarized in Fig. 1.1, where the growth in energy demand is basically modulated by the growth of the population and gross domestic product (GDP), mainly due to the contribution of the non-OECD countries. By 2030, 1.3 billion more people will need energy; and the world income is expected to roughly double the 2011 level.

The raise of fossil fuel prices to record levels in real terms over the past decade inevitably lead to supply responses, by development and deployment of new technologies across a range of energy sources. Thus, the "shale revolution", first for gas and then for oil, will allow account for almost a fifth of the increase in global energy supply to 2030 [2]. Simultaneously, high prices for fossil fuels will also support the expansion of biomass renewable energy supply, accounting for 17 % of the increase in global energy supply by 2030. Hydro and nuclear together will contribute for another 17 % of the growth. Nevertheless, conventional fossil fuel supplies are still required to satisfy expanding energy demand, providing almost half the growth in energy supply till 2030. It is worth remarking that energy demand would increase almost three times if the projections had not taken into consideration a declining of the energy intensity (the amount of energy consumed per unit of GDP), probably in response to the statement by Serge Latouche[2] that "anyone who thinks that infinite growth is consistent with a finite planet is either crazy, or an economist".

The projections till 2050 [3] show a decline in the use of oil and gas after 2030, compensated by a large contribution from coal, biomass and renewables. It is expected that by 2050 the global coal industry be two and a half times as large as in 2000. But intensive coal use by USA, China and India in the next decades could face restrictions due to environmental pressure groups, which are very active in developed countries and protest's pockets started to be visible in China, the main user of coal in the world and, consequently, the major emissor of CO_2.

The increasing use of biofuels to follow demand growth is not free of problems. First-generation of biofuels compete with food production, driving up world market prices, especially in those countries that use maize as a staple [3]. The demand of biofuels by developed countries indirectly encouraged poorer nations to destroy large areas of rainforests and habitats in order to grow palm oil and sugar cane. Moreover, the potential benefits of the use of biofuels on the environment is compensated because these land use produce significant releases of CO_2 stored in the soils, and because the extra N_2O entering the atmosphere as a result of using N-fertilizers to produce crops for biofuels. As a source for NO_x, N_2O plays a major role in stratospheric ozone chemistry (296 times more greenhouse effect than CO_2). Crutzen et al. [5] have concluded that the use of N-fertilizers to grow crops for the production of biofuels could exacerbate the already huge challenge of getting global warming under control.

A second generation of biofuels, produced from the woody parts of plants, including waste products such as stalks and leaves from plants grown for food production, could help to avoid partially the above-mentioned problems and it is estimated that this second generation of biofuels contribution will surpass the first generation beyond 2030 [3].

1.2 Methanol Versus Hydrogen Economy

Hydrogen is thought as the fuel of the future. Although it is the most abundant element in cosmos, it is not available in Earth in its stable elementary form, H_2. That means one have to spend energy to obtain it from water or fossil fuels.

The way we produce H_2 determines if it is "green" or "black" hydrogen. Green hydrogen refers to that obtained from renewable energy sources, like wind or solar,

[2] French economist and sociologist born in 1940, author of "Degrowth Economics".

by splitting water in its components. Black hydrogen, on the other hand, is that obtained from hydrogen-containing hydrocarbons or biomass by chemical reactions (reforming). Nowadays reforming of natural gas is the less expensive and preferred method to produce molecular hydrogen.

Therefore, a really zero-emission hydrogen economy would be true once the cost of splitting water in electrolysers could compete with the actual reforming processes. Obviously, the economical balance will depend not only on technical issues such as increasing the efficiency of photovoltaic conversors or wind-power generators, and developing new water electrolysis technologies, but on political and environmental factors.

With this in mind a "methanol economy" has been proposed by Olah and coworkers [6] as an alternative, or a previous step, to a future "hydrogen economy". Methanol (CH_3OH), is the simplest, safest, and easiest way to store and transport hydrogen as a liquid hydrocarbon. It is prepared almost exclusively from a mixture of CO and H_2 (syn-gas) resulting from the incomplete combustion of natural gas or coal. Methanol economy involves not only the use of methanol as a fuel and gasoline additive, but also its conversion to synthetic hydrocarbons and their products and materials, which are essential part of our life [6].

In fact, the use of methanol as additive for gasoline has attracted interest over a number of years during times of critical shortage. However, much more attention is paid to the use of ethanol obtained from corn and sugar cane which, for instance, satisfy a small part of the transportation fuel requirements in United States and Brazil. The fundamental difference between the production of bioethanol and biomethanol is that the latter does not rely on agriculture or on diminishing fossil fuels [6].

As quoted by Carl Winter [7]: technologies compete not fuels!. In relation to methanol economy, the associated technologies are: (1) new and more efficient ways of producing methanol, such as the oxidative conversion of natural gas, or the hydrogenative recycling of CO_2 to methanol from industrial exhausts (including power plants) or even from the air itself; (2) use of methanol in a new generation of direct methanol fuel cells (DMFC) (mainly for transportation); (3) the use of methanol as the raw material for producing synthetic hydrocarbons and their products.

The production of methanol by recycling excess CO_2 (from fossil power plants or atmospheric) will contribute to mitigate global warming due to the greenhouse effect.

The price of methanol produced from syn-gas is shown in Fig. 1.2. The peaks observed in 2009–2010 are attributed to the global financial crisis. Increasing methanol worldwide demand during the last years (55 million metric tons in 2011) determines the averaged raise in price.

The use of methanol or ethanol as fuels started with the replacement of steam engines for farm machinery and train locomotives and, at the end of the nineteenth century, alcohols were used in internal combustion engines (ICE). During the first decade of the twentieth century, alcohols lost the race against gasoline-powered automobiles, even when they released less contaminants, a fact that deserved not concern at that time. The main reason for that was economic: alcohols could not

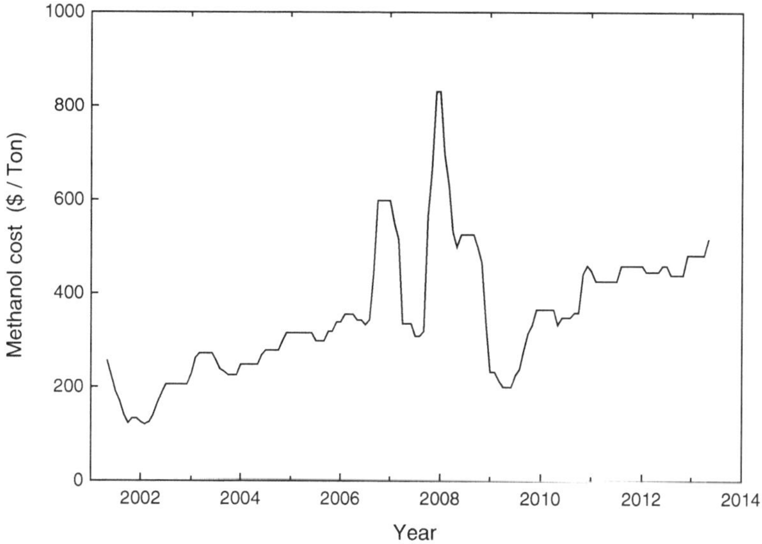

Fig. 1.2 Evolution of the methanol price during the last 12 years (Source: Methanex)

compete with gasoline, especially in U.S., where its price was much lower since it was obtained from the abundant reserves of oil.

Despite the energy density of methanol is half than that of gasoline, it has a higher octane rating and its mixture with air can be compressed to a smaller volume before ignition. As a result methanol as a fuel for ICE is more efficient than gasoline. Methanol higher heat of vaporization, along with a faster and more complete combustion in an ICE, as compared to gasoline, compensates in part its lower energy density.

Regarding the use of methanol as a fuel for transportation in DMFC, it could have the same driving distance as gasoline in an ICE because the fuel cell works with an efficiency that is twice that of the ICE or diesel engine [8].

Plausibly and perspectively, during twenty-first century will take place the transition toward the "second solar civilization".[3] The first civilization, when mankind used exclusively renewable energies, end at the beginning of the industrial revolution. During the second solar civilization, starting this century, mankind would benefit from efficient and sustainable energies, employing all sort of solar-based renewable technologies. Probably there will be no silver bullets in terms of fuels or energy technologies, but a combination of new technologies currently under development (including fuel cells).

[3] This concept was introduced in 1975 by M. K. Hubbert, who claims that the fossil era (coal, oil, gas) is just a "blink of eyes between the first and second solar civilizations".

1.3 Fuel Cells Technology

Fuel cell technology is one of the key issues for the development of hydrogen or methanol economies, as claim several authors [6, 7]. While fuel cell is a nineteenth century invention by William Grove (1839), its evolution toward a technology starts in the twentieth century.

Alkaline fuel cells (AFC) using aqueous KOH as electrolyte were the first type of fuel cells with practical applications at the beginning of the last century [9] but the formation of carbonate in the liquid electrolyte due to the CO_2 contamination in the oxidant gas stream has limited its application to systems running with pure oxygen, such as the fuel cells used by the NASA in the 1960' Apollo space program, and currently used in the shuttle missions [10].

Other types of cells emerged also in the past century, classified mainly for the electrolyte used to conduct current inside the cell. Phosphoric acid fuel cells (PAFC), which operate at about 200 °C, reached its technology maturity at the end of twentieth century but the cost per kW is higher than the target expected by the developers, so its applications are restricted to the military sector in units of 200 kW. Higher temperature fuel cells using molten carbonates (MCFC) and solid oxide (SOFC) electrolytes are attractive for stationary uses because they could use low purity hydrogen as a fuel (CO is a reactant in these cell, not a catalyst poison) and the waste heat can be used for cogeneration, yielding higher efficiencies.

Proton exchange membrane fuel cells (PEMFC) are low temperature fuel cells, operating at temperatures below 100°C, being Nafion the sulfonated perfluopolymer commonly used as electrolyte. The development of low temperature fuel cells (AFC and PEMFC) showed periods of exponential growth, measured in terms of publications, as described by Lamy et al. [8]. The space era, during the 1960s, marked the maturity of the AFC technology, while the energy era, stimulated by the energy crisis in 1973, was the starting of the PEMFC technology. Then it follows the military era in 1980s and 1990s, where the drop in the oil price shifted the interest of PEMFC applied research toward specific military applications. Finally at the early 1990s the incentives for fuel cells were triggered by the need to reduce environmental pollution.

1.4 Research History of the DMFCs

Müller reported the first detailed studies of the electrochemical oxidation of methanol and other organic compounds at platinum anodes in aqueous alkaline electrolytes in 1922 [11]. Some years later Tanaka [12] extended these studies for methanol electroxidation on Pt, Pd, Rh, and Au in a sodium hydroxide electrolyte. Based on these studies Kordesch and Marko [13] first recognized in 1951 the possibility of building an alkaline DMFC, as that schematized in Fig. 1.3, whose major trouble was that the cross-leakage of the methanol to the air cathode damages the noble metal

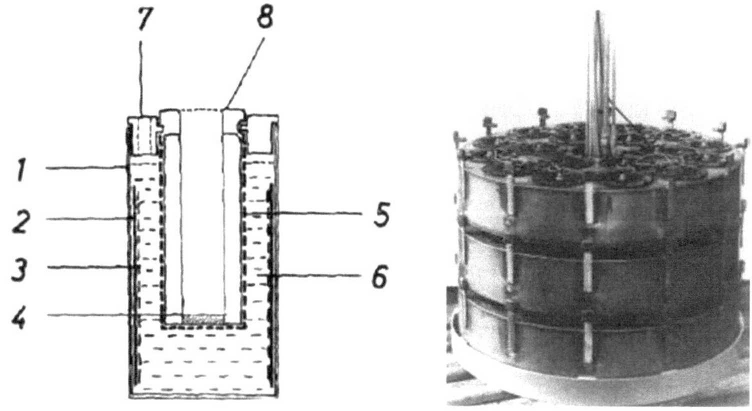

Fig. 1.3 *Left*: Scheme of the first alkaline DMFC built by Kordesch and Marko showing Can (−) (*1*), Pt-fuel electrode (*2*), Screen (*3*), Pt-cat. Carbon Tube (*4*), Separator (*5*), CH_3OH + KOH (*6*), Filler (*7*), Metal Cap (+) (*8*). *Right*: part of a 60-W methanol–air battery with cylindrical air diffusion electrodes built by Brown Boveri (Reproduced from Refs. [13] and [18] with permission

catalysts. Several prototypes of alkaline DMFC were reported during the 1950s, using porous nickel [14], platinized porous carbon impregnated with Ag-Co-Al mixed oxides [15] anodes, yielding maximum power below 3 mW.cm^{-2}. The first methanol-fed PEM fuel cell was that reported by Hunger [16] with MEAs built by pressing porous catalysts on an anionic exchange membrane.

During the 1960s, alkaline DMFC were built with powers in the range of 60–750 W [17–19]. Vielstich [17] reported some laboratory cells fed with 10N KOH + 4.5 M methanol using Pt as anodic catalyst and polyethylene hydrophobized active carbon electrodes as cathode. It was the base of the alkaline DMFC built by Brown Boveri at Switzerland. The 6 V–10 A cell (Fig. 1.3), that powered a flashing sea buoy, contained ten cylindrical cells, each cell had 18 pairs of electrodes connected in parallel [18]. Murray and Grimes [19] at Allis-Chambers built a 40-cell DMFC stack having nickel sheet anode coated with a Pt-Pd catalyst and a porous nickel cathode impregnated with Ag, and it was fed with 5 M KOH. The cell deliver a maximum power of 750 W at 9 V and 83 A. Exxon-Alsthom in France developed alkaline and buffer electrolyte DMFC technology during the 1960s, but the problems associated with carbonation of the electrolyte, due to the complete methanol oxidation to CO_2, forced to pulled out the research in the late 1970s [20, 21].

Pioneering studies of the methanol oxidation reaction (MOR) were conducted under acidic conditions since the late 1960s and 1970s probing that the kinetics of the MOR is slower in acid media compared with the alkaline electrolyte [21]. Cathro [22] studied bimetallic Pt-Sn catalyst, while Janssen and Moolhuysen [23] recognized Sn and Ru ad-atoms on Pt as promising anodic catalyst for DMFC. Finally, Watanabe and Motoo [24] showed that Pt-Ru solid alloys were the catalysts with larger potentialities.

Shell Research Centre (England) and Hitachi Research Laboratories (Japan) developed DMFC with concentrated sulfuric acid in the late 1950s and early 1980s, respectively. The interest of Shell in methanol fuel cells was focused on its application to road transportation [20], and the first approach was to use new thin electrodes in alkaline media and then they operated a 5 kW stack which run with hydrogen generated by methanol reforming. In the 1960s Shell researchers realized that direct oxidation of methanol in sulfuric acid would be a better option and discovered that Pt-Ru electrodes were much more active than pure Pt. By using high Pt loading (10 mg.cm^{-2}) they were able to construct in 1963 a 300 W stack having a quite low performance and an excessive cost per kW.

During the 1970s the efforts of Shell were devoted to reduce costs by increasing catalysts efficiency. Thus, improved electrodes were developed, two orders of magnitude more active, reaching maximum power densities of 25 mW.cm^{-2}. Hitachi built larger stacks, up to 5 kW, with similar maximum power densities. In 1981, after the oil industry recovered from the 1973 crisis, Shell abandoned the development of DMFC because the drop of the oil prices moved the costs target for DMFC even further.

Other acid DMFC systems have been investigated during the late 1970s and early 1980s for different applications, such as military communication systems (US Army) and electric wheelchairs (Royal Institute of Technology, Stockholm), and golf carts driven by DMFC in conjunction with lead acid batteries (Hitachi, Japan) [25].

After that, advances in DMFC mostly hibernated until 1992, when DuPont's Nafion™ solid polymer electrolyte membrane was found to be an excellent proton conducting media [26]. Thus, in 1994, a collaborative work involving the Jet Propulsion Laboratory, University of Southern California, and Giner Inc., demonstrated a Nafion-based DMFC [27], which could deliver power outputs up to 150 mW.cm^{-2} at 90 °C.

The aim of this book is to discuss the development, characterization, fuel cell performance, and applications of the PEM direct methanol fuel cells. Regarding the last issue, the lower power density and higher costs of DMFCs compared to conventional hydrogen-feed PEM fuel cells, have reoriented their short-term applications towards portable electronics.

The change of paradigm due to the development of efficient proton-exchange membranes is well documented by the exponential increases in the number of publications related to DMFC and other others alcohols since 1994, as illustrated in Fig. 1.4.

The growth of interest in DMFC and DEFC seems to have reached a steady state during the last years; with almost a factor 5 more works on methanol respect to ethanol. It can be also observed an incipient interest for ethylene glycol and propanol as a direct fuel in PEM fuel cell since 2000, and a small but raising number of studies exploring glycerol as a new liquid fuel since 5 years ago.

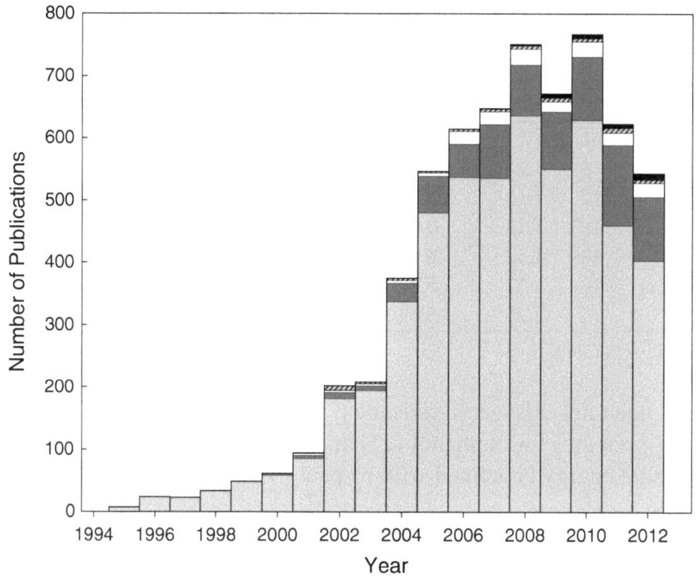

Fig. 1.4 Number of publications on DAFC according to Scopus database. From bottom to top: methanol (*gray*), ethanol (*dark gray*), ethylene glycol (*white*), propanol (*crossed gray*), glycerol (*black*)

1.5 Operating Principle of DAFCs

1.5.1 Thermodynamics of Alcohol Oxidation

The general hemi-reactions for a $C_nH_{2n+1}OH$ mono-alcohol that oxidizes completely to CO_2 in a DAFC employing an acid electrolyte are:

$$\text{Anode}: C_nH_{2n+1}OH + +2n - 1\ H_2O \rightarrow n\ CO_2 + 6n\ H^+ + 6n\ e^-$$
$$\text{Cathode}: 3n/2\ O_2 + 6n\ H^+ + 6n\ e^- \rightarrow 3n\ H_2O$$

and the global reaction is: $C_nH_{2n+1}OH + 3n/2\ O_2 \rightarrow nCO_2 + n + 1\ H_2O$. In the case of $C_2H_4(OH)_2$ (ethylene glycol) and $C_3H_5(OH)_3$ (glycerol) the number of electrons involved in the reaction are 10 and 14, respectively.

The standard potential of the cell, E^0, can be calculated by resorting to Nernst's equation:

$$E^0 = -\frac{\Delta G^0}{nF} \tag{1.1}$$

where ΔG^0 is the Gibbs free energy of the reaction with pure oxygen at 1 bar.

Table 1.1 summarizes the values of the standard potential of DAFCs at 25 °C (referred to the NHE), which are close to the values for the hydrogen/oxygen fuel

Table 1.1 Number of electrons, standard potential, theoretical specific energy and density energy, pure compound capacity, and theoretical energy conversion efficiency for alcohol oxidation in DAFC

Fuel	n	E^0 (V)	W_s (kWh.kg^{-1})	W_e (kWh.l^{-1})	C (Ah.kg^{-1})	η (%)
Hydrogen	1	1.23	32.80	0.18[a]	26,668	83
Methanol	6	1.21	6.07	4.82	5,019	97
Ethanol	12	1.15	8,03	6.28	6,981	97
1-Propanol	18	1.13	9.07	7.28	8,027	97
2-Propanol	18	1.12	8.99	7.07	8,027	97
Ethylene glycol	10	1.22	5.27	5.87	4,318	99
Glycerol	14	1.25	5.09	6.42	4,074	

[a] H_2 storage at 70 bar and 25 °C

cell, and it does not change significantly in the operating range 20–130 °C and 1–3 bar (O_2 pressure) for methanol [21] and other alcohols.

The specific energy is defined with respect to the mass of fuel, that is,

$$W_s = -\frac{\Delta G^0}{M} \tag{1.2}$$

where M is the molecular weight of the alcohol. The values of E_{sp}, shown in Table 1.1, are higher for more energetic (high n or ΔG^0) and light alcohols [28], leading to the order 1- and 2-propanol > ethanol > methanol > ethylene glycol > glycerol. E_{sp} for hydrogen is much higher than for alcohols, but this advantage disappears when energy density, storage and distribution issues are considered. Energy density is defined as the energy per unit of volume,

$$W_e = -\frac{\rho \Delta G^0}{M} \tag{1.3}$$

ρ being the alcohol density. All the alcohols exhibit energy densities much higher than that of natural gas compressed at 200 bar, hydrogen storage as a liquid or in metal hydride systems ($\approx$ 2 kWh.l^{-1}), and the state of the art Li-ion batteries ($\approx$ 0.3 kWh.kg^{-1}), although they lie below the energy density of gasoline and diesel (between 7 and 9 kWh.l^{-1}).

The pure compound capacity accounts for the amount of charge that can be released by the fuel, it is independent of E^0 and proportional to the ratio n/M. Therefore exhibits the same trend as E_{sp}. The theoretical energy conversion efficiency is the ratio between the reversible (maximum) electric work that can be obtained by electrochemical oxidation of the fuel and the heat released by direct combustion with oxygen, that is:

$$\varepsilon_{th} = \frac{\Delta G^0}{\Delta H^0} \tag{1.4}$$

where ΔH^0 is the standard enthalpy of combustion of the fuel. The average ε_{th} for alcohols is 97 %, much higher than the value for the energy conversion using hydrogen in a PEM fuel cell.

1.5.2 Anatomy of a PEM DAFC and Alcohol Crossover

Figure 1.5 schematizes a DAFC using a proton exchange membrane (PEM) as electrolyte. Both sides of the proton conducting membrane (1) are in contact with a thin layer of the catalysts (2, 3) supported on a electron conducting carbon material. The gas diffusion layers (GDL) (4) allow the liquid fuel and the oxygen (air) to distribute homogeneously over the whole catalyst area, while it should permit the vent of CO_2 in the anode and water in the cathode. Graphite current collector plates with flowing channels (5) are used to collect the electrons generated in the anode and inject electrons to the cathode, and direct the fuel/oxidant to the corresponding GDL. Adequate seals (6) avoid fuel and oxidant leakage from the cell and a load (7), which limit the current circulation through the cell, completes the system.

One of the main differences of DAFC with H_2/O_2 PEM fuel cells is the biphasic flow in the anode exhaust due to the feed of aqueous alcohol and the production of CO_2, which is only partially soluble in the liquid.

The flow of protons and the crossover of alcohol (ROH) from the anode to the cathode are also indicated in Fig. 1.5. Alcohol crossover decreases the current through the cell because part of the fuel permeates through the membrane without reacting in the anode. Thus, a fuel efficiency (or fuel utilization) can be defined as:

$$\varepsilon_f = \frac{i}{i + i_{cr}} \tag{1.5}$$

where i is the actual current density in the fuel cell, and i_{cr} is the crossover current, that is the current which is not generated because of the alcohol crossover. In other words, $i + i_{cr}$ represents the theoretical total current calculated from the Faraday law on the basis of complete alcohol consumption, that is $n = 6$ for methanol or $n = 12$ for ethanol.

The actual current could be lower than the theoretical one if the alcohol oxidation does not proceed completely toward the CO_2 formation. Thus, the electrooxidation of methanol could stop at the formaldehyde stage:

$$CH_3OH \rightarrow HCHO + 2H^+ + 2e^- \quad (n = 2)$$

or at the formic acid stage:

$$CH_3OH \rightarrow HCOOH + 4H^+ + 4e^- \quad (n = 4)$$

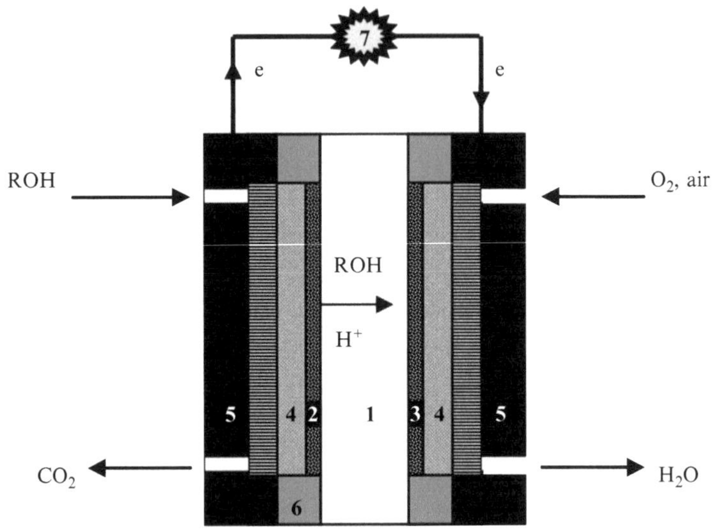

Fig. 1.5 Scheme of a PEM direct alcohol fuel monocell showing: (1) proton conducting membrane: (2) anodic catalyst layer; (3) cathodic catalyst layer; (4) gas diffusion layers; (5) current collector with flow channels; (6) seals; (7) electric load

Therefore, the experimental number of electrons involved in the methanol (or any other alcohol) oxidation, n_{exp}, can be lower that the theoretical, n_{th}, and a Coulombic efficiency is defined as [8]:

$$\varepsilon_c = \frac{n_{\mathrm{exp}}}{n_{th}} \tag{1.6}$$

In practice is not simple to determine whether the current reduction in the cell is due to alcohol crossover, incomplete oxidation, or both. A way to quantify the contribution of both effects is to measure the methanol concentration in the anode exhaust (for determining the amount of methanol oxidized and permeated), and the concentration of methanol and CO_2 in the cathode exit. Part of the CO_2 at the cathode exhaust is due to parasitic oxidation of methanol at the cathode, while the rest is a consequence of CO_2 crossover from the anode. The last can be determined by a half-cell experiment by flowing hydrogen through the cathode, to avoid methanol oxidation, in such a way that all the CO_2 measured in the cathode outlet stream must have crossed the membrane from the anode [29].

1.5.3 Alkaline DAFC

As mentioned above the primitive direct methanol fuel cells were built with aqueous alkaline hydroxide solutions as electrolyte, but the production of CO_2

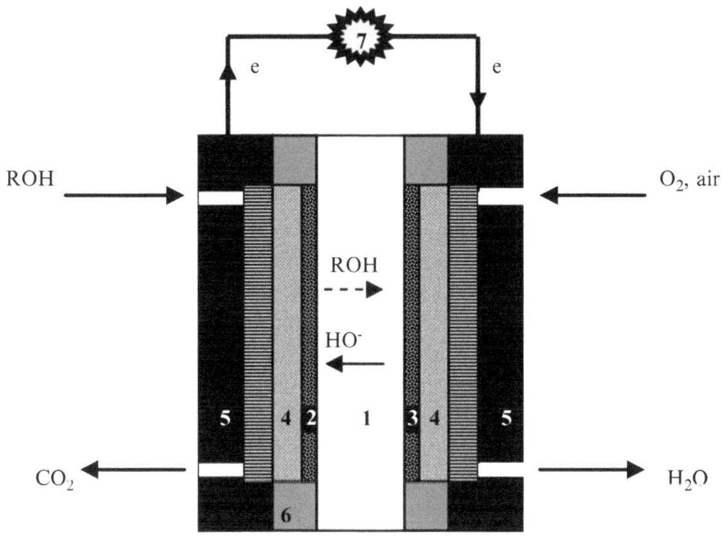

Fig. 1.6 Scheme of a AEM direct alcohol fuel monocell. The numbers have the same meaning that in Fig. 1.5

leading to carbonate/bicarbonate formation can be a problem in this kind of alkaline direct alcohol fuel cells (ADAFC). The use of anion exchange membranes (AEM) could be a solution for the development of alkaline direct alcohol fuel cells (ADAFC), considering that the anodic oxidation of methanol and other alcohols and the oxygen reduction is more feasible in alkaline media than in acidic ones [30].

Figure 1.6 shows a DAFC using an anion exchange membrane as electrolyte, where the hemi-reactions are:

$$\text{Anode}: C_nH_{2n+1}OH + 6n\ HO^- \rightarrow n\ CO_2 + 4n + 1H_2O + 6n\ e^-$$
$$\text{Cathode}: 3n/2\ O_2 + 3n\ H_2O + 6n\ e^- \rightarrow 6n\ HO^-$$

and the global reaction is: $C_nH_{2n+1}OH + 3n/2\ O_2 \rightarrow nCO_2 + n + 1\ H_2O$, as in the case of DAFC in acidic media. In this case the charge through the membranes is transported by the hydroxide ions from the cathode to the anode and, consequently, there is no electro-osmotic flow of alcohol, and the only contributions to alcohol crossover are diffusion and hydraulic permeation due to concentration and pressure gradients, respectively.

The less corrosive nature of an alkaline environment ensures a longer durability of the ADAFC, and the faster kinetics of the ORR allows the use of non-noble, low-cost, metal electrocatalysts. Thus, ADAFC meets a number of potential advantages compared to PEM DAFC, which in turn triggers the resurgence of interest in this kind of fuel cell technology.

1.5.4 Membrane Electrode Assembly and Three Phase Region

The membrane-electrode assembly (MEA) comprises the PEM and both catalyst layers deposited on it, as shown in Fig. 1.7 (left). Some authors include the GDLs as part of the MEA when this is prepared by depositing the catalysts on the GDL, forming the so-called gas diffusion electrodes (GDE), as shown in Fig. 1.7 (right).[4]

When the MEA is prepared by fixing the catalyst ink (containing Nafion as a binder) directly on the membrane, several procedures have been described [31], including rolling [32], spraying [33] or printing [34]. The final hot-pressing is essential to get a good contact between the electrode and the membrane or between the electrode and the GDL, according to the preparation procedure, in order to minimize resistance losses. The "three phases region" (TPR), that is, the boundaries where reactant (alcohol in the anode and oxygen in the cathode), electrolyte (PEM), and supported metallic catalyst are in contact is the hearth of the cell, because these three phases should be in contact for allowing the electrochemical reactions.

Figure 1.8 shows a scheme of the microstructure of the catalyst layer, where the presence of TPRs allows the simultaneous flows of H^+ and electrons toward the PEM and GDL, respectively. Note that the thickness of the proton conducting membrane covering the carbon-supported catalyst nanoparticles is of the order of nanometers, depending of the concentration of proton conducting binder (usually Nafion) in the ink used to prepare the MEA.

The effective electro-catalytic area of the electrode is the sum of all the nano-sized TPRs in the catalyst layer, and it should be as large as possible in order to obtain an optimal relation between the current and the load of catalyst (usually expressed as mA/mg Pt).

1.5.5 Operation Modes and Cell Efficiency

There are different kinds of DAFC operation conditions depending of the way the fuel and the oxidant (oxygen/air) are fed into the cell. In complete "active" fuel cells the liquid fuel (neat alcohol or aqueous solution) is pumped and gas is compressed, using auxiliary pumps and blowers, in order to improve mass transport and reduce concentration polarization losses in the system. On the other hand, in complete "passive" DAFC the alcohol reaches the anode catalyst layer by natural convection and the cathode breathes oxygen directly from the air. A number of intermediate options have been also studied and tested.

While the applications of DAFC in transportation is not so critical in relation to the use of auxiliary components for active operation, micro-DAFC aiming to

[4] In this book we use both definitions for MEA, according to the authors preferences.

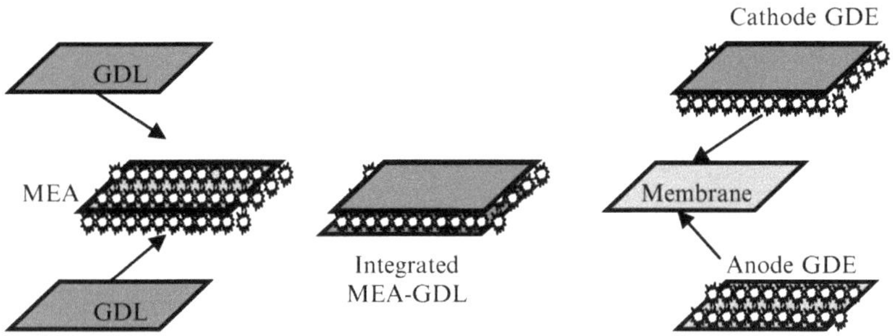

Fig. 1.7 Preparation of MEA by depositing the catalyst layers directly on the membrane (*left*) or on the GDL (*right*), followed by hot pressing of the components

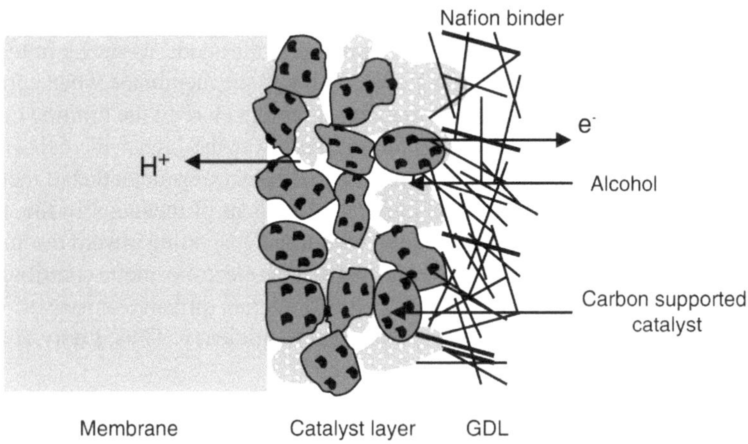

Fig. 1.8 Scheme (not at scale) of the catalyst layer showing TPRs where coexist fuel (alcohol), PEM and catalyst

replace Li-ion battery packs require the minimization of any auxiliary component in order to optimize the energy density of the fuel + fuel cell + auxiliaries pack.

The three main barriers to the reduction of superior energy density of existing DMFC have been [35]: (1) high methanol permeability of the commonly used ionomeric membranes; (2) the balance of water challenge, due to the need of removing water produced in the cathode and needed in the anode; (3) the moderate power density of the DMFC, due to the slow kinetics of methanol oxidation.

The first barrier, related to methanol permeation across the membrane, can be quantified by the fuel utilization (Eq. 1.5). Thus, for a Nafion membrane, the rate of methanol permeation is equivalent to 100–200 $mA.cm^{-2}$, which is similar to the current density normally achievable, meaning that fuel utilization could be as low as 0.5. Higher fuel utilization can be obtained with less permeable membranes, but usually the price to be paid is a higher membrane resistance, which reduce the voltage efficiency defined by:

$$\varepsilon_E = \frac{E(i)}{E_r} \tag{1.7}$$

where $E(i)$ is the cell potential under working conditions, which is lower than the reversible cell potential, E_r, due to the cathodic and anodic overpotentials and the ohmic drop as a consequence of the cell resistance, R_c, mainly due to the membrane:

$$E(i) = E_r - (|\eta_c| + |\eta_a| + R_c i) \tag{1.8}$$

It has been pointed out [35] that the crossover current density in Eq. 1.5 is not a constant, but it is rather sensitive function of the cell current density, decreasing at higher cell current due to the lowering of methanol crossover. This can be explained by resorting to the methanol concentration profiles in the anode, as shown in Fig. 1.9.

Methanol (alcohol) concentration gradient through the membrane when current is close to zero is much higher than the gradient at current close to the limiting current, and methanol crossover due to diffusion will be lower the higher the cell current. However, part of the methanol crossover is due to the electro-osmotic flux of methanol due to proton solvation. At high cell current the amount of methanol in the anode-membrane interface is low, but the amount of proton transporting toward the anode is proportional to the current, hence it is expected that the electro-osmotic contribution to alcohol crossover be less sensitive to the cell current than diffusive crossover.

There is no single definition of the overall cell efficiency. Thus, Lamy et al. [8] defined the cell efficiency, $\varepsilon_{\mathrm{cell}}$, as the product:

$$\varepsilon_{cell} = \varepsilon_{th}\varepsilon_c\varepsilon_E \tag{1.9}$$

while Aricó et al. [21], consider the overall cell efficiency in terms of the fuel utilization, instead of the Coulombic efficiency:

$$\varepsilon'_{cell} = \varepsilon_{th}\varepsilon_c\varepsilon_E \tag{1.10}$$

Indeed, a more complete definition of overall cell efficiency should include the four efficiencies defined by Eqs. 1.4, 1.5, 1.6 and 1.7. An additional term should be included for active cells accounting for the amount of energy consumed by auxiliaries, such as pumps, blowers, etc., if the system efficiency is calculated.

1.5.6 Other DAFC Configuration and Operation Modes

If catalysts can be developed that are active only for the MOR and ORR then a mixed-reactants DMFC (MR-DMFC) could be designed where aqueous methanol

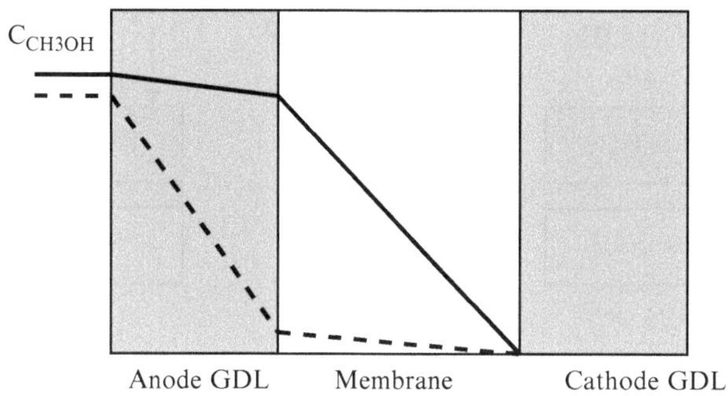

Fig. 1.9 Methanol concentration profile at zero cell current (*solid line*), and close to the anode limiting current (*dashed line*) (Adapted from Ref. [35])

fuel and oxygen or air are mixed together before entering the cell, and the oxidation of methanol and oxygen reduction take place on the respective electrodes due to catalyst selectivity [36, 37], as shown in Fig. 1.10a. The advantage of MR-DMFCs is that can use a simple porous electrolyte instead of Nafion membranes, and simplify the sealing of reactants/products delivery structure [38].

Shukla et al. [39] developed a MR-DMFC using carbon supported RuSe cathode and PtRu anode, which reached 45 $mW.cm^{-2}$ at 0.2 V. However, there were little efforts in the open literature reporting improvements in this type of cells, probably because of the difficulty of obtaining selective catalysts for ORR and MOR and also for the oxidation of higher alcohols.

Wilkinson and coworkers [40, 41] introduced the concept of a mixed-reactant direct liquid fuel cell where the air cathode was substituted with a metal-ion redox couple. This type of cell, which in the case of methanol is called mixed-reactant direct methanol redox fuel cell (MR-DMRFC) has the advantage of cathode selectivity, avoiding Pt group metals as cathodic catalysts, minimize flooding in the cathode and allows the use of larger fuel concentrations.

A scheme of the MR-DMRFC is shown in Fig. 1.10b, where the redox couple is Fe^{2+}/Fe^{3+}, which is mixed with methanol and water and feed to the cathode, while supplying nothing to the anode.

The cathodic reaction in the MR-DMRFC is: $6\ Fe^{3+} + 6\ e^- \rightarrow 6\ Fe^{2+}$, and the overall reaction: $CH_3OH + H_2O + 6\ Fe^{3+} \rightarrow CO_2 + 6\ H^+ + 6\ Fe^{2+}$. The redox couple could be regenerated electrochemically by flowing oxygen through the cathode to oxidize Fe^{2+} to Fe^{3+}, while forming water.

Wilkinson and coworkers also tested a membraneless direct methanol fuel cell, built by replacing the PEM by an open spacer filled with liquid electrolyte (aqueous H_2SO_4). Reasonable power densities (close to 10 $mW.cm^{-2}$) were reached, and the main aim of the new architecture, to control the total power output of the cell, was achieved by disrupting the TPR with a physical guard on or within the electrode assembly [42].

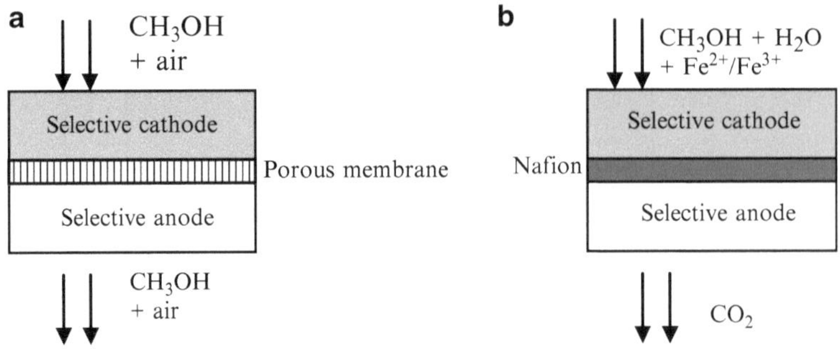

Fig. 1.10 Scheme of a MR-DMFC (**a**) and a MR-DMRFC (**b**)

1.6 DAFC Components

This section is devoted to a brief description of the main components of DAFC as an introduction to the most exhaustive analysis in Chaps. 2, 3, 4, and 5 for electrocatalysts for methanol, ethanol, and higher alcohols, in Chap. 6 for proton exchange and alkaline membranes, and Chap. 7 for carbonous materials used as catalysts support, gas diffusion layers and bipolar plates.

1.6.1 Catalysts

The catalysts used to oxidize alcohols take as a first consideration the mechanism of the oxidation reaction. The mechanism will be illustrated with reference to the methanol oxidation on platinum but, with different stoichiometries, the principles can be extended to other alcohols. The first step is believed to be the adsorption of the alcohol:

$$CH_3OH + Pt \rightarrow Pt\text{-}CH_3OH_{ads}$$

followed by platinum catalyzed dehydrogenation reactions leading to:

$$Pt\text{-}CH_3OH_{ads} \rightarrow Pt\text{-}CO_{ads} + 4\,H^+ + 4\,e^-$$

On the same catalyst, but at different potential, the dissociative adsorption of water takes place with the formation of oxygenated species:

$$Pt\text{-}H_2O \rightarrow Pt\text{-}H_2O_{ads} \rightarrow Pt\text{-}OH_{ads} + H^+ + e^-$$

this is followed by a Langmuir-Hinshelwood type of mechanism leading to the formation of carbon dioxide:

$$\text{Pt-CO}_{ads} + \text{Pt-OH}_{ads} \rightarrow 2\,\text{Pt} + CO_2 + H^+ + e^-$$

The problem with this mechanism is that the formation of Pt-OH_{ads} requires a high anodic potential (>0.75 V) which reduces significantly the potential difference between anode and cathode. In order to overcome this problem, several approaches have been investigated but the most common is to use alloys Pt with other metals to form bimetallic or even multimetallic materials.

Two modes of action are expected from metals alloyed to platinum: (1) the so-called bifunctional effect. When the second metal is more oxophilic it will form the oxygenated species from water at lower potentials, thus diminishing the operational potential of the anode and (2) the intrinsic effect. The second metal may diminish the electron density of the 5d band of platinum weakening the adsorption strength of poisonous species (like CO) and making them easier to oxidize.

The oxidation of methanol on platinum has been widely studied for several decades. In acid media platinum is still the preferred catalyst, although it is seldom used alone. In order to increase the benefits of the two effects just discussed nanostructured catalysts usually involve one or two other metals besides platinum. These metals are chosen to increase the bifunctional effect by forming oxygenated species from water at lower potentials or to increase favorably the intrinsic electronic effect. However, in order that this last effect is present the second metal must form an alloy with platinum, a fact that is disregarded in many works in electrocatalysis. This is an important consideration because many works study the effect of submonolayers of foreign metals or device screening methods to test quickly a large number of combinations leaving undetermined the question of whether there is, or there is not, alloy formation. The bottom line is that many works give the atomic metal composition of the catalysts without giving information on the formation of alloys and/or segregated phases, which is essential to discuss the activity of the catalyst. Needless to say, if an alloy is not formed it will be difficult to observe the benefits of the intrinsic mechanism.

Concerning bimetallic platinum based catalysts platinum-ruthenium has been the most investigated material for the oxidation of alcohols and other small molecules. Since the early work by Watanabe and Motoo [24] Pt-Ru has been widely investigated in all possible physical forms from bulk alloys to nanoparticles. Despite the fact that most works attribute the highest activity for oxidation to a 50 at.% composition it is important to bear in mind that other factors may be crucial in determining the activity. Among these it may be mentioned the degree of alloy, the homogeneity in the distribution of the nanoparticles on the support and the nature of the support itself. The main effect of the presence of ruthenium seems to be lower potential needed to form oxygenated species in comparison to platinum. But it is interesting to remark that the kinetics of the reaction to form those oxygenated species is slower on ruthenium.

One important aspect that must be considered is the meaning of activity of a catalyst used for the oxidation of alcohols. Very often the current obtained with a given electrochemical technique, or even in fuel cell experiments, is taken as indicative of the activity of the catalyst. However, other aspects must be considered. It is crucial that the electrode potential at which the current is evaluated be sufficiently low so that, when combined with the cathodic reaction the resulting potential is meaningful for fuel cell operation. Another important consideration is that of the products formed in the oxidation reaction particularly for the oxidation of alcohols containing two or more carbon atoms. The oxidation of ethanol is a good example of this problem. Ideally, the oxidation should proceed to carbon dioxide as final product with the production of 12 electrons. However, several catalysts for ethanol oxidation give acetic acid as the final product with the production of only 4 electrons. Even if these catalysts are considered good because they produce high enough current densities it remains the problem that it will be difficult to implement an operational fuel cell that produces acetic acid as the final product. The problem becomes even more complex when the oxidation of higher alcohols is considered, glycerol being a good example of this. The variety of oxidation products may become too large. True, a justification for the R&D of these systems may be to use the fuel cell to obtain products that may be of interest in several applied areas. Thus, the fuel cell becomes at the same time an energy converter and a chemical reactor.

Although electrocatalysis is a very important factor in determining the feasibility of a direct alcohol fuel cell a problem of the same magnitude, if not bigger, is the crossover of the alcohol through the membrane to the cathode side. Due to the high potential of the cathode the crossover alcohol gets oxidized and this creates a mixed potential lower than the normal cathode potential. So efficiency is reduced because part of the anodic reagent is lost and also the cathode operates at a lower potential. To this we may add that the crossover alcohol may interfere with the cathodic reaction. This effect depends on the magnitude of the crossover and the nature of the alcohol, so generalizations cannot be made. But as a general rule, metals that when alloyed to platinum reduce the performance for alcohol oxidation are good candidates to reduce the effect on the cathode of the crossover alcohol.

The faster kinetics of alcohol oxidation and oxygen reduction reactions in alkaline direct alcohol fuel cells opens up the possibility of using less expensive Pt-free catalysts, as nickel, gold, palladium and their alloys [30]. Thus, the cost of ADAFC could be potentially lower compared to the acid DAFC technology if non-precious metal alloys are used for the alcohol electrooxidation, being the nanoparticulated Ni-Fe-Co alloys developed by Acta (Italy) with the trade name of HYPERMEC a good example.

1.6.2 Catalysts Support

Another aspect that has emerged in the last few years in the electrocatalysis of alcohol oxidation is a consideration of the support used for the catalyst nanoparticles. The support may influence the electrocatalysis of the reaction through more than one effect. The support may influence the dispersion of the catalyst nanoparticles and usually an increase in the dispersion will increase the current density because of the increase in the catalyst utilization. The support may also have an effect on the electronic density of the catalyst, affecting the adsorption strength of poisons and intermediates of the reaction. If the support is an oxide it may take part directly in the electrocatalytic process by providing oxygenated species that will act through the bifunctional mechanism. Traditionally, carbon black has been the preferred support used almost universally. It has a sufficiently small particle size, it disperses well the catalyst nanoparticles and it is relatively cheap. However, it has some problems, the most important of which is perhaps the fact that it suffers oxidation and degradation, particularly under the high potential of the cathode. Various treatments have been proposed to improve the properties of carbon black supports. Usually, the treatments are oxidative, seeking an increase of the oxygenated species on the surface. Alternative supports usually involve oxides or carbides, but these may have less electrical conductivity than the carbon blacks, so often a mixture of the alternative support and carbon black is used.

Considering that the electrocatalysis of the oxidation of low molecular weight alcohols in acid media on platinum and platinum group metals has been studied for almost a century it may be said that dramatic improvements in this area should not be expected. However, further studies are necessary because even incremental improvements may have a significant impact on the pursue of making direct alcohol fuel cells a practical reality.

1.6.3 Membranes

Nafion, a sulfonated perfluorinated polymer developed by Dupont, which triggered the development of PEM fuel cells fed with hydrogen was also the PEM used in all the first generation of DMFC and DEFC. Very early in the development of DAFC it was recognized that alcohol crossover due to the relatively large permeability of Nafion to alcohols was a severe limitation to be overcome. In the case of methanol the high permeability can be understood by the preference of Nafion for sorbing methanol instead of water in the binary mixtures.

The first efforts aimed to reduce the methanol permeability of Nafion using inorganic and organic fillers. Among the most common inorganic fillers in Nafion composite membranes are Pd, Pt and Pt/Ru nanoparticles, SiO_2 and functionalized silica, zeolites, clays, and zirconium phosphate. The composites are prepared by the recast method, dispersing filler nanoparticles and Nafion in an appropriate solvent and casting on a plane surface. Alternatively, nanoparticles can be incorporated by in situ sol–gel reaction, where the Nafion membrane is swollen in water and methanol and immersed in a methanol solution containing the nanoparticle precursor, for instance silicon tetraetoxysilane (TEOS) in the case of SiO_2. Less common is the use of layered Nafion membranes where films or nanoparticles are deposited on the surface of a Nafion membrane by in situ reduction of a metal salt precursor, colloid adsorption or by sputtering.

The most studied Nafion composite membranes with organic fillers include blends of Nafion with polypyrrole, polybenzimidazole, poly(vinyl alcohol), polyvinylidene fluoride, polyaniline, sulfonated poly(ether ether ketone), and poly(tetrafluoroethylene).

The strategy of adding inorganic and organic fillers to Nafion was to retard the methanol (alcohol) diffusion by blocking the water channels with nanoparticles or by reducing the alcohol solubility with polymer units with less affinity for alcohols. Nevertheless, as it will be discussed in Chap. 6, although the alcohol relative permeability, that is the ratio between the alcohol permeability (P) through the composite and pristine Nafion is reduced in same cases for a factor greater than 10, this effect is in most of the cases linked to a corresponding reduction in the relative proton conductivity (σ). Therefore, the alcohol membrane selectivity, β, defined as the ratio between the proton and alcohol mobility, that is $\beta = \sigma/P$ [43], for the composites is usually close to the selectivity of Nafion.

The decrease of alcohol permeability and, consequently, of alcohol crossover, even if accompanied by a reduction of the proton conductivity, open the possibility to new strategies of MEAs preparation by choosing the optimal membrane thickness and alcohol concentration, among several parameters, in order to increase DAFC performance. Other beneficial effect of incorporating fillers in Nafion-based membranes, is the chance of increasing the operation temperature of the fuel cell, due to the retention of water, avoiding the dramatic drop of proton conductivity taking place in Nafion above 100 °C.

In the search for PEMs with lower alcohol permeability than Nafion and other perfuorinated membranes, without degradation of the proton conductivity, a number of new polymeric membranes were synthetized and characterized, such as sulfonated polyimides, poly(arylene ether)s, polysulfones, poly(vinyl alcohol), polystyrenes, and acid-doped polybenzimidazoles. A comprehensive discussion of the properties of these alternative membranes is given in Chap. 6, along with those of Nafion and Nafion composites.

It should be mention here that even when the alcohol membrane selectivity of many of the alternative membranes are higher than that for Nafion, the correlation between this parameter and the DAFC performance using those membranes is very poor. The reason for this could be found in the complex mass and charge transport processes taking place in the three phase region, where ion proton conduction through thin films of the PEM is essential for a good electrochemical efficiency.

Anion-exchange membranes are also discussed in Chap. 6 because of the facile electro-oxidation of alcohols in alkaline media and because of the minimization of alcohol crossover in alkaline direct alcohol fuel cells.

1.6.4 Gas Diffusion Layers and Two Phase Flow Phenomena

The role of the gas diffusion or backing layers[5] (DL) in a DAFC (see Fig. 1.5) is to provide current transport from the catalyst layer to the ribs of the current collector as well as optimize the fuel and oxidant distribution to the catalysts. As discussed in

[5] The anode GDL in a DAFC should be named DL because the alcohol is usually fed as an aqueous solution.

Chap. 7, this task is achieved by the use of non-woven carbon paper or woven carbon cloth with very good electronic conductivity.

However, CO_2 is produced in the anode of a DAFC as final product of the alcohol electro-oxidation and, if the CO_2 bubbles cannot be efficiently removed from the surface of the DL, they block the access of alcohol to the catalyst layer decreasing the cell efficiency. On the cathode, the formation of water on the catalyst layer and the electro-osmotic drag of water through the membrane can lead to GDL flooding if the excess water cannot be eliminated, reducing the transport of oxygen/air to the catalyst.

To avoid the problems associated with the biphasic (gas + liquid) flow in the backing layers it is common to form on them a micro-porous layer (MPL), having thickness around 10–30 μm. The purpose of the MPL is to improve the mass transport in the MEA by modifying is hydrophobicity/hydrophilicity, which becomes one of the fundamental parameters to achieve an appropriate mass transport in the DAFC, along with other characteristics such as morphology, porosity, and gas/liquid permeability. Thus, the CO_2 bubble growth and detachment from the DL is strongly correlated with the surface wettability and it has been visualized using transparent DAFC and digital cameras [44, 45].

The elimination of water from the cathode GDL is facilitated by adding a coating of poly(tetrafluoroethylene) to the DL The hydrophobicity of PTFE can expell excess water from the GDL to the air/oxygen flowing inside the channels, avoiding flooding. Park et al. [46] have also shown that the MPL in the cathode DL could contribute to decrease methanol crossover, probably as a consequence of inducing hydraulic pressure on the cathode size [47].

The role of MPL in the anode of DMFC is no so clear. Several authors have tested wet-proofed PTFE and hydrophilic MPL on the anode GDL and attributed the improved performance to enhanced gas transfer in the liquid phase and methanol transport, respectively [48]. Zhang et al. [48] analyzed the effect of hydrophobic MPL (Vulcan carbon black + PTFE) and hydrophilic MPL (Vulcan carbon black + Nafion) on Toray carbon papers, concluding that in low-temperature DMFC the higher power are reached with the hydrophilic MPL, in agreement with the direct visualization of smaller and uniform CO_2 bubbles on the hydrophilic GDLs. On the other hand, Kang et al. [49] reported a high-performance DMFC (78 $mW.cm^{-2}$ at 60 °C) working at high methanol concentration with a hydrophobic (Vulcan carbon black + PTFE) anodic MPL and cathode air humidification.

The pore size distribution of the mesoporous carbon used to fabricate anode GDL play a fundamental role on the enhancing of mass transport. Xing et al. [50] suggested that the significant improvement of the cell performance using a GDL fabricated with mesoporous carbon having hydrophilic small pores and hydrophobic middle pores is the result of benefit the liquid and gas transport simultaneously.

1.6.5 Current Collectors and Flow Fields

Current collector plates in large DMFC stacks are commonly designed and fabricated following the same strategies as for hydrogen fed PEM fuel cells, using graphite as substrate [51, 52], although stainless steel anode and cathode current collectors are used in passive stacks [53–55]. However, the main interest in the current collector materials and flow fields design is triggered during the last decade due to the increasing interest for the development of micro direct alcohol fuel cells (μDAFC), after Kelley et al. fabricated a miniature DMFC [56].

The miniaturization of DMFC is much easier than hydrogen fed PEMFC, because of the lack of fuel storage limitations, and it has a lot a advantages derived from the so-called scaling law. Assuming a characteristic scaling factor, L, the surface/volume ratio scales as L^{-1}, meaning that miniaturization (low L) enhances surface effects. Moreover, at constant flow rate, the pressure drop scale as L^{-3} leading to a high pressure gradient upon miniaturization, which improves the transport of fuel and oxygen to the MEA. The Peclet number, Pe, defined as the ratio of convective and molecular diffusive flows, scales as L, assuming that the fluid mean velocity and diffusion coefficient are constants. A relatively large Pe in a μDAFC may prevent fuel from diffusing toward the MEA, while a small Pe may lead to insufficient water removal in the cathode [57]. Thus, for a given μDAFC design there is an optimal channel dimension for fuel/air transport, which according to Cha et al. [58] corresponds to a cross section around 100 μm × 100 μm.

The current collector is probably the most important component of the μDAFC because not only include the flow channels for the transport of reactants, but also collect electrical charges and provides the structural support for the MEA. The choice of the material for the current collector of small DAFC is crucial due the fact that accounts for about 80 % of the total weight and it must exhibit high electronic conductivity, good chemical stability, high thermal conductivity, high-mechanical strength, and it should be fabricated in large scale at low cost [59]. Three kinds of materials are currently being investigated for use as current collectors in μDAFC: silicon, metals and polymers [44, 59], although low-temperature co-fired ceramic (LTCC) has also been used in μDMFC prototypes [60].

Silicon-based μDMFC and μDEFC first prototypes were fabricated using MEMS micromachining technology, which allows micro flow channels having width and deepness in the order of 100–200 μm and deliver a maximum power density of up to 14 $mW.cm^{-2}$ for active methanol feed [61], and 8 $mW.cm^{-2}$ for passive ethanol feed [62] at room temperature.

Due to the fragility of the silicon substrate there is a severe limitation to compress the cell for sealing and minimize contact resistance between MEA and current collectors. For this reason, Lu and Chang have also used photochemical etching of stainless steel to replace silicon in the fabrication of flow plate/current collector [63], reaching a power density of 34 $mW.cm^{-2}$ at room temperature in an active μDMFC. A 8-μDMFC air-breathing stack feed with pure methanol could deliver up to 20–25 $mW.cm^{-2}$ at room temperature for more than 10 h.

Zhang et al. [59] also used stainless steel to generate current collectors by micro-stamping technology, followed by a deposition of a 500 nm thick TiN layer by magnetron sputtering ion plating technology. A maximum power density of 66 $mW.cm^{-2}$ was obtained at 40 °C in an active μDMFCoperating with 1.5 M

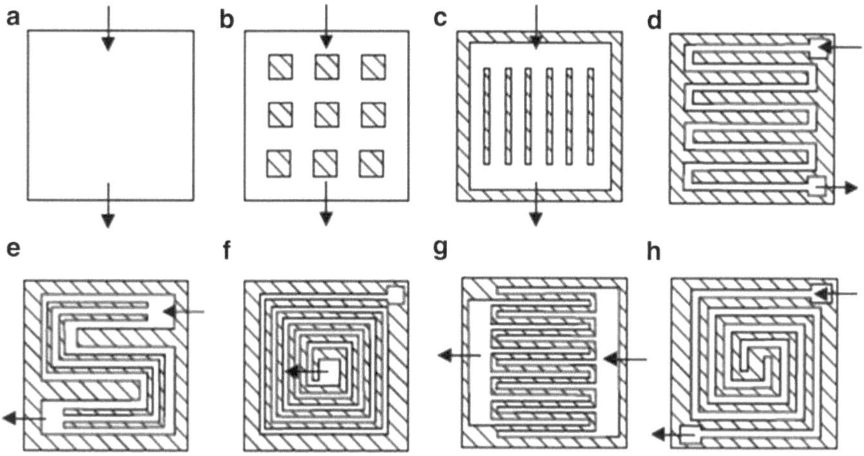

Fig. 1.11 Typical flow field designs: (**a**) direct supply, (**b**) pillars, (**c**) parallel, (**d**) serpentine, (**e**) parallel/serpentine, (**f**) spiral, (**g**) interdigitated, (**h**) spiral/interdigitated (Reproduced from Ref. [57] with permission)

methanol solution. More recently, Yuan et al. [64] reported an active μDMFC fabricated with gold-coated Al current collectors, which exhibited lower weight and excellent conductivity and corrosion resistance. A 6-cell mono-polar DMFC stack with a total volume of 13 cm^3 and a weight of 24 g. delivers 568 mW (33 mW. cm^{-2}) at room temperature for several days.

Litterst et al. [65] developed a μDMFC using poly(methyl methacrylate) (PMMA) and graphite-filled polypropylene as substrates. The authors used an original T-shaped channel design, with defined tapering angles over their cross section and along their axes, which facilitates the removal of CO_2 bubbles. At room temperature a maximum power density of 8 mW.cm^{-2} was achieved in passive mode.

The fuel delivery system reported for μDAFC are similar to that used in larger DAFC and are summarized in Fig. 1.11 [57].

The most used flow field design is the parallel microchannels (c), because it reduce the supplying pressure and decrease the fuel velocity, which leads to longer residence time and better fuel distribution. The serpentine microchannels (d) have large lengths which increase the pressure drop and allows a better fuel permeability. However, the high fuel consumption at the entrance induces fuel starvation and poor current density distribution at the end of the channel. The parallel/serpentine design (e) reduce the pressure drop and mitigate that problem. The spiral microchannel (f) keeps the fuel starvation region at the end of the channel close to the fuel entrance, allowing a better current distribution. The interdigitated microchannels (g) is a dead-ended design, so the fuel is almost depleted while permeating through the exit side. Other flow field designs have been analyzed by Sundarrajan et al. [66].

The state of the art regarding mass transport in DMFC has been discussed by Zhao and coworkers in two recent reviews, including the effect of operating conditions, flow field design and diffusion layer morphology on the transport of methanol and water [67], and the transport of methanol, water and oxygen in small DMFC with passive supply of reactants [68].

1.7 State of the Art of DMFC and Other DAFCs

Several reviews on DMFC have been written since 2000, dealing with several aspects of their fundamentals and technology [8, 21, 66, 69, 70]. More recent reviews focused on certain aspects of DMFC, such as catalysts [71], catalysts support [72], durability [73], modeling [74], and technology and applications [75]. These last three subjects will be addressed in detail in Chaps. 8 and 9.

Reviews on DEFC become available since 2006, describing the state of the art of catalysts for ethanol oxidation [31, 76, 77], while more recent works focused on alkaline DEFC [78–80]. A comprehensive review on alkaline direct alcohol fuel cells was recently published by one of us [30] over viewing catalysts, membranes and cell performance of ADAFC fuelled with methanol, ethanol, and ethylene glycol.

Tables showing the main characteristics of DMFC, including type and load of anodic and cathodic catalyst, membrane type, methanol concentration, reactants flowrates, and maximum power density (MPD) have been provided in previous works [8, 38, 57, 69, 75], for stacks made by companies or laboratories, and for micro fuel cells. The summary of DMFC and DEFC performances shown in Table 1.2 is just a brief upgrading of all the previous literature data [27, 81–102].

Finally, it is worth mentioning the microfluidic fuel cells concept [103] introduced by Whitesides in 2002 [104], based in a membraneless fuel cell design which exploit the laminar flow1 that occurs in liquids flowing at low Reynolds number to eliminate convective mixing of fuels. Using this concept on-chip, membraneless, air-breathing monolithic μDAFC has been constructed by Osaka and coworkers [105, 106] which operate with methanol, ethanol and 2-propanol solution containing sulphuric acid or phosphate buffer. The cell consists of two cathodes at the top of the channel, and the liquid fuel is supplied by capillary force to the anode formed on the bottom of the channel, as indicated in Fig. 1.12a, b.

One single cell can generate only power in the range of few μW, but integration based on silicon microfabrication would permit stacking the integrated cells (see Fig. 1.12c) for application as power sources for portable electronics. According to the authors, methanol is the fuel suitable for applications requiring long life, while neutral ethanol is preferred for safety operation, and 2-propanol for applications prioritizing power.

Table 1.2 Summary of DMFC and DEFC electrical performance for commercial and laboratory prototypes feed with oxygen

Company/Res. group	Type of FC*	Catalysts/load ($mg.cm^{-2}$)	Electrolyte	T (°C)	$c_{alcohol}$ (M)	MPD ($mW.cm^{-2}$)	Refs.
Esso	M-active	Pt	H_2SO_4	60	0.5	40	[81]
Hitachi	M-active	PtRu/8	H_2SO_4	60	1	24	[82]
Siemens	M-active	PtRu/4	Nafion 117	130	Vapor	210	[83]
Univ. Newcastle	M-active	PtRu/5	Nafion 117	98	Vapor	200	[84]
JPL/Giner	M-active	PtRu/2.5	Nafion 117	90	1	150	[85]
LANL	M-active	PtRu/2.2	Nafion 112	110	1	230	[86]
IFC	M-active	PtRu/4	Nafion 117	77	1	75	[87]
Aricó et al.	M-active	PtRu/2	Nafion 112	130	2	390	[88]
Baldauf-Preidel	M-active	PtRu/0.7	Nafion 117	110	1	380	[89]
Ren et al.	M-active	PtRu/2.2	Nafion 112	130	1	380	[90]
Scott et al.	M-active	PtRu/1	Nafion 117	80	1	210	[91]
Shukla et al.	M-active	PtRu/1	Nafion 117	90	1	180	[92]
Surampudi et al.	M-active	PtRu/2.5	Nafion 117	95	2	135	[27]
Witham et al.	M-active	PtRu/0.9	Nafion 117	90	1	100	[93]
Arico et al.	M-active	PtRu/2	Nafion/SiO_2	145	2	350	[94]
			Nafion/ZnO_2			320	
Yang et al.	M-active	PtRu/2.3	Nafion/$ZrPO_4$	145	2	375	[95]
KIER	M-active	PtRu/	Nafion 115	50	2.5	207	[96]
KIST	M-passive	PtRu/	Nafion 115	25	4	40[a]	[97]
Kim et al.	M-passive	PtRu/	Nafion 115	25	4	40	[98]
Yen et al.	M-active	PtRu/	Nafion 112	25	1	47	[61]
Wong et al.	M-active	PtRu/	Nafion	60	2	100	[99]
Kelley et al.	M-active	PtRu/	Nafion 117	70	0.5	90	[56]
Lu et al.	M-active	PtRu/	Nafion 112	60	2	100	[100]
Hou et al.	E-active	PtRu/	PBI/KOH	90		61	[101]
Fujiwara et al.	E-active	PtRu/	AAEM (Tokuyama)	20		58	[102]

[a]Cathode feed with air

*M-active: DMFC active mode; M-passive: DMFC passive mode; E-active: DEFC active mode

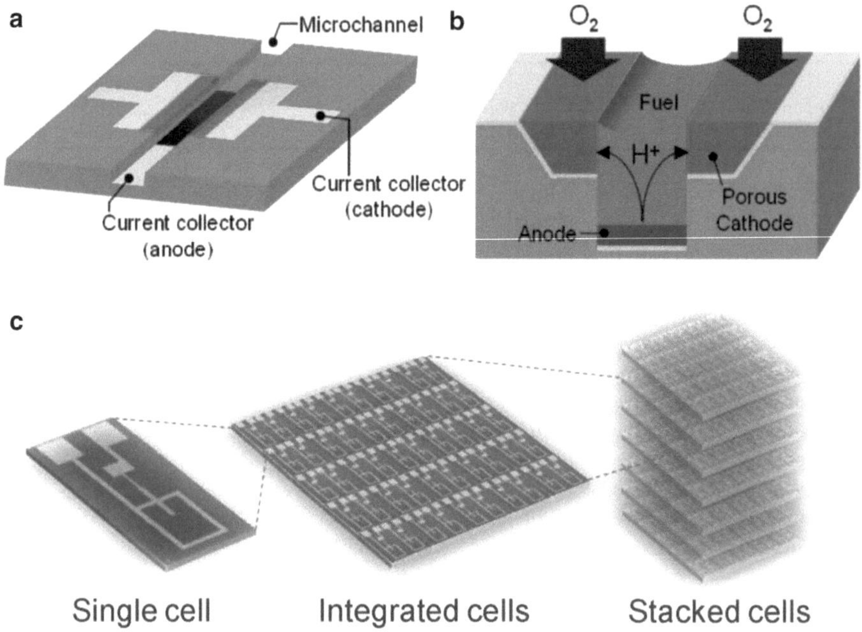

Fig. 1.12 Scheme of the air-breathing, membraneless on-chip fuel cell: (**a**) layout of current collectors and a microchannel, (**b**) cross-sectional view of the electrodes and the microchannel filled with a fuel solution, (**c**) integration of a single cell into multiple cells and snack ((a) and (b) reprinted from Ref. [105] with permission of The Royal Society of Chemistry; (c) reprinted from Ref. [106], copyright (2009) American Chemical Society)

References

1. BP Statistical Review of World Energy (2011). Available in: bp.com/statisticalreview
2. The BP Energy Outlook 2030 (2013). Available in: http://www.bp.com/liveassets/bp_internet/globalbp/globalbp_uk_english/reports_and_publications/statistical_energy_review_2011/STAGING/local_assets/pdf/BP_World_Energy_Outlook_booklet_2013.pdf
3. Shell energy scenarios to 2050 (2008). Available in: http://s08.static-shell.com/content/dam/shell/static/future-energy/downloads/shell-scenarios/shell-energy-scenarios2050.pdf
4. Rühl C, Appleby P, Fennema J, Naumov A, Schaffer ME (2012) Economic development and the demand for energy: a historical perspective on the next 20 years. Energy Policy 50:109–116, Smith Bits S.T.A.T.S
5. Crutzen P, Mosier AR, Smith KA, Winiwarter W (2007) N_2O release from agro-biofuel production negates global warming reduction by replacing fossil fuels. Atmos Chem Phys Discuss 7:11191–11205
6. Olah GA, Goeppert A, Surya Prakash GK (2006) Beyond oil and gas: the methanol economy. Wiley-VCH, Weinheim
7. Winter CJ (2000) In: Winter C-J (ed) On energies-of-change, the hydrogen solution: policy, business, and technology decisions ahead. Gerling Akademie Verlag, Munich
8. Lamy C, Léger JM, Srinavasan S (2001) Direct methanol fuel cells: from a twentieth century electrochemical's dream to a twenty-first century emerging technology. In: Bockris JO'M et al (eds) Modern aspect of electrochemistry, vol 34. Kluwer/Plenum, New York, pp 53–118

9. Reid J (1903) Process of generating electricity. US Patent 736,016
10. Kordesch KV, Simader GR (1995) Environmental impact of fuel cell technology. Chem Rev 95:191–207
11. Miiller E (1922) Die elektrochemische oxydation organischer verbindungen. Z Elektrochem 28:101–106
12. Tanaka S (1929) Z Elektrochem 35:38–42
13. Kordesch K, Marko A (1950) Oesterr Chem Ztg 52:125–130
14. Justi EW, Winsel AW (1955) British Patent 821,688
15. Wynn JE (1960) Proc Ann Power Sources Conf 14:52–57
16. Hunger HE (1960) Proc Ann Power Sources Conf 14:55–59
17. Vielstich W (1965) In: Baker BS (ed) Hydrocarbon fuel cell technology. Academic, New York, p 79
18. Koscher GA, Kordesch KV (2003) Alkaline methanol–air system. J Solid State Electrochem 7:632–636
19. Murray JN, Grimes PG (1963) Fuel cells. American Institute of Chemical Engineers, New York, pp 57
20. McNicol BD, Rand DAJ, Williams KR (1999) Direct methanol-air fuel cells for road transportation. J Power Sources 83:15–31
21. Aricó AS, Baglio V, Antonucci V (2009) Direct methanol fuel cells: history, status and perspectives. In: Liu H, Zhang J (eds) Electrocatalysis for direct methanol fuel cells. Wiley, Weinheim, pp 1–78
22. Cathro KJ (1969) The oxidation of water-soluble organic fuels using platinum-tin catalysts. J Electrochem Soc 116:1608–1611
23. Janssen MMP, Moolhuysen J (1976) Platinum-tin catalysts for methanol fuel cells prepared by a novel immersion technique, by electrocodeposition and by alloying. Electrochim Acta 21:861–868
24. Watanabe M, Motoo S (1975) Electrocatalysis by ad-atoms: Part III. Enhancement of the oxidation of carbon monoxide on platinum by ruthenium ad-atoms. J Electroanal Chem 60:275–283
25. Cameron DS, Hards GA, Harrison B, Potter RJ (1987) Direct methanol fuel cells. Recent developments in the search for improved performance. Platinum Metals Rev 31:173–181
26. Apanel G, Johnson E (2004) Direct methanol fuel cells – ready to go commercial? Fuel Cells Bull 2004:12–17
27. Surampudi L, Narayanan SI, Vamos F, Frank H, Halpert G, LaConti A, Kosek J, Surya Prakash GK, Olah GA (1994) Advances in direct oxidation methanol fuel cells. J Power Sources 47:377–385
28. Demirci UB (2007) Direct liquid-feed fuel cells: thermodynamic and environmental concerns. J Power Sources 169:239–246
29. Eccarius S, Garcia BL, Hebling C, Weidner JW (2008) Experimental validation of a methanol crossover model in DMFC applications. J Power Sources 179:723–733
30. Antolini E, Gonzalez ER (2010) Alkaline direct alcohol fuel cells. J Power Sources 195:3431–3450
31. Song S, Tsiakaras P (2006) Recent progress in direct ethanol proton exchange membrane fuel cells (DE-PEMFCs). Appl Catal B Environm 63:187–193
32. Bever D, Wagner N, Von Bradke M (1998) Innovative production procedure for low cost PEFC electrodes and electrode/membrane structures. Int J Hydrogen Energy 23:57–63
33. Giorgi L, Antolini E, Pozio A, Passalacqua E (1998) Influence of the PTFE content in the diffusion layer of low-Pt loading electrodes for polymer electrolyte fuel cells. Electrochim Acta 43:3675–3680
34. Ralph TR, Hards GA, Keating JE, Campbell SA, Wilkinson DP, Davis H, St. Pierre J, Johnson MC (1997) Low cost electrodes for proton exchange membrane fuel cells: performance in single cells and Ballard stacks. J Electrochem Soc 144:3845–3857
35. Gottesfeld S, Minas C (2008) Optimization of direct methanol fuel cell systems and their mode of operation. In: Kakaç S, Pramuanjaroenkij A, Vasiliev L (eds) Mini-micro fuel cells. Springer, Dordrecht, pp 257–268

36. Barton SC, Patterson T, Wang E, Fuller TF, West AC (2001) Mixed-reactant, strip-cell direct methanol fuel cells. J Power Sources 96:329–336
37. Priestnall MA, Kotzeva VP, Fish DJ, Nilsson EM (2002) Compact mixed-reactant fuel cells. J Power Sources 106:21–30
38. Scott K, Shukla AK (2007) Direct methanol fuel cells: fundamentals, problems and perspectives. In: White RE (ed) Modern aspects of electrochemistry, vol 40. Springer, New York, pp 127–227
39. Shukla AK, Raman RK (2003) Methanol-resistant oxygen-reduction catalyst for direct methanol fuel cells. Annu Rev Mater Res 33:155–168
40. Ilicic AB, Wilkinson DP, Fatih K, Girard F (2008) High fuel concentration direct-liquid fuel cell with a redox couple cathode. J Electrochem Soc 155:B1322–B1327
41. Ilicic AB, Wilkinson DP, Fatih K (2010) Advancing direct liquid redox fuel cells: mixed-reactant and in situ regeneration opportunities. J Electrochem Soc 157:B529–B535
42. Lam A, Wilkinson DP, Zhang J (2009) Control of variable power conditions for a membraneless direct methanol fuel cell. J Power Sources 194:991–996
43. Kim YS, Pivovar BS (2007) Chapter 4: Polymer electrolyte membranes for direct methanol fuel cells. In: Zhao TS, Kreuer KD, Van Nguyen T (eds) Advances in fuel cells. Elsevier, San Diego, pp 187–234
44. Lu G, Wang CY (2005) Chapter 9: Two-phase microfluidics, heat and mass transport in direct methanol fuel cells. In: Sundén B, Faghri M (eds) Transport phenomena in fuel cells. WIT Press, Southampton/Boston, pp 317–358
45. Liao Q, Zhu X, Zheng X, Ding Y (2007) Visualization study on the dynamics of CO_2 bubbles in anode channels and performance of a DMFC. J Power Sources 171:644–651
46. Park YJ, Lee JH, Kang S, Sauk JH, Song I (2008) Mass balance research for high electrochemical performance direct methanol fuel cells with reduced methanol crossover at various operating conditions. J Power Sources 178:181–187
47. Pasaogullari U, Wang CY, Chen KS (2005) Two-phase transport in polymer electrolyte fuel cells with bilayer cathode gas diffusion media. J Electrochem Soc 152:A1574–A1582
48. Zhang J, Yin GP, Lai QZ, Wang ZB, Cai KD, Liu P (2007) The influence of anode gas diffusion layer on the performance of low-temperature DMFC. J Power Sources 168:453–458
49. Kang K, Lee G, Gwak G, Choi Y, Ju H (2012) Development of an advanced MEA to use high-concentration methanol fuel in a direct methanol fuel cell system. Int J Hydrogen Energy 37:6285–6291
50. Xing LH, Gao YZ, Wang ZB, Du CY, Yin GP (2011) Effect of anode diffusion layer fabricated with mesoporous carbon on the performance of direct dimethyl ether fuel cell. Int J Hydrogen Energy 36:11102–11107
51. Aricó AS, Cretí P, Baglio V, Modica E, Antonucci V (2000) Influence of flow field design on the performance of a direct methanol fuel cell. J Power Sources 91:202–209
52. Vijayakumar R, Rajkumar M, Sridhar P, Pitchumani S (2012) Effect of anode and cathode flow field depths on the performance of liquid feed direct methanol fuel cells (DMFCs). J Appl Electrochem 42:319–324
53. Martin JJ, Qian W, Wang H, Neburchilov V, Zhang J, Wilkinson DP, Chang Z (2007) Design and testing of a passive planar three-cell DMFC. J Power Sources 164:287–292
54. Chan YH, Zhao TS, Chen R, Xu C (2008) A self-regulated fuel-feed system for passive direct methanol fuel cells. J Power Sources 176:183–190
55. Chan YH, Zhao TS, Chen R, Xu C (2008) A small mono-planar direct methanol fuel cell stack with passive operation. J Power Sources 178:118–124
56. Kelley SC, Deluga GA, Smyrl WH (2000) A miniature methanol/air polymer electrolyte fuel cell. Electrochem Solid State Lett 3:407–409
57. Nguyen NT, Chan SH (2006) Micromachined polymer electrolyte membrane and direct methanol fuel cells – a review. J Micromech Microeng 16:R1–R12
58. Cha SW, O'Hayre R, Prinz FB (2004) The influence of size scale on the performance of fuel cells. J Power Sources 175:789–795

59. Zhang B, Zhang Y, He H, Li J, Yuan Z, Na C, Liu X (2010) Development and performance analysis of a metallic micro-direct methanol fuel cell for high-performance applications. J Power Sources 195:7338–7348
60. Pavio J, Bostaph J, Fisher A, Hallmark J, Mylan BJ, Xie CG (2002) LTCC fuel cell system for portable wireless electronics. Adv Microelectr 29:1–8
61. Yen TJ, Fang N, Zhang X, Lu GQ, Wang CY (2003) A micro-methanol fuel cell operating at near room temperature. Appl Phys Lett 83:4056–4058
62. Aravamudhan S, Rahman ARA, Bhansali S (2005) Porous silicon based orientation independent, self-priming microdirect ethanol fuel cell. Sens Actuat A 123–124:497–504
63. Lu GQ, Wang CY (2006) Development of high performance micro DMFCs and a DMFC stack. J Fuel Cell Technol 3:131–136
64. Yuan Z, Zhang Y, Fu W, Li Z, Liu X (2013) Investigation of a small-volume direct methanol fuel cell stack for portable application. Energy 51:462–467
65. Litterst C, Eccarius S, Hebling C, Zengerle R, Koltay P (2006) Increasing μDMFC efficiency by passive CO_2 bubble removal and discontinuous operation. J Micromech Microeng 16: S248–S253
66. Sundarrajan S, Allakhverdiev SI, Ramakrishna S (2012) Progress and perspectives in micro direct methanol fuel cell. Int J Hydrogen Energy 37:8765–8786
67. Zhao TS, Xu C, Chen R, Yang WW (2009) Mass transport phenomena in direct methanol fuel cells. Prog Energy Comb Sci 35:275–292
68. Zhao TS, Chen R, Yang WW, Xu C (2009) Small direct methanol fuel cells with passive supply of reactants. J Power Sources 191:185–202
69. Garcia BL, Weidner JW (2007) Review of direct methanol fuel cells. In: White RE (ed) Modern aspects of electrochemistry, vol 40. Springer, New York, pp 229–284
70. Kamarudin SK, Daud WRW, Ho SL, Hasran UA (2007) Overview on the challenges and developments of micro-direct methanol fuel cells (DMFC). J Power Sources 163:743–754
71. Zhao X, Yin M, Ma L, Liang L, Liu C, Liao J, Lu T, Xing W (2011) Recent advances in catalysts for direct methanol fuel cells. Energy Environ Sci 4:2736–2753
72. Sharma S, Poleet BG (2012) Support materials for PEMFC and DMFC electrocatalysts – a review. J Power Sources 208:96–119
73. Kim YS, Zelenay P (2009) Direct methanol fuel cell durability. In: Büchi FN et al (eds) Polymer electrolyte fuel cells durability. Springer, New York, pp 223–240
74. Bahrami H, Faghri A (2012) Review and advances of direct methanol fuel cells: Part II: Modeling and numerical simulation. J Power Sources 230:286–296
75. Dillon R, Srinivasan S, Aricó AS, Antonucci V (2004) International activities in DMFC R&D: status of technologies and potential applications. J Power Sources 127:112–126
76. Antolini E (2007) Catalysts for direct ethanol fuel cells. J Power Sources 170:1–12
77. Friedl J, Stimming U (2013) Model catalyst studies on hydrogen and ethanol oxidation for fuel cells. Electrchim Acta 101:41–58
78. Zhao TS, Li YS, Shen SY (2010) Anion-exchange membrane direct ethanol fuel cells: status and perspective. Front Energy Power Eng China 4:443–458
79. Brouzgou A, Podias A, Tsiakaras P (2013) PEMFCs and AEMFCs directly fed with ethanol: a current status comparative review. J Appl Electrochem 43:119–136
80. Kamarudin MZF, Kamarudin SK, Masdar MS, Daud WRW (2013, in press) Review: direct ethanol fuel cells. Int J Hydrogen Energy. doi:10.1016/j.ijhydene.2012.07.59
81. Heath CE (1964) Proc Ann Power Sources Conf 18:33
82. Tamura K, Tsukui T, Kamo T, Kudo T (1984) Hitachi Hyoron 66:49
83. Waidhas M, Drenckhahn W, Preidea W, Landes H (1996) Direct-fuelled fuel cells. J Power Sources 61:91–97
84. Hogarth MP, Hards GA (1996) Direct methanol fuel cells. Platinum Metal Rev 40:150–159
85. Narayanam SR, Halpert G, Chun W, Jeffries-Nakamura B, Valdez TI, Frank H, Surampudi S (1996) Proceedings of 37th Power Sources Conference, Cherry Hill, NJ (USA), pp 96–99

86. Gottesfeld S, Cleghom SJC, Ren X, Springer TE Wilson MS, Zawodzinski T (1996) In: Courtesy Associates (ed) Fuel cell seminar. Washington, DC, pp 521–524
87. Fuller TF, Murach BL, Maricle DL (1997) 191th meeting of the electrochemical society, vol 97-1, abstract 620. The Electrochemical Society, Pennington, p 812
88. Aricó AS, Antonucci PL, Modica E, Baglio V, Kim H, Antonucci V (2002) Effect of Pt-Ru alloy composition on high-temperature methanol electro-oxidation. Electrochim Acta 47:3723–3732
89. Baldauf M, Preidel W (2001) Experimental results on the direct electrochemical oxidation of methanol in PEM fuel cells. J Appl Electrochem 31:781–786
90. Ren X, Wilson MS, Gottesfeld S (1996) High performance direct methanol polymer electrolyte fuel cells. J Electrochem Soc 143:L12–L15
91. Scott K, Taama W, Cruickshank J (1998) Performance of a direct methanol fuel cell. J Appl Electrochem 28:289–297
92. Shukla AK (2002) An improved-performance liquid-feed solid-polymer-electrolyte direct methanol fuel cell operating at near-ambient conditions. Electrochim Acta 47:3401–3407
93. Witham CK, Chun W, Valdez TI, Narayanan SR (2000) Performance of direct methanol fuel cells with sputter-deposited anode catalyst layers. Electrochem Solid State Lett 3:497–500
94. Antonucci PL, Aricó AS, Creti P, Ramunni E, Antonucci V (1999) Investigation of a direct methanol fuel cell based on a composite Nafion-silica electrolyte for high temperature operation. Solid State Ion 125:431–437
95. Yang C, Srinivasan S, Aricó AS, Creti P, Baglio V, Antonucci V (2001) Composite Nafion/zirconium phosphate membranes for direct methanol fuel cell operation at high temperature. Electrochem Solid State Lett 4:A31–A34
96. Jung DH, Jo YH, Jung JH, Cho SH, Kim CS, Shin DR (2000) Proceedings of fuel cell seminar, Portland, pp 420–423
97. Kim D, Cho EA, Hong SA, Oh IH, Ha IH (2004) Recent progress in passive direct methanol fuel cells at KIST. J Power Sources 130:172–177
98. Kim C, Kim YJ, Yanagisawa T, Park KC, Endo M (2004) High-performance of cup-stacked-type carbon nanotubes as a Pt-Ru catalyst support for fuel cell applications. J Appl Phys 96:5903–5905
99. Wong CW, Zhao TS, Ye Q, Liu JG (2006) Experimental investigations of the anode flow field of a micro direct methanol fuel cell. J Power Sources 155:291–296
100. Lu CQ, Wang CY (2005) Development of micro direct methanol fuel cells for high power applications. J Power Sources 144:141–145
101. Hou H, Sun G, He R, Wu Z, Sun B (2008) Alkali doped polybenzimidazole membrane for high performance alkaline direct methanol fuel cell. J Power Sources 182:95–99
102. Fujiwara N, Siroma Z, Yamazaki S, Ioroi T, Senoh H, Yasuda K (2008) Direct ethanol fuel cells using an anion exchange membrane. J Power Sources 185:621–626
103. Kjeang E, Djilali N, Sinton D (2009) Chapter 3: Advances in microfluidic fuel cells. In: Zhao TS (ed) Micro fuel cells. Academic, Burlington, pp 99–139
104. Ferrigno R, Stroock AD, Clark TD, Mayer M, Whitesides GM (2002) Membraneless vanadium redox fuel cell using laminar flow. J Am Chem Soc 124:12930–12931
105. Tominaka S, Nishizeko H, Ohta S, Osaka T (2009) On-chip fuel cells for safe and high-power operation: investigation of alcohol fuel solutions. Energy Environ Sci 2:849–852
106. Tominaka S, Ohta S, Obata H, Momma T, Osaka T (2008) On-chip fuel cell: micro direct methanol fuel cell of an air-breathing, membraneless, and monolithic design. J Am Chem Soc 130:10456–10457

Chapter 2
Catalysts for Methanol Oxidation

Ernesto R. Gonzalez and Andressa Mota-Lima

Abstract Methanol electroxidation proceeds via a multistep reaction. Herein, mechanisms of reaction and reaction rates of sub-set of the mechanism on different surfaces are discussed. Platinum is a reasonable catalyst for the first methanol electroxidation steps (dehydrogenation), but not for the last (CO electroxidation). Hence, alloying Platinum with a second metal was used as a strategy to enhance the rate of the last step. Accordingly, enhanced rates of CO electroxidation are attained by modifying the bonding energy of the adsorbate to the catalyst or by promoting the formation of oxygenated species at lower overvoltage. Finally, new perspectives on the field of methanol catalysts are commented including the search for catalysts that promote early onset of oscillations, i.e. under lower overvoltage.

2.1 Introduction

Methanol is the typical fuel with one carbon (C1) atom for fuel cells. Methanol was one of the first small molecules chosen to study the oxidation on platinum group metals in the very early beginning of electrocatalysis. In that time, the oxidation of other C1 molecules such as formic acid and formaldehyde (interest in CO oxidation came latter with the oxidation of reformatted gases) were investigated as a model oxidation because their elementary steps were supposedly present in the mechanism of methanol oxidation. From the point of view of CO_2 emission, methanol has, among the other small molecules, the highest energy production per unit of produced

E.R. Gonzalez (✉) • A. Mota-Lima
Instituto de Química de São Carlos-USP, CP 780, São Carlos, SP 13560-970, Brazil
e-mail: ernesto@iqsc.usp.br

H.R. Corti and E.R. Gonzalez (eds.), *Direct Alcohol Fuel Cells: Materials, Performance, Durability and Applications*, DOI 10.1007/978-94-007-7708-8_2,

CO_2 considering the complete reaction. The complete oxidation of methanol proceeds anodically in acid medium according to the reaction:

$$CH_3OH + H_2O \rightarrow CO_2 + 6H^+ + 6e^- \quad (2.1)$$

Which is normally a multistep reaction, as discussed below. The interest in methanol as a fuel may have been generated because of its much larger energy density (4.82 $kWh.l^{-1}$ or 6.1 $kWh.kg^{-1}$) than hydrogen.

In the following, the catalysts that have been investigated for the methanol oxidation reaction (MOR) will be presented. No attempt will be made to be exhaustive neither in terms of all the materials that have been studied nor in terms of the historical development of those materials. Most of the initial studies of the electrocatalysis of the MOR were carried out on massive electrodes and using electrochemical techniques. Later, the feasibility of the direct methanol fuel cell (DMFC) precluded the use of massive electrodes. Electrochemical reactions are surface reactions, so it is apparent that there is much to be gained by using large surface area electrodes, which led to the development of diffusion electrodes where the catalyst is in the form of nanoparticles. These electrodes have large specific surface areas which not only favor intrinsically the reaction but also allow for the use of minimal amounts of catalyst metals, usually rather expensive and, in some cases, scarce.

There are many challenges to produce an ideal DMFC, but it may be said that the electrocatalytic aspects of both MOR and oxygen reduction reaction (ORR) are emphasized in order to increase the power density. Also, much effort is devoted to reduce the crossover, i.e. the methanol from the anode crosses the electrolyte, migrates to the cathode and thus promotes both the diminution of potentially useful fuel and the creation of a mixed potential at the cathode. Because of the high potential of the cathode the methanol that reaches this electrode is oxidized and the consequence is a mixed potential, lower than the normal cathode potential, which contributes to the reduction of the power density. Obviously, this issue is a technological challenge specific for the PEMFCs which are not relevant for other sorts of fuel cell. Despite the development of MOR catalysts had been done together with the development of methanol tolerant ORR catalysts, this chapter is mainly concerned with the catalysts and mechanism of the MOR widely investigated in the last five decades as well as the recent findings which could drive this field forward.

2.2 Methanol Electroxidation

2.2.1 Mechanism of Methanol Electroxidation

The complete oxidation of methanol is given by the reaction:

$$CH_3OH + 3/2\ O_2 \rightarrow CO_2 + 2\ H_2O \quad (2.2)$$

For which $\Delta G^\circ = -702$ kJ.mol^{-1}. Therefore, the standard reversible potential for this reaction is

$$E_\gamma^\circ = \frac{702 \cdot 10^3}{nF} = 1.21\,V \tag{2.3}$$

This value is very near the value of the standard reversible potential for the reaction $H_2 + ½\ O_2 = H_2O$ (1.23 V).

The literature before 1989 on the oxidation of methanol and other small organic molecules has been covered in the excellent review of Parsons and Vander Noot (1988). The complete oxidation of methanol on platinum involves six electrons, which indicates that the mechanism has multiple elementary steps, as shown in a relatively recent study [1].

In fact several products or intermediates have been identified like CO, $HCOOCH_3$, HCOOH and H_2CO, but eventually, and depending on the potential, the final product is CO_2.

Presently, it is believed that the first step is the adsorption of methanol:

$$CH_3OH + Pt \rightarrow Pt\text{-}CH_3OH_{ads} \tag{2.4}$$

followed by platinum catalyzed electrochemical dehydrogenation reactions leading to:

$$Pt\text{-}CH_3OH_{ads} \rightarrow Pt\text{-}CO_{ads} + 4H^+ + 4e^- \tag{2.5}$$

The oxidation of CO_{ads} requires the participation of oxygenated species formed from the dissociative adsorption of water:

$$Pt + H_2O \rightarrow Pt\text{-}H_2O_{ads} \rightarrow Pt\text{-}OH_{ads} + H^+ + e^- \tag{2.6}$$

Finally, the two adsorbed species react in a Langmuir-Hinshelwood (LH) type of mechanism:

$$Pt\text{-}CO_{ads} + Pt\text{-}OH_{ads} \rightarrow 2Pt + CO_2 + H^+ + e^- \tag{2.7}$$

The problem with this mechanism is that Reaction (2.6) requires a high potential (>0.75 V vs. SHE on platinum). If the objective of oxidizing methanol is to run a methanol/oxygen fuel cell, an anode potential so high makes the cell impractical because the potential approaches the potential necessary to reduce oxygen.

In this context, it is not surprising to find that most work in electrocatalysis is devoted to reduce the anodic potential and to increase the potential of the cathode at current densities of practical interest.

By far the most widely explored approach to decrease the anodic potential has been the formation of bimetallic, trimetallic or multimetallic platinum based catalysts. In a bimetallic platinum based catalyst, the second metal will be chosen to favor one (or both) of two effects: (1) the second metal is an oxophilic element, with which Reaction (2.6) proceeds at a lower potential than on platinum (sometimes called a promoted mechanism) or (2) the formation of an alloy with the second metal modifies the electronic structure of platinum so that the adsorption of CO attached to an adjacent platinum becomes weaker (sometimes called an intrinsic mechanism). Another effect claimed in the literature is the (3) third body effect which is of less importance for the MOR.

In spite of all the efforts to reduce the anodic overpotential of the methanol oxidation, the success has been modest, so major advances are still necessary in order to achieve significant power densities with a DMFC. In part, this is the reason for still large number of papers dedicated to this issue found in the literature nowadays.

2.2.2 The Oxidation of Methanol on Platinum

The oxidation of methanol on platinum has been studied for more than 50 years. Most studies have been devoted to the identification of adsorbed species and the nature of intermediates [2]. Detailed pathways for methanol oxidation in acid and alkaline electrolytes were compiled by Cohen et al. [3] and depicted in Fig. 2.1. On platinum, methanol oxidation is a slow reaction that proceeds through multiple parallel mechanisms. Reaction (2.5) was written as global reaction to elucidate that the set of dehydrogenation elementary steps leads to dual pathways. So far, formaldehyde (HCHO), formic acid (HCOOH), formate ($HCOO^-$) and methyl formate ($HCOOCH_3$) were identified as soluble intermediates which conduct the discussion on the pathways of methanol oxidation. In case the first stripped hydrogen of the adsorbed methanol originates from the hydroxyl, the resulting adsorbed species, H_3CO_{ad} (methoxide), eliminates a second hydrogen from methoxy, forming formaldehyde which then desorbs [4]. In case the first eliminated hydrogen proceeds from methoxy group, CH_2OH_{ad} (hydroxymethyl) is the adsorbed intermediate whose further dehydrogenation would lead to formaldehyde and formic acid formation. The other soluble intermediates derive from further interaction of adsorbed formic acid, for instance, methyl formate is formed via chemical reaction with the bulk MeOH.

According to the current understanding, methanol oxidation proceeds via a dual-path mechanism: formaldehyde and formic acid routes. Due to short lifetime most methanolic intermediates are invisible to the current surface spectroscopy techniques; however, adsorbed CO and adsorbed formate are well confirmed by IR spectroscopy studies [5, 6]. With that, all weakly adsorbed intermediates (therefore, soluble) and stronger adsorbed intermediates (carbon monoxide) seem to be recognized. Using a special approach to quantify simultaneously all soluble intermediates by DEMS [7], Zhao et al. [8] confirmed undoubtedly that only three soluble intermediates account for all the faradaic current observed. Hence, in the potential region from 0.65 to 0.80 V (vs. RHE), CO_2 formation accounts for about

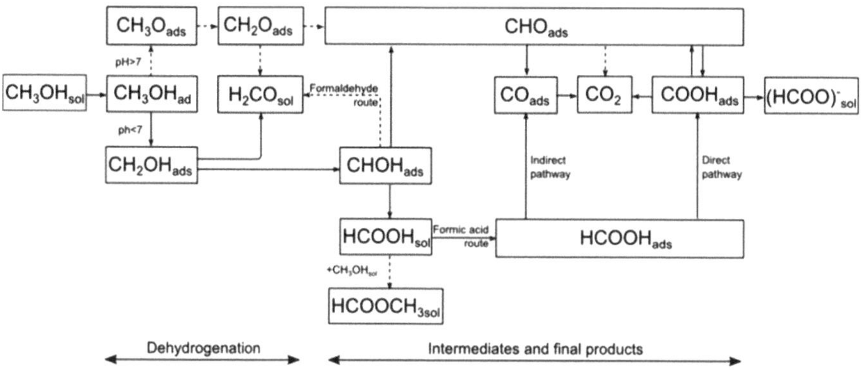

Fig. 2.1 Detailed pathways for methanol oxidation in acid (pH < 7) and alkaline (pH > 7) electrolyte. *Dashed lines* represent pathways that have been suggested in the literature, but are unlikely to occur under typical experimental conditions

half of the overall reaction products, while methanol conversion to formic acid and formaldehyde is only a quarter of the overall amount, or in other words, product yields of 53 ± 5 % for CO_2, 23 ± 2 % for formic acid, and 24 ± 2 % for formaldehyde formation.

The main soluble intermediates could be readsorbed and oxidized to form CO_2 or extracted from the surface under configuration of continuous flow rate. The last situation represents a loss of energy due to an incomplete methanol oxidation. This is well elucidated in the experiments where the extraction of solution in front of the electrode results in lower current than in experiments without sample collection [9]. For supported platinum, Jusys et al. [10] observed that an increasing conversion to CO_2 would be attained with increasing Pt load by the cost of faster consumption of formaldehyde; facts that are attributed to an increased readsoption rate on electrodes with enlarged electrochemical surface area.

Presently, the oxidation of methanol on pure platinum has more academic interest than practical application once DMFC universally employs platinum based materials having two or more metals as an anodic catalyst. In absence of methanolic intermediate readsorption, the maximum reaction rate for CO oxidation is 100-fold smaller than maximum reaction rate for CO adsorption from methanol dehydrogenation steps [11]. Indeed, the mechanism of methanol oxidation on platinum is expected to be equal to that on its alloys despite different kinetics which would result in a selection of pathway. In terms of complex activation theory, “alloyed Pt” is intend to lower the Ea barrier for CO adsorption, thus driving methanol oxidation to completion. As previously established [3], there are several factors that affect the calculated activation energy for the MOR at a given potential, such as coverage of methanolic intermediates and anion adsorption from the electrolyte as well as pH and oxide formation processes.

The MOR is surface sensitive [12, 13], reflecting either the behavior of carbon monoxide oxidation [14] or the active intermediate reaction path [13] on Pt single crystal electrodes. On Pt(110), dissociative chemisorption occurs primarily at low potential on defect sites such as steps [15], and yields initial currents much higher than on any other low index planes, but the decay is also fast due to the strong adsorption of CO [16–18]. Nevertheless, only low CO coverage ($\theta_{CO} \leq 0.3$) was obtained on all three Pt faces from methanol dissociative chemisorptions according to FTIR spectroscopy [19]. Moreover, terminally bonded (i.e., on-top) CO is the major form of this adsorbate detected under most conditions from the characteristic C–O stretching frequencies at ca. 2,030–2,060 cm^{-1}, although bridging CO was also observed in some cases. More recently, oriented nanoparticles, i.e. prepared from processes able to control the nanoparticle shape and orientation, demonstrated to have the same surface sensitive activity [20].

Another aspect is the effect of the adsorbing sub-monolayers of ad-atoms on the oxidative current of the MOR, the targeted issue in the scientific research mainly in the second half of the twentieth century [21–23]. In essence, the sub-monolayer of an adsorbed atom does not form a bulk alloy with platinum but actually a surface segregated phase such as in the case of an upd (under potential deposition) layer. Moreover, this deposit might be unstable toward potential cycling and the ad-atom is gradually desorbing by cycling which creates a multiple scenario to evaluate the impact of ad-atoms toward the MOR. The oxidation of methanol is inhibited by both tin and bismuth adsorbed on all sorts of electrodes, i.e. single-crystal, polycrystalline and nanoparticle platinum [22] given that the anodic current increases according to the evolution of ad-atom desorption. Conversely, the enhancement of methanol oxidation on any ad-atom modified platinum is discussed based in three effects: (1) the electronic effect which modifies the adsorption properties of platinum, (2) the bifunctional mechanism promotion (the second metal is more oxidizable than Pt which increases the oxidation rate of adsorbed carbonaceous intermediates) and (3) third body effect (decreasing of the electrode poisoning by preventing the formation of a strongly bonded intermediate). Nevertheless, an additional interpretation which comes from the necessity to keep the upd metallic ion in the solution is that a chemical step between the dissolved metal and the surface is possible as argued for experiments with tin [23].

In the very beginning, the upd sub-monolayers of ad-atoms were used as a first approach to measure the impact of the bimetallic materials. However, these platinum modified surfaces do not predict fully the behavior of the bulk bimetallic material.

2.2.3 *The Oxidation of Methanol on Bimetallic Platinum Materials*

As discussed above, there is much to be gained by using bimetallic platinum materials for the methanol oxidation. Nevertheless, electrochemists preparing Pt-alloys are facing the challenges of the material science field with respect to the preparation of the metal mixture in nano-scale range. Very frequently, the literature is misleading concerning the term "alloys". The metal mixture is a macrostate (physical mixture) which may be composed of a variety of microstates (segregated pure metals, oxides, different intermetallic phases, etc.) and a given metal mixture is rarely found as a unique solid solution, i.e. forming a solely 100 % alloyed state. For this reason, great care must be taken when discussing electro-activity in terms of the composition effect. Actually, these compositions are usually given in terms of the global composition, i.e. atoms%, which are not useful for characterizing a surface unless the composition in each of the different microstates is given.

This scenario calls for a deep characterization of Pt based materials in terms of how many alloyed phases and segregated phases are present in the mixture as well as their composition. The more accurately the bulk of the Pt based materials is characterized the more precise the characteristics of surface are because the bulk and surface should not be so dissimilar, above the micrometric dimension, in terms of phases and composition. However, the same should not be strictly expected for nanoscale dimensions. Platinum nanoparticle based materials may not form a true alloy but a surface composition much dissimilar from the core given the equalized quantity of both the bulk and surface' free energy. To complicate the picture even more, the electrochemical results often depend on the technique used to evaluate the activity of the catalysts.

2.2.3.1 The Oxidation of Methanol on Platinum-Ruthenium

Many studies carried out with bimetallic materials show that Pt-Ru is today one of the best options to oxidize methanol. Thus, a simple way to prepare active Pt-Ru catalysts involves the deposition of metallic nanoparticles from a suspention onto the carbon microparticles by the method known as formic acid method [24]. Considering that the crystal structures of Pt and Ru are different, Pt being fcc and Ru hcp, the final crystal structure of the alloy depends on the composition. For Ru atomic fractions up to 0.6–0.7, the two metals form solid solutions in which Ru atoms replace Pt lattice points in the fcc structure. The opposite situation, Pt atoms replacing Ru atoms in the hcp structure is found for Ru atomic fractions higher than about 0.7. However, the crystal structure seems to depend also on the physical state of the material. When nanoparticles are prepared by reduction of ionic metal species, there is at least one report [25] claiming that the fcc structure prevails up to 80 at.%Ru. On the other hand in sputtered films the hcp structure is predominant even at low Ru fractions [26].

Pt-Ru materials have been studied as bulk alloys, dispersed nanoparticles, etc. and Watanabe et al. presented the first report pointing to the practical applications [27]. Independently of the physical state there seems to be agreement that the best composition in terms of activity is 50 at.%Ru. However, this may not be the case in an actual fuel cell [28] where the homogeneity of the nanoparticle distribution on the support seems to be more important. Early works with Pt-Ru materials tried to establish a correlation between activity and composition, as in the work of Chu and

Gilman [29] where they examined a wide range of compositions. Notwithstanding, and in view of the discussions above, it may be risky to draw conclusions from results obtained with materials that have not been fully characterized.

As mentioned above, the morphology of the Pt-Ru catalysts plays an important role in the activity toward methanol oxidation [30]. To give a global overview concerning this, we recall the work of Villullas' group that describes with great detail the link between microstructure (dispersion, degree of alloying and the second phase formation) of catalysts prepared by different methods and the catalytical properties. For instance, synthesis of PtRu/C nanocatalysts with controlled particle sizes was attained using a microemulsion method that simply manipulates values of water/surfactant molar ratio [31]. Even with similar particle size, significant differences during the oxidation of CO-monolayer and methanol were observed for different-atmospheres-heating-treated PtRu/C catalysts [32]. For the CO-monolayer oxidation, the peak potentials are shifted to lower values as the amount of the alloyed phase increases; however, the same tendency is not observed for methanol oxidation. Godoi et al. [32] ascribed this fact to different amounts of residual oxides and alloyed phases, i.e. when preparing a Ru-alloyed Pt, neither all ruthenium content is introduced into the Pt lattice and the residual amount remains segregated as oxide. Godoi et al. observed higher methanol oxidation current for catalysts that combine segregated residual oxide with low-degree-of-alloyed metal.

On the one hand, the Ru presence helps the oxidation of the last and strongest adsorbed intermediate, the carbon monoxide. Ru helps the bifunctional mechanism because the formation of oxygenated species (Reaction (2.6)) takes place at lower potentials than on Pt [33]. However, Reaction (2.6) has slower kinetics on Ru than on Pt as proven by electrochemical impedance spectroscopy [34]. For this reason, the currents for Pt-Ru in potential scan techniques are larger than for Pt below 0.65 V, but become smaller above 0.65 V [34].

On the other hand, the Ru presence would lead to specific characteristics of the surface toward the adsorption and oxidation of other CO-like intermediates. In this regard, alloying degree and amount of residual oxides would be employed to alter selectively different reaction rates of the two parallel pathways of methanol oxidation: the formic acid and the formaldehyde route. The first pathway was confirmed for PtRu surface given the presence of prominent DEMS signal $m/e = 60$ (methylformate) [35]. And the second pathway confirmed by the huge production of formaldehyde observed in aliquots analyzed by HPLC [36]. It is known that the increase in the Pt 5d-band vacancy can be promoted by the incorporation of Ru into the Pt lattice as well as by the presence of oxide species. The contribution of the oxide species to emptying the Pt band is more pronounced than the effect of the Ru-alloyed Pt. As a result, catalysts that combine residual oxides with lower-degree-of-Ru-alloyed Pt enhance the Pt 5d-band vacancy leading to increased CO_2 production. Presumably, the enhanced Pt 5d-band vacancy affect both parallel pathways by respectively changing the adsorption energy of formaldehyde and methoxy intermediates [36]. This evidences would be extended to explain the enhancement of methanol oxidation observed for PtRu nanoparticles supported on TiO_2 [37, 38] as well as Pt supported on RuO_2 [39].

2.2.3.2 The Oxidation of Methanol on Platinum-Tin Catalysts

Early in 1989, Kuznetsov et al. [40] suggested that nearly all alloys between Pt and Sn are possible. However, the situation is complicated because Pt and Sn also form the intermetallic phases Pt_3Sn, PtSn, Pt_2Sn_3, $PtSn_2$ and $PtSn_4$. These intermetallic phases are identified by definite crystalline structures as revealed by X-ray diffraction. Therefore, according to Radmilovic et al. [41] lattice parameters determined from DRX may be the result of mixtures of different phases containing Pt and Sn.

As shown by Gonzalez et al. [42], carbon supported nanocatalysts have been studied for the oxidation of methanol and other fuels. There is, however, a problem in the comparison of results, because the catalysts have been prepared by a variety of different methods.

At this point, a general comment is relevant. It is common to find in the literature that different methods of preparation lead to catalysts with different activities and almost continuously new methods of preparation, or modifications of methods already used, are being proposed. This may be classified as a bad scientific approach. What determine the activity of a catalyst are intrinsic properties like the atomic composition, the presence of residual oxide, electronic effects, etc., but not something undefined as the method of preparation. Therefore, it is not surprising that several works report an enhancement of the activity for methanol oxidation due to the presence of Sn [43–48] while others report that there is a very small enhancement or even no effect of the presence of Sn [21–23, 49–52]. It may be concluded that the effect of incorporating tin to the Pt catalyst for methanol oxidation varies from strong to moderate to mild or even negative and, for the time being, this will have to be attributed to the "method of preparation", although the real reason lies on the intrinsic properties of the improperly characterized nano-surface.

Actually, those controversies in technological nanocatalysts could partially be clarified by investigations with well characterized surfaces. To give the reader an idea, ordered alloy Pt_3Sn was securely created by using the Bridgeman technique and the faces of Pt_3Sn single crystal investigated toward the MOR. Astonishingly, none of the alloy surfaces were more effective catalysts than any of the pure platinum surfaces under the conditions of measurement employed in that work [23]. Translating the discussion to the field of nanoparticle catalyst, it is assumed that the "method of preparation" is, actually, a procedure to generate nanosurfaces with different reproductivity concerning the amount of segregate oxide, type of intermetallic phases and its relative proportion. In review [53], the presence of tin oxide/hydrodyde and the degree of alloying are all factors that have impact on MOR.

On the evaluation of alloyed degree at nanoscale range, X-ray diffraction is used to measure the interplanar distance (ID) and correlate the increasing ID to the extension of incorporating-alloying atom to the Pt lattice. Thus, Colmati et al. [54] observed a positive correlation between the MOR activity and the Sn-Pt alloying degree, and pointed $Pt_{75}Sn_{25}/C$ (or Pt_3Sn/C) as slightly better catalyst for MOR than $Pt_{90}Sn_{10}/C$ or Pt/C; Further increasing in Sn-alloying content from Pt_3Sn/C to PtSn/C, however, does not increase activity toward MOR [55] because it leads to a phase change from fcc (Pt_3Sn) to hcp (PtSn) being hcp phase less active toward MOR.

All types of in situ techniques are of great interest to infer about how the surface affects the mechanism of the MOR. In particular, the techniques which probe the local structure of the catalyst open a perspective unexplored so far: the structural changes of the catalyst during the progress of the reaction. Here is a brief example. Very fresh investigations [56] with in situ time-resolved energy-dispersive X-ray Absorption Fine Structure (DXAFS) and quick XAFS (QXAFS) techniques discover structural changes of the Pt_xSn nanocatalyst during the progress of the oxidation reaction. The oxidation of Pt to PtO in Pt_3Sn/C proceeded via two successive processes, while the oxidation of Sn to SnO_2 in Pt_3Sn/C proceeded as a one step process. Accordingly, the authors proposed the structural changes which include an oxide formation whose density depends on the nanoparticle composition. The SnO_2 on Pt_3Sn/C is porous which allows a second oxidation of the Pt located in the core while the compact SnO_2 layer on PtSn/C precludes further oxidation of the Pt-core and preserves, in this way, a metallic core.

On the methanol pathways, there is no information unless that the formic acid is neglectfully formed on Sn-alloyed Pt while CO_2 is largely produced [55].

2.2.4 The Oxidation of Methanol on Trimetallic Platinum Materials

On a basis of trial and error it was noticed that a practical fuel cell attains higher performance employing ternary platinum based materials than employing the binary catalysts. During the last decade, the global observation reveals an increasing of performance for the H_2/CO oxidation as well as for the MOR when a third element was added to the best bimetallic catalyst, the Pt-Ru [57] or Pt-Sn [58] based material. An overview of the preparation and structural characteristics of Pt-based ternary catalysts [59] and their electrochemical performance [60] was presented by Antolini. Therein, all the relevant works before 2007 are found. In summary, many ternary Pt–Ru–M catalysts ($M = W_{in}W_{Ox}$ or W_2C form, Mo, Ir, Ni, Co, Rh, Os, V) perform better than commercial standard Pt–Ru catalysts and/or Pt–Ru catalysts prepared by the same method than the ternary.

So far, the role of the third metallic additive is not clear. While a significant enhanced catalytic activity for MOR is attributed to the presence of hydroxyl Ru oxide, as discussed in the Sect. 2.2.3.1, a further activity increase due to the third additive (M) is only to some extension discussed in the literature and most of the time with controversy [61]. For instance, the promotion effect of Ni would be on the Pt–Ni electronic interaction rather than on the ability of the sample to nucleate OH species. Conversely, Ni is claimed to act as an agent which promotes the more proper physical dispersion of the ternary catalyst but the electrocatalytic effect still remains attributed to the hydroxyl Ru oxide [62].

2.2.5 *Catalysts for Methanol Oxidation in Alkaline Environment*

So far, kinetics of MOR were discussed throughout this chapter considering the acid medium, however the kinetics are, in average, halffold faster in alkaline medium. The same reads for ORR. Concisely, the enhanced activity in alkaline media results from the lack of specifically-adsorbed spectator ions in alkaline solutions, and the higher coverage of adsorbed OH at low potential [63]. The fastest kinetics allows to use low catalyst loadings and to select Pt-free catalysts.

The beneficial effect of catalysis by Pd addition to Pt is ascribed to the facile adsorption of OH^- on this surface which is further reflected by the high yield of formate as well as low yield of carbonate obtained during the oxidation of methanol on Pt e Pd/C catalyst [64]. An overview of catalysts and membranes for alkaline DAFC is found in Ref. [65].

2.3 Electroxidation of Methanol Intermediaries

In Sects. 2.1 and 2.2 of this chapter the mechanism of methanol oxidation on platinum was discussed. This section is dedicated to clarify the impact of the main Pt-alloys, as already mentioned in Sects. 2.2.3 and 2.2.4, on the kinetics of the most relevant oxidative steps belonging to the MOR mechanism. Baring this in mind, the CO oxidation, the last step for complete methanol oxidation, is analyzed as well as the oxidation of other MOR intermediates such as formic acid and formaldehyde.

2.3.1 *Residues of Methanol Adsorption: Dehydrogenation*

The presence of an adsorbed COH was detected as a residue of the methanol adsorption using in situ infra-red spectroscopy [66], hence confirming the threefold dehydration in methanol oxidation. However, the precise pathway remains on debate, partially because the stretching frequencies of many postulated intermediates are similar and therefore the precise path cannot be fully elucidated by in situ infra-red spectroscopy. Facing this scenario, the most promising way to gain insights comes from theory. On pure platinum, the dehydrogenation of surface-adsorbed methanol, CH_3OH_{ads}, is energetically favorable, with a hydrogen atom first stripped off the carbon and further decomposition goes mainly through a hydroxymethyl (CH_2OH_{ads}) intermediate as predicted from quantum chemical theory [67, 68]. Figure 2.2 depicts all possible geometries of fragments of methanol adsorbed on platinum during the dehydrogenation.

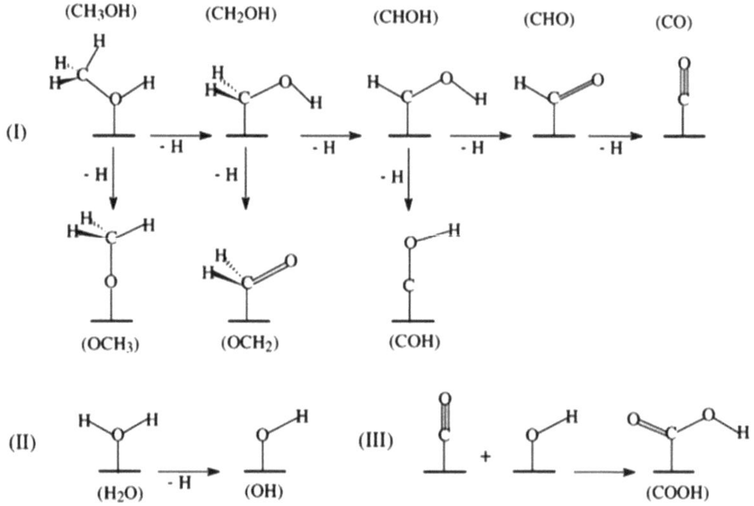

Fig. 2.2 Geometries of the possible intermediate fragments involved in CH_3OH dehydrogenation on metal surface: intermediates for (**I**) CH_3OH dehydrogenation, (**II**) H_2O dissociation, and (**III**) CO oxidation (Reprinted from Ref. [67], Copyright (2011), with permission from Elsevier)

Conversely, the gas/solid surfaces present essential differences regarding methanol adsorption [69]. There are three pathways in UHV, a simple decomposition via a methoxonium ($CH_3O\text{–}_{ads}$) intermediate, an SN1 pathway via a methoxoniumcation ($[CH_3OH_2]^+$), and an SN2 pathway via a methoxonium intermediate.

The adsorption energy of CO on Pt(111)/C is relatively higher than the adsorption energy of CH_2OH on the same surface [70]. Therefore, Pt(111)/C is favorable for the oxidation but loses the activity by continuous CO poisoning. This scenario is depicted in Fig. 2.3. There, a potential energy diagram is shown for each elementary step involved in the dehydrogenation as well as the water splitting and CO oxidation by LH mechanism. All stages of dehydrogenation are exothermic, i.e. spontaneous, and therefore the complete dehydrogenation proceeds till the production of CO attached on the surface. Afterward, the barrier to the proceeding reactions demands an endothermic oxide formation followed by the exothermic CO oxidation. The main intend of alloying a metal to platinum is exemplified also here for the case of Ru as a decreasing endothermic free energy for the oxide formation. However, a less commented effect is the alloying effect on the methanol dehydrogenation. For PtRu cluster this step is slightly favorable compared to the pure platinum. Similarly, PtMo(111)/C was found to facilitate methanol decomposition compared to Pt(111)/C [70]. Moreover, a top-Pt site is predicted to be the favorite adsorption site for methanol on Pt-Ru and Pt-Mo alloy [70].

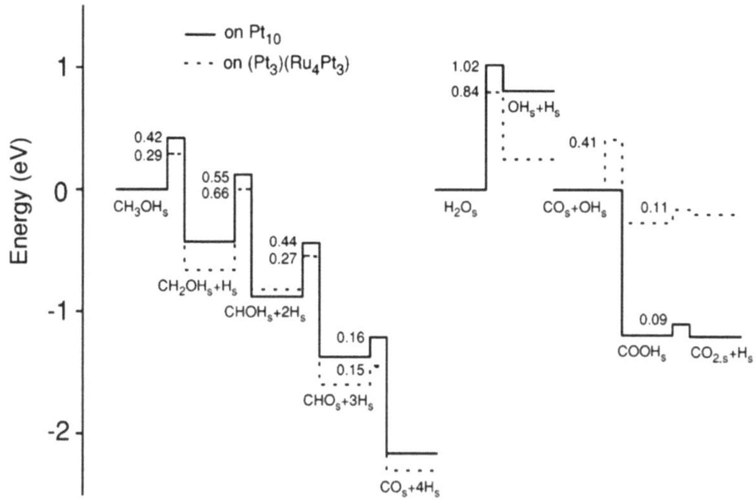

Fig. 2.3 Potential energy diagram for the CH_3OH dehydrogenation, water dissociation and CO_{ad} + OH_{ad} combination on pure platinum and mixed Pt-Ru surfaces (Reprinted from Ref. [67], Copyright (2011), with permission from Elsevier)

2.3.2 CO Oxidation

2.3.2.1 CO Oxidation on Palladium and Pt-Pd

Palladium is a preeminent catalyst for many catalytic reactions given its ability to oxidize carbon monoxide. However, the carbon monoxide electroxidation reaction on pure palladium as well as palladium based materials was mainly studied recently in the context of the catalysts for fuel cells. So, most part of the experimental knowledge about the CO oxidation mechanism is found for this technological device, especially in the context of binary/ternary platinum based materials. Among other interesting aspects about palladium, its ability to solubilize hydrogen into its lattice network [71] rendered it, in the earliest times, technological application as membrane to separate hydrogen out of gaseous mixtures [72]. Its fast hydrogen permeation is, however, an aspect which plays roles in the mechanism of CO oxidation which, so far, had been hypothesized as suffering impact only from species dissolved in the liquid phase.

The sparse investigation of CO oxidation on pure palladium surfaces before 2000 is compiled in the work of Yépez and Scharifker [73]. These authors argue about the possibility of assisting the oxidation of C1 molecules to CO_2 by reaction of CO with hydrogen occluded in the Pd lattice. Indeed, this proposal is based on infrared reflectance spectroscopy (IRS) results shown in Fig. 2.4a for oxidation of a CO monolayer which displays the intensity of the peak due to CO_2 as a function of electrode potential. For a palladium hydrogen free surface, the onset potential of

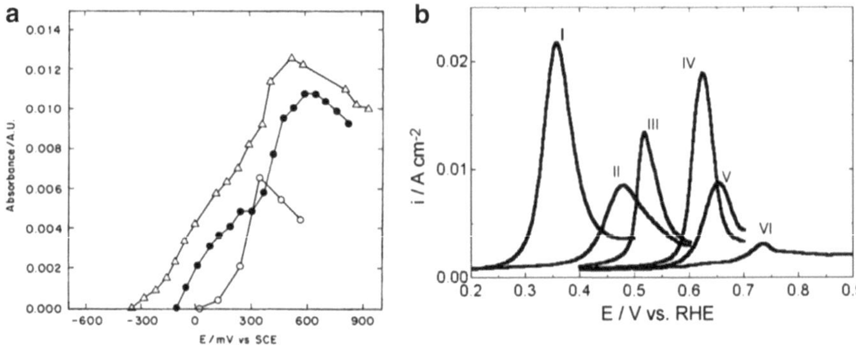

Fig. 2.4 (**a**) Amount of CO_2 produced as a function of electrode potential on palladium loaded with 0.0 (*circles*), 0.1 (*solid circles*) and 2.2 Coulombs (*triangles*) of Hydrogen (With kind permission from Springer: Ref. [74], Fig. 4″). (**b**) CO stripping scan for anodes formed by: (*I*) $Pd_{25}Pt_{25}Ru_{50}/C$, (*II*) $Pd_{65}Pt_{25}Ru_{10}/C$, (*III*) Pt/C, (*IV*) $Pd_{60}Pt_{40}/C$, (*V*) $Pd_{90}Pt_{10}/C$, (*VI*) Pd/C. Others details: electrosorption at 100 mV sec^{-1} (vs. RHE), Nafion 115 membrane, T = 85 °C (Reprinted from [75], Copyright (2011), with permission from Elsevier)

CO_2 production is very low, ca. 0.25 V (vs. SHE), and the rate of CO_2 production increases with the potential up to a maximum, ca. 0.68 V (vs. SHE). The two main impacts of different amount of sub-surface hydrogen on the CO oxidation on palladium are the decrease of the onset CO oxidation potential and the increase of the CO oxidation rate at a given potential. Additionally, oxygenated species seem to inhibit this CO oxidation mechanism once the oxidation rate decreases after the oxide formation potential around 0.8 V (vs. SHE) as seen in Fig. 2.4a. Taking this insight into consideration, the effect predicted for PtPd alloy compared to Pt toward the MOR mechanism should not be based on the bifunctional mechanism but rather in the electronic effect. Also, it may be based in lowering the fraction of poisoned surface as a result of further CO oxidation in the palladium sites which should have higher amounts of subsurface hydrogen originated from the step 5, the methanol dehydrogenation.

With this background, the CO oxidation mechanism on palladium in the presence of sub-surface hydrogen is assumed to be

$$CO_{ads} + 2H^{\bullet} \rightarrow H_2CO_{ads} \quad (2.8)$$

followed by the reaction

$$H_2CO_{ads} + H_2O \rightarrow CO_2 + 4H^+ + 4e^- \quad (2.9)$$

Note that Reaction (2.9) is not an elementary step and for this reason formaldehyde is a candidate for such molecule whose oxidation mechanism could be employed to gain insights over CO oxidation on palladium.

Considering the technological device, the CO bonding on Pd is stronger than on Pt considering that the main peak potential moves to more anodic values in a CO stripping CV [75, 76], see Fig. 2.4b. As a consequence, a much larger activation energy for the CO oxidation on Pd compared to Pt should be predicted, which in turn, indicates that the current of CO oxidation on Pd is much less sensitive to the variation of temperature than on Pt.

There are several experimental indications which support the hypothesis that the palladium surface admits low CO saturation coverage compared to platinum. Therefore this could be the reason behind the CO tolerant effect of the palladium-platinum based materials [77]. For instance, bimetallic Pt-Pd based materials prepared as nanoparticles supported on carbon [75] admit a high level of current at the same potential in a fuel cell when exposed to poisoned hydrogen (with 100 ppm of CO) which is explained by the lower CO coverage on palladium containing platinum compared to pure platinum. Pd/C has CO saturation coverage 5 % smaller than Pt/C when comparing the reduction charge in the hydrogen region due to the CO carpet [76].

2.3.2.2 CO Oxidation on Pt-Ru

According to the bifunctional mechanism at least three Pt atoms are needed to adsorb one methanol molecule and activate it to form a Pt-$(CO)_{ad}$ while only one Ru atom is needed to activate one water molecule to form Ru–OH. Thus, the best atomic ratio of Pt to Ru should be 4:1. However, as mentioned above, the best MOR activity is found for 1:1 Pt-Ru atomic ratio. Actually, the proportion 1:1 Pt-Ru provides the smaller CO oxidation onset potential [78]. Therefore, this scenario indicates that the CO oxidation is the rate determining step of MOR on Pt-Ru catalyst. So, it is not a surprise to find the best Pt-Ru proportion for MOR being equal to the best Pt-Ru proportion for the CO oxidation.

The role of Ru species within the Pt–Ru system is to form hydrous Ru oxides at smaller potentials than hydrous Pt oxides. However, the alloying of Pt should also exhibit a certain impact on the CO binding energy. So, both effects, the 'bifunctional mechanism' and 'ligand effect', should be present in an alloy although with different relative importance. Experimental data with temperature programmed desorption (TPD) taken by adsorbed CO on Pt(110) show a reduction in the binding energy of CO equivalent to 2 kcal/mol [79] in the presence of a Ru-deposit which only produces a 1 kcal/mol change in the activation barrier of Reaction (2.6). However, the barrier for OH recombination was reduced around 3–5 kcal/mol in the presence of ruthenium. Taken both effects together, the Reaction (2.7) is predicted to have a 4–6 kcal/mol reduction in the activation energy for the electrochemical CO removal; only about 1 kcal/mol is associated with the ligand effect, whereas 3–5 kcal/mol are associated with the bifunctional mechanism showing that the bifunctional effect is four times larger than the ligand effect.

The bifunctional mechanism assumes the formation of the oxygenated species, by Reaction (2.6), at a lower potential than for pure platinum. However, the elementary steps involved in the CO oxidation are

$$CO_{ads} + OH_{ad} \rightarrow COOH_{ads} \quad (2.10)$$

$$COOH_{ads} \rightarrow CO_2 + H^+ + e^- \quad (2.11)$$

Given the relevance of the oxygenated species formation for the bifunctional mechanism, the fundamental knowledge of this process is a crucial issue. Much of the actual belief about this concern comes from the assumption made in a Conway paper [80] which in summary established that any charges passed between 0.4 V and 0.7–0.8 V (vs. SHE) are solely due to the double layer and that hydroxyls are typically formed on the Pt surface after 0.7–0.8 V which in turn become irreversibly adsorbed above about 0.95 V, with a coverage of approximately 0.25, as well as reaches the monolayer limit of one hydroxyl per Pt at 1.1 V. Beyond this value, oxide formation is supposed to take place. For a comprehensive review of oxide formation up to the mid-1990s the reader is referred to Conway and references therein [81]. More recently, a concise investigation discusses the formation of hydroxyl at potential lower than 0.7–0.8 V [82] supporting a collection of other papers.

2.3.2.3 CO Oxidation on Pt-Sn

The single crystal, $Pt_3Sn(110)$ alloy surface [49] is very effective catalyst for CO electroxidation which does not reflected in an expressive enhancement for methanol electroxidation as compared to Ru-alloyed Pt, shown in Fig. 2.5. The potentiostatic current from MOR is one-order magnitude smaller for $Pt_3Sn(110)$ than those observed for Pt-Ru alloys. Conversely, $Pt_3Sn(110)$ is very active for CO-bulk oxidation given that $Pt_3Sn(110)$ has approximately 300 mV lower potential for current onset than the most active Pt-Ru alloy surface, and more than 500 mV lower than polycrystalline Pt, as seen in Fig. 2.5b. The superior CO oxidation ability can be ascribed to the facile formation of Sn–OH_{ad}. However, the relative inactivity of Pt_3Sn for methanol oxidation was explained by Wang et al. [49] based on a "crowding" state of CO on Pt_3Sn, i.e. the low potentiostatic current found for methanol oxidation is due to a much lower CO coverage formed by the dehydrogenation of methanol than formed by CO bulk adsorption, since the required Pt clusters for methanol dehydrogenation are already blocked by CO at low coverage.

On nanoparticle catalysts CO electroxidation develops faster charge transfer rate on Pt_3Sn/C than on Pt/C or Pt_3Ru/C [83] on solution free of methanol. Moreover, Ciapina and Gonzalez [83] suggest that Pt_3Sn/C should have adsorbed or activated water on the surface rather than the simple oxygenated species given the dominant capacitive behavior down to 10 mHz on EIS experiments.

2.3.2.4 CO Oxidation on Pt-Pb

Among the several intermetallic phases (PtPb, PtBi, Pt_3Ti, Pt_3Zr, Pt_3Ta and Pt_2Ta) PtPb and Pt_2Ta exhibited CO stripping potentials less positive than bulk Pt whereas

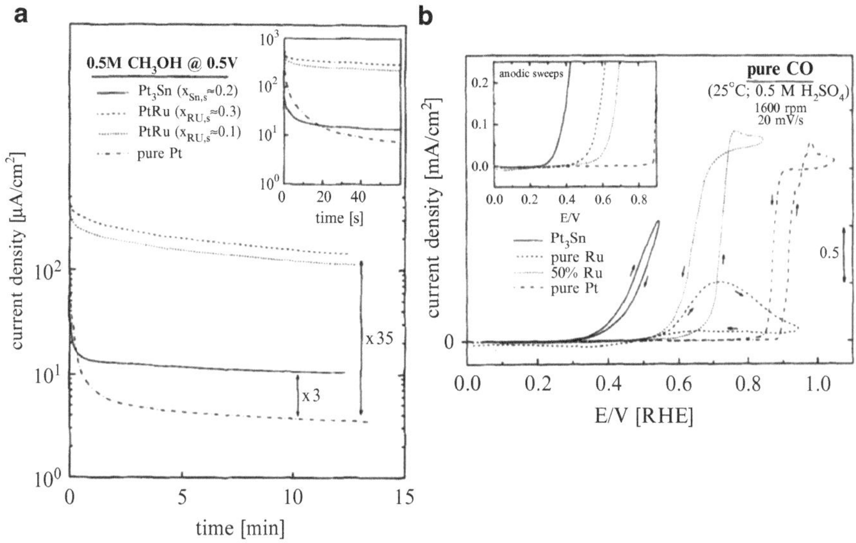

Fig. 2.5 (**a**) Potentiostatic methanol oxidation current densities at 500 mV in 0.5 M methanol and (**b**) Potentiodynamic (20 mV/s) CO oxidation current at rotation rate of 1,600 rpm. Others: 0.5 M H_2SO_4 (Reprinted from [49], Copyright (2011), with permission from Elsevier)

PtBi exhibited virtual immunity to CO poisoning, i.e., CO probably does not adsorb [84]. Globally, the other intermetallic phases exhibited little or no improvement over Pt.

2.3.2.5 DFT Theoretical Prediction

A periodic density-functional theory study of the adsorption of CO and hydroxyl (OH) on different clusters of metal was computed to evaluate the binding energies of these adsorbates [85]. Alloying Pt with Mo leads to weakly adsorbed CO on both Pt and Mo sites, and OH strongly adsorbed only on Mo sites. In the case of alloying Pt with Ru, a weaker bond of both CO and OH to the Pt sites occurs whereas alloying Ru with Pt causes a stronger bond of CO and OH to the Ru sites. This suggests that PtMo could be a better bifunctional catalyst for CO oxidation than PtRu. On $Pt_3Sn(111)$ the calculations show that CO binds only to Pt and not to the Sn, whereas OH has an energetic preference for the Sn sites. This also implies that PtSn should be a better CO oxidation catalyst than PtRu.

Concerning the alloying of platinum with Pb and Bi, the CO adsorption energies show stronger surface orientation dependence [86]. The CO adsorption energy decrease due to downward shift of the Pt d-band center (electronic effect) in the sequence: PtBi (100)B and PtBi (110) <PtPb (100)B and PtPb (110) << Pt(111). The same global trend was observed for other orientations, for instance, CO binding energies on $Pt_3Pb(111)$ are computed to be generally smaller than binding energies on Pt(111) [87].

Additionally, Christoffersen et al. [88] observed that a weaker CO–Pt bond also makes the CO more reactive at higher overpotentials, i.e. the LH mechanism is favorable. The effect of the alloying atom (M) on mixed Pt-M metals is to promote the formation of OHs combined with a favored oxidative removal of COs and formation of COOHs. The relative contributions of ligand effect and bifunctional effect are discussed in several quantum chemistry calculations [67, 89]. A ligand effect plays some role in the oxidation of CO on Pt-Ru, but it is unimportant on Pt-Sn [67].

2.3.3 *Formic Acid Oxidation*

The interest in formic acid oxidation (FAO) rose up in the 1970s with the aim of shedding light on the mechanism of methanol oxidation beyond the commercial interest in direct formic acid oxidation in fuel cells [90]. The FAO in acid solution was extensively investigated on surfaces of platinum [91–100]; The FAO on other pure metallic surfaces seems to have been restricted to the palladium surface [98, 101–104]. In the 1980s, the remarkable contribution was done by the studies on the influence of the ad-atom in the activity of the platinum electrode [91–94]. In the 1990s, superficial spectroscopic techniques were employed to describe the electrochemical mechanism on palladium surface [98, 101–103] as well as platinum surface [97, 98, 105]. In the last 10 years, there was a triplication of publications about the FAO, specially driven by the use of nanoparticles.

A generally accepted dual-path mechanism for formic acid oxidation on platinum was proposed by Capon and Parsons [90] and continuously revised by the new spectroscopic insights mainly due to the discussion of the nature of the poison species. According to the current accepted mechanism, the direct path of FAO occurs via a reactive intermediate described

$$\mathrm{HCOOH} \rightarrow {}^{*}\mathrm{COOH} + \mathrm{H}^{+} + \mathrm{e}^{-} \quad (2.12)$$

$${}^{*}\mathrm{COOH} \rightarrow \mathrm{CO}_2 + \mathrm{H}^{+} + \mathrm{e}^{-} \quad (2.13)$$

where * indicates the number of platinum sites bonded to the carbon atom of the organic species. Surface-blocking residues, the poisons, are formed respectively by the parallel reaction

$${}^{*}\mathrm{COOH} + \mathrm{HCOOH} \rightarrow {}^{***}\mathrm{COH} + \mathrm{CO}_2 + \mathrm{H}_2\mathrm{O} \quad (2.14)$$

and by the indirect pathway

$${}^{*}\mathrm{COOH} + {}^{*}\mathrm{H} \rightarrow {}^{*}\mathrm{CO} + \mathrm{H}_2\mathrm{O} + \mathrm{e}^{-} \quad (2.15)$$

The adsorbed carbon monoxide is assumed to be the strongly adsorbed intermediate and the formate the weakly adsorbed which belong respectively to the direct pathway (active species) and indirect pathway (assumed to occur via poisonous species).

Prediction from DFT calculation shows that HCOOH oxidation under a water-covered surface behaves substantially differently than in the gas phase or using a solvation model involving only a few water molecules [106].

2.3.3.1 Formic Acid Oxidation on Palladium

An astonishing aspect of formic acid oxidation on palladium is the formation of CO_2 at potentials of hydrogen oxidation and oxygen evolution. On-line mass spectroscopy analysis of volatile products reveals the production of CO_2 not only in the double-layer region but also near 0.25 and 1.75 V [101]. The fact that the current and the CO mass signal during a potential scan do not follow the same pattern indicates that HCOOH oxidation proceeds through parallel mechanisms.

Another aspect is the several experimental indications that palladium admits a fast reaction rate for the direct pathway of small molecules oxidation, specially the C1 molecules. Experiments with IR spectroscopy on Pd-Au alloy detected the lack of CO signal during formate electroxidation [102]. Anodic stripping voltammetry of adsorbed formic acid on Pt ad Pd nanoparticles indicated a presence of a sharp peak due to CO oxidation in the former and a lack in the latter [107]. Actually, adsorbed CO is confirmed to be the strongly adsorbed residue for FAO on platinum single crystals [108]. Taken all together, these results suggest two interpretations. First that CO saturation coverage is low under FAO more for matter of peculiar condition of palladium surface (such type of "tolerance") than solely because the presence of the fastest direct way. And second, the CO oxidation rate is faster than the direct oxidation rate that the latter turns into rate determining step, and in turn, make room for other parallel reactions. Despite this concerns, the original proposal [101] of the so called catalytic mechanism under the peak 0.25 V

$$HCOOH \rightarrow CO_2 + 2\,H_{ads} \tag{2.16}$$

followed by

$$2\,H_{ads} \rightarrow 2\,H^+ + e^- \tag{2.17}$$

explains the CO_2 production at the hydrogen region and a homogeneous reaction

$$HCOOH + Pd^{+2} \rightarrow Pd + CO_2 + 2\,H^+ \tag{2.18}$$

explains CO_2 production beyond *ca.* 1.4 V. Baring in mind that Pd has a relevant electrochemical corrosion rate [109], for instance 5 μg cm^{-2} h^{-1} at 750 mV, then an analytical amount of palladium complex should be present in the solution during the electrochemical experiments.

Investigations with Surface-enhanced infrared absorption spectroscopy in the attenuated total reflection mode (ATR-SEIRAS) give great detail of the poisoning species on FAO on palladium [98]. They reported the presence of different poisons: adsorbed CO and bridge-bonded formate as well as carbonate (CO_3^{2-}), bicarbonate (HCO_{3-}) and supporting anions. Globally, poisoning of the Pd surface by CO formed by step (2.13) is very slow and scarcely affects formic acid oxidation. According to the results, formate should be a short-lived reactive intermediate in formic acid oxidation and is hence detected when its decomposition yielding CO_2 is suppressed.

2.3.3.2 Formic Acid Oxidation on Pt-Sn

The effects of antimony, tin, and lead additions to the palladium black catalyst was analyzed [110]. Accordingly, each adatom strongly promotes formic acid oxidation in an electrochemical cell and reduces the amount of CO poison that develops on the catalyst surface after 1 h of oxidation. The authors attributed this effect to the third body effect (steric effect) but did not discard an electronic effect regarding that a decrease in the CO binding energy on palladium due to the presence of the adatoms, using XPS technique, was observed.

2.3.3.3 Formic Acid Oxidation on Pt-Pb

Stabilized upd lead on platinum was prepared and employed as electrode to measure simultaneously the rate of CO_2 production from both bulk and adsorbed formic acid [111]. In the experiments of Fig. 2.6, the strongly adsorbed residue is formed previously in a solution with ^{13}C-labeled formic acid by applying 0.25 V per 3 min, then the solution is replaced by ^{12}C-formic acid and the potentiodynamic is recorded simultaneously with the mass signals (using the DEMS system). Hence the CO_2 molecules coming from bulk formic acid oxidation (m/e = 44) and resulting from the oxidation of the strongly adsorbed residue (m/e = 45) are distinguished. On pure platinum, the surface is poisoned by the previously adsorbed ^{13}C residue and the V-I profile is in good agreement with the mass signal. In the presence of lead, the bulk HCOOH oxidation starts at 200 mV (monitored by m/e = 44) while the onset of residue oxidation occurs at the same potential as in pure platinum. Despite the 100 mV displacement of the peak to lower potentials in the case of the residue oxidation on upd-Pt compared to on pure platinum, the residue is oxidized by the platinum oxide. These results clearly demonstrate that the direct oxidation of bulk formic acid is accelerated by the presence of updPb adatoms even though in the presence of stuck residues on the surface. Experiments with electrochemical quartz crystal microbalance (EQCM) reveal that the majority of the current is delivered by formic acid oxidation but that the mass response is dominated by the changes in UPD

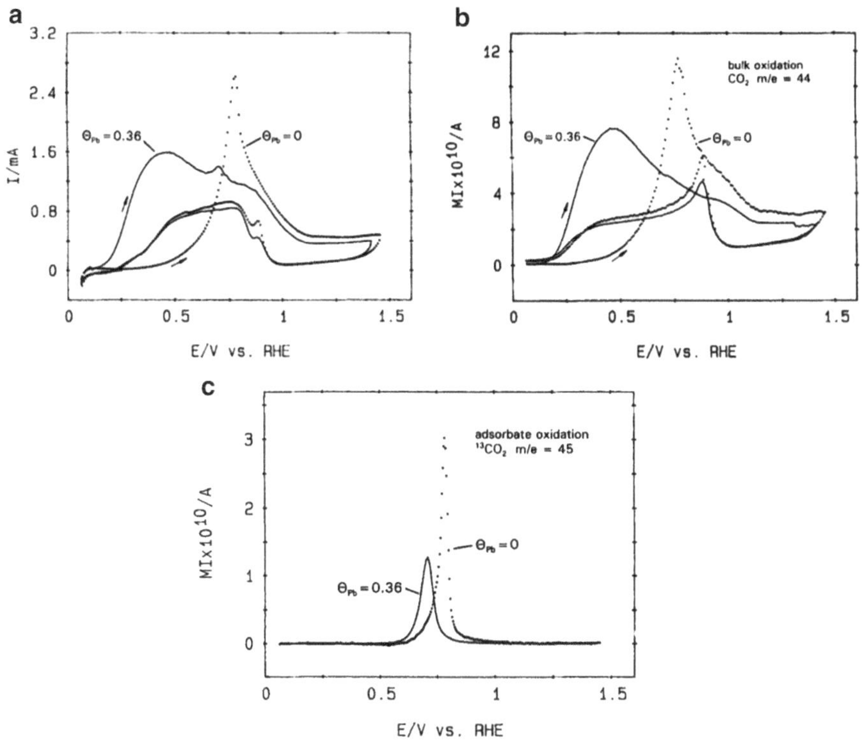

Fig. 2.6 Oxidation of bulk formic acid on sputtered platinum previously modified with 0.005 M ^{13}C-formic. (**a**) Cyclic voltammetry as well as (**b**) CV mass signal/potential (MSCV) for CO_2 from bulk $H^{12}COOH$, m/e = 44 and (**c**) for CO_2 from adsorbed ^{13}C residue, m/e = 45. Others: the *dotted lines* show the results without Pb and *solid line* with updPb (Θ_{Pb} = 0.36) (Reproduced by permission of The Electrochemical Society)

coverage [94]. Thus examination of mass responses reveals both variations in UPD coverage and the manner in which the underpotential deposits are affected by adsorbates derived from formic acid. At low concentrations of formic acid there is some suppression of the underpotential deposit and data suggest that strongly adsorbing intermediates form most rapidly in the hydrogen adsorption region of potential. The high current efficiency of CO_2 formation for methanol and formaldehyde oxidation at a PtPb electrode can be ascribed to the complete dehydrogenation of formaldehyde and formic acid due to electronic effects [86].

Several authors [111, 112] suggest two catalytic effects: third-body effect and electronic effect. However, the third-body effect cannot account solely for the observed catalytic effects. Thus, an electronic interaction between the updPb and the Pt substrate is likely as well. Other aspect concerning the lead deposits is the microstructural change during electrochemical experiments. Electrodeposited layers, near stoichiometric PtPb, yields a smooth compact surface that suffers microstructure changes after cyclic electrochemical experiments [113]. The dealloyed structure exhibited significant electrocatalytic behavior which was presumably related to Pb_{upd} on the available Pt surface.

Beyond the investigations with thin films, the study with nanodendritic materials suggest that nanoparticle morphology is responsible for the significant improvement of electrocatalytic activities toward the FAO mainly because PtPbnanodendrites could expose preferential crystalline planes from the intermetallic phase evolution from a Pt-based face-centered cubic to a PtPb hexagonal [114]. With respect to the fractal dimension of the dendritic surface, an improvement of surface area could be achieved.

In agreement with the flat surfaces, the PtPb nanoparticles exhibited significantly CO tolerance and a stable oxidation current for FAO was observed over 9 h, compared to a large decrease in oxidation current, ostensibly due to CO poisoning, on Pt, Pd, and Pt-Ru/Vulcan alloy nanoparticles [115].

2.4 Methanol Oxidation Far from Equilibrium: An Outlook

Standard investigations evaluate electrochemical proprieties of a given surface according to its steady stationary state; however, CO oxidation undergoes kinetic instabilities in the catalytic environment [116–120] whose electrochemical version were, so far, identified [121–123] even that sparsely. Among the kinetic instabilities the most broadly report in the electrochemical field are bistability and oscillations. It happens that theoretical approach treats oscillating CO oxidation [124] as part of the mechanism for the oscillating methanol oxidation [125] as well as the oscillating formic acid oxidation [126]. Predictions that match fair agreement with experimental reports on oscillating electroxidation of methanol [127–131] and formic acid [95, 132–134] using a conventional-three-electrode cell. In this section, two new perspectives concerning catalysts for DMFC are commented: (a) evaluation of kinetics and mechanism on different catalysts by analyzing its far from equilibrium behavior, and (b) taking advantages of the better conversion exhibited by the oscillatory state [135].

The mechanism of the methanol oxidation defines the time evolution of species coverage on the electrode surface, hence, defines also the set of ordinary differential equations which describes the time evolution for each coverage species [136] which is known as chemical network stability analyses (SNA). Accordingly, the unique instability which allows for oscillatory behavior is the Hopf bifurcation (HB) which most of the time is proofed by numerical solution. However, the methodology to solve analytically the set of differential equations was very recently applied to chemistry [137].

Figure 2.7 joins information from theory and experiments. The stationary CV in potentiostatic sweep is shown during the methanol oxidation. Conversely, under galvanostatic sweep mode, a region of instabilities appears with the beginning at the

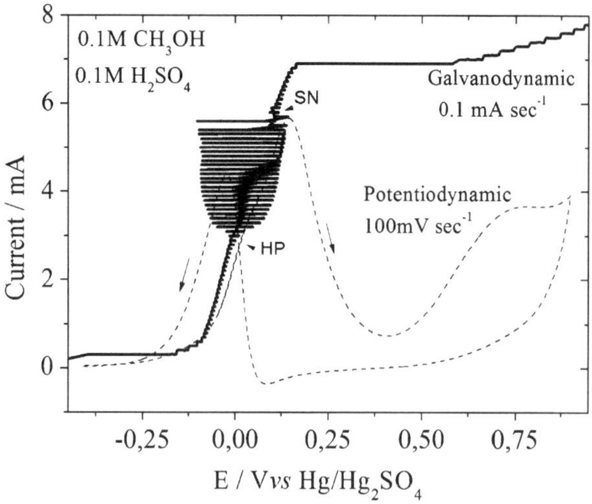

Fig. 2.7 Methanol oxidation on platinum under potentiostatic and galvanostatic sweep. *HB* means hopf bifurcation point and *SN* the saddle node bifurcation point. Both predicted by theory (Courtesy of the authors)

Hopf bifurcation point and the ending on Saddle-node bifurcation point, just as predicted by analytical solution of the set of EDOs as well as numerical simulations. Both saddle point and hopf bifurcation point are defined for a certain electrochemical mechanism as a function of a set of kinetic rate constants belonging to the given mechanism. Therefore, the location of the HB as well as SN is defined by a set of kinetic parameters which are mainly determined by the sort of catalyst. The time scale of poisoning process could be evaluated, for instance, by analyzing different current sweeps as well as by the corresponding time series. The lack or the appearance of the instability region at different range of current leads to a direct way to evaluate the rate constants for a given surface. Likewise, the feasibility of a certain mechanism could be evaluated leading to new perspectives or proposition to understand with more detail the aspects behind an electroxidation of small molecules.

Finally, oscillatory state render better conversion than stationary state considering the electrochemical class of oscillator in which the methanol is included [135]. The promise of lunching more efficient methanol conversion using oscillations is in its infancy, but there are evidences that point toward good prospects. Current oscillation on DMFC emerges [138] depending on the flow rate of methanol at the anode side and on the fixed potential. Another possibility for oscillations in the practical DMFC operation was reported by Du et al. as a result of the methanol crossover; in this case, the presence of methanol at the cathode led to superposition of the anodic MOR on the cathodic ORR [139]. Together, both evidences reinforce the parallel between findings using a conventional-three-electrode cell and practical fuel cells. On the bases of our experience, the feasibility of using the benefits from oscillations will run a searching for methanol catalysts that promote oscillation onset at lower overvoltage. To elucidate this we recall the understanding gained with the electroxidation of CO-containing hydrogen [123] which works as mimetic for the methanol electroxidation mechanism [140]. Kadyk et al. [141] reported an anodic overvoltage around 600 mV for the oscillations onset on Pt/C which is around 300 mV larger than for oscillations onset on PtRu/C [142]. According to the energy balance recommended in reference [135] that 300 mV higher overvoltage represents dispel of energy conversion.

References

1. Batista EA, Malpass GRP, Motheo AJ et al (2004) New mechanistic aspects of methanol oxidation. J Electroanal Chem 571:273–282
2. Iwasita T (2002) Electrocatalysis of methanol oxidation. Electrochim Acta 47:3663–3674
3. Cohen JL, Volpe DJ, Abruña HD (2007) Electrochemical determination of activation energies for methanol oxidation on polycrystalline platinum in acidic and alkaline electrolytes. Phys Chem Chem Phys 9:49–77
4. Batista EA, Malpass GRP, Motheo AJ et al (2003) New insight into the pathways of methanol oxidation. Electrochem Commun 5:843–846
5. Chen YX, Miki A, Ye S et al (2003) Formate, an active intermediate for direct oxidation of methanol on Pt electrode. J Am Chem Soc 125:3680–3681
6. Liu SX, Liao LW, Tao Q et al (2011) The kinetics of CO pathway in methanol oxidation at Pt electrodes, a quantitative study by ATR-FTIR spectroscopy. Phys Chem Chem Phys 13:9725–9735
7. Zhao W, Jusys Z, Behm RJ (2010) Quantitative online analysis of liquid-phase products of methanol oxidation in aqueous sulfuric acid solutions using electrospray ionization mass spectrometry. Anal Chem 82:2472–2479
8. Zhao W, Jusys Z, Behm RJ (2012) Complete quantitative online analysis of methanol electrooxidation products via electron impact and electrospray ionization mass spectrometry. Anal Chem 84:5479–5483
9. Santasalo-Aarnio A, Kwon Y, Ahlberg E et al (2011) Comparison of methanol, ethanol and iso-propanol oxidation on Pt and Pd electrodes in alkaline media studied by HPLC. Electrochem Commun 13:466–469
10. Jusys Z, Kaiser J, Behm RJ (2003) Methanol electrooxidation over Pt/C fuel cell catalysts: dependence of product yields on catalyst loading. Langmuir 19:6759–6769
11. Liao LW, Liu SX, Tao Q et al (2011) A method for kinetic study of methanol oxidation at Pt electrodes by electrochemical in situ infrared spectroscopy. J Electroanal Chem 650:233–240
12. Parsons R, Vandernoot T (1988) The oxidation of small organic molecules: a survey of recent fuel cell related research. J Electroanal Chem Interfacial Electrochem 257:9–45
13. Grozovski V, Climent V, Herrero E et al (2011) The role of the surface structure in the oxidation mechanism of methanol. J Electroanal Chem 662:43–51
14. García G, Koper MTM (2011) Carbon monoxide oxidation on Pt single crystal electrodes: understanding the catalysis for low temperature fuel cells. Chem Phys Chem 12:2064–2072
15. Vidal F, Tadjeddine A, Humbert C et al (2012) The influence of surface defects in methanol dissociative adsorption and CO oxidation on Pt(1 1 0) probed by nonlinear vibrational SFG spectroscopy. J Electroanal Chem 672:1–6
16. Juanto S, Beden B, Hahn F et al (1987) Infrared spectroscopic study of the methanol adsorbates at a platinum electrode: Part II. The Pt (100) surface in an acid medium. J Electroanal Chem Interfacial Electrochem 237:119–129
17. Herrero E, Franaszczuk K, Wieckowski A (1994) Electrochemistry of methanol at low index crystal planes of platinum: an integrated voltammetric and chronoamperometric study. J Phys Chem 98:5074–5083
18. Papoutsis A, Léger JM, Lamy C (1993) Study of the kinetics of adsorption and electro-oxidation of MeOH on Pt(100) in an acid medium by programmed potential voltammetry. J Electroanal Chem 359:141–160

19. Chang SC, Leung LWH, Weaver MJ (1990) Metal crystallinity effects in electrocatalysis as probed by real-time FTIR spectroscopy: electrooxidation of formic acid, methanol, and ethanol on ordered low-index platinum surfaces. J Phys Chem 94:6013–6021
20. Solla-Gullon J, Vidal-Iglesias FJ, Lopez-Cudero A et al (2008) Shape-dependent electrocatalysis: methanol and formic acid electrooxidation on preferentially oriented Pt nanoparticles. Phys Chem Chem Phys 10:3689–3698
21. Beden B, Kadirgan F, Lamy C et al (1981) Electrocatalytic oxidation of methanol on platinum-based binary electrodes. J Electroanal Chem Interfacial Electrochem 127:75–85
22. Campbell SA, Parsons R (1992) Effect of Bi and Sn adatoms on formic acid and methanol oxidation at well defined platinum surfaces. J Chem Soc Faraday Trans 88:833–841
23. Haner AN, Ross PN (1991) Electrochemical oxidation of methanol on tin-modified platinum single-crystal surfaces. J Phys Chem 95:3740–3746
24. Lizcano-Valbuena WH, Paganin VA, Gonzalez ER (2002) Methanol electro-oxidation on gas diffusion electrodes prepared with PtRu/C catalysts. Electrochim Acta 47:3715–3722
25. Zhang X, Chan K-Y (2002) Water-in-oil microemulsion synthesis of platinum–ruthenium nanoparticles, their characterization and electrocatalytic properties. Chem Mater 15:451–459
26. Kim T-W, Park S-J, Jones LE et al (2005) Structure and electrocatalysis of sputtered RuPt thin-film electrodes. J Phys Chem B 109:12845–12849
27. Watanabe M, Uchida M, Motoo S (1987) Preparation of highly dispersed Pt + Ru alloy clusters and the activity for the electrooxidation of methanol. J Electroanal Chem Interfacial Electrochem 229:395–406
28. Lizcano-Valbuena WH, De Souza A, Paganin VA et al (2002) Performance of a DMFC using Pt-Ru/C catalysts prepared by reduction with formic acid. Fuel Cells 2:159–165
29. Chu D, Gilman S (1996) Methanol electro-oxidation on unsupported Pt-Ru alloys at different temperatures. J Electrochem Soc 143:1685–1690
30. Lizcano-Valbuena WH, Paganin VA, CaP L et al (2003) Catalysts for DMFC: relation between morphology and electrochemical performance. Electrochim Acta 48:3869–3878
31. Godoi DRM, Perez J, Mercedes Villullas H (2007) Influence of particle size on the properties of Pt–Ru/C catalysts prepared by a microemulsion method. J Electrochem Soc 154: B474–B479
32. Godoi DRM, Perez J, Villullas HM (2009) Effects of alloyed and oxide phases on methanol oxidation of Pt – Ru/C nanocatalysts of the same particle size. J Phys Chem C 113:8518–8525
33. Frelink T, Visscher W, Van Veen JAR (1995) On the role of Ru and Sn as promotors of methanol electro-oxidation over Pt. Surf Sci 335:353–360
34. Azevedo DC, Lizcano-Valbuena WH, Gonzalez ER (2004) An impedance study of the rate determining step for methanol oxidation on platinum and platinum-ruthenium supported on high surface area carbon. J New Mater Electrochem Syst 7:191–196
35. Krausa M, Vielstich W (1994) Study of the electrocatalytic influence of Pt/Ru and Ru on the oxidation of residues of small organic molecules. J Electroanal Chem 379:307–314
36. Godoi DRM, Villullas HM (2012) Relevance of electronic effects on the yield of CO_2 from fethanol oxidation. Langmuir 28:1064–1067
37. Salgado JRC, Paganin VA, Gonzalez ER et al (2013) Characterization and performance evaluation of Pt–Ru electrocatalysts supported on different carbon materials for direct methanol fuel cells. Int J Hydrogen Energy 38:910–920
38. Hepel M, Kumarihamy I, Zhong CJ (2006) Nanoporous TiO_2-supported bimetallic catalysts for methanol oxidation in acidic media. Electrochem Commun 8:1439–1444
39. Villullas HM, Mattos-Costa FI, Bulhões LOS (2004) Electrochemical oxidation of methanol on Pt nanoparticles dispersed on RuO_2. J Phys Chem B 108:12898–12903
40. Kuznetsov VI, Belyi AS, Yurchenko EN et al (1986) Mössbauer spectroscopic and chemical analysis of the composition of Sn-containing components of Pt-Sn/Al_2O_3(Cl) reforming catalyst. J Catal 99:159–170

41. Radmilovic V, Richardson TJ, Chen SJ et al (2005) Carbon-supported Pt–Sn electrocatalysts for the anodic oxidation of H_2, CO, and H_2/CO mixtures. Part I. Microstructural characterization. J Catal 232:199–209
42. González MJ, Hable CT, Wrighton MS (1998) Electrocatalytic oxidation of small carbohydrate fuels at Pt–Sn modified electrodes. J Phys Chem B 102:9881–9890
43. Janssen MMP, Moolhuysen J (1976) Platinum–tin catalysts for methanol fuel cells prepared by a novel immersion technique, by electrocodeposition and by alloying. Electrochim Acta 21:861–868
44. Watanabe M, Furuuchi Y, Motoo S (1985) Electrocatalysis by ad-atoms: Part XIII. Preparation of ad-electrodes with tin ad-atoms for methanol, formaldehyde and formic acid fuel cells. J Electroanal Chem Interfacial Electrochem 191:367–375
45. Mcnicol BD, Short RT, Chapman AG (1976) Methanol electro-oxidation catalysts. Platinum promoted by tin. J Chem Soc Faraday Trans 1 72:2735–2743
46. Rahim MAA, Khalil MW, Hassan HB (2000) Platinum–tin alloy electrodes for direct methanol fuel cells. J Appl Electrochem 30:1151–1155
47. Honma I, Toda T (2003) Temperature dependence of kinetics of methanol electro-oxidation on PtSn alloys. J Electrochem Soc 150:A1689–A1692
48. Cathro KJ (1969) The oxidation of water-soluble organic fuels using platinum-tin catalysts. J Electrochem Soc 116:1608–1611
49. Wang K, Gasteiger HA, Markovic NM et al (1996) On the reaction pathway for methanol and carbon monoxide electrooxidation on Pt-Sn alloy versus Pt-Ru alloy surfaces. Electrochim Acta 41:2587–2593
50. Frelink T, Visscher W, Van Veen JAR (1994) The effect of Sn on Pt/C catalysts for the methanol electro-oxidation. Electrochim Acta 39:1871–1875
51. Bittins-Cattaneo B, Iwasita T (1987) Electrocatalysis of methanol oxidation by adsorbed tin on platinum. J Electroanal Chem Interfacial Electrochem 238:151–161
52. Morimoto Y, Yeager EB (1998) Comparison of methanol oxidations on Pt, Pt|Ru and Pt|Sn electrodes. J Electroanal Chem 444:95–100
53. Antolini E, Gonzalez ER (2011) Effect of synthesis method and structural characteristics of Pt–Sn fuel cell catalysts on the electro-oxidation of CH_3OH and CH_3CH_2OH in acid medium. Catal Today 160:28–38
54. Colmati F, Antolini E, Gonzalez ER (2005) Pt–Sn/C electrocatalysts for methanol oxidation synthesized by reduction with formic acid. Electrochim Acta 50:5496–5503
55. Herranz T, García S, Martínez-Huerta MV et al (2012) Electrooxidation of CO and methanol on well-characterized carbon supported PtxSn electrodes. Effect of crystal structure. Int J Hydrogen Energy 37:7109–7118
56. Uemura Y, Inada Y, Bando KK et al (2011) In situ time-resolved XAFS study on the structural transformation and phase separation of Pt_3Sn and PtSn alloy nanoparticles on carbon in the oxidation process. Phys Chem Chem Phys 13:15833–15844
57. Arikan T, Kannan AM, Kadirgan F (2013) Binary Pt–Pd and ternary Pt–Pd–Ru nanoelectrocatalysts for direct methanol fuel cells. Int J Hydrogen Energy 38:2900–2907
58. Ramos SG, Calafiore A, Bonesi AR et al (2012) Supported catalysts for alcohol oxidation synthesis and analysis of their catalytic activity. Int J Hydrogen Energy 37:14849–14853
59. Antolini E (2007) Platinum-based ternary catalysts for low temperature fuel cells. Part I. Preparation methods and structural characteristics. Appl Catal B Environ 74:324–336
60. Antolini E (2007) Platinum-based ternary catalysts for low temperature fuel cells. Part II. Electrochemical properties. Appl Catal B Environ 74:337–350
61. Martínez-Huerta MV, Rojas S, Gómez De La Fuente JL et al (2006) Effect of Ni addition over PtRu/C based electrocatalysts for fuel cell applications. Appl Catal B Environm 69:75–84
62. Liu J, Cao J, Huang Q et al (2008) Methanol oxidation on carbon-supported Pt-Ru-Ni ternary nanoparticle electrocatalysts. J Power Sources 175:159–165

63. Spendelow JS, Wieckowski A (2007) Electrocatalysis of oxygen reduction and small alcohol oxidation in alkaline media. Phys Chem Chem Phys 9:2654–2675
64. Mahapatra SS, Dutta A, Datta J (2011) Temperature dependence on methanol oxidation and product formation on Pt and Pd modified Pt electrodes in alkaline medium. Int J Hydrogen Energy 36:14873–14883
65. Antolini E, Gonzalez ER (2010) Alkaline direct alcohol fuel cells. J Power Sources 195:3431–3450
66. Iwasita T, Nart FC, Lopez B et al (1992) On the study of adsorbed species at platinum from methanol, formic acid and reduced carbon dioxide via in situ FT-ir spectroscopy. Electrochim Acta 37:2361–2367
67. Ishikawa Y, Liao MS, Cabrera CR (2000) Oxidation of methanol on platinum, ruthenium and mixed Pt-M metals (M = Ru, Sn): a theoretical study. Surf Sci 463:66–80
68. Vilekar SA, Fishtik I, Datta R (2007) Topological analysis of catalytic reaction networks: methanol decomposition on Pt(111). J Catal 252:258–270
69. Waszczuk P, Lu GQ, Wieckowski A et al (2002) UHV and electrochemical studies of CO and methanol adsorbed at platinum/ruthenium surfaces, and reference to fuel cell catalysis. Electrochim Acta 47:3637–3652
70. Li LC, Wang YW, Tian AM (2008) Adsorption of methanol on the Pt-Mo(111)/C surface. Wuli Huaxue Xuebao//Acta Physico – Chimi Sin 24:2013–2018
71. Grdeń M, Łukaszewski M, Jerkiewicz G et al (2008) Electrochemical behaviour of palladium electrode: oxidation, electrodissolution and ionic adsorption. Electrochim Acta 53:7583–7598
72. Lewis FA (1967) The palladium hydrogen system. Academic, London
73. Yépez O, Scharifker BR (2002) Oxidation of formate on hydrogen-loaded palladium. Int J Hydrogen Energy 27:99–105
74. Yépez O, Scharifker BR (1999) Oxidation of CO on hydrogen-loaded palladium. J Appl Electrochem 29:1185–1190
75. Garcia AC, Paganin VA, Ticianelli EA (2008) CO tolerance of PdPt/C and PdPtRu/C anodes for PEMFC. Electrochim Acta 53:4309–4315
76. Papageorgopoulos DC, Keijzer M, Veldhuis JBJ et al (2002) CO tolerance of Pd-rich platinum palladium carbon-supported electrocatalysts. J Electrochem Soc 149: A1400–A1404
77. Hoshi N, Kida K, Nakamura M et al (2006) Structural effects of electrochemical oxidation of formic acid on single crystal electrodes of palladium. J Phys Chem B 110:12480–12484
78. Gasteiger HA, Markovic N, Ross PN et al (1994) Carbon monoxide electrooxidation on well-characterized platinum-ruthenium alloys. J Phys Chem 98:617–625
79. Lu C, Masel RI (2001) The effect of ruthenium on the binding of CO, H_2, and H_2O on Pt (110). J Phys Chem B 105:9793–9797
80. Angerstein-Kozlowska H, Conway BE, Sharp WBA (1973) The real condition of electrochemically oxidized platinum surfaces: Part I. Resolution of component processes. J Electroanal Chem Interfacial Electrochem 43:9–36
81. Conway BE (1995) Electrochemical oxide film formation at noble metals as a surface-chemical process. Prog Surf Sci 49:331–452
82. Kucernak AR, Offer GJ (2008) The role of adsorbed hydroxyl species in the electrocatalytic carbon monoxide oxidation reaction on platinum. Phys Chem Chem Phys 10:3699–3711
83. Ciapina EG, Gonzalez ER (2009) Investigation of the electro-oxidation of CO on Pt-based carbon supported catalysts (Pt75Sn25/C, Pt65Ru35/C and Pt/C) by electrochemical impedance spectroscopy. J Electroanal Chem 626:130–142
84. de los Santos-Álvarez N, Alden LR, Rus E et al (2009) CO tolerance of ordered intermetallic phases. J Electroanal Chem 626:14–22
85. Shubina TE, Koper MTM (2002) Quantum-chemical calculations of CO and OH interacting with bimetallic surfaces. Electrochim Acta 47:3621–3628

86. Wang H, Alden L, Disalvo FJ et al (2008) Electrocatalytic mechanism and kinetics of SOMs oxidation on ordered PtPb and PtBi intermetallic compounds: DEMS and FTIRS study. Phys Chem Chem Phys 10:3739–3751
87. Ranjan C, Hoffmann R, Disalvo FJ et al (2007) Electronic effects in CO chemisorption on Pt-Pb intermetallic surfaces: a theoretical study. J Phys Chem C 111:17357–17369
88. Christoffersen E, Liu P, Ruban A et al (2001) Anode materials for low-temperature fuel cells: a density functional theory study. J Catal 199:123–131
89. Liao M-S, Cabrera CR, Ishikawa Y (2000) A theoretical study of CO adsorption on Pt, Ru and Pt–M (M = Ru, Sn, Ge) clusters. Surf Sci 445:267–282
90. Capon A, Parson R (1973) The oxidation of formic acid at noble metal electrodes: I. Review of previous work. J Electroanal Chem Interfacial Electrochem 44:1–7
91. Fernandez-Vega A, Feliu JM, Aldaz A et al (1991) Heterogeneous electrocatalysis on well-defined platinum surfaces modified by controlled amounts of irreversibly adsorbed adatoms: Part IV. Formic acid oxidation on the Pt(111)-As system. J Electroanal Chem Interfacial Electrochem 305:229–240
92. Llorca MJ, Herrero E, Feliu JM et al (1994) Formic acid oxidation on Pt(111) electrodes modified by irreversibly adsorbed selenium. J Electroanal Chem 373:217–225
93. Herrero E, Llorca MJ, Feliu JM et al (1995) Oxidation of formic acid on Pt(100) electrodes modified by irreversibly adsorbed tellurium. J Electroanal Chem 383:145–154
94. Zhang M, Wilde CP (1995) The influence of organic adsorbates on the UPD process. Oxidation of formic acid at UPD lead-modified platinum electrodes. J Electroanal Chem 390:59–68
95. Seland F, Tunold R, Harrington DA (2008) Impedance study of formic acid oxidation on platinum electrodes. Electrochim Acta 53:6851–6864
96. Batista BC, Varela H (2010) Open circuit interaction of formic acid with oxidized Pt surfaces: experiments, modeling, and simulations. J Phys Chem C 114:18494–18500
97. Samjeské G, Miki A, Osawa M (2007) Electrocatalytic oxidation of formaldehyde on platinum under galvanostatic and potential sweep conditions studied by time-resolved surface-enhanced infrared spectroscopy. J Phys Chem C 111:15074–15083
98. Miyake H, Okada T, Samjeske G et al (2008) Formic acid electrooxidation on Pd in acidic solutions studied by surface enhanced infrared absorption spectroscopy. Phys Chem Chem Phys 10:3662–3669
99. Clavilier J, Parsons R, Durand R et al (1981) Formic acid oxidation on single crystal platinum electrodes. Comparison with polycrystalline platinum. J Electroanal Chem Interfacial Electrochem 124:321–326
100. Iwasita T, Xia X, Herrero E et al (1996) Early stages during the oxidation of HCOOH on single-crystal Pt electrodes as characterized by infrared spectroscopy. Langmuir 12:4260–4265
101. Solis V, Iwasita T, Pavese A et al (1988) Investigation of formic acid oxidation on palladium in acidic solutions by on-line mass spectroscopy. J Electroanal Chem Interfacial Electrochem 255:155–162
102. Nishimura K, Kunimatsu K, Machida K-I et al (1989) Electrocatalysis of Pd + Au alloy electrodes: Part IV. IR spectroscopic studies on the surface species derived from formaldehyde and formate in alkaline solutions. J Electroanal Chem Interfacial Electrochem 260:181–192
103. Pavese A, Solís V (1991) Comparative investigation of formic acid and formaldehyde oxidation on palladium by a rotating ring-disc electrode and on-line mass spectroscopy in acidic solutions. J Electroanal Chem Interfacial Electrochem 301:117–127
104. Pavese AG, Solis VM, Giordano MC (1987) Oxidation of formic acid on palladium anodes in acidic medium. Effect of Pd(II) ions. Electrochim Acta 32:1213–1216
105. Wolter O, Willsau J, Heitbaum J (1985) Reaction pathways of the anodic oxidation of formic acid on Pt evidenced by ^{18}O labeling – a DEMS study. J Electrochem Soc 132:1635–1638

106. Gao W, Keith JA, Anton J et al (2010) Theoretical elucidation of the competitive electro-oxidation mechanisms of formic acid on Pt(111). J Am Chem Soc 132:18377–18385
107. Zhang J, Qiu C, Ma H et al (2008) Facile fabrication and unexpected electrocatalytic activity of palladium thin films with hierarchical architectures. J Phys Chem C 112:13970–13975
108. Clavilier J, Sun SG (1986) Electrochemical study of the chemisorbed species formed from formic acid dissociation at platinum single crystal electrodes. J Electroanal Chem Interfacial Electrochem 199:471–480
109. Llopis JF, Gamboa JM, Victori L (1972) Radiochemical study of the anodic behaviour of palladium. Electrochim Acta 17:2225–2230
110. Haan JL, Stafford KM, Masel RI (2010) Effects of the addition of antimony, tin, and lead to palladium catalyst formulations for the direct formic acid fuel cell. J Phys Chem C 114:11665–11672
111. Xia XH, Iwasita T (1993) Influence of underpotential deposited lead upon the oxidation of HCOOH in $HClO_4$ at platinum electrodes. J Electrochem Soc 140:2559–2565
112. Lei H-W, Hattori H, Kita H (1996) Electrocatalysis by Pb adatoms of HCOOH oxidation at Pt (111) in acidic solution. Electrochim Acta 41:1619–1628
113. Hwang S-M, Bonevich JE, Kim JJ et al (2011) Formic acid oxidation on $Pt_{100-x}Pb_x$ thin films electrodeposited on Au. J Electrochem Soc 158:B1019–B1028
114. Wang J, Asmussen RM, Adams B et al (2009) Facile synthesis and electrochemical properties of intermetallic PtPb nanodendrites. Chem Mater 21:1716–1724
115. Matsumoto F, Roychowdhury C, Disalvo FJ et al (2008) Electrocatalytic activity of ordered intermetallic PtPb nanoparticles prepared by borohydride reduction toward formic acid oxidation. J Electrochem Soc 155:B148–B154
116. Eiswirth M, Ertl G (1986) Kinetic oscillations in the catalytic CO oxidation on a Pt(110) surface. Surf Sci 177:90–100
117. Eiswirth RM (1987) Phänomene der selbstorganisation bei der oxidation von CO an Pt(110). Thesis, Ludwig-Maximilians Universität, München
118. Eiswirth M, Bürger J, Strasser P et al (1996) Oscillating Langmuir–Hinshelwood mechanisms. J Phys Chem 100:19118–19123
119. Krischer K (1990) Nichtlineare Dynamik zweier Grenzflächenreaktionen – kinetische Oszillationen, Bifurkationen und deterministisches Chaos. Thesis, Free University of Berlin, Berlin, p 249
120. Nettesheim S (1993) Reaction diffusion patterns in the catalytic CO oxidation on Pt(110): front propagation and spiral waves. J Chem Phys 98:9977
121. Koper MTM, Schmidt TJ, Marković NM et al (2001) Potential oscillations and S-shaped polarization curve in the continuous electro-oxidation of CO on platinum single-crystal electrodes. J Phys Chem B 105:8381–8386
122. Azevedo DC, Pinheiro ALN, Gonzalez ER (2002) Current oscillations during CO electro-oxidation on smooth platinum. Electrochem Solid State Lett 5:A51–A54
123. Lima ABDM (2012) Kinect instabilities in the electro-oxidation of CO-containing hydrogen. Thesis, Physical Chemistry USP-IQSC, São Carlos
124. Sauerbrei S, Sensse A, Eiswirth M (2011) Application of quotient rings for stability analysis in chemical systems. Z Naturforsch A 66a:231–241
125. Sauerbrei S (2010) Mechanism and model of the oscillatory electro-oxidation of methanol. J Chem Phys 132:154901
126. Sensse A, Gatermann K, Eiswirth M (2005) Analytic solution for the electrocatalytic oxidation of formic acid. J Electroanal Chem 577:35–46
127. Li L, Wei Z, Qi X et al (2008) Chemical oscillation in electrochemical oxidation of methanol on Pt surface. Sci China B Chem 51:322–332
128. Krausa M, Vielstich W (1995) Potential oscillations during methanol oxidation at Pt-electrodes. 1. Experimental conditions. J Electroanal Chem 399:7–12
129. Schell M (1998) Mechanistic and fuel-cell implications of a tristable response in the electrochemical oxidation of methanol. J Electroanal Chem 457:221–228

130. Seland F, Tunold R, Harrington DA (2010) Activating and deactivating mass transport effects in methanol and formic acid oxidation on platinum electrodes. Electrochim Acta 55:3384–3391
131. Dahlstrom PK, Harrington DA, Seland F (2012) A study of methanol oxidation by dynamic electrochemical impedance spectroscopy. ECS Trans 41:35–47
132. Strasser P, Eiswirth M, Ertl G (1997) Oscillatory instabilities during formic acid oxidation on Pt(100), Pt(110) and Pt(111) under potentiostatic control. II. Model calculations. J Chem Phys 107:991–1003
133. Chen S, Noles T, Schell M (2000) Effects of anions on chemical instabilities in the oxidation of formic acid. Electrochem Commun 2:171–174
134. Raspel F, Eiswirth M (1994) Current oscillations in the electrochemical oxidation of formic acid at Pt single crystal surfaces. J Phys Chem 98:7613–7618
135. Mota A, Gonzalez ER, Eiswirth M (2013) Enhanced efficiency of CO-containing hydrogen electroxidation with autonomous oscillations. J Phys Chem C 117:12495–12501
136. Eiswirth M (1994) Instability and oscillations in chemistry. Suuri kagaku 372:59–64
137. Gatermann K, Eiswirth M, Sensse A (2005) Toric ideals and graph theory to analyze Hopf bifurcations in mass action systems. J Symb Comput 40:1361–1382
138. Hacquard A (2005) Improving and understanding direc methanol fuel cell (DMFC) performance. Thesis, Worcester Polytechnic Institute
139. Du CY, Zhao TS, Yang WW (2007) Effect of methanol crossover on the cathode behavior of a DMFC: a half-cell investigation. Electrochim Acta 52:5266–5271
140. Mota A, Gonzalez ER, Eiswirth M (2013). Continuous CO electroxidation: bistability and oscillations. Phys Chem Chem Phys (to Submitted)
141. Kadyk T, Kirsch S, Hanke-Rauschenbach R et al (2011) Autonomous potential oscillations at the Pt anode of a PEM fuel cell under CO poisoning. Electrochim Acta 56:10593–10602
142. Mota A, Gonzalez ER, Eiswirth M (2011) Kinetic insights over a PEMFC operating on stationary and oscillatory states. J Phys Chem A 115:13773–13782

Chapter 3
Pt and Pd-Based Electrocatalysts for Ethanol and Ethylene Glycol Fuel Cells

O. Solorza-Feria and F. Javier Rodríguez Varela

Abstract Direct Oxidation Fuel Cells (DOFCs) are power systems that can replace H_2/O_2 fuel cells in different applications where the use of hydrogen is a major problem. The use of liquid fuels can be of great advantage due to the easiness of their transport and handling. A considerable number of small organic molecules have been considered as fuels in DOFCs. Methanol is the most studied liquid fuel, but the main problem of this molecule is its high toxicity score. Some alternative liquid fuels are taking an important role and are being considered as replacements of methanol, mainly because they are prone to be electro-oxidized at low temperatures at suitable electrocatalysts. C_2-fuels such as ethanol (EtOH, C_2H_5OH) and ethylene glycol (EG, $C_2H_6O_2$) are some of the most interesting molecules for DOFCs, because of their high energy density and due to the fact that only one C–C bond scission occurs during the dissociative adsorption of the molecule to form CO_2. In this chapter we present a description of the Direct Ethanol Fuel Cells (DEFC) and the Direct Ethylene Glycol Fuel Cell (DEGFC). We describe the reaction mechanism of the electro-oxidation of these fuels, the problems related to their crossover and the development of EtOH and EG tolerant cathodes.

3.1 Introduction

Polymer Electrolyte Membrane Fuel Cells (PEMFCs) have a number of advantages over other types of fuel cells. For example, due to their configuration, large power densities can be obtained in stacks of compact size. Moreover, PEMFCs can be fueled with a variety of fuels [1]. Besides the use of H_2, it has been demonstrated that liquid fuels such as methanol, ethanol and ethylene glycol can work as reliable

---

O. Solorza-Feria (✉) • F.J. Rodríguez Varela (✉)
Center for Research and Advanced Studies of the National Polytechnic Institute (CINVESTAV-IPN), Mexico City, Mexico
e-mail: osolorza@cinvestav.mx; javier.varela@cinvestav.edu.mx

H.R. Corti and E.R. Gonzalez (eds.), *Direct Alcohol Fuel Cells: Materials, Performance, Durability and Applications*, DOI 10.1007/978-94-007-7708-8_3,

source of chemical energy to react with O_2, to have a Direct Oxidation Fuel Cell (DOFC) configuration. Methanol has been largely studied and the complex oxidation mechanism of this alcohol on Pt-alloys anodes has been proposed [2–5]. It has been demonstrated that higher catalytic activities for the methanol oxidation reaction can be obtained by using Pt-Ru nanostructures over a wide range of operating temperatures [1].

Unfortunately, methanol is a toxic substance. Among the organic molecules used in fuel cell applications, it has one of the highest ecotoxicity scores, only behind hydrazine [6]. Therefore, alternative organic substances like C_2H_5O and $C_2H_6O_2$ are being tested in DOFCs. Due to the fact that the oxidation of these C_2-fuels form reaction intermediates such as CO or oxalic acid, Pt-alone anodes electrocatalysts become rapidly depolarized. Electrocatalysts with alternative chemical compositions are needed in order to sustain high current densities, maintaining an excellent stability. Pt-alloys have demonstrated a high catalytic activity for the oxidation of EtOH and EG [7–10]. The high performance of this type of materials relies on the bifunctional mechanism and the ligand effect. The first one describes the synergetic effect of adding an alloying metal M into the electrocatalyst's structure, which actively participates in the water discharge process to form (–OH) intermediates [11]:

$$H_2O_{ads(M)} \rightarrow OH_{ads(M)} + H^+ + e^- \tag{3.1}$$

The (–OH) molecules take part in the oxidation of CO_{ads} in the neighboring Pt sites:

$$CO_{ads(Pt)} + OH_{ads(M)} \rightarrow COOH_{ads(Pt)} \rightarrow CO_2 + H^+ + e^- \tag{3.2}$$

Reactions 3.1 and 3.2 take place at lower potentials (200–300 mV) on Pt-alloys compared to Pt-alone anodes [11–12]. The ligand effect has been related to an increase in the Pt d-band vacancies, modifying the adsorption energy of alcoholic intermediates on Pt, thus, the reaction rate is affected by electronic effects due to the interaction between Pt and the alloying element [1].

3.2 Direct Ethanol Fuel Cells (DEFC)

In a DEFC the anodic, cathodic and overall reactions are:

$$\text{Anode} : C_2H_5OH + 3H_2O \rightarrow 2CO_2 + 12H^+ + 12e^- \tag{3.3}$$

$$\text{Cathode} : 3O_2 + 12H^+ + 12e^- \rightarrow 6H_2O \tag{3.4}$$

$$\text{Overall reaction} : C_2H_5OH + 3O_2 \rightarrow 2CO_2 + 3H_2O \tag{3.5}$$

In several reports, Lamy et al. have developed in detail the different relationships that allow us to calculate several important thermodynamic parameters of DEFCs [13, 14]. The Gibbs energy change of ethanol (under standard conditions) is −1,326.7 kJ/mol. Therefore, the electromotive force (EMF) of a DEFC is:

$$E^{o}_{FC} = 1.145\,V \tag{3.6}$$

The thermodynamic efficiency of the cell, considering a value of $\Delta H^{o} = -1{,}367.9$ kJ/mol for ethanol, is:

$$\varphi_{rev} = 0.97 \tag{3.7}$$

The energy density in the case of C_2H_5OH is:

$$W_e = 8.01\,\frac{kWh}{kg} \tag{3.8}$$

3.2.1 Ethanol Oxidation Reaction (EOR) on Pt and Pt-Alloys

The understanding of the EOR mechanism on Pt-based electrocatalysts at low temperatures is relevant because of the importance of the DEFC technology. Several studies indicate that the electrocatalysts with higher catalytic activity for the EOR are Pt-Sn/C alloys [1, 15–19]. Even though, recently novel anode chemical compositions have been proposed. For example, the modification of Pt with rare earth oxides seems to improve the catalytic activity for this reaction. Cerium oxide is an interesting material for this kind of applications because it has been reported that can be used as co-catalyst or as co-support [20]. CeO_2 has the purpose of acting as a buffer for intermediates that serve to oxidize CO species into CO_2. Anodes of composition Pt-CeO_2/C have demonstrated an enhanced catalytic activity for the EOR, in some cases higher than that of Pt-Ru/C materials [21]. Also, CeO_2 has been used as co-support along with Vulcan. The addition of cerium oxide to disperse Pt-Sn lead to higher catalytic activity of a Pt-Sn/CeO_2-C anode, related to a Pt-Sn/C material [22]. At the same time, tri-metallic electrocatalyst have shown a high catalytic activity for the EOR. Basu et al. demonstrated that the Pt-Ir-Sn/C (20:5:15) anode increased the power density obtained from a DEFC [23].

The EOR mechanism on Pt-based anodes has been studied by different methods. The proposed path in acid media involves the first steps of ethanol adsorption via an O-adsorption or a C-adsorption. From these dissociative reactions, acetaldehyde (AAL) is formed [17, 24–26]:

$$Pt + CH_3CH_2OH \rightarrow Pt - (OCH_2CH_3)_{ads} + e^- + H^+ \tag{3.9}$$

$$Pt + CH_3CH_2OH \rightarrow Pt - (CHOHCH_3)_{ads} + e^- + H^+ \tag{3.10}$$

The formation of AAL at potentials lower than 0.6 V vs. RHE has been reported [17]. Acetic acid also forms as intermediate of this reaction, so the following mechanisms can be proposed [17]:

$E < 0.6$ V *vs* RHE

$$\mathrm{Pt} - (\mathrm{OCH_2CH_3})_{\mathrm{ads}} \rightarrow \mathrm{Pt} + \mathrm{CHOCH_3} + \mathrm{e}^- + \mathrm{H}^+ \quad (3.11)$$

$$\mathrm{Pt} - (\mathrm{CHOHCH_3})_{\mathrm{ads}} \rightarrow \mathrm{Pt} + \mathrm{CHOCH_3} + \mathrm{e}^- + \mathrm{H}^+ \quad (3.12)$$

The next step is the formation of Pt-$COCH_3$ species according to the reaction [17]:

E < 0.4 V/RHE

$$\mathrm{Pt} + \mathrm{CHOCH_3} \rightarrow \mathrm{Pt} - (\mathrm{COCH_3})_{\mathrm{ads}} + \mathrm{H}^+ + \mathrm{e}^- \quad (3.13)$$

Studies carried out by Lamy et al. show that CO species may adsorb on Pt at potentials as low as 0.3 V vs. RHE [17]. Meanwhile, Iwasita et al. found CH_4 traces at potentials lower than 0.4 V vs. RHE [24].

E > 0.3V/RHE

$$\mathrm{Pt} - (\mathrm{COCH_3})_{\mathrm{ads}} + \mathrm{Pt} \rightarrow \mathrm{Pt} - (\mathrm{CO})_{\mathrm{ads}} + \mathrm{Pt} - (\mathrm{CH_3}) \quad (3.14)$$

E < 0.4 V/RHE

$$\mathrm{Pt} - (\mathrm{CH_3})_{\mathrm{ads}} + \mathrm{Pt} - (\mathrm{H})_{\mathrm{ads}} \rightarrow 2\mathrm{Pt} + \mathrm{CH_4} \quad (3.15)$$

Moreover, Lamy et al. reported that at relatively high potentials (0.6 V vs. RHE) the dissociative adsorption of water molecules occurs on Pt, forming –OH species (Reaction 3.16) that further oxidize the alcoholic residues [17]. In this step, several reactions take place, including the oxidation of CO species (Reaction 3.17) and AAL (Reaction 3.18):

E > 0.6 V/RHE

$$\mathrm{Pt} + \mathrm{H_2O} \rightarrow \mathrm{Pt} - (\mathrm{OH})_{\mathrm{ads}} + \mathrm{H}^+ + \mathrm{e}^- \quad (3.16)$$

$$\mathrm{Pt} - (\mathrm{CO})_{\mathrm{ads}} + \mathrm{Pt} - (\mathrm{OH})_{\mathrm{ads}} \rightarrow 2\mathrm{Pt} + \mathrm{CO_2} + \mathrm{H}^+ + \mathrm{e}^- \quad (3.17)$$

$$\mathrm{Pt} - (\mathrm{CHOCH_3})_{\mathrm{ads}} + \mathrm{Pt} - (\mathrm{OH})_{\mathrm{ads}} \rightarrow 2\mathrm{Pt} + \mathrm{CH_3COOH} + \mathrm{H}^+ + \mathrm{e}^- \quad (3.18)$$

The dissociative adsorption of ethanol occurs at lower potentials on Pt-Sn/C electrocatalysts than on Pt/C anodes. OH species are formed on Sn sites at low potentials, leading to the oxidation of $(CO)_{ads}$ into CO_2, in agreement with the bifunctional mechanism.

3.2.2 DEFC Performances

Power densities obtained from DEFCs have been increased by using Pt-alloys anodes. Testing bi-metallic and tri-metallic alloys, Xin et al. showed that the

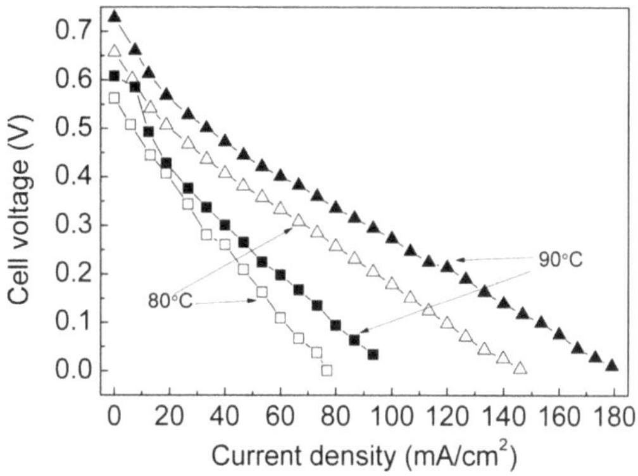

Fig. 3.1 Polarization curves of a DEFC at 80 °C (*open symbols*) and 90 °C (*closed symbols*). Anodes: PtSn/C (*triangles*) and PtRu/C (*squares*). Cathodes: Pt/C in both cases. Catalyst loading on all electrodes: 1 mg/cm^2. Membrane: Nafion® 117. Ethanol concentration and flow rate: 1 M and 3 mL/min. Oxygen pressure and flow rate: 1 atm and 0.4 L/min

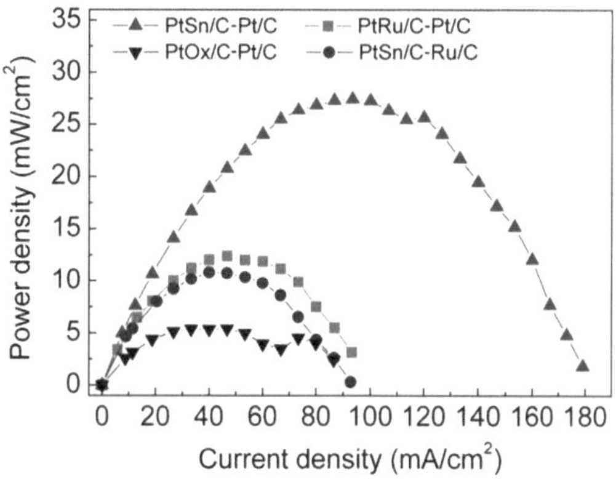

Fig. 3.2 Power density curves of a DEFC at operating at 90 °C with four different anode-cathode MEA configurations (shown in the figure). Catalyst loading on all electrodes: 1 mg/cm^2. membrane: Nafion® 117. Ethanol concentration and flow rate: 1 M and 3 mL/min. Oxygen pressure and flow rate: 1 atm and 0.4 L/min

catalytic activity decreases in the order Pt-Sn/C>Pt-Ru/C>Pt-W/C>Pt-Pd/C>Pt/C [16]. In a single cell, the $Pt_1Ru_1Sn_1/C$ anode showed a better performance than the Pt_1Ru_1/C electrocatalyst. However, the performance of the Pt_1Sn_1/C anode remained higher than that of $Pt_1Ru_1Sn_1/C$ [16]. Studies with different Pt:Sn ratios carried out by the same group indicate that the best performance is reached by using the Pt_2Sn_1/C composition as anode in a single DEFC [15].

Figure 3.1 shows the performance of a DEFC equipped with Pt_3Sn_1/C (triangles) and Pt_1Ru_1/C (squares) anodes, and Pt/C cathodes in both cases. The fuel cell operated at 80 °C and 90 °C. Under these experimental conditions, a comparison of the polarization curves at both temperatures shows that higher performances in current density and open circuit potential can be obtained with the PtSn/C alloy. For instance, there is a twofold increase in current density at 0.4 V when the PtSn–Pt MEA is used in the fuel cell, compared to the current density obtained at the same voltage with the PtRu–Pt based MEA, at any of the two operating temperatures (Fig. 3.1).

Figure 3.2 shows the power density attained by the DEFC of Fig. 3.1, this time with four different anode-cathode configurations (shown in the Fig. 3.2). The higher power density (27 mW/cm^2) is reached by a PtSn/C-Pt/C MEA configuration, followed by the PtRu/C-Pt/C MEA. Interestingly, the MEA composed of PtSn/C-Ru/C (anode-cathode) attains a power density similar to that of the PtRu/C-Pt/C MEA, demonstrating the high contribution of the PtSn/C anode to the current density – voltage characteristics of a DEFC integrated by a low-performance cathode (Ru/C).

3.3 Direct Ethylene Glycol Fuel Cells (DEGFC)

DEGFCs are being considered as an alternative to DMFCs. Peled et al. are among the workers that have extensively investigated the use of EG in fuel cells [10, 27–29]. Ethylene glycol (EG) is the simplest aliphatic diol. It has a high solubility in aqueous solutions [9].

The anode, cathode and overall reactions in a DEGFC are [30]:

$$\text{Anode}: C_2H_6O_2 + 2H_2O \rightarrow 2CO_2 + 10H^+ + 10e^- \quad (3.19)$$

$$\text{Cathode}: 5/2O_2 + 10H^+ + 10e^- \rightarrow 5H_2O \quad (3.20)$$

$$\text{Overall reaction}: C_2H_6O_2 + 5/2O_2 \rightarrow 2CO_2 + 3H_2O \quad (3.21)$$

Under standard conditions, the Gibbs energy change of EG is −1,181 kJ/mol, therefore the EMF of a DEGFC is:

$$E^{\circ}_{FC} = 1.22V \quad (3.22)$$

The thermodynamic efficiency of a DEGFC, considering a value of $\Delta H^{\circ} = -1,181$ kJ/mol for ethylene glycol, is:

$$\varphi_{rev} = 0.99 \quad (3.23)$$

The energy density of EG is:

$$W_e = 5.29\frac{kWh}{kg} \quad (3.24)$$

3.3.1 *Electro-oxidation of Ethylene Glycol on Pt-Based Anodes*

The study of the direct electro-oxidation of EG is a very active area of research. The development of anode electrocatalysts is of interest in order to find a suitable material for this reaction. In his work, Peled's group found that the intermediates formed during the electro-oxidation of EG include mainly oxalic acid and glycolic acid [29]. Therefore, it is necessary to use Pt-alloys or Pt-composite electrodes that may sustain a high catalytic activity in long-term tests. Peled et al. studied the performance of a DEGFC with Pt-Ru/C anodes and Pt/C cathodes [27–31]. They developed for the first time a high performance 10 cm^2 DEGFC stack based on Pt-Ru/C anodes [10]. At the same time, the studies by Selvaraj et al. on the electrocatalytic oxidation of ethylene glycol showed that Pt-Ru/C anodes have a higher catalytic activity than Pt/C anodes for this reaction [9]. A comparison of the performance of Pt-Ru/Vulcan versus Pt-Ru/CNTs showed that the use carbon nanotubes produced larger anodic current densities [9]. Spinacé et al. compared the performance of Pt-Sn/C and Pt-Sn-Ni/C anodes for the electro-oxidation of EG. They reported higher current densities for the trimetallic Pt-Sn-Ni/C (with 50:40:10 composition) related to the Pt-Ru/C anode [32]. Chetty and Scott showed a superior performance of the ternary PtRuW/Ti anode electrocatalyst related to PtRu/Ti, PtRuNi/Ti and PtRuPd/Ti anodes (where Ti was a titanium mesh) [33].

The EG oxidation reaction mechanism on bimetallic anodes (Pt-Ru) has been proposed as [9, 34]:

$$(CH_2OH)_2 \rightarrow (CH_2OH)_{2(ads)} \tag{3.25}$$

$$(CH_2OH)_{2(ads)} \rightarrow (:COH)_2 \tag{3.26}$$

$$(:COH) + 2H_2O \rightarrow (HCOOH)_{ads} + 2H^+ + 2e^- \tag{3.27}$$

$$Pt(HCOOH)_{ads} \rightarrow Pt(CO)_{ads} + H_2O \tag{3.28}$$

$$Pt + H_2O \rightarrow Pt(OH)_{ads} \tag{3.29}$$

$$Pt(OH)_{ads} + Pt(CO)_{ads} \rightarrow Pt + CO_2 + H^+ \tag{3.30}$$

$$Ru + H_2O \rightarrow Ru(OH)_{ads} + H^+ + e^- \tag{3.31}$$

$$Ru(OH)_{ads} + Pt(CO)_{ads} \rightarrow Pt + Ru + CO_2 + H^+ + e^- \tag{3.32}$$

Equations 3.31 and 3.32 describe the capability of Ru atoms for the water discharge process at lower potentials (0.35 V) than Pt (0.6 V) [34].

3.3.2 DEGFC Performances

Peled's group demonstrated the capacity of $C_2H_6O_2$ as a real fuel, using it to fuel single cells [27, 28, 31]. The DEGFC delivered power densities as high as 320 mW/cm^2 at 130 °C with a nanoporous proton-conducting membrane (NP-PCM) [27].

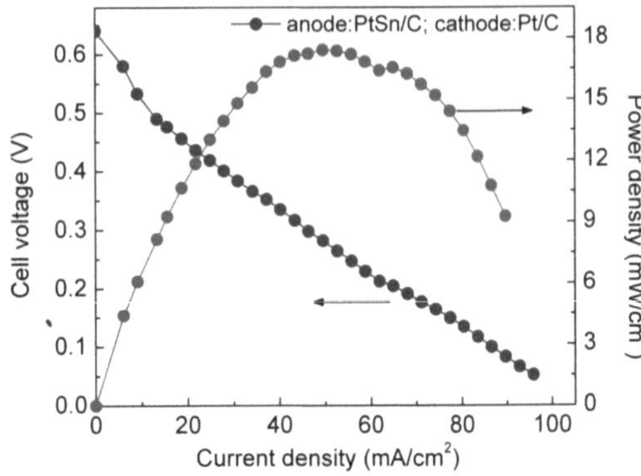

Fig. 3.3 Polarization curves of a DEGFC. Cell temperature: 80 °C. Anode: 20 % PtSn/C, cathode 20 % Pt/C. Catalyst loading in both electrodes: 2 mg/cm^2. Membrane: Nafion® 117. EG concentration and flow rate: 1 M and 2 mL/min. Oxygen flow rate: 0.5 L/min without backpressure

Figure 3.3 shows the polarization curve of a DEGFC with a 20 % Pt-Sn/C anode and a 20 % Pt/C cathode obtained by Rodríguez Varela and Savadogo. The operating temperature is 80 °C with a Nafion 117 membrane. The maximum power density of the cell is 17.5 mW/cm^2. The same authors tested a DEGFC cell equipped with a Pt-Sn/C anode and a Pd/C cathode, which delivered a maximum power density of 13.4 mW/cm^2 [35].

3.4 ORR at Ethanol Tolerant Electrocatalysts

The DOFC offers several advantages in comparison to hydrogen fed polymer electrolyte membrane PEM fuel cells, because the alcohol based aqueous solution can be processes inside and stored easier than gases. However, it should be acknowledged that the DOFC shows significantly lower power density than PEMFC due to slow oxidation kinetics and mass transport problem, i.e., transport of carbon dioxide out of the system and alcohol crossover through the membrane to the cathode side, where the alcohol oxidation on the cathodic compartment leads to a significant performance loss and lowering of the efficiency of the DOFC, situation which could be alleviated by use of alcohol tolerant electrocatalysts.

Electrocatalysts used for low temperature fuel cells have been extensively studied aiming the improvement of the catalytic activity, selectivity and greater stability at low cost. Beside intense studies of the ethanol and ethylene glycol penetration due to crossover effect into the cathode space of a fuel cell operation, there are some problems related to the cathodic reaction. The oxygen reduction reaction (ORR) is a multi-electron process which occurs at the fuel cell cathode side and has been the focus of attention because of the sluggish kinetics which limits the energy conversion efficiency of PEM fuel cells. The significant overpotential for the oxygen reduction reaction and the alcohol tolerance electrocatalyst required to overcome this electrochemical process, which could be attained by modification of the electronic structure of the electrode surface, reducing the tendency to bond to oxygen-containing species and hence increasing the number of available sites for oxygen adsorption and reduction processes. Although significant progress has been made in understanding how the ORR occurs, the exact reaction mechanism is still the subject of extensive discussions and remains indefinable. Recent publications give a good overview of the current state of the art in the development of ethanol [36] and ethylene glycol fuel cells [34], where a description mechanism of the ethylene glycol electrooxidation in acid media on Pt and different Pt-based alloys are analyzed.

The oxygen reduction reaction is of great importance in the development of novel cathode electrodes in PEM fuel cells. It is a complex process and includes several individual reactions where a desired discrete step involves the water formation through a fourth-electron pathway and the other is production of hydrogen peroxide as intermediate. Recent detailed review articles devoted to the present state and advances made in recent year on the ORR appear reported in literature [37–39]. In these articles authors emphasize that one of the target on cathode electrodes is to reduce the platinum content or the most expensive metallic content in electrocatalysts. Ruthenium-based chalcogenides have been considered as an alternative to Pt for the ORR because with the incorporation of Se the activity and effective selectivity improves significantly, modifying the electronic superficial structure of the catalyst weakening the O–O bond at the interface, making the catalyst more stable in the fuel cell operation. Furthermore, the action of adding a second metal to ruthenium chalcogenides have also been investigated as thin film electrodes in acid media and membrane-electrode assemblies, conducting to alcohol tolerance oxygen reduction electrocatalysts [39–42]. Lately, investigations on Pd-based alloys as cathode catalysts have been provided better understanding of the relationship between the size, shape, nanostructure, composition, and activity for the oxygen reduction reaction. The performances exhibited by these catalysts are equivalent to that reported for Pt and the enhanced activity of the alloy surface is assigned to bi-functional effects in which the unique catalytic properties of each element in the alloy combine in synergetic manner to yield a surface which improve the stability and enhance the catalytic activity compared to each element alone.

Most of the recent theoretical studies on ORR are based on Density Functional Theory (DFT) calculations, relating on the correlation of the electronic structure of the outermost layer of the surface composition and the catalytic activities of Pt- and Pd-based alloys [43–45]. It was found that there is a strong linear correlation between the adsorption energy of O and OH which are governing the rate of ORR on metallic catalyst surfaces. The ORR involves the sequential addition of protons and electrons to adsorbed oxygen. An approach to understand the electrochemistry of the ORR taking into account three possible mechanisms (oxygen dissociation, peroxyl dissociation and hydrogen peroxide dissociation) have been proposed by Nilekar and Mavrikakis [46], to describe how an oxygen molecule is reduced to form water molecule involving different intermediate species. The ORR reactivity of different surfaces is dictated by the strength of the oxygen adsorption, with the OH removal via hydrogenation and O–O bond scission. The detailed series of protonation elementary steps of the three ORR mechanisms is shown in Table 3.1, where intermediate species such as atomic hydrogen (H), atomic oxygen

Table 3.1 Three possible mechanisms for oxygen reduction reaction [43]

	1. Oxygen dissociation	2. Peroxyl dissociation	3. Hydrogen peroxide dissociation
(a)	$O_2 + * \rightarrow O_2^*$	$O_2 + * \rightarrow O_2^*$	$O_2 + * \rightarrow O_2^*$
(b)	$O_2^* + * \rightarrow 2\,O^*$	$O_2^* + H^+ + e^- \rightarrow OOH^*$	$O_2^* + H^+ + e^- \rightarrow OOH^*$
(c)	$O^* + H^+ + e^- \rightarrow OH^*$	$OOH^* \rightarrow O^* + OH^*$	$OOH^* + H^+ + e^- \rightarrow H_2O_2^*$
(d)	$OH^* + H^+ + e^- \rightarrow H_2O$	$O^* + H^+ + e^- \rightarrow OH^*$	$H_2O_2^* \rightarrow 2\,OH^*$
(e)		$OH^* + H^+ + e^- \rightarrow H_2O$	$OH^* + H^+ + e^- \rightarrow H_2O$

(O), hydroxyl (OH), peroxyl (OOH), and hydrogen peroxide (H_2O_2) are being the rate-determining step with surfaces with stronger and weaker oxygen binding.

Since carbon supported nanoparticles of Pt is the model of fuel cell electrocatalyst, almost all theoretical and experimental publications are referred to Pt and Pt-based alloys that exhibit significant ORR activity, even in presence of methanol [47, 48]. Experiments have shown that Pt monolayer deposited on a number of metals (Au, Rh, Pd, Ru and Ir) has significant changes in ORR electrocatalytic activity compared to pure Pt [49] and the authors reported a Volcano-type dependence of monolayer catalytic activity on the substrate was novel in the field of electrocatalysis and catalysis in general. For pure Pt (111) surface the ORR has been reported to follow a peroxyl dissociation mechanism [45] for its rate-determining O protonation reaction and on Pt (111) modified by a subsurface of transition metal containing Ni, Co and Fe, the ORR adopt a hydrogen peroxide dissociation mechanism for their rate-determining O_2 protonation reaction. The theoretical study reported by Wang et al. [45] reveals also how subsurface transition metals would modify the electronic structure and the catalytic activity of the outermost Pt monolayer surface.

Some alloys of carbon supported nanoparticles of Pt-Pd catalyst and Pt-based cathode catalysts modified with S, P and Bi have been reported as highly tolerant to ethanol for direct ethanol fuel cells (DEFC) [41, 50], observing a decreasing in the overpotential of the ORR on the bimetallic alloy in relation to that observed on pure Pt, ascribed to a reduced ethanol adsorption on Pt-Pd. Recently, Savadogo and Rodríguez Varela, have published that Pd and Pd-Co cathode catalysts exhibit high degree of tolerance to ethanol during the ORR in acid medium, observing a selective cathodic process attributed to a slow rate of adsorption of the ethanol and the reaction intermediates on the catalytic surface [51, 52]. Also, a high catalytic activity is observed originated from the synergistic effect between the supports and the bimetallic catalysts, making those material promising Pt-free alternative for cathodic electrocatalysts in direct alcohol fuel cells (DAFC). The enhanced catalytic activity of bimetallic surfaces in comparison with pure metal surfaces is usually ascribed to two effects: the bifunctional effect in which the unique catalytic properties of each of the elements in the compound are combined to yield a more active surface than each of the elements alone, and the electronic effect in which one of the elements alters the electronic properties of the other [53, 54].

Figure 3.4 shows the polarization curves of the ORR at the 10 % Pt_1Co_1/C cathode. The scans were performed in a non-stirred 0.5 M H_2SO_4 solution with and without 0.5 M EtOH at 2 mV/s. The purpose of such test is to evaluate the detrimental effect of C_2H_5OH in the catalytic activity for the ORR of the Pt-alloy.

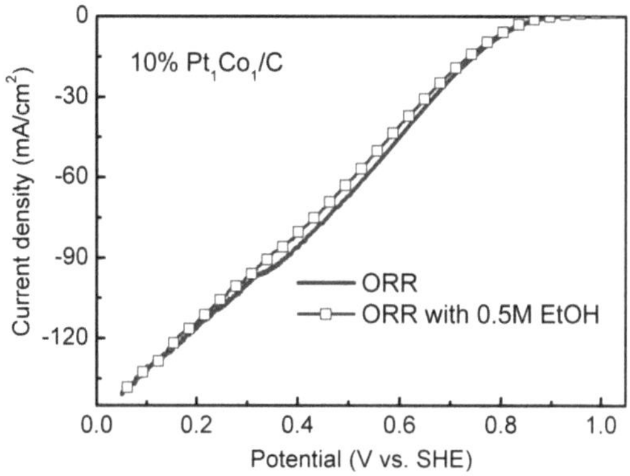

Fig. 3.4 Polarization curves of the ORR at Pt_1Co_1/C without and with 0.5 M C_2H_5OH. Support electrolyte: 0.5 M H_2SO_4. Scan rate: 2 mV/s

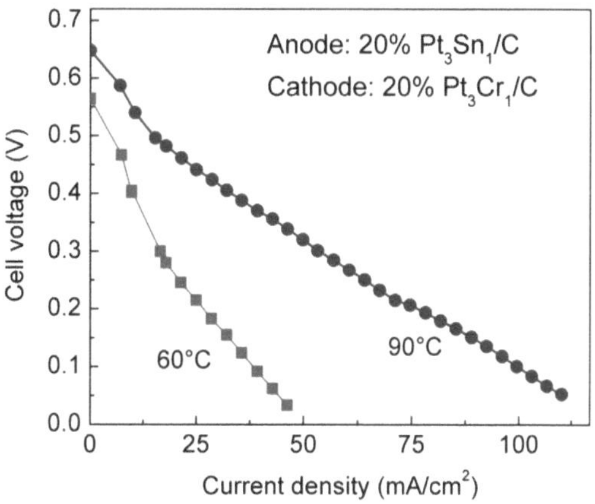

Fig. 3.5 Polarization curves of the DEFC based on a 20 % Pt_3Cr_1/C cathode at 60 °C and 90 °C, with a 20 % PtSn/C anode. Catalyst loading: 1 mg/cm^2. Membrane: Nafion® 117. Ethanol concentration: 1 M

The most important parameter for these evaluations is the shift of the onset potential for the ORR when ethanol is present in the solution and therefore, the increase in cathode overpotential [19]. The Pt_1Co_1/C cathode clearly shows a high selectivity for the ORR. The characteristics of the curves indicate the high insensibility of the PtCo material to the presence of the organic molecule. The Pt_1Co_1/C cathode also showed a higher mass activity for the ORR when compared to a Pt_3Cr_1/C cathode [19].

The performance characteristic of a DEFC employing 1 mg/cm^2 20 % Pt_3Cr_1/C ethanol tolerant ORR catalyst at the cathode and 20 % PtSn/C at the anode is shown in Fig. 3.5. Non-preheated ethanol was pumped to the anode side without backpressure at a flow rate of 2 ml/min. The performance curves at 60 °C and 90 °C are presented. The expected enhancement in *cell voltage vs. current density* performances is observed with the increased temperature in Ref. [19].

3.5 ORR at Ethylene Glycol Tolerant Electrocatalysts

Active cathode catalysts are required to eliminate the adverse effect brought along by the crossover of EG. The electrocatalytic activity of the catalysts toward the ORR, both in presence and absence of $C_2H_6O_2$ is evaluated for application in DEGFCs. Carbon supported Pd electrocatalysts synthesized by the formic acid method, and $Pt_{40}Pd_{60}$/MWCNT produced by reduction with $NaBH_4$, have been recently reported by Rodríguez Varela et al. as active and EG tolerant cathode electrocatalyst [35, 55]. Mixed-potential due to cathodic ethylene glycol oxidation is detrimental to the performance of the cathode catalyst in DEGFCs and this effect was observed when Pt was used as cathode catalyst. However, good performances were observed when PtCoSn/C and Pd/C catalyst were used as cathodes in DEGFCs [29, 56]. Elsewhere, a 30 % Pt_{70}-Co_{30}/MWCNT material was evaluated as cathode in 0.5 M H_2SO_4. The alloy showed a good behavior as ORR selective electrocatalyst and a high tolerance to EG [47]. The EIS results of the ORR at Pt-Co/MWCNT in the absence and presence of the liquid fuel showed one-arc spectra, indicating that the ORR mechanism at the cathode may not change when that substance was in the solution. The behavior of Pt-Co/MWCNT for the ORR with EG, EtOH or 2-Prop was similar. Thus, the ORR mechanism with the three organic molecules was proposed to be analogous [47].

Figure 3.6a depicts the linear scan voltammograms (LSVs) of the ORR on 20 % Pd/C without ethylene glycol at $\omega = 400, 800, 1{,}200$ and 1,600 rpm. The LSVs show a clear dependence of the ORR current densities with potential and rotation rate. The ORR on the Pd electrocatalyst seems to be under kinetic and mixed control in the potential scanned, not reaching a well defined limiting current. Levich-Koutecky plots 1/j vs. $1/\omega^{1/2}$ at different potentials corresponding to the experimental values are shown in the insert in Fig. 3.6a. Clearly, these plots show linearity and parallelism, indicating first order kinetics with respect to molecular oxygen.

Figure 3.6b shows the LSVs of the ORR on the Pd/C electrocatalyst in 0.5 M H_2SO_4 solution containing 0.125, 0.25, 0.5 or 1 M EG. The polarization curve of the ORR without EG at 1,600 rpm (from Fig. 3.6a) is shown as well in Fig. 3.6b for comparison purposes. The high selectivity for the ORR and tolerance of Pd/C to the presence of $C_2H_6O_2$ concentrations as high as 1 M, can be clearly seen in Fig. 3.6b. The values with EG remain practically unchanged with respect to the value without organic molecule.

No organic molecule oxidation peak, normally observed at Pt/C, appear in the case of Pd/C, making the negative effect of the liquid fuel on the cathode overpotential and in the onset potential for the ORR exceptionally limited.

The ORR was also analyzed on perovskite-type oxide $La_{1-x}Sr_x MnO_3$ in direct ethylene glycol alkaline fuel cells (DEGAFC) observing a high tolerance to EG since the cathodic polarization curves were not affected by the concentration of EG supplied to the anodic side [57].

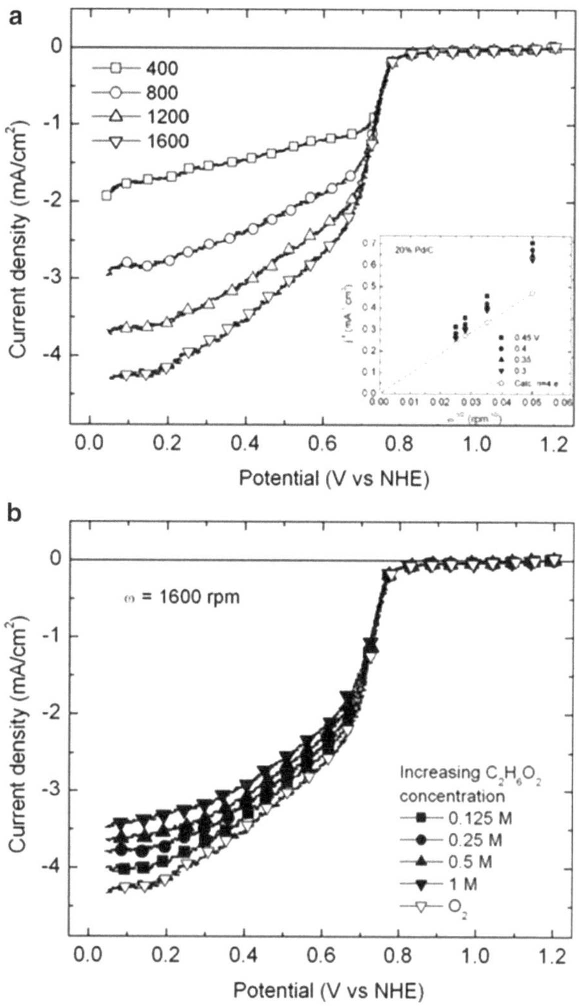

Fig. 3.6 (**a**) LSVs of the ORR at different rotation rates without EG. (**b**) LSVs of the ORR in the presence of increasing EG concentrations (indicated). The polarization curve of the ORR without EG at 1,600 rpm (from Fig. 3.6a) is also shown in Fig. 3.6b. Electrocatalyst: 20 % Pd/C. Electrolyte: O_2-saturated 0.5 M H_2SO_4 at 25 °C. Scan rate: 10 mV/s [56] (Reproduced with permission of the publisher)

3.6 Conclusions

The use of C_2-fuels in DOFCs has proven to be advantageous from a lower toxicity point of view, related to methanol. Moreover, high energy densities have been attained from ethanol and ethylene glycol fuel cells. Due to the fact that only one C–C scission occurs during the dissociative adsorption of these fuels, their oxidation at low temperatures (i.e., 80 °C, the operating temperature of DOFCs) at Pt-based electrocatalysts to form CO_2 is possible. Although the reaction mechanism of EtOH and EG can be complex, most of the reaction intermediates have been identified. In the case of C_2H_5OH, acetic acid, acetaldehyde and CO are the main intermediates. Pt-Sn/C alloys have shown the best performance for this reaction. On the other hand, during the oxidation of $C_2H_6O_2$, oxalic and glycolic acid are the most important intermediates. The definition of the most active catalysts for the EG oxidation reaction is still an on-going line of research. The studies indicate that Pt-Ru/C and Pt-Sn/C materials have a high catalytic activity of this anodic reaction, although the most active chemical composition is yet to be clearly defined. Novel anode compositions, such as ceria-modified Pt (Pt-CeO_2/C), are attractive alternatives for this reaction. A combination of theoretical and experimental studies have been performed to achieve a desirable design of stable nanocatalysts that have higher activity and alcohol tolerance ORR, with lower Pt loading and increased resistance to OH and diverse intermediate poisoning.

References

1. Rodríguez Varela FJ, Savadogo O (2008) Catalytic activity of carbon-supported electrocatalysts for direct ethanol fuel cell applications. J Electrochem Soc 155:B618–B624
2. Lamy C, Léger J-M, Srinivasan S (2001) Direct methanol fuel cells: from a twentieth century electrochemist's dream to a twenty-first century emerging technology. In: Bockris J-O'M, Conway B, White RE (eds) Modem aspects of electrochemistry, vol 34. Kluwer Academic/ Plenum Publishers, New York
3. Lamy C, Lima A, LeRhun V, Delime F, Coutanceau C, Léger J-M (2002) Recent advances in the development of Direct Alcohol Fuel Cells (DAFC). J Power Sources 105:283–296
4. Aricò AS, Srinivasan S, Antonucci V (2001) DMFCs: from fundamental aspects to technology development. Fuel Cells 1:133–161
5. Neergat M, Leveratto D, Stimming U (2002) Catalysts for direct methanol fuel cells. Fuel Cells 2:25–30
6. Demirci UB (2009) How green are the chemicals used as liquid fuels in direct liquid-feed fuel cells? Environ Int 35:626–631
7. Stevanovic S, Tripkovic D, Rogan J, Minic D, Gavrilovic A, Tripkovic A, Jovanovic VM (2011) Enhanced activity in ethanol oxidation of Pt_3Sn electrocatalysts synthesized by microwave irradiation. Russ J Phys Chem A 85:2299–2304
8. Silva JCM, Parreira LS, De Souza RFB, Calegaro ML, Spinacé EV, Neto AO, Santos MC (2011) PtSn/C alloyed and non-alloyed materials: differences in the ethanol electro-oxidation reaction pathways. Appl Catal B Environ 110:141–147
9. Selvaraj V, Vinoba M, Alagar M (2008) Electrocatalytic oxidation of ethylene glycol on Pt and Pt–Ru nanoparticles modified multi-walled carbon nanotubes. J Colloid Interf Sci 322:537–544
10. Livshits V, Philosoph M, Peled E (2008) Direct ethylene glycol fuel-cell stack — study of oxidation intermediate products. J Power Sources 178:687–691
11. Kim HJ, Kim DY, Han H, Shul YG (2006) PtRu/C-Au/TiO_2 electrocatalyst for a direct methanol fuel cell. J Power Sources 159:484–490
12. Suk YJ, Tae KH, Joh H-I, Kim H, Heup MS (2011) Preparation of a CO-tolerant $PtRu_xSn_y$/C electrocatalyst with an optimal Ru/Sn ratio by selective Sn-deposition on the surfaces of Pt and Ru. Int J Hydrogen Energy 36:1930–1938
13. Lamy C, Belgsir EM, Léger J-M (2001) Electrocatalytic oxidation of aliphatic alcohols: application to the direct alcohol fuel cell (DAFC). J Appl Electrochem 31:799–809
14. Vigier F, Coutanceau C, Perrard A, Belgsir EM, Lamy C (2004) Development of anode catalysts for direct ethanol fuel cell. J Appl Electrochem 34:439–446
15. Zhou W, Zhou Z, Song S, Li W, Sun G, Tsiakaras P, Xin Q (2003) Pt based anode catalysts for direct ethanol fuel cells. Appl Catal B Environ 46:273–285

16. Zhou WJ, Li WZ, Song SQ, Zhou ZH, Jiang LH, Sun GQ, Xin Q, Poulianitis K, Kontou S, Tsiakaras P (2004) Bi- and tri-metallic Pt-based anode catalysts for direct ethanol fuel cells. J Power Sources 131:217–223
17. Vigier F, Coutanceau C, Hahn F, Belgsir EM, Lamy C (2004) On the mechanism of ethanol electro-oxidation on Pt and PtSn catalysts: electrochemical and in situ IR reflectance spectroscopy studies. J Electroanal Chem 563:81–89
18. Lamy C, Rousseau S, Belgsir EM, Coutanceau C, Léger J-M (2004) Recent progress in the direct ethanol fuel cell: development of new platinum–tin electrocatalysts. Electrochim Acta 49:3901–3908
19. Rodríguez Varela FJ, Savadogo O (2009) Ethanol-tolerant Pt-alloy cathodes for direct ethanol fuel cell (DEFC) applications. Asia-Pac J Chem Eng 4:17–24
20. Antolini E, Perez J (2011) The use of rare earth-based materials in low-temperature fuel cells. Int J Hydrogen Energy 36:15752–15765
21. De Souza RFB, Flausino AEA, Rascio DC, Oliveira RTS, Teixeira Neto E, Calegaro ML, Santos MC (2009) Ethanol oxidation reaction on $PtCeO_2/C$ electrocatalysts prepared by the polymeric precursor method. Appl Catal B Environ 91:516–523
22. Oliveira Neto A, Linardi M, dos Anjos DM, Tremiliosi-Filho G, Spinacé EV (2009) Electro-oxidation of ethanol on $PtSn/CeO_2$–C electrocatalyst. J Appl Electrochem 39:1153–1156
23. Tayal J, Rawat B, Basu S (2011) Bi-metallic and tri-metallic Pt-Sn/C, Pt-Ir/C, Pt-Ir-Sn/C catalysts for electro-oxidation of ethanol in direct ethanol fuel cell. Int J Hydrogen Energy 36:14884–14897
24. Iwasita T, Pastor E (1994) A DEMS and FTIR spectroscopic investigation of adsorbed ethanol on polycrystalline platinum. Electrochim Acta 39:531–537
25. Rightmire RA, Rowland RL, Boos DL, Beals DL (1964) Ethyl alcohol oxidation at platinum electrodes. J Electrochem Soc 111:242–247
26. Lamy C, Coutanceau C, Leger J-M (2009) The direct ethanol fuel cell: a challenge to convert bioethanol cleanly into electric energy. In: Barbaro P, Bianchini C (eds) Catalysis for sustainable energy production. Wiley, New York
27. Peled E, Livshits V, Duvdevani T (2002) High-power direct ethylene glycol fuel cell (DEGFC) based on nanoporous proton-conducting membrane (NP-PCM). J Power Sources 106:245–248
28. Livshits V, Peled E (2006) Progress in the development of a high-power, direct ethylene glycol fuel cell (DEGFC). J Power Sources 161:1187–1191
29. Travitsky N, Burstein L, Rosenberg Y, Peled E (2009). Effect of methanol, ethylene glycol and their oxidation by-products on the activity of Pt-based oxygen-reduction catalysts. J Power Sources 194:161–167
30. Rodriguez Varela J, Savadogo O (2010). Celdas de combustible de consumo directo de moléculas orgánicas, In: Rodriguez Varela FJ, Solorza Feria O, Hernández Pacheco E (eds) Celdas de Combustible. yld books, Canada
31. Peled E, Duvdevani T, Aharon A, Melman A (2001) New fuels as alternatives to methanol for direct oxidation fuel cells. Electrochem Solid State Lett 4:A38–A41
32. Oliveira Neto A, Linardi M, Spinacé EV (2006) Electro-oxidation of ethylene glycol on PtSn/C and PtSnNi/C electrocatalysts. Ionics 12:309–313
33. Chetty R, Scott K (2007) Catalysed titanium mesh electrodes for ethylene glycol fuel cells. J Appl Electrochem 37:1077–1084
34. Serov A, Kwak C (2010) Recent achievements in direct ethylene glycol fuel cells (DEGFC). Appl Catal B Environ 97:1–12
35. Rodríguez Varela FJ, Fraire Luna S, Savadogo O (2009) Synthesis and evaluation of highly tolerant Pd electrocatalysts cathodes in direct ethylene glycol fuel cells. Energies 2:944–956
36. Antolini E, Gonzalez ER (2010) Alkaline direct alcohol fuel cells. J Power Sources 195:3431–3450
37. Morozan A, Jousselene B, Palacin S (2011) Low-platinum and platinum free catalysts for the oxygen reduction reaction at fuel cell cathodes. Energy Environ Sci 4:1238–1254
38. Keith JA, Jerskiewicz G, Jacob T (2010) Theoretical investigations of the oxygen reduction reaction. Chem Phys Chem 11:2714–2731

39. Feng Y, Gago A, Timperman L, Alonso-Vante N (2011) Chalcogenide metal centers for oxygen reduction reaction: activity and tolerance. Electrochim Acta 56:1009–1022
40. Tsivadze AY, Tarasevich MR, Kuzov AV, Romanova IA, Pripadchev DA (2008) New nanosized cathode electrocatalysts tolerant to ethanol. Doklady Phys Chem 421:166–169
41. Suárez-Alcántara K, Solorza-Feria O (2009) Comparative study of oxygen reduction on $Ru_xM_ySe_z$ (M = Cr, Mo, W) electrocatalysts for polymer exchange membrane fuel cell. J Power Sources 192:165–169
42. Ramos-Sánchez G, Solorza-Feria O (2010) Synthesis and characterization of $Pd_{0.5}Ni_xSe_{(0.5-x)}$. Int J Hydrogen Energy 35:12105–12110
43. Norskov JK, Rossmeisl J, Logadottir A, Lindquist L, Kitchin JR, Bligaard T, Jónsson H (2004) Origin of the overpotential for the oxygen reduction reaction at a fuel cell cathode. J Phys Chem B 108:17886–17892
44. Tripkovic V, Skulason E, Siahrostami S, Norskov JK, Rossmeisl J (2010) The oxygen reduction mechanism on Pd(111) from density functional theory calculations. Electrochim Acta 55:7975–7981
45. Duan Z, Wang G (2011) A first principle study of oxygen reduction reaction on a Pt(111) surface modified a subsurface transition metal M (M = Ni, Co or Fe). Phys Chem Chem Phys 13:20178–20187
46. Nilekar AU, Mavrikakis M (2008) Improved oxygen reduction reactivity of platinum monolayers on transition metal surfaces. Surf Sci 602:L89–L94
47. Morales-Acosta D, López de la Fuente D, Arriaga LG, Vargas Gutierrez G, Rodríguez Varela FJ (2011) Electrochemical investigation of Pt-Co/MWCNT as an alcohol-tolerant ORR catalyst for direct oxidation fuel cells. Int J Electrochem Sci 6:1835–1854
48. Rodríguez Varela FJ, Gaona Coronado AA, Jiang Q-Z, Bartolo Pérez P (2011) Pt-CeO_x/MWCNT electrocatalysts as ethanol-tolerant ORR cathodes for direct alcohol fuel cells. J New Mater Electrochem Syst 14:75–80
49. Zhang JL, Vukmirovic MB, Xu Y, Mavrikakis M, Adzic RR (2005) Controlling the catalytic activity of platinum monolayer electrocatalysts for oxygen reduction with different substrates. Angew Chem Int Ed 44:2132–2135
50. Lopes T, Antolini E, Gonzalez ER (2008) Carbon supported Pt-Pd alloys as an ethanol tolerant oxygen reduction electrocatalyst for direct ethanol fuel cells. Int J Hydrogen Energy 33:5563–5570
51. Savadogo O, Rodriguez Varela FJ (2008) Palladium-alloy catalysts as ethanol tolerant cathodes for direct alcohol fuel cell applications. J New Mater Electrochem Syst 11:69–74
52. Savadogo O, Rodríguez Varela FJ (2006) Palladium-alloy catalysts as ethanol tolerant cathodes for direct alcohol fuel cell (DEFC) applications. ECS Trans 1:247–254
53. Salvador-Pascual JJ, Collins-Martínez V, López-Ortiz A, Solorza-Feria O (2010) Low Pt content on the $Pd_{45}Pt_5Sn_{50}$ cathode catalyst for PEM fuel cells. J Power Sources 195:3374–3379
54. Ramos-Sánchez G, Yee-Madeira H, Solorza-Feria O (2008) PdNi electrocatalyst for oxygen reduction in acid media. Int J Hydrogen Energy 33:3596–3600
55. Morales-Acosta D, Arriaga LG, Alvarez-Contreras L, Fraire Luna S, Rodríguez Varela FJ (2009) Evaluation of $Pt_{40}Pd_{60}$/MWCNT electrocatalyst as ethylene glycol-tolerant oxygen reduction cathodes. Electrochem Commun 11:1414–1417
56. Rodríguez Varela FJ, Fraire Luna S, Dabek Klapco R (2009) Evaluation of Pd/C electrocatalyst as ORR cathodes tolerant to ethylene glycol. J New Mater Electrochem Syst 12:3–8
57. Miyazaki K, Sugimura N, Matsuoka K, Iriyama Y, Abe T, Matsuoka M, Ogumi Z (2008) Perovskite-type oxides $La_{1-x}Sr_xMnO_3$ for cathode catalysts in direct ethylene glycol alkaline fuel cells. J Power Sources 178:683–686

Chapter 4
Electro-oxidation of 3-Carbon Alcohols and Its Viability for Fuel Cell Application

Janaina Fernandes Gomes, Patricia Maria Patrizi Pratta, and Germano Tremiliosi-Filho

Abstract N-propanol, 2-propanol, 1,2-propanediol, 1,3-propanediol, glycerol, allyl and propargyl are all aliphatic alcohols containing three carbon atoms.

The difference between them is the number of hydroxyl groups, their positions in the molecular chain and the saturation degree. The interaction between these alcohols and model surfaces (such as platinum and gold) and the further reaction has been widely investigated in the last three decades. Previous studies have revealed that the electrosorption properties of alcohols with three carbon atoms chain depend on: (1) the number of hydrogen atoms bonded to the α-carbon atom; (2) the position of the OH in the hydrocarbon chain; (3) the number of OH groups; and (4) the relative position of the OH group. One of the motivations for studying alcohols with three carbon atoms chain is the possibility of application of these alcohols in direct alcohol fuel cells for electric energy generation, co-generation of heat and chemicals. In particular, glycerol is an interesting alcohol, since it is abundantly produced as a co-product of the biodiesel fabrication and it has high energy content. In this chapter, we concentrate attention on the reactivity of glycerol over Pt, Pd and Au-based electrodes in acidic and alkaline medium and on the application of glycerol in direct glycerol fuel cell, highlighting the main challenges and perspectives on that.

J.F. Gomes (✉) • G. Tremiliosi-Filho
Instituto de Química de São Carlos, Universidade de São Paulo, São Carlos, SP C.P. 780-13560-970, Brazil
e-mail: janainafg@iqsc.usp.br

P.M.P. Pratta
Departamento de Engenharia Química, Universidade Federal de São Carlos, Rodovia Washington Luiz, km 235–SP 310, São Carlos, SP CEP 13565-905, Brazil

H.R. Corti and E.R. Gonzalez (eds.), *Direct Alcohol Fuel Cells: Materials, Performance, Durability and Applications*, DOI 10.1007/978-94-007-7708-8_4,

4.1 Introduction

N-propanol, 2-propanol, 1,2-propanediol, 1,3-propanediol and glycerol are all aliphatic alcohols containing three carbon atoms. Additionally, allyl and propargyl are unsaturated alcohols. The difference between them is the number of hydroxyl groups, their positions in the molecular chain and the saturation degree, as shown in Table 4.1. The interaction between alcohols with three carbon atoms chain and model surfaces (such as platinum and gold) and the further reaction has been widely investigated in the last three decades [1–17]. From the fundamental point of view, these studies are important in order to reach an understanding about model systems. Fundamental studies provide us with information about the role of OH groups and their location on the interaction and reaction of these alcohols with surfaces. Previous studies have revealed that the electrosorption properties of alcohols with three carbon atoms chain depend on: (1) the number of hydrogen atoms bonded to the α-carbon atom; (2) the position of the OH in the hydrocarbon chain; (3) the number of OH groups; and (4) the relative position of the OH group [18, 19].

One of the motivations for studying alcohols with three carbon atoms chain is the possibility of application of these alcohols in direct alcohol fuel cells for electric energy generation, co-generation of heat and chemicals. In particular, glycerol is an interesting alcohol, since it is abundantly produced as a co-product of the biodiesel fabrication and it has high energy content. On the other hand, the industrial production of the other alcohols pointed out in Table 4.1 is either complicated or somewhat expensive. For this reason, their large scale production is technically unfeasible and represents an unviable alternative to fuel cell application.

Generally, the electro-oxidation reactions of alcohols involve different steps: alcohol adsorption, breaking of the inter-atomic bonds, electronic charge transfer, reaction between oxygenated species and fragments from the alcohol, and reaction products desorption. As a result, the anode performance depends on the: (1) interaction between the catalyst surface and the alcohol molecules, (2) interaction between the catalyst surface and the resulting adsorbed fragments from the original

Table 4.1 Structural similarities and particularities of alcohols with three carbon atoms chain

Three carbon atoms alcohol	Schematic representation	Number of OH	OH group position
n-propanol		1	C1
2-propanol	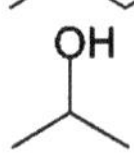	1	C2
1,2-propanediol	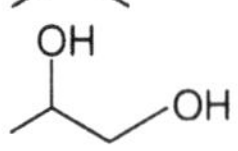	2	C1, C2
1,3-propanediol		2	C1, C3
Glycerol	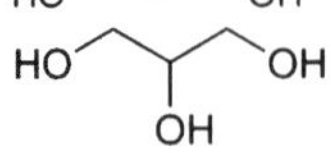	3	C1, C2, C3
Allyl alcohol	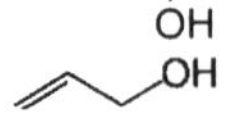	1	C1
Propargyl alcohol	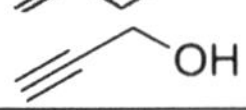	1	C1

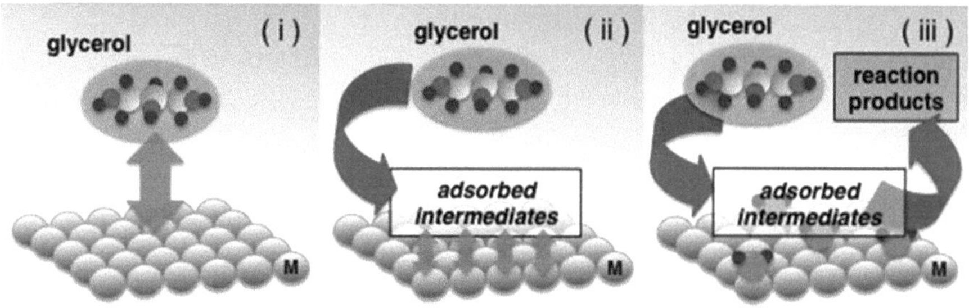

Fig. 4.1 Important aspects for the anode performance towards the alcohol electro-oxidation: (**i**) interaction between the catalyst surface and the alcohol molecules, (**ii**) interaction between the catalyst surface and the resulting adsorbed fragments from the original alcohol molecules, and (**iii**) surface oxides formation from the water splitting and reaction products desorption. Glycerol is taken as an example of alcohol molecule; *M* metal atom

alcohol molecules, and (3) surface oxides formation from water splitting and reaction products desorption (as schematically presented in Fig. 4.1). In this context, from the technological point of view, an efficient catalyst regarding alcohol electro-oxidation would possibly contain multiple elements with distinct roles and not a single element such as Au and Pt. On the other hand, from the fundamental point of view, Au and Pt are interesting model catalysts for elementary studies of the electrochemical oxidation of alcohols.

In this chapter, we concentrate attention on the reactivity of glycerol on Pt, Pd and Au-based electrodes in acidic and alkaline medium. Firstly, in Sect. 4.2, we present an overview about glycerol, including its current availability, some possible applications and the main challenges. Next, in Sect. 4.3, we present the most important results reported in the literature on the electrochemical reaction of glycerol over Pt, Pd and Au-based catalysts at different electrochemical environments. In the sequence, in Sect. 4.4, we summarize some evidences of application of glycerol in direct glycerol fuel cell prototypes, and some patents on that. And finally, in Sect. 4.5, we present some perspectives concerning the application of glycerol in direct glycerol fuel cells.

4.2 Glycerol: Availability, Possible Applications and Challenges

Glycerol, popularly known as glycerin, is an alcohol with three carbon atoms and three OH groups, each one of them attached to one carbon atom. The traditional trades of glycerol are cosmetics, medicaments, food, and chemistry industries, as shown in Fig. 4.2 [20].

Until the middle of last century, glycerol was mostly obtained as a co-product of soap fabrication. Afterwards, glycerol was also produced from propene synthesis (route represented in Fig. 4.3) and more recently, when many countries started to produce biodiesel in large scales to partially substitute the traditional diesel of

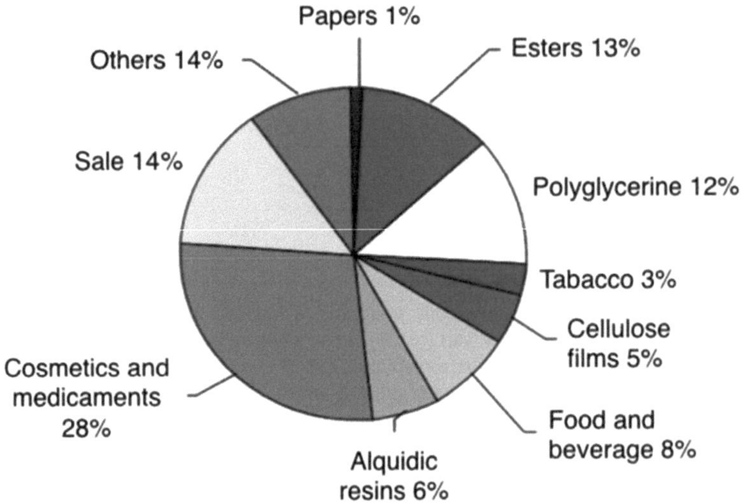

Fig. 4.2 Main industrial sectors of glycerol applications (Reproduced from Ref. [20], with kind permission of © Sociedade Brasileira de Química)

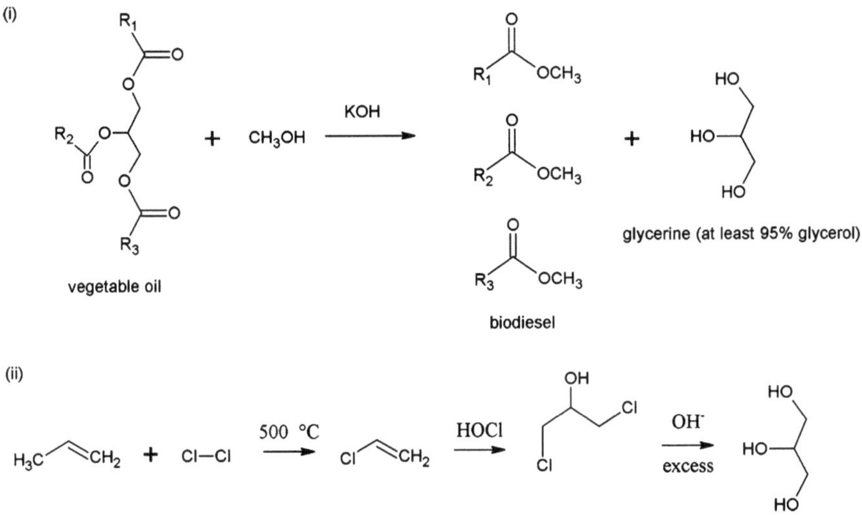

Fig. 4.3 Routes of glycerol production. (**i**) Production of biodiesel and glycerol from transesterification reaction; (**ii**) Industrial route for production of glycerol from propene synthesis

petroleum, glycerol is largely generated as co-product of the biodiesel fabrication in the proportion of 1 kg of glycerol to 10 kg of biodiesel. The main route of biodiesel production is through transesterification reaction between vegetable oils or animal fat and alcohols (methanol and ethanol) under acid or base catalysis, as represented in Fig. 4.3. It is estimated that nowadays 1.5 millions of tons of glycerol come from biodiesel industries around the world. Based on data from the Association of Producers of Biodiesel of Brazil (Aprobio), Argentine is currently one of the main producers of biodiesel, surpassing USA, Brazil and Germany. With the advance of the biodiesel production: what to do with the additional glycerol, which exceeds the demand of traditional trades for glycerol? To exemplify, in 2011, about 260,000 tons of glycerol were produced only as co-product of the biodiesel fabrication in Brazil. This amount corresponds to approximately eight times the Brazilian demand for glycerol, which is estimated in 40,000 tons. The exceeding glycerol is partly burned in order to generate heat that is used in biodiesel production and in many other industrial processes. This is an environmentally correct practice since glycerol substitutes wood and fossil fuels, such as diesel and coal. Additionally, another fraction of the exceeding glycerol is exported to other countries, as China, where it is used as raw material in the traditional trades of glycerol. Despite some application, huge volumes of glycerol are being accumulated in Brazil and other countries. For this reason, at present, there is the opportunity for new applications for the glycerol, which is currently an abundant raw material, consequently, of low cost and in addition to that presents chemical potential for the development of new products and processes with high added value.

One of the possibilities is the application of glycerol in direct glycerol fuel cells for generation of clean electric energy and co-generation of heat and commercially interesting chemicals [21–23]. Another possible and more noble application for the glycerol is the use of it (in alkaline medium) as reducing agent in the synthesis of gold [24] and silver [25] particles, which have numerous applications. Particularly, gold particles have been employed in studies of alternative detection and treatment of cancer [26–28] and silver particles have been largely used as anti-bactericide [29–31]. In addition to that, gold and silver particles are applicable in electro-catalysis of important reactions occurring in fuel cell, such as borohydride oxidation [24] and oxygen reduction [25]. In this chapter, we will focus on the first abovementioned possible application of glycerol. Information concerning direct glycerol fuel cells, challenges and existing patents on that are given in Sect. 4.4.

Considering the complete oxidation to CO_2, which for glycerol involves the generation of 14 electrons/glycerol molecule, the theoretical energy density of glycerol is comparable to those of methanol and ethanol. Specifically, the theoretical energy density of glycerol corresponds to 5.0 kWh kg^{-1} while those of methanol and ethanol are 6.1 kWh kg^{-1} and 8.1 kWh kg^{-1}, respectively [32]. On the other hand, the theoretical energy density of glycerol is relatively lower than the theoretical energy density of hydrogen that is 33 kWh kg^{-1} [33]. This means that, theoretically, the mass of glycerol required to achieve a given power in a glycerol/oxygen fuel cell is about seven times higher than the mass of hydrogen required to reach the same power in a hydrogen/oxygen fuel cell. As previously mentioned, in countries like Argentine, USA, Brazil and Germany, where glycerol is abundantly produced, its commercial value is very low. In this manner, in these countries, the need of high consumption of glycerol in glycerol/oxygen fuel cells would be irrelevant. Nevertheless, in principle, these devices would demand large reservoirs for glycerol and, hence, they would be more appropriately employed for stationary (not moving) fuel cell applications.

Concerning the maximum conversion of glycerol in energy, the viability of using glycerol as fuel in direct glycerol fuel cells depends on the improvement of the energetic yield of the oxidation reaction. In this aspect, nowadays, two main difficulties make the application of glycerol in fuel cells impracticable. One of the difficulties is the breaking of the C–C–C bonds. As a consequence of that, glycerol oxidizes by different reaction pathways, leading to the formation of partially oxidized products, in addition to CO_2. While the production of CO_2 involves the generation of 14 electrons/glycerol molecule, the formation of partially oxidized products involves less than this. Therefore, the formation of the partially oxidized products decreases the total efficiency of the system. On the other hand, the generation of specific products of the partial oxidation, such as dihydroxyacetone, ethylene glycol and propanediol, would be of great interest. Dihydroxyacetone is used as an ingredient in sunless tanning products, ethylene glycol is employed as anti-freezing in automotive radiators and propanediol is utilized in the fabrication of many industrial products, such as composite materials, aliphatic polyesters and solvents. In this context, one of the great challenges of glycerol catalysis is to control the selectivity of its oxidation towards the formation of the desired product (CO_2 or any other product), once the direct glycerol fuel cell can also operates as a chemical reactor. Going back to the difficulties related to the application of glycerol in direct glycerol fuel cells, another tricky point is the formation of poisoning adsorbed intermediates, such as CO, that block the catalyst surface at low overpotentials. In order to overcome these difficulties, important issues related with the identification of the adsorbed intermediates and final products of the glycerol oxidation as well as their relation with: (1) the geometric and electronic properties of the electrodes, (2) the glycerol concentration, and (3) the working conditions (electrolyte solution, temperature, etc.) need to be widely explored.

An atomic level understanding of the interaction between glycerol molecules and the catalyst active sites is absolutely important for designing catalysts more tolerant with respect to the poisoning adsorbed intermediates and, in addition, more selective towards the formation of CO_2, which is the product of highest energetic yield, or another product of high added value. Catalysts with these characteristics would allow us to improve the glycerol electro-oxidation process for more efficient fuel cell applications.

4.3 Fundamental Studies on the Glycerol Oxidation and Advances on the Reaction Understanding

Previous studies about the glycerol oxidation in heterogeneous liquid phase [34–39] as well as in electrochemical environment [13, 40–42] show that glycerol react on metal surfaces through a complex mechanism that lead to the formation of a large variety of final reaction products, including compounds with one, two and three carbon atoms. There are evidences that the pH [13, 40, 41, 43], the nature of the catalyst [13, 32, 40, 44, 45], and the presence of a support [36, 46] influences the catalyst activity and the reaction selectivity. Furthermore, it is shown that the selectivity of this reaction also depends on the electrode potential [13, 32, 40, 41]. Details about the reaction mechanism of glycerol on metal surfaces, however, are still largely unknown. The main advances on the understanding of this reaction will be summarized below.

Lamy and co-workers [41] proposed that glycerol initially adsorb on Pt in acidic medium by the terminal oxygen atom, forming an adsorbed intermediate like $(CH_2OH–CHOH–CH_2OH)_{ads}$:

$$(\mathrm{CH_2OH - CHOH - CH_2OH})_{ads} \rightarrow (\mathrm{CH_2OH - CHOH - CH_2\bar{O}})_{ads} + 1\mathrm{H^+} + 1\mathrm{e^-} \quad (4.1)$$

where Ō represents the oxygen atom by which the reaction intermediate is adsorbed on the electrode surface.

In addition to this possibility, Lamy and co-workers [41] also proposed that glycerol can interact with platinum oxides, formed at high overpotentials, by simultaneous attack of two carbon atoms:

$$\mathrm{CH_2OH - CHOH - CH_2OH + 2PtOH} \rightarrow \left(\mathrm{\dot{C}HOH - CHOH - \dot{C}HOH}\right)_{ads} + 4\mathrm{H^+} + 4\mathrm{e^-} \quad (4.2)$$

$$\mathrm{CH_2OH - CHOH - CH_2OH + 2PtO} \rightarrow \left(\mathrm{CH_2OH - \dot{C}OH - \dot{C}HOH}\right)_{ads} + 2\mathrm{H^+} + 2\mathrm{e^-} \quad (4.3)$$

where Ċ represents the carbon atoms by which the reaction intermediates are adsorbed on the electrode surface.

The formation of species such as $(CH_2OH–CHOH–CH_2OH)_{ads}$, $(\dot{C}HOH–CHOH–\dot{C}HOH)_{ads}$ and $(CH_2OH–\dot{C}OH–\dot{C}HOH)_{ads}$ was not yet confirmed by spectro-electrochemical studies. It is possible that these adsorbed intermediates are formed and rapidly reacts towards the production of more stable species. So far, only 2,3-dihydroxy-propionyl species (also named as glyceroyl), glycerate $[(CH_2OH–CHOH–COO)^-]$, linearly and multiply bonded CO species were detected as adsorbed intermediates of the glycerol reaction on Pt in acidic aqueous solution [47]. The formation of adsorbed 2,3-dihydroxy-propionyl and glycerate species indicates that at least one alcohol group of glycerol has been oxidized, most likely without scission of the C–C bond. On the other hand, the formation of adsorbed intermediates with one carbon atom, as linearly and multiply bonded CO, shows that glycerol undergo dissociative adsorption on Pt. Another evidence of the C–C bond breaking during the glycerol adsorption was reported by Gootzen et al. [48]. They applied FTIR spectroscopy to study the mechanism of glycerol adsorption/oxidation at 0.4 V versus reversible hydrogen electrode (RHE) on a Pt electrode in perchloric acid electrolyte. Based on adsorbate-stripping measurements (in the absence of glycerol in the solution bulk), they concluded that glycerol is completely dehydrogenated to form adsorbed CO, since no methane desorption was found at low potentials and CO_{ad} was the only adsorbate detected via IR spectroscopy. Leung and Weaver estimated the CO_{ad} coverage from dissociative adsorption of glycerol on Pt in perchloric acid solution [49]. By comparing the intensity of the CO_{ad} band resulting from glycerol dissociative adsorption and the CO_{ad} intensity measured when gas phase CO was adsorbed on the electrode, they found that the CO_{ad} coverage was about 0.55 monolayers, comparable to the CO_{ad} coverage from dissociative adsorption of 1,2-propanediol on Pt at the same experimental conditions [49]. These results indicate that the presence of an additional vicinal OH group in the glycerol molecule (with respect to the 1,2-propanediol molecule) does not significantly affect the formation of CO on the Pt surface.

Glyceraldehyde (CH_2OH–CHOH–CHO), dihydroxyacetone [CH_2OH–(CO)–CH_2OH], glyceric acid (CH_2OH–CHOH–COOH), tartronic acid (COOH–CHOH–COOH), glycolic acid (COOH–CH_2OH), glyoxylic acid (COOH–CHO), formic acid (HCOOH) and carbon dioxide (CO_2) were detected as solution phase reaction products of the glycerol oxidation on Pt in acidic medium [13, 40, 41, 47]. The variety of incomplete oxidation products with the C–C–C bonds remaining intact or only partly broken points to pronounced kinetic limitations for the dissociative adsorption and subsequent complete oxidation of glycerol to CO_2.

A detailed analysis of the reaction products formed during long term continuous electrolysis of glycerol on Pt in acidic solution was performed by Roquet et al. [41], using high performance liquid chromatography (HPLC). At 0.75 V versus RHE, the main reaction product was glyceraldehyde, and glyceric acid and tartronic acid were detected as side products. At 1.3 V, differently, glyceraldehyde, glycolic acid and formic acid were the major products and glyceric acid was detected as a minor product. Therefore, in an acid medium, the electrochemical oxidation of glycerol leading to formic acid and glycolic acid is mainly observed when the potential of electrolysis is set in the so called "oxygen region". It has been demonstrated that at positive potentials corresponding to the formation of a Pt surface oxide, the PtO_x surface oxide catalyzes the conversion of glyceraldehyde finally to formic acid and glycolic acid [40]. Conversely, the selectivity towards the formation of glyceraldehyde is very high when the applied potential is chosen before the beginning of the oxidation of the electrode surface [41].

For potentials above 0.5 V versus RHE, CO_2 formation was detected upon potentiodynamic and potentiostatic oxidation of glycerol on a platinum thin film in acidic medium under continuous flow conditions [47]. The maximum current efficiency for CO_2 formation was observed at around 0.7 V and corresponds to only 3–5 %, indicating that incomplete oxidation of glycerol prevails under these conditions. On the other hand, in a recent FTIR spectroscopy study on the glycerol electro-oxidation over polycrystalline Pt under stagnant flow conditions, performed in an external reflection configuration, massive CO_2 production was observed at potentials above 1 V versus RHE [42]. In this potential range, the CO_2 and glyceric acid formation rates coincide, as seen by differentiating the corresponding integrated FTIR bands with respect to the potential. Thus, possibly, further reaction

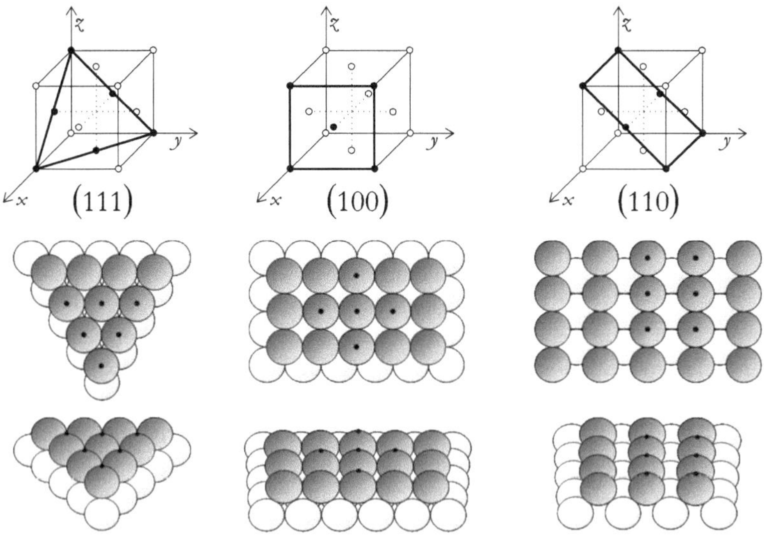

Fig. 4.4 Schematic representation of the planes referred to the (111), (100) and (110) low Miller index faces of face-centered cubic (fcc) crystals and respective balls models

of partial oxidation products (such as glyceric acid) to CO_2 may explain the relatively large CO_2 production under stagnant electrolyte conditions. In addition to this pathway occurring at high potentials, another pathway leading to CO_2 was also evidenced at the same experimental conditions [42, 50]. This second pathway arises at low potentials and involves CO adsorbed on Pt. By means of experiments with isotopically labeled glycerol ($^{13}CH_2OH$–CHOH–$^{13}CH_2OH$), it was observed the co-existence of $^{13}CO_2$ and $^{12}CO_2$ as glycerol electro-oxidation products [51]. These results demonstrate the C–C–C bonds breaking and the formation of CO_2 from both the central and the terminal carbon atoms.

An important aspect of the glycerol electro-oxidation reaction is its dependence on the electrode surface structure [5, 52, 53]. Particularly in our group, we recently investigated the electrochemical oxidation of glycerol on (111), (100) and (110) low Miller index Pt single crystals in acidic media (H_2SO_4 and $HClO_4$) by cyclic voltammetry and Fourier Transform Infrared (FTIR) spectroscopy and we verified that this is a surface sensitive reaction. Figure 4.4 shows the schematic representation of (111), (100) and (110) low Miller index faces of face-centered cubic (fcc) crystals, such as platinum. It can be noted that the (111) face presents the highest surface atomic density (hexagonal array) in comparison with the (100) and (110) faces. On the other hand, the (110) face has the less compact surface structure and the highest number of atoms low coordinated to neighboring atoms compared to the other faces. We have verified that Pt(100) and Pt(110) surface structures favor the breaking of the C–C–C bond at low potentials (say 0.05 V), as seen by the formation of CO. Pt(111) surface structure does not favor the C–C–C bond breaking at potentials as low as 0.05 V. However, Pt(111) is less poisoned by residues of glycerol dissociation and, for this reason, it is more active for glycerol oxidation than Pt(100) and Pt(110) at low potentials. Carbonyl containing compounds and CO_2 were detected as reaction products of the glycerol oxidation on all investigated single-crystal Pt surfaces, but the ratio between CO_2 and carbonyl containing compounds is clearly much higher for Pt(100) and Pt(110) than for Pt(111) [53]. Effect of the electrode surface crystallographic orientation has been also previously evidenced on the electro-oxidation of other alcohols, such as methanol and ethanol, on metallic single crystals [54–62].

Besides the catalyst surface structure, other key points are the nature of the catalyst and the pH of the electrolytic solution. Platinum and gold, which are generally used as model catalysts, present distinct catalytic performances towards glycerol electro-oxidation. Platinum and gold are adjacent elements in the periodic table. Both of them are transition metals. Their distinct electronic configurations (Pt: $[Xe]4f^{14}5d^{9}6s^{1}$; Au: $[Xe]4f^{14}5d^{10}6s^{1}$), however, make them very different with respect to the catalytic properties and, for this reason, they interact in a different way with glycerol and other alcohols. Notoriously, platinum has strong interaction with hydrogen, oxygen and organic radicals. As a consequence, the breaking of O–H and C–H bonds and the subsequent interaction of radicals and H atoms with active Pt sites constitutes the first step in the alcohols electro-oxidation mechanism on Pt, as previously mentioned at the beginning of this section. In opposition, gold is a much less active metal in relation to the abovementioned reactions. This could be due to the presence of completely filled d-bands, so that the formation of covalent bonds with the radicals resulting from dissociation of the alcohols would no longer be possible. Studies on the electro-oxidation of glycerol [13, 40] and other alcohols [63, 64] indicate that under specific conditions, however, the interaction between gold and alcohols is significantly improved. Particularly, in alkaline media, gold interacts much more with glycerol and other alcohols than in acidic medium. It has been shown that glycerol oxidation reaction on gold is significantly favored with increasing the pH [13, 43]. This positive effect of the high pH on the catalytic activity of gold towards the glycerol oxidation was also evidenced for platinum [13, 43]. Under acidic conditions, however, only platinum exhibits some catalytic activity, while Au loses its catalytic properties completely. Therefore, gold catalyzes the oxidation of glycerol and other alcohols only under alkaline conditions, in contrast to a “real catalyst”, such as Pt, which catalyzes these reactions over the entire pH range [40]. One of the explanations for the distinct behavior observed for gold in alkaline medium is that it becomes covered by OH^- species above the zero-charge potential and, through the formation of hydrogen bridges, Au-OH interacts with alcohol molecules, that can be further oxidized [64]. In addition to that, the acid–base equilibrium involving the alcoholic group of alcohols may also contribute to the high catalytic activities in alkaline media. At high pH, alcohol molecules deprotonate and the deprotonated species are believed to be the main reactant in alkaline media [40]. Further results of the electro-oxidation of a series of similar alcohols with varying pK_a on gold electrodes in alkaline solution revealed that the first alcohol deprotonation is base catalyzed, and the second one is fast but gold catalyzed. The base catalysis follows a Hammett-type correlation with pK_a. These results indicate that the base catalysis is the main driver force behind the high oxidation rate of alcohols, including glycerol, in alkaline medium, and not the catalyst interaction with hydroxide [16].

Other recent studies focused on the understanding of the role of ad-atoms on the catalytic performance (in terms of both catalytic activity and glycerol oxidation overpotential) of Pt-based [11, 65] and Pd-based [32, 66] materials towards glycerol electro-oxidation in acidic and alkaline medium.

Electrocatalytic oxidation of glycerol was studied over a carbon-supported PtRuSn catalyst (PtRuSn/C), which was prepared with a Pt:Ru:Sn atomic ratio of 5:4:1 by using a colloidal method combined with a freeze-drying procedure at room temperature. The ternary PtRuSn/C catalyst was characterized by various physicochemical analyses, such as: X-ray diffraction (XRD), transmission electron microscopy (TEM), energy dispersive X-ray spectroscopy (EDS), thermogravimetric analysis (TGA), X-ray photoelectron spectroscopy (XPS) and X-ray absorption-near-edge spectroscopy (XANES). The PtRuSn/C catalyst presented noticeable modifications of the geometric and electronic structures in comparison with the Pt/C and PtRu/C catalysts. The structurally modified PtRuSn/C catalyst presented better electrocatalytic performance for glycerol oxidation than Pt/C and PtRu/C catalysts, as seen by larger peak currents and lower onset potential. In addition, with basis on results of CO stripping experiments, it was suggested that the PtRuSn/C catalyst is more efficient than Pt/C and PtRu/C catalysts for oxidative removals of CO or CO-like carbonaceous intermediate species produced during the glycerol oxidation reaction [11].

It has been verified that bimetallic PdAu catalysts present better catalytic performance than that of monometallic Pd and Au materials, being close to that of platinum [32, 50]. Such synergetic effect between Pd and Au is still not completely understood, and further studies with different PdAu catalyst structures may help to bring new insights on that. The same scientific group has shown that Bi-modified Pt/C and Pd/C present higher catalytic activity and lower overpotentials to the glycerol oxidation than non-modified materials [65]. In particular, the order of activity of catalysts was $Pd/C < Pt/C = Pd_{90}Bi_{10}/C < Pt_{90}Bi_{10}/C = Pt_{45}Pd_{45}Bi_{10}/C$. Therefore, the replacement of half of the platinum atoms by palladium atoms resulted in the same catalytic activity. The improvement of the catalytic performance of the Bi-modified materials with respect to non-modified materials was explained in terms of a bifuctional mechanism, since no evidence for either bismuth-platinum or bismuth-palladium alloy formation was found in that work. In the proposed bifunctional mechanism, the adsorption of glycerol occurs on platinum and/or palladium and the formation of oxygenated species occurs on bismuth [65].

With a different approach, the influence of bismuth on the electrocatalysis of glycerol was also investigated by Koper and co-workers [67]. They observed that a carbon supported platinum electrode in a bismuth-saturated solution is highly selective to the electro-oxidation of the secondary alcohol of the glycerol, leading to 100 % of dihydroxyacetone at a carefully chosen potential. Using a combination of online HPLC and in situ FTIR, the authors have shown that bismuth not only blocks the pathway towards the primary alcohol oxidation but also provides a specific Pt-Bi surface site for secondary alcohol oxidation. On the other hand, in the absence of bismuth, at the same potential condition, the primary alcohol oxidation is dominant with respect to the secondary alcohol oxidation, leading to the formation of other species.

4.4 Direct Glycerol Fuel Cell: State-of-Art, Challenges and Patents

4.4.1 Overview on Fuel Cells

In general, a fuel cell is an electrochemical device that converts chemical energy of a fuel (such as: hydrogen, methanol, ethanol and glycerol) into electrical energy. Fuel cells present several advantages in comparison with internal combustion engines. Fuel cells are more efficient in extracting energy from fuel (e.g., 60–70 % efficiency as compared to 40 % for turbodiesel engines and 30 % for gasoline engines). Furthermore, fuel cells are noiseless and emit negligible amount of pollutants.

Typically, a fuel cell consists of an anode (a positively charged electrode), a cathode (a negatively charged electrode) and an electrolyte in between the two electrodes, as schematically represented in Fig. 4.5. Each electrode is coated with a catalyst layer. At the anode, a fuel, such as glycerol, is converted catalytically to form products, protons and electrons. The electrons travel in the form of an electric current through an external circuit to the cathode. At the same time, the protons diffuse through the electrolyte to the cathode, where they react exothermically with oxygen to produce water, thus completing the overall process. The electric current that circulates externally can be used for electrical work. Additionally, the co-generated heat can be employed, for example, in water heating for residence and hospital proposals. Finally, the glycerol fuel cell is also an interesting chemical reactor once produces chemicals of higher market price than glycerol.

The current produced in a fuel cell is proportional to the geometric area of the electrodes. A single fuel cell typically produces a relatively small voltage (about 1 V). In order to produce a higher voltage, several fuel cells are connected in series or in parallel, through bipolar plates separating adjacent fuel cells (i.e., "stacked").

In direct glycerol fuel cells, the fuel and the oxidant used are glycerol (from biodiesel fabrication, for example) and oxygen from the air, respectively. In such fuel cells, the complete reaction taking place at the anode and cathode are represented by the equations:

$$\text{Anode reaction: } C_3H_8O_3 + 3\ H_2O \rightarrow 3\ CO_2 + 14\ H^+ + 14\ e^- \quad (4.4)$$

$$\text{Cathode reaction: } 3.5\ O_2 + 14\ H^+ + 14\ e^- \rightarrow 7\ H_2O \quad (4.5)$$

$$\text{Overall reaction: } C_3H_8O_3 + 3.5\ O_2 + \rightarrow 3\ CO_2 + 4\ H_2O \quad (4.6)$$

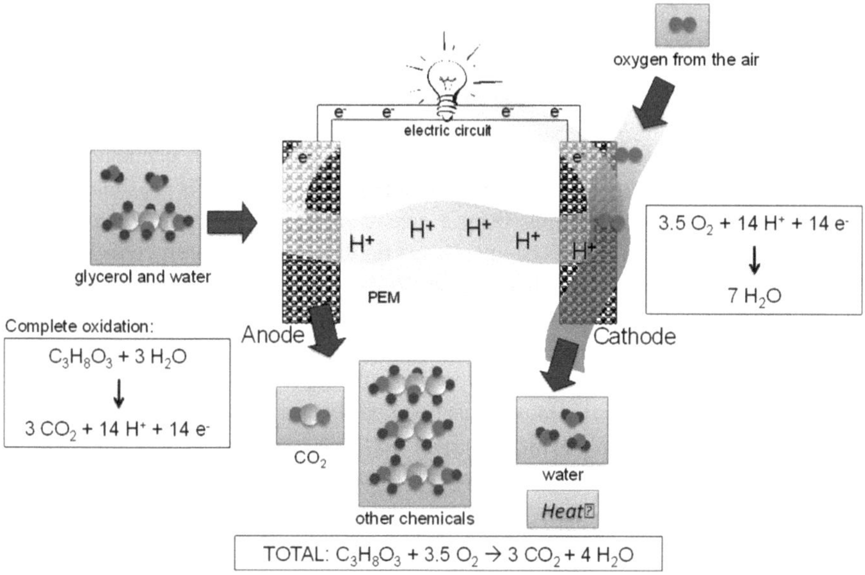

Fig. 4.5 Representation of a direct glycerol fuel cell for generation of clean electric energy and co-generation of heat and commercially interesting chemicals, such as propanediol, ethylene glycol and dihydroxyacetone

Currently, there are five types of fuel cells, categorized by their electrolyte (solid or liquid), operating temperature, and fuel characteristics. Glycerol could be employed in any one of them as fuel. The categories of fuel cells include: proton exchange membrane fuel cell (PEMFC), phosphoric acid fuel cell (PAFC), molten carbonate fuel cell (MCFC), solid oxide fuel cell (SOFC) and alkaline fuel cell (AFC). See details about the different types of fuel cells in Chap. 1.

Based on our fundamental studies [13, 53, 68], we believe that the most convenient cell for employing glycerol as fuel is the AFC because the kinetics of glycerol oxidation and oxygen reduction reactions are favored in alkaline media with respect to those in acidic media and, in addition, the variety of non-noble metal-based catalysts which may be used in alkaline media is large. On the other hand, AFCs should work at relatively higher temperatures in order to compensate kinetic problems resulting from the use of non-noble metal-based catalysts.

In general, the electrochemical reactions taking place in a fuel cell assembly are exothermic. However, the catalyst employed in these reactions is normally sensitive to heat. To perform optimally, fuel cells should be maintained at a certain temperature that is nearly uniform across each cell in the stack. For example, at high temperatures, the catalyst may be destroyed, while at low temperatures, ice may form within the fuel cell assembly. Thus, to accommodate such temperature requirements, heat transfer compositions are needed.

4.4.2 Direct Glycerol Fuel Cells: Comparison with Hydrogen and Direct Ethanol Fuel Cells

The power of an ethanol/oxygen fuel cell operating at temperatures between 80 °C and 90 °C is about 60–70 mW cm^{-2} [69]. However, 10 years ago, an ethanol/oxygen fuel cell did not generate more than 15 mW cm^{-2}. In this way, it is possible to affirm that, in the last years, there has been a considerable progress with respect to the increase of the yield of these systems. Based on that, there are perspectives of more significant improvements. Similarly, glycerol/oxygen fuel cells with non-platinum-based multimetallic nanocatalysts as anode, operating at low temperatures (80–90 °C), generate up to 80 mW cm^{-2} [70], power slightly higher than that of an ethanol/oxygen fuel cell. Like the power of an ethanol/oxygen fuel cell, the power of a glycerol/oxygen fuel cell also presents potentiality to achieve the power levels of a hydrogen/oxygen fuel cell, which is about 500 mW cm^{-2}. It is very important to emphasize that, while hydrogen/oxygen fuel cell operate with efficiency of about 90 %, ethanol/oxygen and glycerol/oxygen fuel cells work nowadays with efficiency of approximately 15 %. For this reason, there is an expectation of improvement of the efficiency of alcohol fuel cells, which theoretically can reach 90 % of efficiency. On the other hand, it is not expected that the internal combustion engines, which operate with about 30–40 % of efficiency, have significant improvement with respect to their efficiency, since they are thermodynamically limited by the Carnot cycle. Therefore, direct glycerol and ethanol fuel cells, which are not thermodynamically limited by the Carnot cycle, are considered promising systems for the efficient production of clean energy and, for this reason, they deserve attention and position in spotlight in the current world scenario, which shows a transition phase from fossil fuels to the biofuels. See more details about direct ethanol fuel cells in Chap. 3. The major bottlenecks to the technologic advances on the development of direct alcohol fuel cell, such as the direct glycerol fuel cell, are: (1) the inefficiency of the catalysts (in general, based on noble metals) towards the oxidation of the alcohol to CO_2 or any other desired product and (2) the diffusion of the alcohol from the anode to the cathode through the ion exchange membrane.

The use of non-noble metal-based catalysts is highly preferred for economic reasons and also for restrict availability of noble metals. Catalysts consisting of Pt and other non-noble metals have been successfully developed. These catalysts present good stability and higher activity than pure Pt catalysts for reactions of practical interest [71].

In order to overcome difficulties related to the existing ion exchange membrane, new materials are being developed. Preferably, the ion exchange membranes may present low permeability to alcohols and, in addition, good mechanical properties at high temperature. These characteristics will allow the fuel cell to be operated at relatively higher temperatures. In this way, phosphoric acid impregnated polybenzimidazoles (PBI) present the following advantages in comparison with Nafion®: (1) high proton conductivity (>0.01 S cm^{-1} at 200 °C), (2) low permeability to fuel, (3) good chemical/thermal stability (>550 °C in air), (4) 1,000 times more tolerant to CO than Nafion® [72].

Recently, in our group, we evaluated the potentiality of a poly(imide) (PI)/ organically-modified montmorillonite (O-MMT) nanocomposite membrane for the use in alkaline fuel cells [73]. Both X-ray diffraction and scanning electron microscopy revealed a good dispersion of O-MMT into the PI matrix and preservation of the O-MMT layered structure. When compared to the pure PI, the addition of O-MMT improved thermal stability and markedly increased the capability of absorbing electrolyte and ionic conductivity of the composite. Based on these results, the PI/ O-MMT nanocomposite is a promising candidate for alkaline fuel cell applications.

4.4.3 Patents

Regarding systems of energy generation from glycerol, some patents were found to be interesting [74–77]. In one of the patents, it is claimed the simultaneous production of electricity and chemical compounds, such as carboxylic acids, ketones, keto-acids, and other oxygenated species, by the oxidation of glycerol and other alcohols in direct alcohols fuel cells (DAFC) comprising an anion-exchange membrane and catalysts based on: Ni-Zn-P-Pd, Ni-Zn-P-Pt, Ni-Co-Fe-Pt mixtures and Pd [74]. In particular, the direct glycerol fuel cell generates approximately 30 mW cm^{-2} of power density, which is considered a respectable performance taking into account the use of Ni-based and other non-noble metals-based catalysts with low amount of noble metal [74]. In another patent, it is claimed a fuel cell system comprising one or more fuel cell assemblies and a heat transfer composition providing high electrical resistance. In addition to that, it is also claimed a method for removing heat from a fuel cell assembly by contacting the fuel cell assembly with a heat transfer composition [75]. Finally, other patent claims an implantable fuel cell assembly containing a device for converting fat to glycerol and fatty acid, devices for converting glycerol and fatty acid to hydrogen, a device for converting a bodily fluid to hydrogen, oxygen and mixtures thereof, and a fuel cell for producing electricity from hydrogen and oxygen [77].

Concerning catalysts employed in system of energy generation from glycerol, a patent is remarkable. It is claimed a double-effect oxygen electrode catalyst slurry for uses in fuel cells [78]. The preparation of catalyst slurry involves preparing metal nanoparticles, adding proton-conductive polymer solution in alcoholic solution, adding soluble precursor with metal active component, stirring, adding alkaline solution and reducing agent, adjusting pH to 8–13, stirring, obtaining colloidal solution, adding in reaction kettle at 90 °C–200 °C for 2 h, obtaining colloidal solution of nanoparticles modified by proton-conductive polymer and dialyzing the obtained colloidal solution. The soluble precursor is chloroplatinic acid, ruthenium chloride, hexachloroiridic acid, palladium chloride, osmium chloride, chromium nitrate, nickel nitrate, cobalt nitrate, manganese chloride, copper chloride, titanium chloride, tin chloride, vanadium chloride, iron chloride, gallium nitrate, molybdenum hexachloride or sodium selenite. The alkaline solution is sodium carbonate, sodium bicarbonate, sodium hydroxide, potassium carbonate, potassium bicarbonate, potassium hydroxide or ammonium hydroxide. The catalyst nanoparticles contain composite nanoparticles and metal nanoparticles as carrier. The carrier is platinum nanoparticles or alloy of platinum nanoparticles, PtM, where M is ruthenium, iridium, palladium, rhodium, osmium, gold, chromium, nickel, cobalt, manganese, copper, tin, titanium, vanadium, iron, gallium, molybdenum or selenium [78].

4.5 Perspectives

Despite the relatively modest power density of direct glycerol fuel cells reported in patents, fundamental and applied studies up till now, there are positive perspectives regarding the enhancement of their efficiencies. A promising direct glycerol fuel cell-type is the AFC because the kinetics of glycerol oxidation and oxygen reduction reactions are favored in alkaline media with respect to those in acidic media and, in addition, the variety of non-noble metal-based catalysts which may be used in alkaline media is large. On the other hand, AFCs should work at relatively higher temperatures in order to compensate kinetic problems resulting from the use of non-noble metal-based catalysts. Many efforts need to be made in order to solve issues related to catalysts, membranes and formation of carboxylates and carbonates. This will allow direct glycerol fuel cells to reach power densities of practical interest.

References

1. Alonso C, Gonzalez-velasco J (1988) Study of the electrooxidation of 1,2-propanediol on an Au electrode in basic medium. J Electroanal Chem 248:193–208
2. Avramovivic ML, Leger JM, Lamy C, Jovic VD, Petrovic SD (1991) The electrooxidation of glycerol on the gold(100)-oriented single-crystal surface and polycrystalline surface in 0.1 M NaOH. J Electroanal Chem 308:309–317
3. Gao P, Lin CH, Shannon C, Salaita GN, White JH, Chaffins SA, Hubbard AT (1991) Studies of adsorbed saturated alcohols at Pt(111) electrodes by vibrational spectroscopy (EELS), Auger-spectroscopy, and electrochemistry. Langmuir 7:1515–1524
4. Elshafei AA, Elmaksoud SAA, Moussa MNH (1992) Electrocatalytic oxidation of the propanol isomers on platinum ad-atom electrodes in alkaline-medium. Z Phys Chem 177:211–223
5. Hamelin A, Ho YH, Chang SC, Gao XP, Weaver MJ (1992) Surface crystallographic dependence of voltammetric oxidation of polyhydric alcohols and related systems at monocrystalline gold acidic aqueous interfaces. Langmuir 8:975–981
6. Luczak T, Beltowska-Brzezinska M, Holze R (1993) Molecular structure effects in the adsorption of terminal and vicinal aliphatic diols on a gold electrode. Electrochim Acta 38:717–720
7. Pastor E, Schmidt VM, Iwasita T, Arevalo MC, Gonzalez S, Arvia AJ (1993) The reactivity of primary C3-alcohols on gold electrodes in acid-media – a comparative-study based on DEMS data. Electrochim Acta 38:1337–1344

8. Venancio EC, Napporn WT, Motheo AJ (2002) Electro-oxidation of glycerol on platinum dispersed in polyaniline matrices. Electrochim Acta 47:1495–1501
9. Sen Gupta S, Datta J (2005) An investigation into the electro-oxidation of ethanol and 2-propanol for application in direct alcohol fuel cells (DAFCs). J Chem Sci 117(4):337–344
10. Rodrigues IA, Nart FC (2006) 2-Propanol oxidation on platinum and platinum-rhodium electrodeposits. J Electroanal Chem 590:145–151
11. Kim HJ, Choi SM, Green S, Tompsett GA, Lee SH, Huber GW, Kim WB (2011) Highly active and stable PtRuSn/C catalyst for electrooxidations of ethylene glycol and glycerol. Appl Catal B Environ 101:366–375
12. Santasalo-Aarnio A, Kwon Y, Ahlberg E, Kontturi K, Kallio T, Koper MTM (2011) Comparison of methanol, ethanol and iso-propanol oxidation on Pt and Pd electrodes in alkaline media studied by HPLC. Electrochem Commun 13:466–469
13. Gomes JF, Tremiliosi-Filho G (2011) Spectroscopic studies of the glycerol electro-oxidation on polycrystalline Au and Pt surfaces in acidic and alkaline media. Electrocatalysis 2:96–105
14. Feng Y, Yin W, Li Z, Huang C, Wang Y (2010) Ethylene glycol, 2-propanol electrooxidation in alkaline medium on the ordered intermetallic PtPb surface. Electrochim Acta 55:6991–6999
15. Giannetti BF, Almeida C, Bonilla SH, Mengod MOA, Raboczkay T (2003) Electrocatalytic effect of Pb and Sn adatoms on the oxidation of 1-propanol on platinized platinum electrodes: determination of the apparent activation energy. Zeits Phys Chem Int 217:35–44
16. Kwon Y, Lai SCS, Rodriguez P, Koper MTM (2011) Electrocatalytic oxidation of alcohols on gold in alkaline media: base or gold catalysis? J Am Chem Soc 133:6914–6917
17. Santasalo A, Vidal-Iglesias FJ, Solla-Gullon J, Berna A, Kallio T, Feliu JM (2009) Electrooxidation of methanol and 2-propanol mixtures at platinum single crystal electrodes. Electrochim Acta 54:6576–6583
18. Sokolova E (1975) Influence of nature of alcohols on mechanisms of their electrochemical oxidation. Electrochim Acta 20:323–330
19. Sokolova E (1979) Zur elektrooxidation zweiwertiger alkohole. Electrochim Acta 24:147–155
20. Mota CJA, da Silva CXA, Gonçalves VLC (2009) Gliceroquímica: novos produtos e processos a partir da glicerina de produção de biodiesel. Química Nova 32:639–648
21. Ilie A, Simoes M, Baranton S, Coutanceau C, Martemianov S (2011) Influence of operational parameters and of catalytic materials on electrical performance of direct glycerol solid alkaline membrane fuel cells. J Power Sources 196:4965–4971
22. Matsuoka K, Iriyama Y, Abe T, Matsuoka M, Ogumi Z (2005) Alkaline direct alcohol fuel cells using an anion exchange membrane. J Power Sources 150:27–31
23. Zhang Z, Xin L, Li W (2012) Electrocatalytic oxidation of glycerol on Pt/C in anion-exchange membrane fuel cell: cogeneration of electricity and valuable chemicals. Appl Catal B Environ 119:40–48
24. Gasparotto LHS, Garcia AC, Gomes JF, Tremiliosi G (2012) Electrocatalytic performance of environmentally friendly synthesized gold nanoparticles towards the borohydride electro-oxidation reaction. J Power Sources 218:73–78
25. Garcia A, Gasparotto LS, Gomes JF, Tremiliosi-Filho G (2012) Straightforward synthesis of carbon-supported Ag nanoparticles and their application for the oxygen reduction reaction. Electrocatalysis 3:147–152
26. Huang X, Jain PK, El-Sayed IH, El-Sayed MA (2007) Gold nanoparticles: interesting optical properties and recent applications in cancer diagnostics and therapy. Nanomedicine 2:681–693
27. Pissuwan D, Valenzuela SM, Cortie MB (2006) Therapeutic possibilities of plasmonically heated gold nanoparticles. Trends Biotechnol 24:62–67
28. El-Sayed IH, Huang XH, El-Sayed MA (2005) Surface plasmon resonance scattering and absorption of anti-EGFR antibody conjugated gold nanoparticles in cancer diagnostics: applications in oral cancer. Nano Lett 5:829–834
29. Furno F, Morley KS, Wong B, Sharp BL, Arnold PL, Howdle SM, Bayston R, Brown PD, Winship PD, Reid HJ (2004) Silver nanoparticles and polymeric medical devices: a new approach to prevention of infection? J Antimicrob Chemother 54:1019–1024
30. Lee HJ, Yeo SY, Jeong SH (2003) Antibacterial effect of nanosized silver colloidal solution on textile fabrics. J Mater Sci 38:2199–2204

31. Sondi I, Salopek-Sondi B (2004) Silver nanoparticles as antimicrobial agent: a case study on *E-coli* as a model for Gram-negative bacteria. J Colloid Interface Sci 275:177–182
32. Simões M, Baranton S, Coutanceau C (2010) Electro-oxidation of glycerol at Pd based nano-catalysts for an application in alkaline fuel cells for chemicals and energy cogeneration. Appl Catal B Environ 93:354–362
33. Lamy C, Lima A, LeRhun V, Delime F, Coutanceau C, Leger JM (2002) Recent advances in the development of direct alcohol fuel cells (DAFC). J Power Sources 105:283–296
34. Carrettin S, McMorn P, Johnston P, Griffin K, Kiely CJ, Hutchings GJ (2003) Oxidation of glycerol using supported Pt, Pd and Au catalysts. Phys Chem Chem Phys 5:1329–1336
35. Bianchi CL, Canton P, Dimitratos N, Porta F, Prati L (2005) Selective oxidation of glycerol with oxygen using mono and bimetallic catalysts based on Au, Pd and Pt metals. Catal Today 102:203–212
36. Demirel S, Lehnert K, Lucas M, Claus P (2007) Use of renewables for the production of chemicals: glycerol oxidation over carbon supported gold catalysts. Appl Catal B Environ 70:637–643
37. Dimitratos N, Lopez-Sanchez JA, Lennon D, Porta F, Prati L, Villa A (2006) Effect of particle size on monometallic and bimetallic (Au, Pd)/C on the liquid phase oxidation of glycerol. Catal Lett 108:147–153
38. Garcia R, Besson M, Gallezot P (1995) Chemoselective catalytic-oxidation of glycerol with air on platinum metals. Appl Catal A Gen 127:165–176
39. Kimura H, Tsuto K, Wakisaka T, Kazumi Y, Inaya Y (1993) Selective oxidation of glycerol on a platinum bismuth catalyst. Appl Catal A Gen 96:217–228
40. Kwon Y, Schouten KJP, Koper MTM (2011) Mechanism of the catalytic oxidation of glycerol on polycrystalline gold and platinum electrodes. ChemCatChem 3:1176–1185
41. Roquet L, Belgsir EM, Leger JM, Lamy C (1994) Kinetics and mechanisms of the electrocatalytic oxidation of glycerol as investigated by chromatographic analysis of the reaction-products – potential and pH effects. Electrochim Acta 39:2387–2394
42. Martins CA, Giz MJ, Camara GA (2011) Generation of carbon dioxide from glycerol: evidences of massive production on polycrystalline platinum. Electrochim Acta 56:4549–4553
43. Kahyaoglu A, Beden B, Lamy C (1984) The electrocatalytic oxidation of glycerol on gold and platinum-electrodes in aqueous-media. Electrochim Acta 29:1489–1492
44. Grace AN, Pandian K (2006) Pt, Pt-Pd and Pt-Pd/Ru nanoparticles entrapped polyaniline electrodes – a potent electrocatalyst towards the oxidation of glycerol. Electrochem Commun 8:1340–1348
45. Kwon Y, Koper MTM (2010) Combining voltammetry with HPLC: application to electro-oxidation of glycerol. Anal Chem 82:5420–5424
46. Demirel-Gulen S, Lucas M, Claus P (2005) Liquid phase oxidation of glycerol over carbon supported gold catalysts. Catal Today 102:166–172
47. Schnaidt J, Heinen M, Denot D, Jusys Z, Jürgen Behm R (2011) Electrooxidation of glycerol studied by combined in situ IR spectroscopy and online mass spectrometry under continuous flow conditions. J Electroanal Chem 661:250–264
48. Gootzen JFE, Wonders AH, Visscher W, vanVeen JAR (1997) Adsorption of C-3 alcohols, 1-butanol, and ethene on platinized platinum as studied with FTIRS and DEMS. Langmuir 13:1659–1667
49. Leung LWH, Weaver MJ (1990) Influence of adsorbed carbon-monoxide on the electrocatalytic oxidation of simple organic-molecules at platinum and palladium electrodes in acidic solution – a survey using real-time FTIR spectroscopy. Langmuir 6:323–333
50. Fernandez PS, Martins ME, Camara GA (2012) New insights about the electro-oxidation of glycerol on platinum nanoparticles supported on multi-walled carbon nanotubes. Electrochim Acta 66:180–187
51. Fernandez PS, Martins ME, Martins CA, Camara GA (2012) The electro-oxidation of isotopically labelled glycerol on platinum: new information on C-C bond cleavage and CO_2 production. Electrochem Commun 15:14–17

52. Avramovivic M, Leger JM, Beden B, Hahn F, Lamy C (1993) Adsorption of glycerol on platinum in alkaline medium – effect of the electrode structure. J Electroanal Chem 351:285–297
53. Gomes JF, de Paula FBC, Gasparotto LHS, Tremiliosi-Filho G (2012) The influence of the Pt crystalline surface orientation on the glycerol electro-oxidation in acidic media. Electrochim Acta 76:88–93
54. Iwasita T (2002) Electrocatalysis of methanol oxidation. Electrochim Acta 47:3663–3674
55. Xia XH, Iwasita T, Ge F, Vielstich W (1996) Structural effects and reactivity in methanol oxidation on polycrystalline and single crystal platinum. Electrochim Acta 41:711–718
56. Herrero E, Franaszczuk K, Wieckowski A (1994) Electrochemistry of methanol at low-index crystal planes of platinum – an integrated voltammetric and chronoamperometric study. J Phys Chem 98:5074–5083
57. Tremiliosi G, Kim H, Chrzanowski W, Wieckowski A, Grzybowska B, Kulesza P (1999) Reactivity and activation parameters in methanol oxidation on platinum single crystal electrodes 'decorated' by ruthenium adlayers. J Electroanal Chem 467:143–156
58. Xia XH, Liess HD, Iwasita T (1997) Early stages in the oxidation of ethanol at low index single crystal platinum electrodes. J Electroanal Chem 437:233–240
59. Morin MC, Lamy C, Leger JM, Vasquez JL, Aldaz A (1990) Structural effects in electrocatalysis – oxidation of ethanol on platinum single-crystal electrodes – effect of pH. J Electroanal Chem 283:287–302
60. Colmati F, Tremiliosi-Filho G, Gonzalez ER, Berna A, Herrero E, Feliu JM (2008) Surface structure effects on the electrochemical oxidation of ethanol on platinum single crystal electrodes. Faraday Discuss 140:379–397
61. Lai SCS, Koper MTM (2008) Electro-oxidation of ethanol and acetaldehyde on platinum single-crystal electrodes. Faraday Discuss 140:399–416
62. Gomes JF, Busson B, Tadjeddine A, Tremiliosi G (2008) Ethanol electro-oxidation over Pt (*hkl*): comparative study on the reaction intermediates probed by FTIR and SFG spectroscopies. Electrochim Acta 53:6899–6905
63. Lai SCS, Kleijn SEF, Ozturk FTZ, Vellinga VCV, Koning J, Rodriguez P, Koper MTM (2010) Effects of electrolyte pH and composition on the ethanol electro-oxidation reaction. Catal Today 154:92–104
64. Ocón P, Alonso C, Celdrán R, González-Velasco J (1986) Study of the electrooxidation of n-propanol on an Au electrode in basic medium. J Electroanal Chem 206:179–196
65. Simões M, Baranton S, Coutanceau C (2011) Enhancement of catalytic properties for glycerol electrooxidation on Pt and Pd nanoparticles induced by Bi surface modification. Appl Catal B Environ 110:40–49
66. Mougenot M, Caillard A, Simoes M, Baranton S, Coutanceau C, Brault P (2011) PdAu/C catalysts prepared by plasma sputtering for the electro-oxidation of glycerol. Appl Catal B Environ 107:372–379
67. Kwon Y, Birdja Y, Spanos I, Rodriguez P, Koper MTM (2012) Highly selective electro-oxidation of glycerol to dihydroxyacetone on platinum in the presence of bismuth. ACS Catal 2:759–764
68. Gomes JF, Gasparotto LHS, Tremiliosi-Filho G (2013) Glycerol electro-oxidation over glassy-carbon-supported Au nanoparticles: direct influence of the carbon support on the electrode catalytic activity. Phys Chem Chem Phys 15:10339–10349
69. Simões FC, dos Anjos DM, Vigier F, Leger JM, Hahn F, Coutanceau C, Gonzalez ER, Tremiliosi-Filho G, de Andrade AR, Olivi P, Kokoh KB (2007) Electroactivity of tin modified platinum electrodes for ethanol electrooxidation. J Power Sources 167:1–10
70. Simões M (2011) Development of multimetallic nanostructured electrocatalysts for an application in a solid alkaline membrane fuel cell (SAMFC). Thesis, Université de Poitiers, Poitiers
71. Gasparotto LHS, Ciapina EG, Ticianelli EA, Tremiliosi-Filho G (2012) Electrodeposition of PVA-protected PtCo electrocatalysts for the oxygen reduction reaction in H_2SO_4. J Power Sources 197:97–101

72. Bose S, Kuila T, Thi XLN, Kim NH, Lau KT, Lee JH (2011) Polymer membranes for high temperature proton exchange membrane fuel cell: recent advances and challenges. Prog Polym Sci 36:813–843
73. Battirola LC, Gasparotto LHS, Rodrigues-Filho UP, Tremiliosi-Filho G (2012) Poly (imide)/organically-modified montmorillonite nanocomposite as a potential membrane for alkaline fuel cells. Membranes 2:430–439
74. Bert P, Bianchini C, Giambastiani G, Marchionni A, Tampucci A, Vizza F (2008) A process for the partial oxidation of alcohols in water by direct alcohol fuel cells. International Patent Number WO 2008/138865 A1
75. Jeffcoate CS, Gershun AV, Woyciesjes PM, Marinho FJ (2003) Heat transfer composition for fuel cell assembly, e.g. proton exchange membrane fuel cell, comprises alcohol, polyalkene oxide, additive, and water. United States Patent Number: US 7,481,948 B2
76. Ragsdale S (2010) Fuel cell using polyhydric mixtures diretly as a fuel. International Patent Number WO/2010/028323
77. Wang X, Chen C, Duan H (2004) Implantable fuel cell. United States. Patent Number: US 2004/0091757 A1
78. Liu BC, Zhang HC, Ma HC, Zhang YC (2010) Double-effect oxygen electrode catalyst slurry for fuel cell, is obtained by using catalyst nanoparticles, proton-conductive polymer and alcoholic solution. China. Patent Number: CN101773825-A

Chapter 5
Nanostrutured Electrocatalysts for Methanol and Ethanol-Tolerant Cathodes

Edson A. Ticianelli and Fabio H.B. Lima

Abstract Polymer electrolyte membrane fuel cells (PEMFC) that use small organic molecules like methanol and ethanol as fuel in the anode, and oxygen in the cathode, are attracting considerable interest for application in portable electronic devices. Carbon-supported platinum has the highest catalytic activity for oxygen reduction when compared to all other pure metals, and serves as state-of-the-art cathode material in low-temperature fuel cells. Regarding the direct methanol or ethanol fuel cells (DMFC or DEFC), one of the major problems is the alcohol crossover through the polymer electrolyte. The mixed potential, which results from the oxygen reduction reaction and the alcohol oxidation occurring simultaneously, reduces the cell voltage, generates additional water and increases the required oxygen stoichiometric ratio. This problem can be, in principle, solved either by using electrolytes with lower methanol or ethanol permeability or by developing new cathode electrocatalysts with both higher alcohol tolerance and higher activity for the oxygen reduction reaction than Pt. This chapter presents an overview of recent developments of platinum and non-platinum-based catalysts as methanol and ethanol-tolerant oxygen reduction materials for direct alcohol fuel cells.

5.1 Introduction

The direct methanol fuel cell (DMFC) is one of most interesting low-temperature fuel cell system for practical application in small or portable electronic devices. This is attributed to its high energy density, feasible operation conditions at ambient temperature, and easy availability of the liquid fuel [1]. In this type of fuel cell, the diffusion of methanol molecules to the three-phase boundary of the oxygen cathode can hardly be avoided. Under working conditions, this leads to a negative shift of

E.A. Ticianelli (✉) • F.H.B. Lima
Institute of Chemistry of Sao Carlos, University of Sao Paulo, Sao Paulo, Brazil
e-mail: edsont@iqsc.usp.br; fabiohbl@iqsc.usp.br

H.R. Corti and E.R. Gonzalez (eds.), *Direct Alcohol Fuel Cells: Materials, Performance, Durability and Applications*, DOI 10.1007/978-94-007-7708-8_5,

the cathode potential and an additional consumption of methanol [2]. Until now, it was assumed that the observed effect results from simultaneous processes of oxygen reduction and methanol oxidation at the Pt surface, the oxygen reduction being only slightly poisoned by the adsorbed methanol. The direct oxidation of methanol in fuel cells has been widely investigated in carbon-supported Pt-Ru bimetallic nanoparticles, which present high activity and high faradaic conversion efficiency to CO_2. As this anode has satisfactory electrochemical conversion efficiency, the main challenge in DMFC is to overcome the problems caused by the methanol crossover.

The Direct Ethanol Fuel Cell (DEFC) is an attractive system since this fuel can be produced by fermentation of sugar-containing raw materials from agriculture. However, despite several advances in recent years, the existing electrocatalyst still possesses very low electrochemical conversion efficiency to CO_2 at ambient temperature, particularly at high concentration values of the alcohol molecule (>0.1 mol L^{-1}) [3]. In addition to this, the ethanol crossover seems to be even more prominent in the DEFC, causing more severe losses in the fuel cell power.

5.2 Oxygen Reduction Reaction (ORR)

The oxygen reduction reaction has remained the focus of considerable attention because of its complex kinetics and the need for better electrocatalysts [4–6]. A further increase in their efficiency critically depends on improving the reaction's kinetics even on Pt, which is the best available electrocatalyst. In addition, the platinum content of electrocatalysts must be lowered before fuel cells can be broadly applied.. Particularly troubling is the large loss in potential of 0.3–0.4 V in the initial part of the polarization curves that is an important source of decline in the fuel cell's efficiency. A part of this polarization is attributed to the inhibition of O_2 reduction caused by OH adsorption on Pt in the potential region 0.75–1 V. Alloying Pt with transition metals was reported to reduce PtOH formation and to produce some improvements in activity [7]. Another problem of existing fuel cell technologies is the high Pt loading in the cathode. Attempts have been made to reduce Pt loadings by using Pt monolayers on less-expensive metal nanoparticles [7, 8].

In acidic electrolyte, the oxygen reduction reaction (ORR) can be represented as follows [4, 6]:

A. "Direct" four-electron pathway:

$$O_2 + 4H^+ + 4e^- \rightarrow 2H_2O \qquad E_0 = 1.229 \text{ V} \tag{5.1}$$

B. "Series" pathway:

$$O_2 + 2H^+ + 2e^- \rightarrow H_2O_2 \qquad E_0 = 0.67 \text{ V} \tag{5.2}$$

followed by:

$$H_2O_2 + 2H^+ + 2e^- \rightarrow 2\,H_2O \qquad E_0 = 1.77\ \text{V} \tag{5.3}$$

or by:

$$2H_2O_2 \rightarrow 2H_2O + O_2 \tag{5.4}$$

5.2.1 Oxygen Reduction Electrocatalysts

On Pt-based electrocatalysts, some evidences suggest that the first electron transfer composed of O_2 adsorption with simultaneous electron transfer and proton addition is the rate determining step at the overall reaction processes. This may be represented by [7]

$$O_2 + Pt \rightarrow Pt\text{---}O_2 \tag{5.5}$$

followed by and electrochemical step:

$$Pt\text{---}O_2 + H^+ + e^- \rightarrow \text{products} \tag{5.6}$$

(the difference to the alkaline electrolyte is that the hydrogen atom for the O–H bond formation comes from the water molecules of the electrolyte, instead of from H_3O^+ species).

From the kinetic point of view, the other steps are less important, since they take place after the rate-determining step. The transfer of the second electron, with the addition of a second proton, generates OH or hydrogen peroxide species. The peroxide can diffuse to the *bulk* of the solution electrolyte and the reaction terminates with a process involving two electrons. One additional transfer of two electrons and two protons involving OH or H_2O_2 completes the reduction via 4 electrons, resulting in H_2O (for the case of acidic media). So, the reaction via 4 electrons, wherein it is of major interest because it involves higher number of electrons per O_2 molecule, has to deal with the O–O bond breaking and the O–H bond formation, as represented below [7]:

$$O_2 \rightarrow O + O \tag{5.7}$$

$$O + H \rightarrow OH \tag{5.8}$$

Thus, electrocatalysts for the ORR have to present electronic structures that result in adsorption forces that strike these two competing steps: while strong adsorption leads to facilitated O–O bond breaking, weak adsorption tend to facilitate the O–H bond formation (hydrogen addition). This produces the so-called "volcano" plot of the activity as a function of the adsorption strength on the catalyst surface [7–9].

A particularly difficult problem of the ORR electrocatalysis is the high loss of potential, which is a substantial source of the decline in the efficiency of fuel cells. As mentioned before, another drawback (from cost point of view) is the high Pt loading in cathode or the low mass-activity (current per mass of active metal) in the regular Pt-based electrocatalysts [10, 11]. Thus, the research in the ORR electrocatalysis aims at developing better electrocatalysts in order to reduce the cathode overpotential and the total mass of Pt [12, 13].

The activity for the ORR of Pt monolayers, deposited on different single-crystal surfaces, using the Cu UPD technique [14], were investigated in acid and in alkaline electrolytes [7, 8]. Figure 5.1a show the typical ORR curves obtained for pure Pt/C and Pt monolayer on Pd/C nanoparticles, and Fig. 5.1b shows the plot of ORR activity versus Pt *d-band* center on different surfaces [7]. As can be seen, the Pt monolayer electrocatalysts exhibited support-induced tunable activity by arising either by structural and/or electronic effects. It can be observed that the most active of all surfaces is Pt_{ML}/Pd(111), and the least active is Pt_{ML}/Ru(0001). The plots of the kinetic current on the platinum monolayers on various substrates at 0.8 V as a function of the calculated *d*-band center, ε_d, generated a volcano-like curve, with Pt_{ML}/Pd(111) showing the maximum activity (Fig. 5.1a).

The modification of the electronic and chemical properties of Pt(111) surfaces by subsurface 3*d* transition metals was studied using density-functional theory [15]. In each case investigated, it was seen that the Pt surface *d*-band was broadened and lowered in energy by interactions with the subsurface 3*d* metals, resulting in weaker dissociative adsorption energies of hydrogen and oxygen on these surfaces. The magnitude of the decrease in adsorption energy was largest for the early 3*d* transition metals and smallest for the late 3*d* transition metals.

The O_2 reduction reaction were also investigated on platinum deposited monolayers, and on polycrystalline alloy films of the type Pt_3M (M = Ni, Co, Fe, and Ti) [9]. The Pt_3M alloys were annealed at 1,000 K under ultrahigh-vacuum (UHV) conditions; analysis of the low-energy ion-scattering spectra indicated that pure Pt constitutes the first surface layer. The surface enrichment of Pt atoms results from a surface segregation phenomenon, whereby one of the alloy's components (in this case Pt) enriches the surface region [16]. UHV experimental analyses and theory revealed that the strong enrichment of Pt in Pt_3M alloy systems is counterbalanced by the depletion of Pt in the first two or three layers beneath the surface, giving a concentration profile that oscillates around the *bulk* value [17]. The *d*-band center of these alloy films was measured by synchrotron-based high-resolution photoemission spectroscopy, a methodology described in a previous publication [18]. Figure 5.2a [9] shows the ORR curves obtained for Pt_3Co and pure Pt electrocatalysts, and Fig. 5.2b presents the plots of the electrocatalytic activity for the ORR at 0.9 V versus the position of the *d*-band center. It exhibits a classical volcano-shaped dependence, agreeing very well with the activity predicted from DFT calculations, and with the results for Pt monolayer deposited on different single-crystal substrates, as discussed above.

As mentioned above, the O_2 reduction reaction via 4 electrons must involve the breaking of the O–O bond (whether in O_2, HO_{2-}, or H_2O_2) and the formation of an O–H bond. So, a more reactive surface, such as one characterized by a

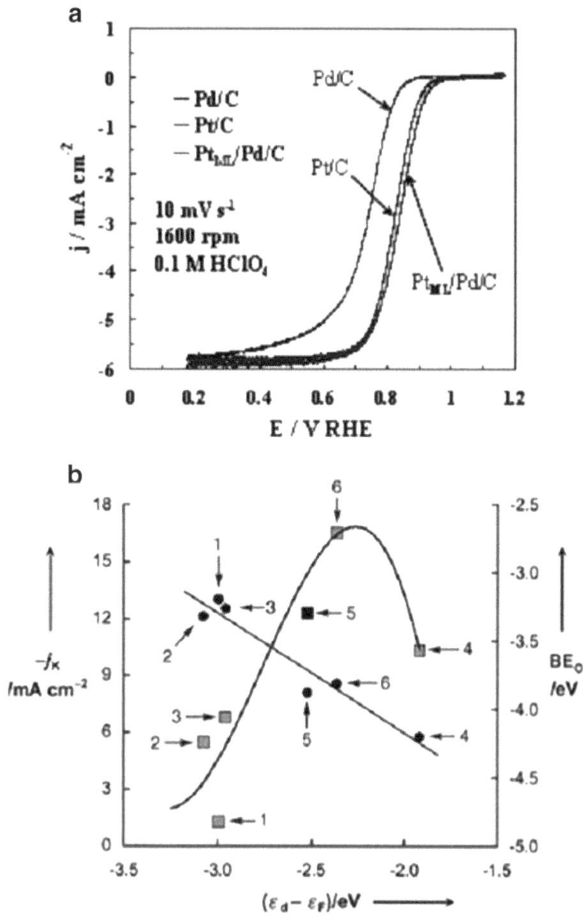

Fig. 5.1 (**a**) Polarization curves for the ORR on Pt/Pd/C material compared to those for pure Pt/C and Pd/C electrocatalysts; (**b**) Kinetic currents (j_K; *square symbols*) at 0.8 V for O_2 reduction on platinum monolayers in a 0.1 mol L^{-1} $HClO_4$ on the different Pt monolayers, as functions of the electrocatalyst *d-band* center. Labels: (*1*) Pt_{ML}/Ru (0001), (*2*) Pt_{ML}/Ir(111), (*3*) Pt_{ML}/Rh(111), (*4*) Pt_{ML}/Au (111), (*5*) Pt(111), (*6*) Pt_{ML}/ Pd(111) (Reproduced from Refs. [7, 13] with kind permission of Ref. [7] © Wiley-VCH Verlag GMBH & CO. KGAA 2012, and Ref. [13] © American Chemical Society 2012)

higher lying ε_d, tends to bind adsorbates more strongly [15], thus enhancing the kinetics of dissociation reactions. On the other hand, a surface with a lower lying ε_d tends to bind adsorbates more weakly, facilitating the formation of bonds in the reaction adsorbates. So, it is expected that the ε_d of the most active platinum monolayer should present an intermediate value.

Regarding the geometric effect, compressive strain tends to down-shift ε_d in energy, whereas tensile strain has the opposite effect, as revealed by DFT studies [7, 15, 16]. The platinum monolayers on Ru(0001), Rh(111), and Ir(111) are compressed compared with Pt(111), whereas Pt_{ML}/Au(111) is stretched by more than 4 %. However, the position of the ε_d depends also on the electronic effect, in which the magnitude of the ε_d shift depends on the intensity of the electronic interaction between the platinum monolayer and its substrate [19]. This indicates that Pt_{ML}/Ru (0001), Pt_{ML}/Ir(111), and Pt_{ML}/Rh(111) are less active for O_2 reduction than platinum because breaking the O–O bond is more difficult on their surfaces than on Pt(111), while the kinetics of hydrogenation of the oxygen atoms may be

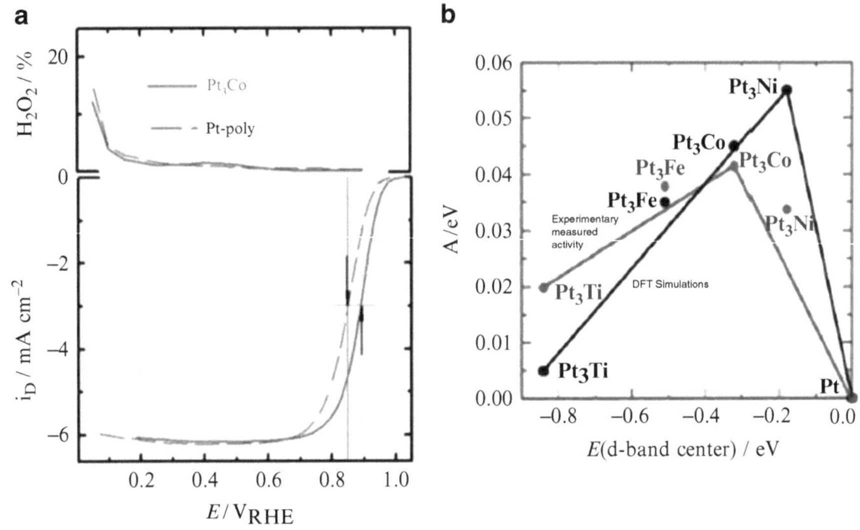

Fig. 5.2 (**a**) Polarization curves for Pt_3Co alloy and Pt at 1,600 rpm and 10 mV s^{-1} and on 0.1 mol L^{-1} $HClO_4$ electrolyte; (**b**) Activity versus the experimentally measured *d*-band center relative to platinum. The activity predicted from DFT simulations is shown in *black*, and the measured activity is in *gray* (Reproduced from Ref. [9] with kind permission of © Wiley-VCH Verlag GMBH & CO. KGAA 2012

hindered on Pt_{ML}/Au(111) due to its stronger binding of oxygen atoms or oxygen-containing fragments compared to Pt(111). These properties of Pt monolayer on different substrates also result in the volcano dependence of kinetic currents as a function of ε_d [7–9]. In addition, slow rates of O or OH hydrogenation cause an increased in the O or OH surface coverage. This results in a blocking of the active sites for the O_2 adsorption, dissociation, or hydrogenation.

Therefore, the results evidence that Pt_{ML}/Pd(111) possesses a better balance between the kinetics of the two opposite steps (Eqs. 5.11 and 5.12) compared to that of Pt(111), suggesting that Pt_{ML}/Pd(111) is close to the position of an optimum compromise or an optimum balance. Experimentally, Adzic and co-authors reported that the superior ORR catalytic activity of Pt_{ML}/Pd(111) could be associated with reduced OH coverage, which was observed by in situ XANES (*X-ray Absorption Near Edge Structure*). This result agrees with DFT findings that the binding energy of OH is weaker on Pt_{ML}/Pd(111) than on Pt(111) [13].

In the case of the Pt_3M polycrystalline alloys [9], the activity versus the position of the metal d states relative to the Fermi level also resulted in a Volcano plot. Pt_3Ni and Pt_3Co presented higher activities when compared to Pt_3Ti, Pt_3Fe or even pure Pt. It was also evidenced that the increased activity of these alloys in relation to Pt was due to a reduced Pt-oxygen species interaction, which in turns accelerates the hydrogenation step. Thus, in both cases, it was observed that Pt binds oxygen a little too strong. So, in order to conduct the Pt-based catalyst closer to the optimum compromise between the two opposite steps, its reactivity for binding oxygen should be slightly reduced.

5.3 Methanol and Ethanol Electrooxidation

The total electrochemical oxidation of methanol, involving 6 electrons, can be represented by:

$$CH_3OH + H_2O \rightarrow CO_2 + 6H^+ + 6e^- \qquad E^0 = 0.019\ \text{V}\ \textit{vs.}\ \text{SHE} \tag{5.9}$$

However, on Pt-based electrocatalysts, the reaction can follow parallel pathways which can, in principle, be formulated as follows [20]:

$$\begin{aligned} CH_3OH \quad \rightarrow \quad & (\text{adsorbed intermediates}) \rightarrow CO_{ad} \rightarrow CO_2 \rightarrow \\ & \rightarrow HCHO, HCOOH, CO_2 \end{aligned} \tag{5.10}$$

It was suggested that methanol adsorption takes place in several steps, forming different species due to dissociation of the molecule [21]:

$$\begin{aligned} CH_3OH \rightarrow \underset{x}{CH_2OH} + H^+ + e^- \rightarrow \underset{xx}{CHOH} + H + + e^- \rightarrow \\ \rightarrow \underset{xxx}{COH} + H^+ + e^- \rightarrow \underset{x}{CO} + H^+ + e^- \end{aligned} \tag{5.11}$$

where x represents a Pt site.

Therefore, in order to follow the CO pathway, forming CO_2, the electrocatalyst needs to have at least three active contiguous sites of Pt.

Similar steps can be drawn for the case of ethanol electro-oxidation [22], in which the total electrochemical oxidation of ethanol, involving 12 electrons, can be represented by:

$$C_2H_5OH + 3H_2O \rightarrow 2CO_2 + 12H^+ + 12e^- \quad E^0 = 0.082\ \text{V}\ \textit{vs.}\ \text{SHE} \tag{5.12}$$

Also, here, the reaction can follow parallel pathways, which can be formulated as follows:

$$\begin{aligned} C_2H_5OH \rightarrow (\text{adsorbed intermediates}) \rightarrow CO_{ad} \rightarrow CO_2 \rightarrow \\ \rightarrow CH_3CHO, CH_3COOH, CO_2 \end{aligned} \tag{5.13}$$

but, now, involving even higher number of contiguous active sites of Pt. However, the exact number of such sites is still unknown, since the ethanol electro-oxidation mechanism remains somewhat unclear.

5.4 Oxygen Reduction in the Presence of Methanol and Ethanol

As mentioned above, the alcohol crossover from the anode to the cathode is a important problems to be overcome to improve the DAFC performance. This is due to the fact that the commonly used Pt-based cathode electrocatalysts are also active for the adsorption and oxidation of methanol [1]. So, in addition to the resulting mixed potential at the cathode, there is a decrease in the fuel utilization. Therefore, considering the above exposed reactions for the alcohol electro-oxidation, and the features that govern the ORR electrocatalytic activity, as discussed in the Sect. 5.2, it is ready to conclude the importance of the modification of the active ORR electrocatalyst surfaces in order to inhibit the methanol or ethanol oxidative adsorption steps. In the next sections, some recent materials being developed to overcome the problems caused by the alcohol crossover will be presented.

5.4.1 *Platinum Nitrides: PtN_x/C*

Kim and Oh [23] have investigated the ORR activity of carbon-supported platinum nitrides (PtN_x/C) in the absence and in the presence of methanol. These materials were synthesized by a pyrolysis procedure. Figure 5.3 shows the obtained linear sweep voltammograms for PtN_x/C and for a commercially available Pt/C catalyst. The molar ratio of N to Pt was changed from 0.5 to 15. As can be observed, the ORR activity of the PtN_x/C catalysts increases as the molar ratio of N:Pt decreases from 15 to 2. With further decrease of the ratio to 0.5, no clear activity change was observed. The ORR activities of the PtN_x/C catalysts, prepared at different annealing temperatures, were also investigated in the absence of methanol. The results showed that the onset potential for PtN_x/C treated at 700 °C is more positive than that treated at 900 °C. It is proposed that this difference is due to the roughness factor, that is, the ratio between the actual and reactive surface area. As discussed by the authors, the surface of Pt is modified and this does not enhance the ORR activity, but rather suppresses the adsorption of methanol. As observed above, at least three contiguous neighbouring Pt atoms are required to initiate the chemisorption of methanol [20].

The methanol-tolerant properties of the PtN_x/C catalysts in terms of polarization in N_2-saturated 0.5 mol L^{-1} H_2SO_4 solution, containing 0.1 mol L^{-1} methanol, is presented in Fig. 5.3b [23]. As can be observed, the methanol oxidation peak for the PtN_x/C is more than ten times smaller than that obtained for the commercial Pt/C catalyst. These curves show that the PtN_x/C materials are less active for methanol oxidation and this might result in a high methanol tolerance during the ORR. With increasing N content, the methanol oxidation is further inhibited and shows almost constant methanol tolerance when the molar ratio of N to Pt is higher than 4.

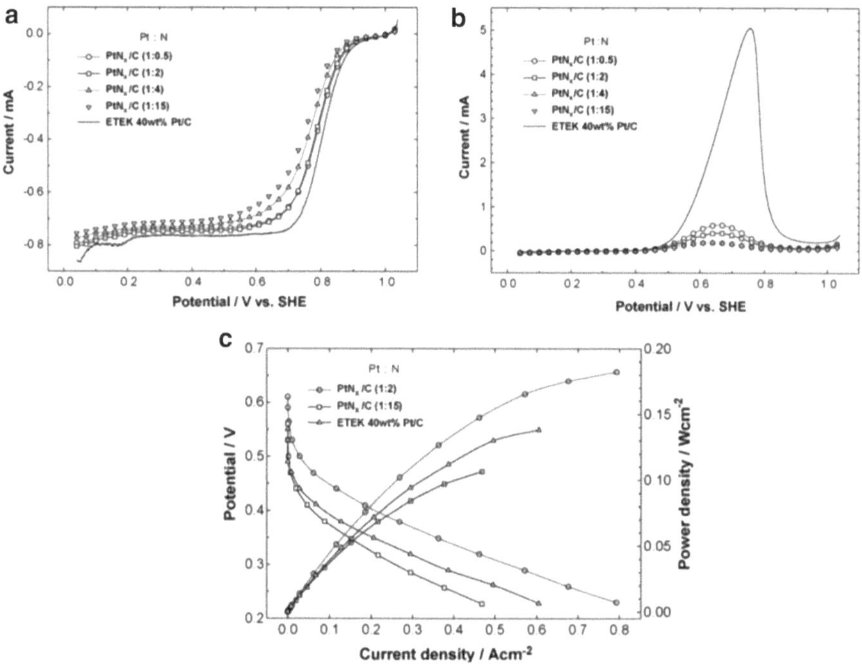

Fig. 5.3 (**a**) Polarization curves obtained on RDE for commercial Pt/C and PtN_x/C catalysts prepared using different molar ratios of Pt to N in precursor solution. Measurements performed in 0.5 mol L^{-1} H_2SO_4 solution saturated with oxygen at scan rate of 5 mV s^{-1} and rotating speed of 1,200 rpm. (**b**) Linear sweep voltammogram for methanol oxidation on PtN_x/C and commercial Pt/C catalysts in 0.5 mol L^{-1} H_2SO_4 + 0.1 mol L^{-1} CH_3OH solution saturated with nitrogen at scan rate of 5 mV s^{-1}. (**c**) Polarization curves in single DMFC using PtN_x/C and Pt/C as cathode electrodes. Anode: PtRu black 4 mg cm^{-2}; cathode: PtN_x/C or commercial Pt/C 1 mg cm^{-2}; electrolyte: Nafion 112; methanol (4 mol L^{-1}) flow rate: 1 mL min^{-1}; O_2 pressure: 1 atm; cell temperature: 70 °C (Reproduced from Ref. [23] with kind permission of © Elsevier 2012)

The researchers also conducted experiments using a unit DMFC. Figure 5.3c shows a comparison of the DMFC polarization curves obtained for the PtN_x/C catalyst and commercial Pt/C at 70 °C and 1 atm. The PtRu black electrocatalyst was used in the anode in all cases. It was seen that the PtN_x/C (Pt:N = 1:2) material gives improved performance, as much as 180 mWcm^{-2}, compared with the commercial Pt/C value of 138 mW cm^{-2}. According to the authors, such improving in the fuel cell performance is attributed to the high oxygen reduction activity and to the lower kinetics of the oxidation of methanol transferred from the anode to the cathode.

5.4.2 *Pt-Mesoporous Carbon/C*

Li and co-authors [24] reported a facile template route for the in situ entrapment of highly distributed *core-shell* Pt/C nanoparticles into the nanochannels of

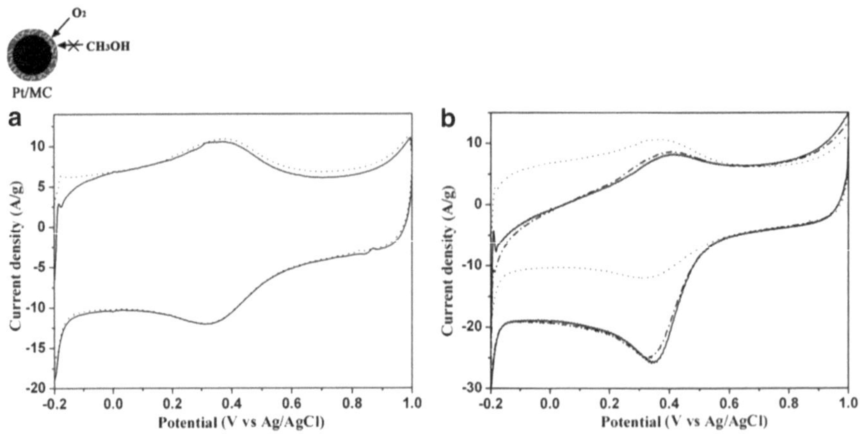

Fig. 5.4 Cyclic voltammograms of Pt@C/MC in: (**a**) a N_2-saturated aqueous solution of 0.5 mol L^{-1} H_2SO_4 with (*dotted line*) and without (*solid line*) 0.5 mol L^{-1} methanol, and (**b**) an O_2-saturated aqueous solution of 0.5 mol L^{-1} H_2SO_4 with (*dash-dot line*) and without (*solid line*) 0.5 mol L^{-1} methanol at a scan rate of 50 mV s^{-1}. Note that the *dotted line* in (**b**) is the CV curve of the Pt@C/MC in the N_2-saturated electrolyte (Reproduced from Ref. [24] with kind permission of © John Wiley and Sons 2012)

mesoporous carbon. The electrocatalysts were synthesized by mixing mesoporous silica and poly(vinylpyrrolidone) in an aqueous solution containing glucose and H_2PtCl_6, followed by a thermal treatment in autoclave, and by a carbonization under argon protection at 750 °C. Figure 5.4a shows the voltammetric curves recorded for a Pt@C/MC material in N_2 saturated 0.5 mol L^{-1} H_2SO_4 solution with and without methanol. The two voltammograms almost completely overlap, suggesting that the Pt@C/MC exhibits a negligible catalytic activity for methanol electrooxidation. The Pt@C/MC material presents two redox peaks at about 0.35 V in the N_2-saturated electrolyte, which was associated to the reduction and oxidation processes of the oxygen-containing groups on the MC surface. Furthermore, the peak shoulder at about −0.195 V and an adsorption feature at −0.184 V was explained in terms of the fact that the exposed active site of the Pt nanoparticles might be predominated by some specified crystallite phase.

In the O_2-saturated electrolyte solution, shown in Fig. 5.4b, the reduction peak at 0.35 V was enhanced and became sharper, which suggests that the Pt@C/MC is active for the ORR. The oxygen reduction peak displayed insignificant changes and no novel peaks were found in the O_2-saturated electrolyte with 0.5 mol L^{-1} methanol. The authors state that the high catalytic activity and methanol tolerance of these Pt@C/MC composites are associated to an overlaid film of carbon, that contain micropores, which allows oxygen to access the active sites of the nanoparticles, but hindering methanol from approaching these sites. As discussed by the authors, the Pt@C/MC composite without carbonization did not show any catalytic behavior for oxygen reduction, suggesting that less-exposed activity sites of Pt nanoparticles could be obtained when no micropores were formed in the carbon layer coating.

This further demonstrates that the micropores formed during the thermal treatment are necessary for oxygen to access the Pt-nanoparticle active sites. Authors also evaluated the durability of the Pt@C/MC hybrids through repeated CV cycles with the appropriate lower and upper potential limits in an O_2-saturated electrolyte containing 0.5 mol L^{-1} of methanol. The results showed that the variation in the current density was only about 4 % after 40 cycles, which means that the Pt@C/MC electrode has a considerable stable electrocatalytic activity for ORR in the presence of methanol [24].

5.4.3 Platinum Alloys

Many ORR experiments have been made on electrocatalysts composed by Pt and Pd with the addition of non-noble metals, such as Co, Fe, and Ni. However, under electrochemical conditions these non-noble metals might leach out from the electrocatalyst, as demonstrated in previous investigations [25]. In order to avoid this problem, Yang and co-authors [26] have investigated PdPt-based electrocatalysts for the ORR in absence and in the presence of methanol, because the long-term stability of Pd in acidic solution is comparable to that of Pt (but this depends on the potential – additionally, Pt may stabilize Pd atoms in the alloy). It was report a novel strategy for surface and structure-controlled synthesis of carbon-supported Pd_3Pt_1 nanoparticles for the ORR as well as for methanol-tolerant ORR electrocatalyst. The influence of the surface composition and structure of the Pd_3Pt_1/C on the ORR activity in the absence and presence of methanol was also reported.

The comparison of the ORR curves on commercial Pt/C and Pd_3Pt_1/C catalysts in the presence of 0.5 mol L^{-1} CH_3OH is made in Fig. 5.5a [26]. Compared to pure acidic solution, all the catalysts show an increase in the overpotential in the presence of methanol. The significant increase in overpotential (ca. 200 mV) of ORR on the Pt/C catalyst was assigned to the competition reaction between oxygen reduction and methanol oxidation. On the other hand, the overpotentials are much lower on the PtPd-based materials. The Pd_3Pt_1/C catalyst showed the highest ORR activity, which was attributed to Pd enrichment on the surface layers. In the presence of methanol, the ORR activity in the kinetic and mixed regions on all the Pd_3Pt_1/C catalysts is much higher than that on Pt/C, evidencing a high methanol tolerance of these materials. Figure 5.5b shows the linear scanning voltammograms obtained for the methanol oxidation for the Pd_3Pt/C materials and of Pt/C, in N_2-saturated solution. As can be noted, the current densities of methanol oxidation on the Pd_3Pt_1/C catalysts (in absence of O_2) are much lower than that of the Pt/C catalyst. So, as discussed in the work [26], the high methanol tolerance of the Pd_3Pt_1/C catalysts can be attributed to the weak competitive reaction of methanol oxidation, which could be induced by composition effects associated with the presence of Pd atoms. The high tolerance of Pd_3Pt_1/C to methanol suggests that this catalyst may be an economically viable candidate to replace Pt in the cathode of a DMFC.

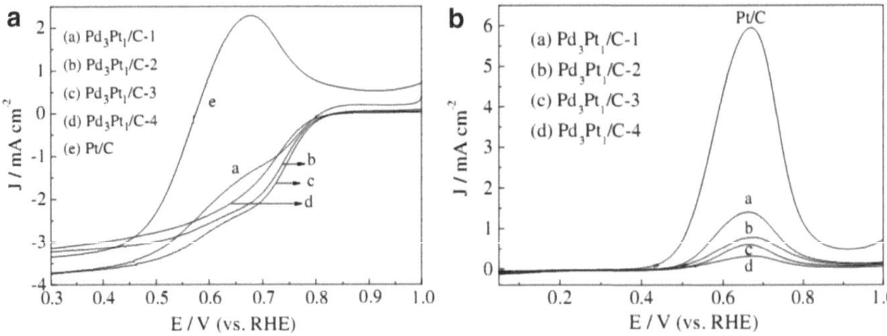

Fig. 5.5 (**a**) Linear sweep voltammograms of the prepared catalysts in oxygen-saturated 0.1 M $HClO_4$ + 0.5 M MeOH at a scan rate of 5 mV s^{-1} with a rotating speed of 1,500 rpm; (**b**) LSVs of methanol oxidation on the Pt/C and Pd_3Pt_1/C catalysts in 0.1 mol L^{-1} $HClO_4$ + 0.5 mol L^{-1} CH_3OH saturated with pure N_2 (Reproduced from Ref. [26] with kind permission of © Elsevier 2012)

In another work, Shukla and co-authors [27] studied the electrocatalytic activity of carbon-supported Pt-Au alloy catalysts, with different atomic ratios, to improve the oxygen reduction reaction (ORR) kinetics and methanol tolerance, in a direct methanol fuel cell. The electrocatalysts were prepared by codeposition of Pt and Au nanoparticles onto a carbon support.

The performance of the Pt-Au/C (2:1) and Pt/C catalysts toward ORR in the presence and absence of methanol was evaluated by RDE experiments and the results are shown in Fig. 5.6a. As can be observed, Pt-Au/C (2:1) and Pt/C possesses similar ORR half-wave potentials, but Pt-Au/C showed higher ORR activity in the presence of methanol. The electrocatalysts have also been tested in a DMFC [27]. The cell polarization data for Pt-Au/C catalysts with different Pt to Au atomic ratios in a methanol/O_2 DMFC are presented in Fig 5.6b. The results obtained for Pt/C were also included for comparison. These results show that the DMFCs with Pt-Au/C cathodes perform better than Pt/C. It was noted that the DMFC with Pt-Au/C (2:1) in the cathode shows an enhanced peak power density of 120 mW cm^{-2}, compared to 80 mW cm^{-2} observed for the DMFC with a Pt/C cathode. As argued by the authors [27], since the anodes in both the DMFCs are identical, the enhanced performance for the DMFC with Pt-Au/C catalyst is clearly due to the synergistic promotion of ORR on the catalyst.

Although the Pt-Au phase diagram shows that these two metals form alloy only in a restricted range of atomic ratios, the formation of alloy seems favorable in particles with nanometric dimensions. Furthermore, DFT calculations show a tendency of Au surface segregation on the particle [16]. So, as the methanol adsorption-dehydrogenation process requires at least three neighboring Pt atoms with appropriate crystallographic arrangement, in the Pt-Au/C samples, the probability of finding this arrangement on the surface decreases for increasing Au contents. This would explain the higher methanol tolerance of the alloy material in relation to that of pure Pt/C.

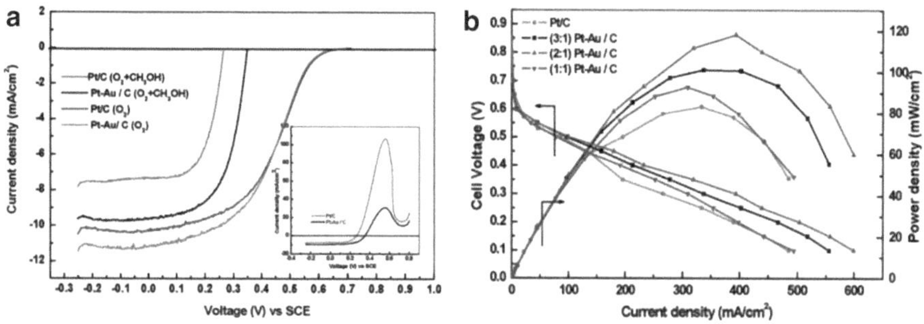

Fig. 5.6 (**a**) Linear sweep voltammetry (LSV) data for ORR on Pt/C and (2:1) Pt Au/C catalysts in O_2-saturated 0.5 mol L^{-1} H_2SO_4 electrolyte in the presence and absence of methanol at 1 mV s^{-1} scan rate (electrode rotation rate: 1,500 rpm). The inset shows the LSV data for ORR on Pt/C and (2:1) Pt Au/C catalyst in O_2-saturated 0.5 mol L^{-1} H_2SO_4 0.5 mol L^{-1} CH_3OH electrolyte. (**b**) Steady-state performances of DMFCs (CH_3OH and O_2) for Pt/C and Pt Au/C with varying atomic ratios (Reproduced from Ref. [27] with kind permission of © American Chemical Society 2012)

In the case of DEFC, Pripadchev and co-authors [28] developed platinum-based (Pt/C) cathode catalysts modified with S, P, and Bi that were highly tolerant to ethanol crossover into the cathode compartment of the fuel cell. In the case of doping Pt/C with these elements, the researchers used triphenylphosphine, bismuth nitrate, and thiourea as dopant sources. The catalysts were synthesized by high-temperature treatment of the catalyst–dopant element mixture in an inert atmosphere (further details see Ref. [28]).

The polarization curves of ethanol oxidation in the presence of oxygen (anodic region) and of oxygen reduction in the presence of ethanol (cathodic region) are presented in the Fig. 5.7a. It is observed that the ethanol oxidation is much slower and occurs at potentials higher than 0.70 V in the modified Pt. The PtBi catalyst is the least active for the ethanol oxidation reaction. In the case of pure Pt/C, the authors explained the significant decrease in the current density of oxygen reduction by a concerted process involving the ethanol oxidation and the poisoning of the platinum surface by oxidation products. This also explained the corresponding increase in ethanol tolerance of the modified catalysts. Figure 5.7b shows polarization curves obtained on PEM single cells, using the investigated cathode material. The results show that the modified platinum catalyst improved the performance of the ethanol–oxygen fuel cell as compared the commercial platinum catalyst. The positive effect of Bi in the fuel cell current characteristics is more evident for relatively high voltages (0.7–0.4 V) since the crossover effect and, consequently, the deterioration of cathode characteristics is much more pronounced at low current densities and high ethanol concentrations.

5.4.4 *Platinum-Free Electrocatalysts*

The activity for the ORR and the tolerance to the presence of methanol of PdCo/C nanoparticles was investigated by Yang and co-authors [29]. The materials were

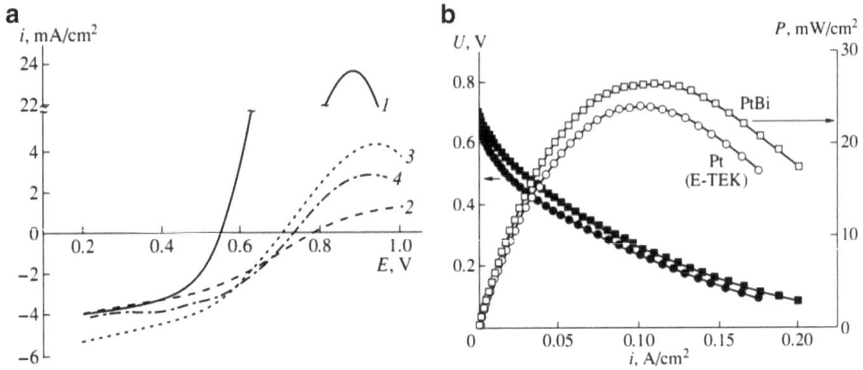

Fig. 5.7 (**a**) Polarization curves of the ethanol oxidation in the presence of oxygen (anodic region) and of the oxygen reduction in the presence of ethanol (cathodic region) measured with a rotating disc electrode in a 0.5 mol L^{-1} H_2SO_4 solution in the presence of 0.2 mol L^{-1} C_2H_5OH at 60 °C in an oxygen atmosphere with a thin layer of catalysts:(*1*) Pt (E-TEK), (*2*) PtBi, (*3*) PtP, and (*4*) PtS. The disc electrode rotation rate is 1,150 rpm, and the potential sweep rate is 5 mV s^{-1}; (**b**) Discharge curves (*solid symbols*) and dependences of power density on current density (*open symbols*) obtained in the tests with the use of MEA with a Nafion 117 membrane for Pt (E–TEK) and PtB catalysts (the Pt loading in the cathode active layer is 1.0 mg cm^{-2}) at 75 °C. The ethanol concentration in the fuel mixture is 1 M, and = pO_2 = 2 atm. The anode catalyst is a HiSPEC 10000 PtRu catalyst (with a Pt loading of 3.5 mg cm^{-2}) in both cases (Reproduced from Ref. [28] with kind permission of © Springer 2012)

synthesized by chemical reduction using sodium borohydride in the presence of NH_4F, as a complexing agent. The ORR curves on the Pd/C and PdCo/C eletrocatalysts obtained by these authors are presented in Fig. 5.8a. For a comparison, the registered curve for commercial Pt/C (E-TEK), with a metal loading of 20 %, was also included. It was observed that in the Tafel region and the mixed kinetics region, that the PdCo/C catalysts present higher ORR mass activities than Pd/C, except for the Pd_1Co_1/C catalyst. According the authors [29], this is probably due to the fact that the high Co content within a Pd_1Co_1/C sample results in a Co surface enrichment which prohibits oxygen molecules to adsorb on the Pd atoms. The onset potential of the ORR on Pd_2Co_1/C and Pd_4Co_1/C catalysts is found to be the most positive among the Pd Co/C catalysts. The higher activities of the Pd_4Co_1/C and Pd_2Co_1/C electrocatalysts, which were close to that of a commercial Pt/C material, was ascribed to the compressed Pd lattice caused by alloying with Co and the downshift of the d-band center of the Pd metal [15].

The effect of the presence of methanol on the ORR activity of commercial Pt/C and as-prepared Pd_4Co_1/C is presented in Fig 5.8b. As compared to the ORR in pure $HClO_4$ solution, both catalysts exhibited an increase in overpotential under the same current density in the presence of methanol. For the ORR on Pt/C in methanol-containing solution, the overpotential increases by ca. 200 mV, while only a small negative shift of ca. 15 mV was observed for Pd_4Co_1/C catalyst. In fact, the results obtained by the researchers indicate that the Pd_4Co_1/C electrocatalyst is very active for the ORR even at a high concentration of methanol.

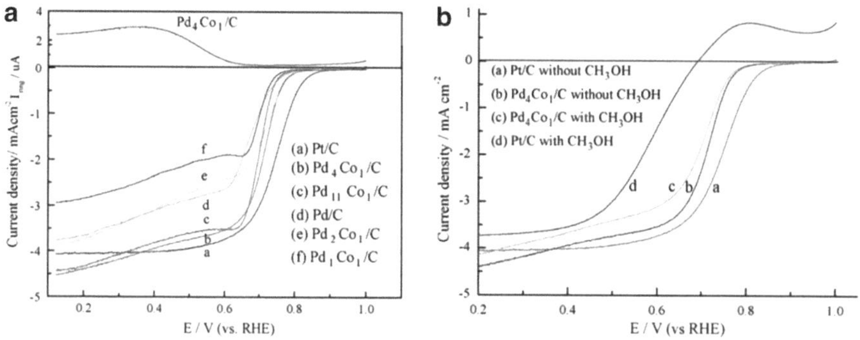

Fig. 5.8 (**a**) Linear scan voltammograms of the Pt/C and Pd Co/C catalysts in 0.1 mol L^{-1} $HClO_4$ saturated with pure oxygen at a scan rate of 5 mV s^{-1} and with a rotation speed of 1,600 rpm. Current density is normalized to the geometrical surface area of the electrode. The ring current, RRDE data, for hydrogen peroxide production is compared; (**b**) LSVs of the Pt/C and Pd_4Co_1/C catalysts in 0.1 mol L^{-1} $HClO_4$ + 0.5 mol L^{-1} CH_3OH saturated with pure oxygen (Reproduced from Ref. [29] with kind permission of © Elsevier 2012)

In general, non-noble metal alloy nanoparticles have shown some methanol tolerance effects, but their activity towards ORR is lower than that of Pt/C. Furthermore, it has been found that the non-precious metal catalysts do not present the required stability in the acidic environment, even in the case of Pd (at high potential). On the other hand, some works have shown that this instability (dissolution), mainly of the non-noble metal, can be overcome by the addition of small amounts of stabilizers like Au. Based on this, Mathiyarasu and Phani [30] examined the effect of the addition of Au, Ag and Pt on the activity and stability of several Pd-Co/C electrocatalysts. Results showed higher ORR activities for Pd-Co-Pt/C, equal to that of a commercial Pt/C electrocatalyst.

Figure 5.9 shows the voltammetric behavior of Pt/C (a) and Pd–Co–Pt/C (b) towards ORR in the presence and absence of methanol. As can be also observed, the presence of 0.5 mol L^{-1} methanol causes a negative shift of 50 mV in the half-wave potential, in contrast to Pt/C for which there is a severe loss of activity. The linear scan voltammograms of the methanol oxidation on all the investigated materials in 0.5 mol L^{-1} H_2SO_4 + 0.5 mol L^{-1} CH_3OH solution, showed that the current densities of the methanol oxidation reaction on Pd–Co–X alloy catalysts (X = Au, Ag, Pt) diminish to values much lower than for Pt/C catalyst, and the onset of methanol oxidation occurs at more positive potentials, demonstrating the lowered MOR activity of the Pd–Co–Pt alloy catalysts.

Conrad and co-workers [31] tested the electrocatalític activity of commercially available 30 wt% Rh_xS_y/C supported on Vulcan XC-72R carbon, and $RuSe_x/C$, as selective oxygen reduction candidates. The methanol electrooxidation curves obtained on the Rh_xS_y/C and $RuSe_x/C$ materials in 0.5 M H_2SO_4 electrolyte with 1 M CH_3OH at 60 °C are presented in Fig. 5.10a. The investigated electrocatalysts showed low activity for methanol electroxidation, when compared to those of Pt/C and PtRu/C. RuSe/C exhibited the lowest methanol oxidation activity, and RhxSy/C

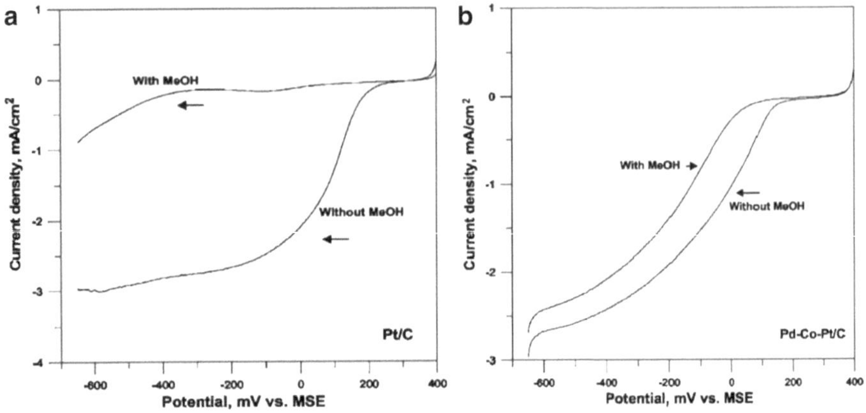

Fig. 5.9 Comparison of polarization curves for ORR on (**a**) Pt/C and (**b**) Pd–Co–Pt in 0.1 mol L^{-1} H_2SO_4 with and without 0.1 mol L^{-1} methanol 1,600 rpm scan rate 5 mV s^{-1} (Reproduced from Ref. [30] with kind permission of © The Electrochemical Society 2012)

displayed a methanol electroxidation peak at 0.85 V with an onset potential of 0.3 V. So, this could have implications relating on the use of this material in fuel cell operations demanding a methanol tolerant cathode.

Figure 5.10b presents RDE curves obtained for the ORR in 0.5 mol L^{-1} H_2SO_4 electrolyte in the absence and presence of 1.0 mol L^{-1} methanol. The ORR curves for Pt/C was included in the absence of methanol for comparison. As can be observed, RuSe/C shows an increase in overpotential of c.a.10 mV, in the presence of methanol which corroborates with its low observed for methanol oxidation. For Rh_xS_y/C, a current contribution from methanol oxidation was seen at potentials above 0.7 V, resulting in a net positive current at above 0.8 V. As discussed by the authors [31], modifying ruthenium with selenium results in a significant enhancement of oxygen reduction activity and almost an entirely suppressing of its methanol oxidation activity, in contrast to rhodium modified by sulfur.

A loss of selenium from the surface was observed upon exposure to potentials greater than 0.85 V, and this can have a detrimental effect on the implementation of RuSe/C as a cathode material in fuel cell applications which should, therefore, be further investigated. While steady-state operation could be confined to low cathode potentials, exposure to higher potentials in transients, e.g. during start-up conditions, could lead to selenium loss with a concomitant drop in fuel cell performance. The commercially available rhodium sulphide underperformed and exhibited higher susceptibility to methanol compared to RuSe/C, but was found to be more stable under similar testing conditions.

Savadogo and co-authors [32] have investigated the tolerance of Pd, PdCo, and Pt electrocatalyts to ethanol, during the ORR. The catalysts were fabricated by using the sputtering method. Linear sweep voltammograms of the oxygen reduction reaction in 0.5 mol L^{-1} H_2SO_4 with 0.5 mol L^{-1} ethanol are shown in Fig. 5.11. The polarization curves of the ORR without ethanol are also shown for comparison.

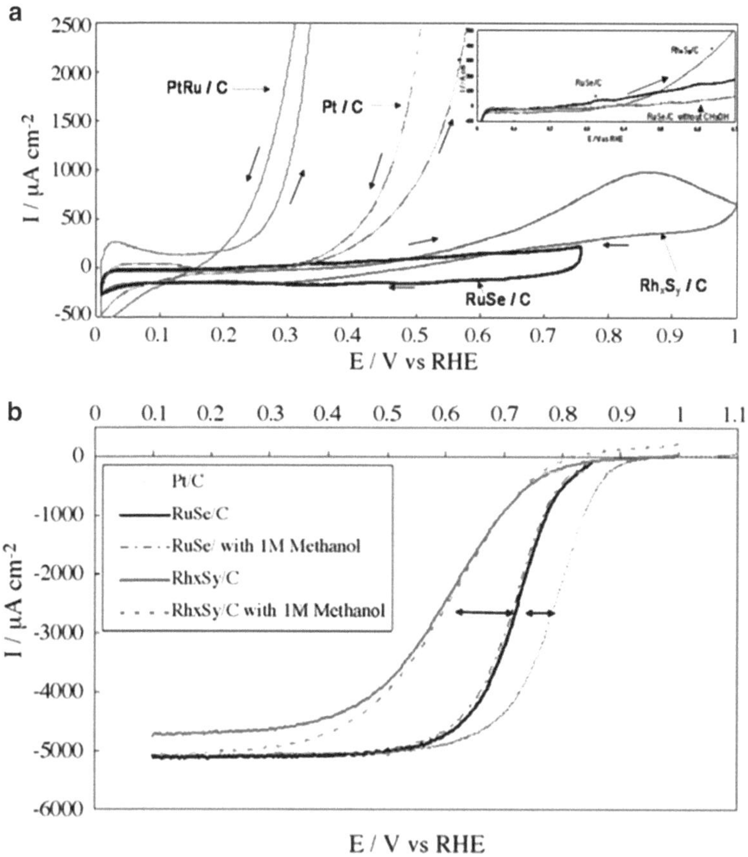

Fig. 5.10 (**a**) Cyclic voltammograms for Rh_xS_y/C, RuSe/C, PtRu/C and Pt/C, samples in 0.5 mol L^{-1} H_2SO_4 + 1.0 mol L^{-1} CH_3OH at 60 °C with a scan rate of 10 mV s^{-1}; (**b**) Polarisation curves for oxygen reduction on RhxSy/C, RuSe/C and Pt/C in oxygen saturated 0.5 mol L^{-1} H_2SO_4 + 1.0 mol L^{-1} CH_3OH at 60 °C and 1,600 rpm. Scan rate $\nu = 2$ mV s^{-1} in the negative direction (Reproduced from Ref. [31] with kind permission of © Elsevier 2012)

Both *bulk*-Pt and sputtered Pt (Fig. 5.11a, b) present a typical current density peak at ca. 0.8 V assigned to the oxidation of ethanol. On the contrary, this peak does not appear in the linear scans for Pd and Pd-Co catalysts. This was assigned to a low rate of adsorption of ethanol molecules and/or reaction intermediate species on the catalytic surface of Pd and Pd-Co.

Different from the results on Pt, the curves demonstrate the very high tolerance of the Pd-based catalysts to ethanol during the ORR. According to the authors, this phenomenon might be related to the inactivity of Pd and/or Co for ethanol oxidation reaction, and this might be due to the absence of adsorption of the alcohol and/or intermediate reaction species onto the catalyst surface. So, these results indicate that the utilization of Pd-alloys as cathode catalysts in a DEFC will help in reducing the high increase in cathode overpotential observed on Pt catalysts due to the ethanol crossover.

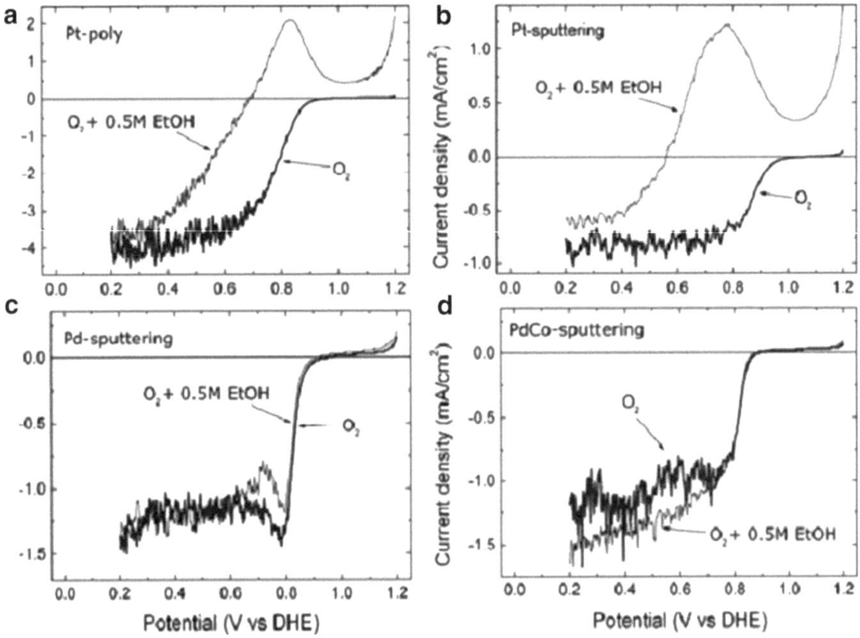

Fig. 5.11 Linear sweep voltammograms of the ORR in the absence and presence of 0.5 mol L^{-1} ethanol for *bulk*-Pt (**a**), sputtered Pt (**b**), sputtered Pd (**c**) and sputtered Pd-Co (**d**). Electrolyte: 0.5 mol L^{-1} H_2SO_4 saturated with oxygen. Scan rate: 5 mV s^{-1} (Reproduced from Ref. [32] with kind permission of © The Electrochemical Society 2012)

5.5 Conclusions

The ORR activity shows a volcano-type dependency on the *d*-band center of different metal catalysts. The electrocatalyst surfaces should exhibit the optimum balance between the kinetics of O–O bonding breaking and the electro-reduction of the oxygenated intermediates or for O–H formation. The highest activity is observed for Pt monolayer deposited on Pd substrate.

PtN_x/C and Pt@C/MC composites exhibit high ORR activity in the presence of methanol. In the last case, this is associated to an overlaid film of carbon, that contain micropores, which allows oxygen to access the active sites of the nanoparticles, but hinders methanol from approaching these sites.

The high methanol tolerance of PtPd/C alloy catalysts is attributed to the weak competitive reaction of methanol oxidation, which could be induced by composition effects associated with the presence of Pd atoms. The methanol adsorption-dehydrogenation process requires at least three neighboring Pt atoms with appropriate crystallographic arrangement, so, in the case of Pt-Au/C materials, the probability this arrangement in the surface decreases for increasing Au contents.

This explains the higher methanol tolerance of the alloy material in relation to that of pure Pt/C. For Pt-free electrocatalysts, Pd_4Co_1/C showed to be very active for the ORR even at a high concentration of methanol. The addition of noble metal such as Au, Ag and Pt onto the PdCo material, in order to increase their stability in acid electrolyte, conducts to a lowered MOR activity and high ORR kinetics. For the RuSe/C and RhS/C materials, the former presents a considerable tolerance to the presence of methanol. However, the observed loss of selenium from the surface, observed upon exposure to potentials greater than 0.85 V, indicates a detrimental effect on the implementation of RuSe/C as a cathode material in fuel cell applications. The commercially available rhodium sulphide underperforms and exhibits higher susceptibility to methanol compared to RuSe/C, but it is more stable under similar testing conditions.

Some presented results demonstrated a very high tolerance of the Pd-based catalysts to ethanol during the ORR. This phenomenon is related to the inactivity of Pd and/or Co for ethanol oxidation reaction, and this might be due to the absence of adsorption of the ethanol and/or intermediate species onto the catalyst surface. So, these results indicate that the utilization of Pd-alloys as cathode catalysts in a DEFC will help in reducing the negative effect due to the ethanol crossover.

Acknowledgments E. A. Ticianelli and F. H. B. Lima acknowledge support from Fundação de Amparo à Pesquisa do Estado de São Paulo – FAPESP, Brazil.

References

1. Vielstich W, Gasteiger HA, Lamm A (2003) Handbook of fuel cells – fundamentals, technology and applications. Wiley, Chichester
2. Vielstich W, Paganin VA, Lima FHB, Ticianelli EA (2001) Nonelectrochemical pathway of methanol oxidation at a platinum-catalyzed oxygen gas diffusion electrode. J Electrochem Soc 148:A502–A505
3. Sun S, Chojak Halseid M, Heinen M, Jusys Z, Behm RJ (2009) Ethanol electrooxidation on a carbon-supported Pt catalyst at elevated temperature and pressure: a high-temperature/high-pressure DEMS study. J Power Sources 190:2–13
4. Yeager E (1984) Electrocatalysts for O_2 reduction. Electrochim Acta 29:1527–1537
5. Kinoshita K (1992) Electrochemical oxygen technology. Wiley-Interscience, New York
6. Adzic R (1998) Recent advances in the kinetics of oxygen reduction. In: Lipkowski J, Ross PN (eds) Recent advances in the kinetics of oxygen reduction. Electrocatalysis. Wiley-VCH, New York, p. 197
7. Zhang J, Vukmirovic MB, Xu Y, Mavrikakis M, Adzic RR (2005) Controlling the catalytic activity of platinum-monolayer electrocatalysts for oxygen reduction with different substrates. Angew Chem Int Ed 44:2132–2135
8. Lima F, Zhang J, Shao M, Sasaki K, Vukmirovic M, Ticianelli E, Adzic R (2007) Catalytic activity-d-band center correlation for the O_2 reduction reaction on platinum in alkaline solutions. J Phys Chem C 111:404–410
9. Stamenkovic V, Mun BS, Mayrhofer KJJ, Ross PN, Markovic NM, Rossmeisl J, Greeley J, Nørskov JK (2006) Changing the activity of electrocatalysts for oxygen reduction by tuning the surface electronic structure. Angew Chem 118:2963–2967

10. Mukerjee S, Srinivasan S, Soriaga MP, McBreen J (1995) Effect of preparation conditions of Pt alloys on their electronic, structural, and electrocatalytic activities for oxygen reduction-XRD, XAS, and electrochemical studies. J Phys Chem 99:4577–4589
11. Lima FHB, Ticianelli EA (2004) Oxygen electrocatalysis on ultra-thin porous coating rotating ring/disk platinum and platinum-cobalt electrodes in alkaline media. Electrochim Acta 49:4091–4099
12. Adzic RR, Zhang J, Sasaki K, Vukmirovic MB, Shao M, Wang J, Nilekar AU, Mavrikakis M, Valerio J, Uribe F (2007) Platinum monolayer fuel cell electrocatalysts. Topics Catal 46:249–262
13. Zhang J, Mo Y, Vukmirovic M, Klie R, Sasaki K, Adzic R (2004) Platinum monolayer electrocatalysts for O_2 reduction: Pt monolayer on Pd (111) and on carbon-supported Pd nanoparticles. J Phys Chem B 108:10955–10964
14. Brankovic S, Wang J, Adzic R (2001) Pt submonolayers on Ru nanoparticles: a novel low Pt loading, high CO tolerance fuel cell electrocatalyst. Electrochem Solid State Lett 4: A217–A220
15. Kitchin JR, Nørskov JK, Barteau MA, Chen J (2004) Role of strain and ligand effects in the modification of the electronic and chemical properties of bimetallic surfaces. Phys Rev Lett 93:156801–156804
16. Ruban A, Skriver HL, Nørskov JK (1999) Surface segregation energies in transition-metal alloys. Phys Rev B 59:15990–16000
17. Bardi U, Atrei A, Zanazzi E, Rovida G, Ross P (1990) Study of the reconstructed (001) surface of the Pt80Co20 alloy. Vacuum 41:437–440
18. Mun BS, Lee C, Stamenkovic V, Markovic NM, Ross PN Jr (2005) Electronic structure of Pd thin films on Re (0001) studied by high-resolution core-level and valence-band photoemission. Phys Rev B 71:115420–115426
19. Sasaki K, Wang J, Balasubramanian M, McBreen J, Uribe F, Adzic R (2004) Ultra-low platinum content fuel cell anode electrocatalyst with a long-term performance stability. Electrochim Acta 49:3873–3877
20. Batista EA, Malpass GRP, Motheo AJ, Iwasita T (2003) New insight into the pathways of methanol oxidation. Electrochem Commun 5:843–846
21. Batista EA, Malpass GRP, Motheo AJ, Iwasita T (2004) New mechanistic aspects of methanol oxidation. J Electroanal Chem 571:273–282
22. Xia XH, Liess HD, Iwasita T (1997) Early stages in the oxidation of ethanol at low index single crystal platinum electrodes. J Electroanal Chem 437:233–240
23. Oh J-G, Kim H (2008) Synthesis and characterization of PtN_x/C as methanol-tolerant oxygen reduction electrocatalysts for a direct methanol fuel cell. J Power Sources 181:74–78
24. Wen Z, Liu J, Li J (2008) Core/Shell Pt/C nanoparticles embedded in mesoporous carbons a methanol-tolerant cathode catalyst in direct methanol fuel cells. Adv Mater 20:743–747
25. Dubau L, Durst J, Maillard F, Guetaz L, Chatenet M, Andre J, Rossinot E (2011) Further insights into the durability of Pt_3Co/C electrocatalysts: formation of "hollow" Pt nano particles induced by the Kirkendall effect. Electrochim Acta 56:10658–10667
26. Wang W, Huang Q, Liu J, Zou Z, Zhao M, Vogel W, Yang H (2009) Surface and structure characteristics of carbon-supported Pd3Pt1 bimetallic nanoparticles for methanol-tolerant oxygen reduction reaction. J Catal 266:156–163
27. Selvarani G, Vinod Selvaganesh S, Krishnamurthy S, Kiruthika GVM, Sridhar S, Pitchumani S, Shukla AK (2009) Methanol-tolerant carbon-supported Pt-Au alloy cathode catalyst for direct methanol fuel cells and its evaluation by DFT. J Phys Chem C 113:7461–7468
28. Tsivadze AY, Tarasevich MR, Kuzov AV, Romanova IA, Pripadchev DA (2008) New nanosized cathode electrocatalysts tolerant to ethanol. Doklady Phys Chem 421:166–169
29. Li X, Huang Q, Zou Z, Xia B, Yang H (2008) Low temperature preparation of carbon-supported Pd Co alloy electrocatalysts for methanol-tolerant oxygen reduction reaction. Electrochim Acta 53:6662–6667

30. Mathiyarasu J, Phani KLN (2007) Carbon-supported palladium-cobalt-noble metal (Au, Ag, Pt) nanocatalysts as methanol tolerant oxygen-reduction cathode materials in DMFCs. J Electrochem Soc 154:B1100–B1105
31. Papageorgopoulos DC, Liu F, Conrad O (2007) Reprint of "A study of Rh*x*S*y*/C and RuSe*x*/C as methanol-tolerant oxygen reduction catalysts for mixed-reactant fuel cell applications". Electrochim Acta 53:1037–1041
32. Savadogo O, Rodríguez Varela FJ (2006) Palladium-alloy catalysts as ethanol tolerant cathodes for direct alcohol fuel cell (DEFC) applications. ECS Trans 1(6):247–254

Chapter 6
Membranes for Direct Alcohol Fuel Cells

Horacio R. Corti

Abstract This chapter is devoted to summarize and discuss the main properties of ionomeric membranes used in direct alcohol PEM fuel cells. Although Nafion is the proton exchange membrane commonly used in methanol and other direct alcohol fuel cells, other proton and alkaline membranes are being investigated in order to improve the efficiency of DAFC. The goals in the development of this critical component of DAFC are: low cost, long durability, low alcohol permeability and high electrical conductivity. The last two properties can be combined in a single parameter, the membrane selectivity that accounts for the ratio between the proton and alcohol transport through the membrane. This parameter can be compared to that measured for Nafion to define a relative selectivity, which is a primary parameter to evaluate the potentiality of a ionomer material to be used in alcohol feed fuel cells.

The vast catalogue of polymeric materials reviewed here included Nafion composite with inorganic and organic fillers, and non-fluorinated proton conducting membranes such as sulfonated polyimides, poly(arylene ether)s, polysulfones, poly (vinyl alcohol), polystyrenes, and acid-doped polybenzimidazoles. Anion-exchange membranes are also discussed because of the facile electro-oxidation of alcohols in alkaline media and because of the minimization of alcohol crossover in alkaline direct alcohol fuel cells.

The performance of different types of membranes in direct alcohol fuel cells, mainly methanol, are summarized and discussed in order to identify the most promissory ones. The lack of correlation between the relative selectivity and fuel cell performance of the membranes indicates that the architecture of the three

H.R. Corti (✉)
Departamento de Física de la Materia Condensada, Centro Atómico Constituyentes, CNEA, and INQUIMAE (Universidad de Buenos Aires – CONICET), Buenos Aires, Argentina
e-mail: hrcorti@cnea.gov.ar

H.R. Corti and E.R. Gonzalez (eds.), *Direct Alcohol Fuel Cells: Materials, Performance, Durability and Applications*, DOI 10.1007/978-94-007-7708-8_6,

phases region is a decisive factor to take into account in the design of enhanced performance membrane-electrode assemblies.

6.1 Introduction

The studies on membranes for DAFC follow the same pattern as reported in Chap. 1 for DAFC. Around 90 % of the most of 4,000 works cited by Scopus till the end of 2012, correspond to direct methanol, while only 10 % deals with membranes for direct ethanol fuel cells. In both case, for DMFC and DEFC membranes, there is a maximum in the number of publications in 2009, as can be seen in Fig. 6.1.

Several reviews on membranes for DMFC fuel cells have been published in the last decade [1–9], starting with that by Kreuer [1], discussing the differences between Nafion and sulfonated polyether ketone membranes. According to Fig. 6.1, reviews published till 2006 [1–4] cover only one third of the ionomeric membranes currently developed for DAFC. More recent reviews deal with polyimide ionomer membranes [5], composite membranes for high temperature DMFC [6], non-perfluorated sulfonic acid membranes [7], modified Nafion membranes [8], and hybrid membranes [9–11].

This chapter aims to cover most of the membranes that have been characterized in relation to alcohol permeability (crossover) or have been tested in single DAFC.

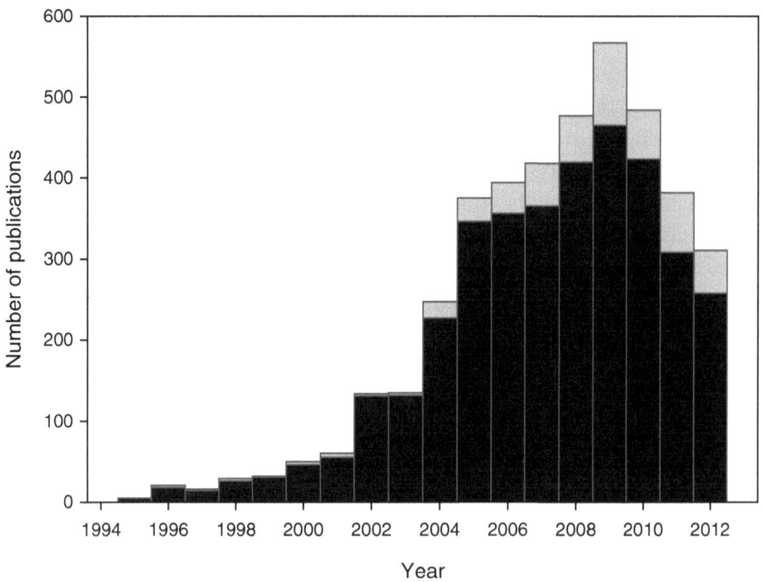

Fig. 6.1 Number of publications on DAFC membranes according to Scopus database: DMFC membranes (*black*) and DEFC (*gray*)

Certainly, it does not intend to be a comprehensive review of the whole literature on membranes for DAFC, but most types of single and composite membranes tested in the last decade will be addressed.

Firstly, Nafion and other perfluorinated sulfonic acid ionomers will be discussed, along with inorganic- and organic-Nafion based composites. Secondly, we will introduce non-fluorinated single and composite membranes, including membranes for high temperature DAFC. Finally we will discuss anion conducting membranes for alkaline DAFC.

6.2 Membrane Requirements for DAFC

Alcohol crossover and cell resistance are the relevant properties determining the DAFC performance, which are closely related to the membrane used in the preparation of the membrane-electrode assembly (MEA). Mechanical properties, as well as the chemical and thermal stability, of the membrane could also be important when durability is considered.

Proton to alcohol membrane selectivity was first defined by Pivovar et al. [12] as the ratio of proton ($\sigma \cdot \nabla \phi$) to alcohol ($P \cdot \nabla$c) flows, which depends on the proton membrane conductivity, σ, and the permeability coefficient of the alcohol, P, and the corresponding potential and concentration gradients through the membrane,

$$\frac{J_p}{J_a} = \frac{\sigma}{P}\left(\frac{\Delta\phi}{Fc}\right) \tag{6.1}$$

where $\Delta\phi$ is the electrostatic potential through the membrane, c the alcohol concentration on the aqueous phase in contact with the anode (assuming the alcohol concentration on the cathode side is zero), and F the Faraday's constant. The quantity $\Delta\phi/Fc$ represent the ratio of the electrical and permeation driving forces, which is independent of the membrane thickness, but it depends on the experimental conditions. For this reason the intrinsic proton to alcohol selectivity of the membrane, β, is defined as the ratio between conductivity and permeability,

$$\beta = \frac{\sigma}{P} \tag{6.2}$$

Membrane selectivity is probably the most important parameter to compare potential polymeric membranes for DAFC because low ohmic resistance and low alcohol crossover are desired. Kim and Pivovar [4] have noted that both of these quantities are independent of membrane thickness, which affect resistance and alcohol crossover in opposite ways. The thinner is a membrane in a DAFC, the lower the resistive losses, but the higher the crossover losses. Thus, a minimum conductivity is required in membranes for DAFC, regardless of how high its

Fig. 6.2 Selectivity plot: relative proton conductivity versus relative alcohol permeability, showing regions with membrane properties better, equal, and worse than Nafion

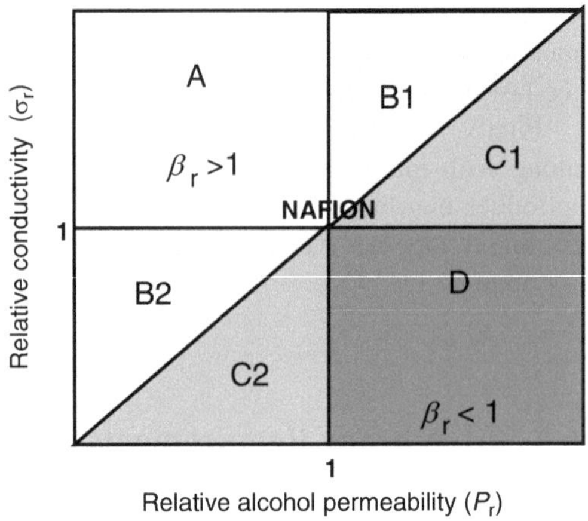

selectivity is, in order to maintain the membrane thickness in a range compatible with good mechanical robustness.

Membrane conductivity and alcohol permeability depend on temperature and degree of hydration of the membrane. Therefore, it is useful to define the relative selectivity, β_r, as the ratio between the selectivity of the membrane to the selectivity of Nafion membrane measured to under similar experimental conditions,

$$\beta_r = \frac{(\sigma/P)}{(\sigma/P)_{Nafion}} \tag{6.3}$$

Thus, in a plot of the membrane relative electrical conductivity *vs.* relative alcohol permeability, the best candidates as membranes for DAFC are those lying in the upper left-hand corner, as shown in the plot of Fig. 6.2.

Membranes having relative conductivities ($\sigma_r = \sigma/\sigma_{Nafion}$) and relative permeability ($P_r = P/P_{Nafion}$) values on the dash line, exhibit the same selectivity that Nafion. The quadrant A in Fig. 6.2 corresponds to membranes with a relative selectivity greater than 1, that is, conductivity and permeability barrier higher than Nafion under the same conditions. Octants B1 and B2, also with $\beta_r > 1$, correspond to membranes having selectivities greater than Nafion due to higher conductivity (B1) or higher permeability barrier (B2), respectively.

On the other hand, octants C1 and C2 correspond to membranes with selectivities lower than Nafion due to lower conductivity (C2) or lower permeability barrier (C1). Finally, in quadrant D lie membranes having properties poorer than Nafion.

We will refer to this generalized selectivity plot later in this Chapter when comparing the performance of DAFC membranes.

Other authors [7] consider the properties of the fuel cell instead of the free-standing membrane to define a MEA-selectivity parameter, α, as the reciprocal of the high frequency resistance times the alcohol crossover limiting current. In this case the cell resistance contains contributions associated with the interfaces electrode-membrane and electronic contact resistances of the fuel cell components. On the other hand, the crossover limiting current has contribution of the drag factor, that is, the coupling between proton and alcohol transport through the membrane, and the mass transport properties of the gas diffusion layer and catalysts layers toward the alcohol.

Independently of the selectivity parameter chosen for comparing the performance of membranes for DAFC, it is clear that a high proton conductance (higher than 0.08 $S{\cdot}cm^{-1}$) and low impermeability to alcohol and oxygen, along with good thermal and oxidation stability and low cost, are desirable properties.

Regarding costs, Neburchilov et al. [3] pointed out that the contribution of the membrane to a PEM stack should be lower than US$ 10 per kW, although this limit could be much higher for mini or micro DAFC, taking into account that the cost of small fuel cells for portable applications would be in the order of US$ 17/W [13].

Other property that could be relevant for the durability of DAFC is the ruthenium crossover, that is, the dissolution of Ru from anode catalysts containing Ru (a typical anode catalyst in DMFC) and its re-precipitation in the cathode [14]. The water and methanol permeability was expected to have an effect on Ru crossover, but the only study available on this seems to indicate that differences is not the case [15].

It should bear in mind that the requirements for active DAFC could be significantly different than the passive DAFC systems used for lower power, portable applications [4]. Water balance and alcohol crossover become more critical in passive cell systems, where information on the performance of advanced ionomer membranes is limited by the confidential nature of the research and development.

6.3 Proton Conducting Membranes for DAFC

In this section the Proton Exchange Membranes (PEM) used or with potential to be used in DAFC will be introduced. The ample spectra of PEM chemical structures, including perfluorinated sulfonic membranes, and non-perfluorinated membranes and their corresponding inorganic and organic composites or blends, will be addressed.

6.3.1 Nafion and Similar Perfluorinated Sulfonic Acid Ionomers

For DAFC operated at room and moderate temperatures (lower than 80 °C) perfluorinated sulfonic acid ionomers (PFSA), mainly Nafion (DuPunt), have been the most studied and used in direct methanol (DMFC) and ethanol fuel cells (DEFC) due to its excellent proton conductivity and chemical stability [16, 17].

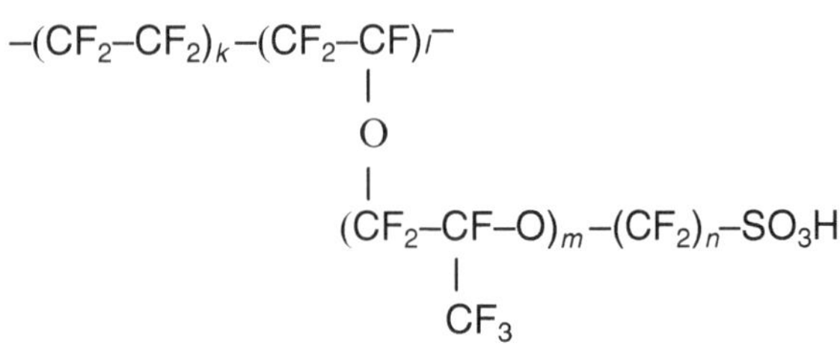

Fig. 6.3 Chemical structure of Nafion and other perfluorinated sulfonic acid polymers

The general chemical structure of Nafion is illustrated in Fig. 6.3, where k =6-7, $l = 1$, $m = 1$, and $n = 1$. Other fluorinated commercial membranes similar to Nafion were developed, such as, Hyflon (Solvay Solexis), Flemion (Asahi Glass Engineering, Japan), 3P membranes (3P Energy, Germany), Gore-Tex and Gore Select (Gore & Assoc, USA), as well as Nafion-like membranes with –COOH weak acid groups instead of $–SO_3H$, like Aciplex (Asahi Kasei, Japan) and membranes prepared with copolymers of tetrafluoroethylene and vinylene, such as the XUS membranes (Dow Chemical, USA). These membranes have chemical structures similar to that shown in Fig. 6.3, but with different values for the indexes k, m, and n.

The nano- and meso-structure of Nafion have been analyzed by small-angle neutron and X-ray scattering [1, 18, 19] and several models were proposed in order to match the scattering data. The Gierke model [18], the most commonly accepted, assumes that the polymer form inverted micelles, about 4 nm diameter, connected by narrow (1 nm diameter) cylindrical pores, as depicted in Fig. 6.4 (top left), forming a network of spherical water clusters. A similar channel networks model is proposed by Kreuer [1], while Rubatat et al. [19] have interpreted the small angle scattering data as suggesting the existence of cylindrical or ribbonlike polymeric aggregates, as shown in Fig. 6.4 (top right), surrounded by the ionic groups and water.

These different models have reanalyzed by Schmidt-Rohr and Chen [20] by numerical simulation of SAXS curves using a new algorithm which consider arguments of symmetry. These authors concluded that the previous models do not match the scattering data, while a model featuring long parallel water channels in cylindrical inverted micelles reproduces all of the SAXS characteristics. Figure 6.4 (bottom) shows the features of the new Nafion model for a hydration level of 11 wt%. The structure is formed by cylindrical water channels (1.8–3.5 nm diameter cylindrical inverted micelles) lined with hydrophilic side groups (Fig. 6.4a, b). A cross section through the cylindrical water channels is shown in Fig. 6.4c, where water channels (white) and Nafion crystallites (black) in the non-crystalline Nafion matrix (grey) are schematized.

As noted by Sanchez and coworkers [10], the inverted micelles model by Schmidt-Rohr and Chem is not a superposition of scattering studies, but takes into account the supramolecular organization of the polymer chains as whole. Even when some aspects of this model, as the finer details of the self-association of the polymer, still remains unclear; it allows to explain a number of features

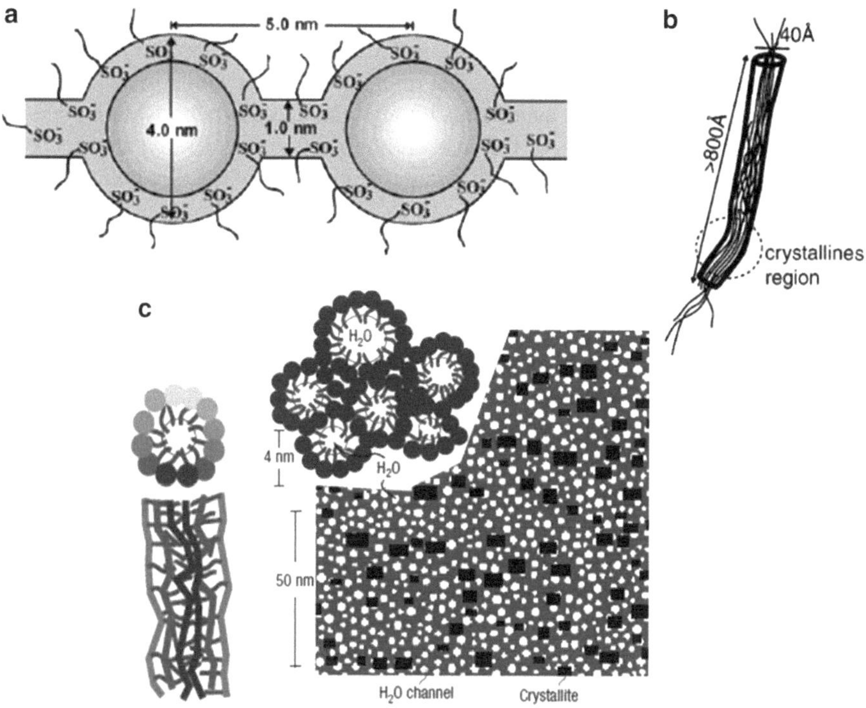

Fig. 6.4 Inverted micelles model by Gierke (*top left*), polymer bundles model (*top right*), and parallel water-channel model (*bottom*) (Reproduced from Refs. [16, 20] with permission of the American Chemical Society (copyright 2004) and Nature Publishing Group)

of Nafion, such as the fast diffusion of protons and water through it and its persistence at low temperatures.

We will return to the discussion on the relationship between structure and properties in Nafion and Nafion-based membranes in a next section. The modification of Nafion with inorganic and organic compounds has been investigated [3, 9] in order to improve the performance of these perfluorinated membranes in DAFC by reducing the methanol crossover, while enhancing water retention and thermal stability. Next subsections are devoted to describe the different types of inorganic/Nafion composites and polymer/Nafion blends that have been reported in the literature in relation to its use as electrolytes in DAFC.

6.3.2 *Inorganic/Nafion Composites*

As quoted by Sanchez and coworkers [10], the properties of hybrid organic–inorganic membranes are largely determined by the interactions in the mineral-organic interface. According with the nature of this interface they classified these hybrid materials as Class I, when the interactions in the interface are weak (Van der Waals, electrostatic, hydrogen bonds), and Class II, for components linked by covalent or ionic covalent bonds. The inorganic/Nafion composites described below correspond to Class I materials, although the Class II hybrids, of increasing development and application in the case on non-perfluorinated polymers, render a number of advantages, such as the minimization of phase separation at the nanometric scale.

Among the inorganic fillers in Nafion composite membranes that have deserved most attention in relation to DAFC (mainly DMFC) are noble metals, such as Au nanoparticles [21], Pd and Pd alloys in the form of nanoparticles, nanowires and thin films [22–27], Pt and Pt/Ru nanoparticles [28, 29], SiO_2 and functionalized silica [30–45], organic silica [46–50], zeolites [51–60], montmorillonite [61–71], silicate and aluminophosphate [72], aluminosilicate [73], calcium phosphate [74], zirconium phosphate [41, 75–80], zirconium sulphophenyl phosphate [81], titanium phosphate [41], titanosilicate [82, 83], titanosulfonate [84], TiO_2 [85], TiO_2 sulfated [86], ZrO_2 sulfated [87], Fe_2O_3 sulfated [88], Fe_2O_3 sulfonated [89], Nd_2O_3 [90], neodymium triflate [91], erbium triflate [92], strontium hydroxide [93], phosphomolybdic acid [33], and phosphotungstic acid [94].

Some Nafion binary inorganic composites have been studied in relation to DMFC, such as silicotungstic acid – SiO_2 [31] phosphotungstic acid-SiO_2 [31, 95], phosphotungstic acid –mesoporous SiO_2 [96], and Pd-SiO_2 [97].

Most of the reported inorganic fillers used to modify Nafion are composite where the inorganic particles (usually nanoparticles) are located in the membrane bulk. Most of them are prepared using the recast method, where the filler nanoparticles dispersed in a solvent are mixed with Nafion ionomer dispersion in the same solvent or a compatible one. The solution is cast on a Petri dish or a plane surface at high temperature to form the recast composite membrane. An alternative method adopted to prepare Nafion composites with silica [31, 32, 41, 95], functionalized silica [35], and zirconium and titanium phosphate [41], is the in situ sol–gel reaction method, schematized in Fig. 6.5.

In the case of SiO_2/Nafion composites the Nafion membrane is swollen in water and methanol and immersed in a methanol solution of silicon tetraetoxysilane (TEOS). Thus, the TEOS hydrolysis take place inside the membrane and the amount of formed SiO_2 inside the membrane depends on the impregnation time. Finally the membrane is dried at high temperature to remove trapped volatiles and promote further silica network condensation [32].

Other reported method to prepare inorganic/Nafion composite membranes is the colloidal [56] or ionic impregnation [75–78, 87, 93], where the inorganic filler is introduced a colloidal dispersion into the membrane or by exchange of H^+ by the salt cation, like Zr^{4+}, followed by salt precipitation with the corresponding acid.

Only few works deal with films or nanoparticles covering the surface of a Nafion membrane. In the case of metal coated Nafion membranes surface nanoparticles were deposited on the surface by in situ reduction of a metal salt precursor [21], by adsorption of a colloidal dispersion of the metal [28], or by sputtering [22, 23].

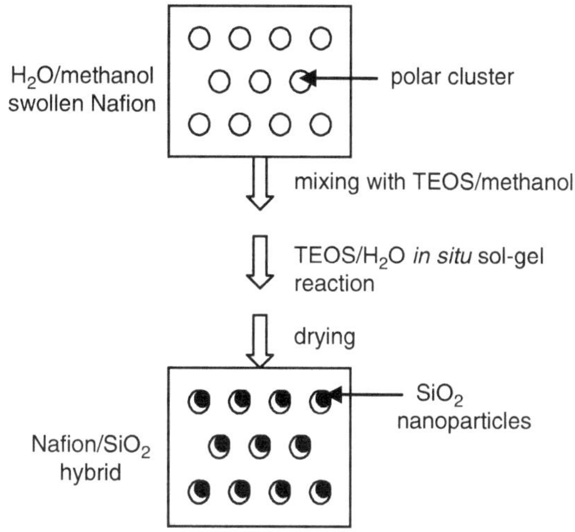

Fig. 6.5 Scheme of preparation of composite silica/Nafion membranes by sol–gel reaction (Adapted from Ref. [9])

SiO_2 surface modified Nafion membranes could be prepared by in situ surfactant-templated sol–gel reaction [38] or by plasma enhanced chemical vapor deposition [45]. TiO_2-coated Nafion membranes were prepared by dip coating from a titania sol [85].

6.3.3 Polymer/Nafion Blends

The acid–base Nafion composite membranes include blends of Nafion with polypyrrole (PPy) [98–104], polybenzimidazole (PBI) [105–107], poly (propyleneoxide) (PPO) [108, 109], polyfurfuryl alcohol (PFA) [110], poly(vinyl alcohol) (PVA) [111–115], sulfonated phenol-formaldehyde (sPF) [116], polyvinylidene fluoride (PVdF) [117–122], poly(p-phenylene vinylene) (PPV) [123], poly(vinyl pyrrolidone) (PVP) [124] polyaniline (PANI) [125–128], polyethylene (PE) [129], poly(ethylene-terephtalate) [130], sulfated β-cyclodextrin (sCD) [131], sulfonated poly(ether ether ketone) (sPEEK) [132–135], sulfonated poly(aryl ether ketone) (sPAEK) [136], poly(arylene ether sulfone) (PAES) [137], poly(vinylimidazole) (PVI) [138], poly(vinyl pyridine) (PVPy) [139], poly (tetrafluoroethylene) (PTFE) [140–142], poly(fluorinated ethylene-propylene) [143], sulfonated polyhedral oligomeric silsesquioxane (sPOSS) [144], poly (3,4-ethylenedioxythiophene) (PEDT) [145, 146], polyrotaxanes (PR) [147], purple membrane [148], sulfonated polystyrene (PSSA) [149, 150], polystyrene-b-poly(ethylene-ran-butylene)-bpolystyrene (SEBS) [151], poly(2-acrylamido-2-methyl-1-propanesulphonic acid-co-1,6-hexanediol propoxylate diacrylate-co-ethyl methacrylate) (AMPS) [152], and chitosan [31]. A binary PVA/chitosan [153] and a ternary Nafion composite with PVA, polyimide (PI) and 8-trimethoxy silylpropyl glycerin ether-1,3,6-pyrenetrisulfonic acid (TSPS) has also been reported [154].

6.3.4 Organic/Inorganic, Grafted, Layer-by-Layer Assembly, and Modified Nafion Membranes

A few examples of Nafion composite membranes with polymers and inorganic compounds has been reported, such as poly(ethylene glycol) (PEG)/SiO_2 [155], PTFE/SiO_2 [156], chitosan/phosphotungstic acid [157], PVI/Pd [158], and PBI/zirconium phosphate [159].

Styrene have been grafted to Nafion using supercritical CO_2 for polymer impregnation [160], or grafted to Nafion surface via plasma-induced polymerization [161]. Recently a poly(glycidyl methacrylate) was grafted to Nafion using a simple chemical initiation system [162].

Some authors superficially modified Nafion membranes in order to reduce methanol crossover. For instance, hydrocarbon films have been deposited on Nafion membranes by plasma deposition [163], and trilayer membranes were prepared by casting Nafion layers on a central Nafion/PVdF composite layer [164]. Nafion treated with supercritical CO_2 seems to increase crystallinity, as well as proton conductivity and DMFC performance [165, 166].

One of the most interesting surface-modified Nafion membranes for DAFC application is the layer-by-layer (LBL) self-assembly of polyelectrolytes onto Nafion, a procedure pioneered by Hammond and coworkers [167] and Moon and Rhim [168], although Tang et al. [25] had previously reported deposited multi-layer Pd nanoparticles onto Nafion using a polycation as a dispersant of the Pd particles. The polyelectrolytes assembled were water soluble polycations, like linear poly (ethylene imine) (LPEI), poly(allylamine hydrochloride) (PAHC), poly (diallyl dimethyl ammonium chloride) (PDAC), and poly(4-vinylpyridine) (P4VP), paired with polyanions, like sulfonated poly(2,6-dimethyl −1,4-phenylene oxide) (sPPO) [167], or PAHC paired with polystyrene sulfonic acid (PSSA) or a fluorosulfonic surfactant [168]. Later, new LBL membranes based on Nafion were prepared with chitosan and sulfonated poly(aryl ether ketone) with carboxyl groups (sPAEK-C) [169], poly(vinylsulfate) (PVS) and PAHC [170], and chitosan and silicotungstic acid [171].

6.3.5 Other Fluorinated Membranes

The significant contribution of Nafion or perfluorosulfonic membranes to the cost of the fuel cells stacks and the high alcohol crossover levels that affect the fuel efficiency, prompted the development of radiation grafted proton exchange membranes based on poly(ethylene-tetrafluoroethylene) (ETFE) [172–178], PVdF [175], and PTFE [179]. The peroxy radicals produced on the base polymer by γ-ray, electron- or proton-beam, react with styrene to form a co-polymer that is then sulphonated.

The preference by ETFE as base polymer was based on its low cost, although the DMFC performance of membranes based on PVdF seems to be superior [175].

Table 6.1 Alternative DMFC membranes and number of publications in open literature

Type	1994–2004 [4]	2005–2012
Polyarylenes	27	21
Polyvinyl alcohols	11	97
Polyimides	9	69
Polystyrenes (grafted)	9	3
Polystyrene (block copolymers)	6	10
Polyphosphazenes	7	6
Polybenzimidazole	–	33
Polyvinylidene fluoride	–	23

6.3.6 Non-Fluorinated Membranes

Several non-fluorinated alternative polymers have been proposed for DAFC, mainly based on sulfonated ionomers with an aromatic or aliphatic hydrocarbon skeleton [7]. Kim and Pivovar [4] have reported the number of DMFC alternative membranes papers appearing in open literature for years 1994–2004, showing that polyarylenes, polyvinyl alcohols, grafted and block polystyrenes copolymers, and polyimides were among the most studies polymer electrolytes. In view of the dramatic increase in the number of publications since 2005 (see Fig. 6.1), the trends have changed, as shown in Table 6.1, which summarized the publications in open literature for the period 2005–2012, as compared to the previous period.

Polyarylenes are still being development and tested for DMFC, although the studies on polyvinyl alcohols (PVA) and polyimides (PI) exhibit a dramatic increment during the last years. As predicted by Kim and Pivovar [4], sulfonated PIs deserved great attention in view of their lower methanol permeability and water uptake compared to polyarylenes, while studies on polyphosphazenes were discouraged in view of the poor DMFC performance reported till 2004. However, these authors failed to assess the huge development of PVA/based membranes for DMFC applications by indicating that this polymer electrolyte has poor chemical stability and DMFC performance had not been reported till 2004. In contrast, a recent review by Maiti et al. [180], reports a large number of PVA blends and composites membranes tested in relation to methanol permeability and DMFC performance.

On the other hand, a number of polymers whose properties in relation with DAFC were scarcely known 8 years ago, have deserved a considerable attention during the last years, particularly membranes based on polybenzoimidazole (PBI) and polyvinylidene fluoride (PVdF).

6.3.6.1 Polyvinyl Alcohols

Because PVA is not a proton-conducting polymer it should be sulfonated using different sulfonic-carboxylic acids forming crosslinked polymers, as shown in Fig. 6.6 [180].

Fig. 6.6 Structure of some sulfonated crosslinked PVAs

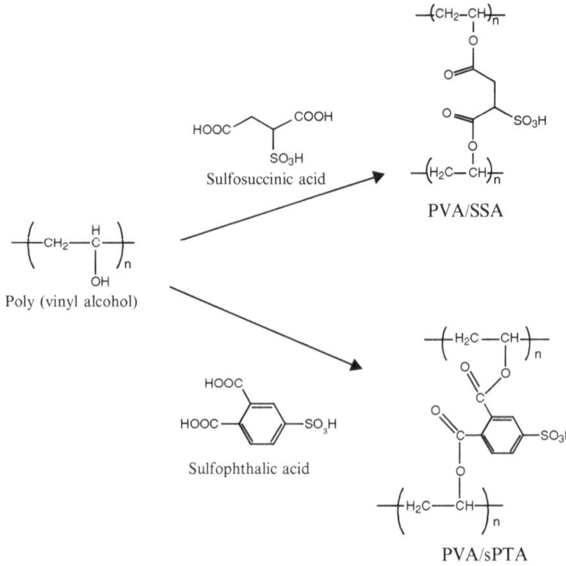

Others crosslinking agent include poly(styrene sulfonic acid-*co*-maleic acid) (PVA/PSSA-MA) and poly(acrylic acid-*co*-maleic acid) (PVA/PAA-MA). Blending of PVA with sulfonated proton-conducting polymers, like sPEEK, Nafion, PSSA, is another approach to achieve the high conductivity requirements [180]. Inorganic/PVA composites have been also considered, including sulfonic or phosphonic modified silica, titania, hydroxyapatite, heteropoly acids, zirconium phosphate, and montmorillonite. Hybrid organic–inorganic PVA composite membranes, like PVA/TiO_2/PSSA, PVA/SiO_2/SSA/, PVA/PWA/sPEEK/PWA, PVA/SiO_2/PPA, etc., have been reported.

Maiti et al. [180] reported the conductivity and methanol permeability more than 50 PVA-based membranes and the readers are referenced to this article for details on different procedures of preparation.

6.3.6.2 Poly(arylene ether)s

Sulfonated poly(arylene ether)s (sPAE)s, such as sulfonated poly(ether ether ketone) (sPEEK) (see Fig. 6.7) are promising ionomer for DMFC [181] due to their chemical stability, possibility of structural variants, and sulfonation procedures, as discussed by Kim et al. [7].

PEEK polymers are characterized by a high crystallinity which can be reduced by PEEK modified with cardo (Latin meaning loop) group which increase the amorphous degree. Sulfonated cardo-group modified PEEK (sPEEK-WC), as shown in Fig. 6.7, has been proposed for DMFC application [182–184]. A comprehensive review of the use of sPEEK and sPEEK-WC membranes in DMFC has been recently performed by Iulianelli and Basile [185].

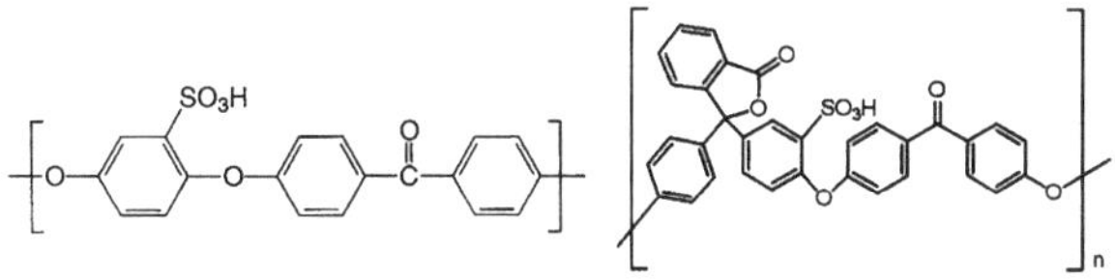

Fig. 6.7 Chemical structure of sPEEK (*left*) and sPEEK-WC (*right*)

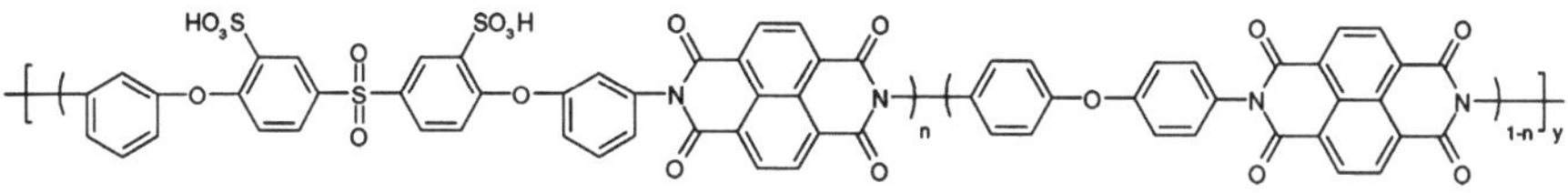

Fig. 6.8 Typical naphthalenic sulphonated polyimide

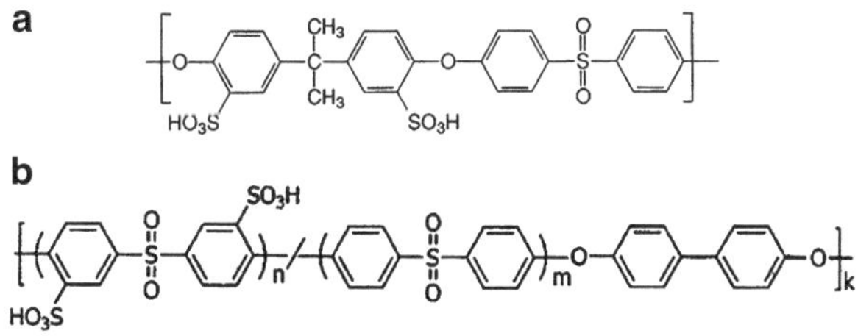

Fig. 6.9 Structure of: (**a**) sulfonated Udel-1700 polysulfone, and (**b**) sulfonated poly(arylene ether sulfone) (sPAES)

6.3.6.3 Polyimides

Sulfonated polyimides (sPI) are another type of ionomers that have been prepared [5] by copolymerization of 4,4′-diamino-biphenyl 2,2′-disulphonic acid (BDSA), 4,4′-oxydianiline (ODA), and naphtalenic or phthalic anhydrides. Polyimide structures based on six-membered rings derived from naphtalenics anhydrides, such as 1,4,5,8-naphtalene tetracarboxylic dianhydride (NTDA), shown in Fig. 6.8, are more hydrolitically stable than the five-membered rings basedoxy-diphthalic anhydrides (ODPA).

Sulfonated polyimide can be also prepared by post-sulfonation of polyimide polymers, but the former procedure is generally adopted because of the existence of commercially available sulfonated aromatic diamines.

6.3.6.4 Polysulfones

Sulfonated polysulfones (sPSf) can be prepared by sulfonation of commercial polysulfones, such as Udel 1700 (Solvay), or poly(arylene ether sulfones) (PAES), leading to structures like that shown in Fig. 6.9a [7, 186]. Also, they can be synthesized by direct copolymerization of sulfonated monomers available commercially [7, 187, 188], to produce polymers with the structure shown in Fig 6.9b.

The stability and mechanical properties depends on the number and position of the sulfonic group on the sulfone-linked rings. The ion exchange capacity of the

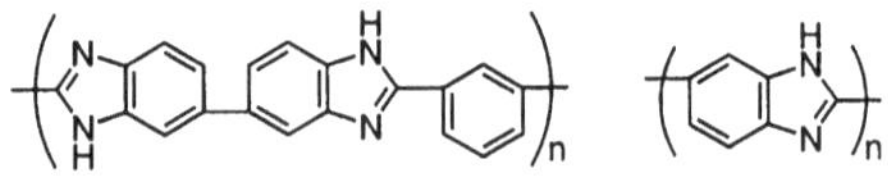

Fig. 6.10 Structure of m-PBI (*left*) and ABPBI (*right*)

polymer can be controlled through the ratio of disulfonated to non-sulfonated + biphenol monomers, in such a way that IEC higher than 2 meq.g^{-1} can be obtained, leading to high proton conducting membranes.

6.3.6.5 Polybenzoimidazoles

Other group of ionomers for PEM fuel cells are based on polymers, such as poly [2,2′-(m-phenylene)-5,5′-bibenzimidazole] (m-PBI), obtained by polymerization of 3,3′-diaminobenzidine, 1,2,4,5-tetraaminobenzene and a variety of aromatic diphenyl dicarboxylates [189], whose structure is shown in Fig. 6.10. This non-ionic polymer becomes a proton conductor when doped with a strong acid such as sulphuric or phosphoric acid.

Commercial PBI is a proprietary product (Celanese) made from 3,3′,4,4′-tetra aminobiphenyl (TAB) and diphenyl isophtalate in a two step, melt/solid polymerization [190]. A modified PBI structure such as poly(2,5)-benzimidazole (ABPBI) (see Fig. 6.10) has been synthesized by condensation of 3,4-diaminobenzoic acid (DABA) monomer in polyphosphoric acid [191].

Similar PBI structures where synthesized by polymerization of DAB or tetraaminobencene with pyridine-dicarboxylic or naphtalene-dicarboxylic acids [190, 192].

Sulfonated polybenzoimidazole (sPBI) membranes have also been prepared by polycondensation of 2-sulphoterephthalic acid with 1,2,4,5-tetraaminobenzene tetrahydrochloride [193] or by grafting alkyl or aryl sulfonates on to the imidazole ring [194]. ABPBI was sulphonated by immersion in sulfuric acid followed by heat treatment at 450 °C [195]. Other methods of sulphonation have been reviewed by Asensio et al. [196], along with a detailed description of the properties of PBI blends with sulfonated and non-sulfonated polymers, and block PBI copolymers.

Pu et al. [197] have prepared and tested N-methyl (PNMBI) and N-ethyl (PNEBI) derivates of PBI, and Chuan and Hsu [198] reported fluorine-containing PBI. A number of PBI membranes modified by ionic or covalent cross-linking and composite PBI membranes have been reviewed by Li et al. [199].

A new process for synthesizing high molecular weight PBI and membrane casting, termed ¨PPA sol–gel process¨, was developed by Xiao et al. [200] based on the chemical and physical transformations illustrated in Fig. 6.11.

After polymerization of TAB and isophtalic acid in polyphosphoric acid (PPA) at 200 °C the PBI solution in PPA (polycondensation agent and polymerization solvent) is cast at directly from the hot mixture. The water uptake by the mixture hydrolyzes the PA to phosphoric acid, which along with the decrease in temperature induce a transition to a gel state. The PBI membranes formed by this method retain

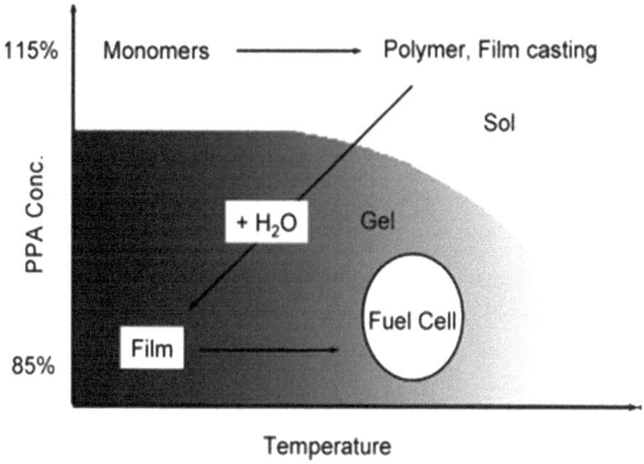

Fig. 6.11 State diagram of the PPA sol–gel process (Reprinted from Ref. [200] with permission Copyright (2005) American Chemical Society)

high levels of phosphoric acid in the gel structure, yielding to high ionic conductivities and stable mechanical properties at high temperatures [200].

A commercial phosphoric acid doped PBI membrane, Celtec V by BASF Fuel Cells, tailored for DMFC is based in a blend of PBI and poly(vinyl phosphonic acid) (PVPA). The PVPA polyacid is immobilized in the PBI matrix by interpenetration, crosslinking and covalent bonding [201].

6.4 Anion Exchange Membranes for Alkaline DAFC

As mention in Chap. 1, anion exchange membranes (AEM) could be a solution for the development of alkaline direct alcohol fuel cells (ADAFC), considering that the anodic oxidation of methanol and other alcohols and the oxygen reduction are more feasible in alkaline media than in acidic ones. In addition, in alkaline media the cathodic catalysts are more methanol tolerant and the methanol crossover due to electro-osmotic drag would be minimized in oxidrile conducting membranes. However, the comparison between PEM and AEM DAFC was unfavorable for alkaline fuel cells ten years ago, and the search for AEM stable at high temperatures, highly conductive and soluble in certain solvents for catalyst ink preparation was strongly suggested [202].

Varcoe and coworkers were pioneers in synthesizing AEM for DMFC by radiation grafting vinylbenzyl chloride (VBC) onto poly (vinylidene fluoride) (PVDF), and poly (tetrafluoroethene-*co*-hexafluoro propylene) (FEP), followed by quaternization in trimethylamine and immersion in KOH solution to obtain the hydroxide ion forms [203], as indicated in Fig. 6.12.

PVDF and FEP based materials were not stable enough to be tested in fuel cells, but the authors prepared a similar AEM by electron-beam grafting of VBC onto poly (ethylene-co-tetrafluoroethylene) (ETFE) that has been used to prepare a MEA for ADMFC [204].

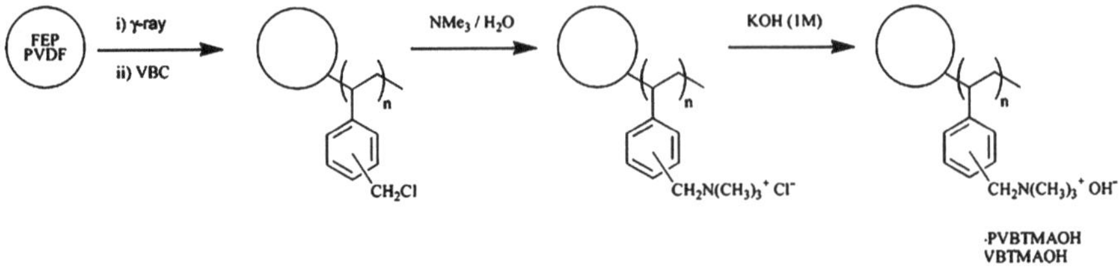

Fig. 6.12 Radiation-grafting of VBC onto PVDF and FEP, and conversion to the anion-exchange membranes [203]

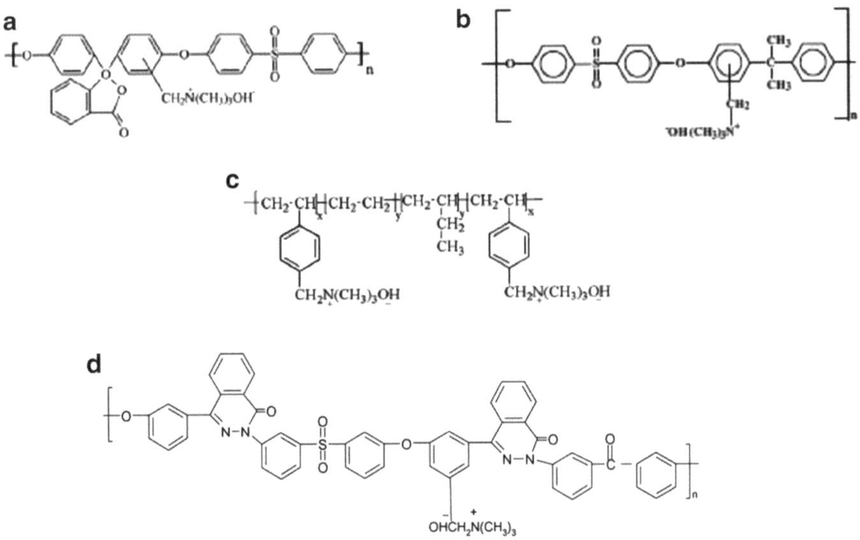

Fig. 6.13 Structure of QPES-C (**a**), QPAES (**b**), QSEBS (**c**), and QPPESK (**d**)

Many kinds of AEMs based on quaternized polymers containing a quaternary ammonium group have been developed and tested in ADAFC, such as polyethersulfone cardo (QPES-C) [205], polyetherketone cardo (QPEK-C) [206], poly (phthalazinone ethersulfone ketone) (QPPESK) [207], poly(arylene ethersulfone) (QPAES) [208–210], QPAES cross-linked with tetraphenylolethane glycidyl ether (QPAES/4EP) [210], poly (arylether oxadiazole) (QPAEO) [211], polystyrene-block-poly(ethylene-ran-butylene)-block-polystyrene (QSEBS) [212], poly(vinyl alcohol) (QPVA) [213], poly(vinyl chloride) (QPVC) [214], and poly (vinylbenzyl chloride) (QPVBC) [215]. The chemical structures of some of these polymers are shown in Fig. 6.13.

Quaternized blends were also prepared as potential membranes for DMFC, such as, quaternized poly(2,6-dimethyl-1,4-phenyleneoxide)/bromomethylated

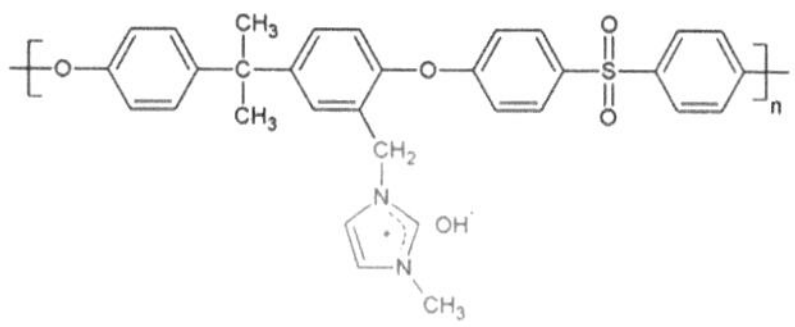

Fig. 6.14 Structure of the imidazolium-functionalized polysulfone (Reproduced from [221] with permission)

poly (2,6-dimethyl 1,4-phenylene oxide) (QCPPO/BPPO) [216, 217], quaternized polyepichlorohydrin (QPECH)/PTFE [218], quaternized copolymers of VBC, methyl metacrylate (MMA), ethylacrylate (EA) (QAPMVE) [219], or VBC, butyl methacrylate (BMA) and hexafluorobutyl methacrylate [220].

Imidazolium-functionalized polysulfone anion exchange membranes have been synthesized by reaction of choloromethylated polysulfone and 1-methyl-imidazole [221], yielding to the structure shown in Fig. 6.14.

The cationic character of the imidazole group has been exploited by alkali-doping of PBI polymers with the structure shown in Fig. 6.10. Hou et al. [222–224] have prepared AEM by doping of PBI with KOH and tested the membranes in a direct methanol single cell, while the permeability of ethanol through PBI has been studied [225], as well as the performance of DEFC using PBI [226, 227].

Nafion membranes, modified by dipping in concentrated NaOH or KOH solution, were used in alkaline DEFC [228, 229].

C.C. Yang and coworkers have studied PVA and quaternized PVA(QPVA)-based membranes in relation to their methanol permeability and DMFC performance, including PVA cross-linked with sulfosuccinic acid (cPVA) [230], and composite membranes, such as cPVA/hydroxyl apatite (HAP) [231, 232], cPVA/TiO_2 [233, 234], cPVA/quaternized silica ($QSiO_2$) [235, 236], cPVA/layered hydroxides [237, 238], cPVA/quaternized hydroxyethylcellulose (QHECE) [239]. Other composite include quaternized PVA, such as QPVA/SiO_2 [240], QPVA/$QSiO_2$ [241], QPVA/Al_2O_3 [242], QPVA/chitosan [243, 244], and QPVA/poly (epichlorohydrin) (PECH) [245].

Other authors have used commercial AEM in DAFC. Thus, several workers have studied methanol, ethanol, ethylene glycol, glycerol, erythritol and xylitol fuel cells using AHA [246], A-006 [247–251], A-010 [252] and A-201 [253–257] membranes from Tokuyama Co. Ltd (Japan), an ammonium copolymer of styrene and divinylbencene (DVB).

Morgane ADP membrane, a cross-linked fluorinated polymer with quaternary ammonium exchange groups, from Solvay has been used in fuel cells fed with methanol, glycerol and ethylene glycol [228, 258–264]. Finally, Fumasep FAA, a commercial quaternized polysulfone membrane by Fumatech, was employed in DMFC [265] and DEFC [266]. A recent review by Yu et al. [267] summarizes the performance of DAFC using commercial and non-commercial AEM.

6.5 Nafion Membranes in DAFC

As mentioned above, Nafion (DuPont) membranes are currently used for hydrogen feed PEM fuel cells due to their excellent chemical stability and high proton conductivity. It is also the most common PEM in DAFC, in spite that alcohol permeability leads to loss of efficiency. In this section we summarized all the properties of Nafion membranes that are relevant to its use in DAFC in order to them with those of alternative membranes.

6.5.1 Alcohol Uptake

The flow, J, of alcohol through the membrane can be described using the first Fick's law,

$$J = -D\left(\frac{\mathrm{d}c}{\mathrm{d}z}\right) \tag{6.4}$$

where D is the diffusion coefficient of the alcohol in the membrane and $\mathrm{d}c/\mathrm{d}z$ is the concentration gradient in the direction z, perpendicular to the membrane surface. The concentration of alcohol in the membrane, c, is related to its concentration in the solution, c', through the apparent partition constant K,

$$c = K\acute{c} \tag{6.5}$$

Therefore, the flow of alcohol can be expressed in terms of the alcohol concentration difference between the solutions in contact with the membrane, $\Delta c'$,

$$J = P\,\frac{\Delta\acute{c}}{\delta} \tag{6.6}$$

where $P = DK$ is the permeability coefficient of the alcohol in the membrane of thickness δ (assuming a linear concentration gradient inside the membrane).

Thus, the uptake of alcohol by Nafion membranes is fundamental to assess their behavior in DAFC because alcohol permeability is proportional to the alcohol content and, consequently, a membrane with low alcohol solubility is preferred.

The uptake of alcohols by Nafion membranes has been much less studied than the water uptake. Nevertheless, several authors determined the sorption of pure methanol from the liquid and the vapor phase in commercial and recast Nafion membranes [268–282] and the reported results are summarized in Table 6.2, expressed as λ, that is, moles of methanol sorbed by sulphonic group. The treatment of the membrane previous to the methanol uptake determination is also indicated.

Table 6.2 Pure methanol sorption by Nafion membranes

Membrane	Treatment	T (°C)	Phase	λ (mol/SO_3group)	Ref.
Nafion 117	A	Ambient	Liquid	27.5	[46]
Cast Nafion	B	Ambient	Liquid	5.2	[49]
Nafion 117	C	Ambient	Liquid	27.8	[76]
Nafion 115	D	Ambient	Liquid	21.1	[102]
Nafion 117	A	Ambient	Liquid	41.2	[125]
Nafion 115	A	?	?	22.2	[127]
Nafion 117	A	Ambient	Liquid	7.6	[168]
Nafion 117	E	Ambient	Liquid	15.4	[268]
Nafion 117	A	Ambient	Liquid	23.0	[269]
Nafion 117	A	Ambient	Liquid	17.2	[270]
Nafion 117	A	60	Liquid	21.5	[271]
			Vapor	7.4	
Nafion 117	None	20	Vapor	16.9	[272]
Nafion 117	A	25	Liquid	24.0	[273]
Nafion 117	A	25	Liquid	18	[274]
			Vapor	11.8	
Nafion 112	A	30	Vapor	11.5	[275]
Cast Nafion	A	30	Vapor	13.1	[276]
Nafion 117	A	25	Liquid	21.5	[277]
Nafion 117	A	18–24	Liquid	27	[278]
Nafion 117	NI	25	Liquid	6.9	[279]
Cast Nafion	None	20	Vapor	6.3	[280]
Nafion	A	40–60	Vapor	13 ($a = 0.93$)	[281]
Nafion 117	NI	Ambient	Liquid	19	[282]

A boiled in H_2O_2/water/acid/water, *B* dried in vacuum at 60 °C, *C* boiled in acid, dried at 80 °C, *D* dried in vacuum at room temperature, *E* boiled in the solvent, *NI* treatment and ionic form not indicated

The observed scatter in the data is in part due to differences in measurement temperature, which was not well controlled in some of the cases, but it is mainly due to different membrane treatment previous to the alcohol uptake. As in the case of water uptake, methanol uptake from the liquid phase is higher than from the vapor phase when compared for the same membrane under the same treatment. This phenomenon, known as Schroeder's paradox, is related to the thermal history of the adsorbing polymer [90].

The expanded form of the Nafion membrane, obtained by boiling the membrane in 3 % H_2O_2, followed by boiling in water, sulphuric acid and finally in water again (treatment A), tends to uptake more methanol than the non-expanded or cast membrane. At temperatures close to ambient, typical liquid methanol uptake (λ) range from 17 to 27, except for a few studies. These values are similar to that observed for the uptake of water by Nafion.

The vapor methanol sorption in expanded membranes range between $\lambda \approx 11$–13 at temperatures in the range 20–30 °C. The case of very-thin cast membrane [280] will be discussed later.

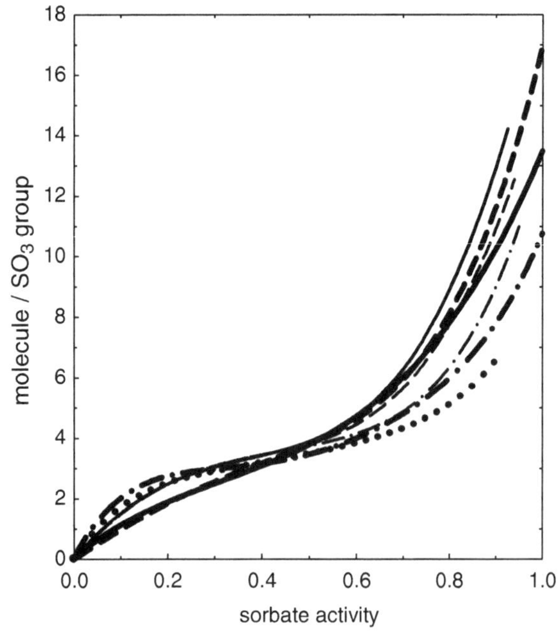

Fig. 6.15 Sorption isotherms of Nafion: water (*solid*), methanol (*dash*), ethanol (*dash-dot*), and propanol (*dot*) from Ref. [272] at 20 °C (*thick curves*) and Ref. [281] at 40–60 °C (*thin curves*)

Datta and coworkers measured the sorption of methanol from vapor by Nafion all over the range of methanol activities at 25 °C [274] and 30 °C [275]. The uptake of ethanol and propanol from vapor by Nafion as a function of the alcohol activities were determined at 20 °C [272] and 30 °C [276]. Zhao et al. [281] reported uptake of methanol and ethanol by Nafion, at 40 and 60 °C, up to alcohol activities close to unity.

In Fig. 6.15 are summarized the sorption isotherms of methanol, ethanol and propanol by Nafion [272, 281], along with the data for water. It can be observed that for alcohol activity higher than 0.5 the alcohol sorption follows the order methanol > ethanol > propanol, as expected according to the size of the sorbate. At low alcohol activities the alcohol uptakes were nearly identical. The behavior of water uptake is different and could be due to the fact that in one case [272] the membrane was used without treatment, while Zhao et al. [281] performed the measurements on the expanded form. In the last case the sorption of water is higher than methanol, while for the untreated membrane more methanol is sorbed in the membrane as compared to water. The data by Jalani et al. [276] for a cast Nafion membrane, not shown in Fig. 6.15, indicates a similar behavior, with the uptake order at high activities being water ≈ methanol > ethanol > propanol.

It is interesting to analyze the effect of inorganic fillers or blending with other polymers on the methanol uptake of Nafion-based membranes. Kim et al. [46] measured methanol uptake in composite of Nafion with organic modified silica (ORMOSIL) and they observed that the methanol uptake decrease proportionally to the amount of inorganic phase in the membrane, reaching a half of its value in Nafion when the filler content is close to 40 wt%. The connectivity between the inorganic phase and the ionic clusters in Nafion is probably responsible for this effect. An inverse effect was found in Nafion composites with diphenylsilicate [49], where the increase of methanol uptake was attributed to the enhanced hydrophobicity of the membrane with the increase of diphenylsilicate beyond 10 wt%.

In the case of Nafion/zirconium phosphate membranes an increase of the water and a decrease of the methanol uptakes is observed with increasing content of the inorganic phase.

Park et al. [102] reported a noticeable reduction of the methanol uptake in Nafion/polypyrrole composite membranes, which is accompanied by a similar decrease in the water sorption. A Nafion/polyaniline membrane containing polyanilines with different oxidation states is reported [127] to reduce around 10 % the methanol uptake, particularly the oxidized state, along with a very small reduction of the water uptake. A similar study on Nafion/polyaniline membrane reported a larger reduction (40 %) of the methanol uptake, which reached 50 % when SiO_2 was added to the composite membrane, without any significant effect on the water uptake [125].

The methanol uptake of a grafted Nafion/poly(glycidyl methacrylate) membrane was studied by Mohy Eldin et al. [162]. The membrane, its amine derivate, and the phosphoric acid –doped membrane exhibit an important enhance of the methanol and water uptake with increasing grafting percentages.

In conclusion, the effect of inorganic and organic fillers or grafting on the methanol (and water) uptake in Nafion is difficult to predict, probably because of several opposite phenomena that could be taking place.

6.5.2 Water-Alcohol Uptake and Partition Constant

Although the former results for the uptake of pure alcohols by Nafion give us an insight of the behavior of this membrane in relation to its use in DAFC, more clear evidence can be obtained by analyzing the uptake from a liquid water-alcohol mixture, and also from the thermodynamic parameter known as the partition coefficient, defined by Eq. 6.5.

Nandan et al. [268] have determined the sorption of an equimolar water–methanol mixture in Nafion 117 and observed that the total uptake of the mixture ($\lambda_w + \lambda_m = 30.8$) was larger than the uptake of the corresponding pure components ($\lambda_w = 14.1$, $\lambda_m = 15.4$). The measured apparent partition constant of methanol, expressed as the ratio of the methanol molar fraction in the membrane, x, and in the solution, x', was $K_x = 0.79$, that is, the membrane sorbs water preferentially. Hietala et al. [270] extended the study of water–methanol sorption in Nafion 117 over all the range of compositions and also observed a maximum at $x = 0.5$ for the sorption of water–methanol in Nafion 117 but the contribution of methanol to the uptake was not determined.

Skou et al. [269] reported that the total uptake goes through a maximum ($\lambda_w + \lambda_m \approx 30$) at $x = 0.6$, while the uptake of the pure components were $\lambda_w = 23.1$ and $\lambda_m = 22.0$. The apparent partition constant is independent of the liquid composition and, contrary to that observed by Nandan et al. [268], was close to one ($K_x = 0.945$). Ren et al. [283] measured the methanol sorption in Nafion 117 and Nafion 1200, at 22 °C, from aqueous solutions up to 13 M. In this composition range ($x < 0.5$) the total uptake increases with x and $K_x \approx 1.0$, in good agreement with the results by Skou et al. [269] and Godino et al. [284].

Gates and Newman [271] have determined the sorption of water–methanol in Nafion 117 at 60 °C, observing a maximum uptake ($\lambda_w + \lambda_m \approx 41$) at $x = 0.5$. The apparent partition constant is higher than one for the methanol dilute solutions ($x < 0.5$), but it tends to one in the concentrated region. Hallinan et al. [278] have studied the sorption of water–methanol in Nafion 117 at temperatures between 18 y 24 °C, observing a maximum ($\lambda_w + \lambda_m \approx 38$) close to $x = 0.5$ and apparent partition coefficients higher than 1, but decreasing with the methanol concentration and reaching a value $K_x \approx 1.2$ at the maximum uptake, in agreement with the results by Gates y Newman [271]. Chaabane et al. [285] also reported K_x slightly higher than one for Nafion 117 water–methanol uptake from solutions up to $x \approx 0.65$, while Diaz et al. [280] found $K_x = 1.9$ for $x = 0.12$.

Saarinen et al. [282] observed that the total uptake by Nafion 117 from a water–methanol solution reaches a maximum at $x = 0.8$, that is, shifted toward more concentrated methanol solutions as compared with former results. However the treatment of the membrane before measurements was not reported in this case. These authors performed uptake measurements for ethanol, 2-propanol and tert-butanol aqueous solutions. The maximum total sorption decrease inversely with the alcohol size ($x \approx 0.7$ for ethanol, $x = 0.25$ for 2-propanol, and $x = 0.10$ for t-butanol), as can be observed in Fig. 6.16. The total mass up taken from the ethanol-water mixture is similar to that reported by Song et al. [286] for dilute solutions ($x < 0.2$). Godino et al. [287] have studied the liquid uptake of methanol, ethanol, 1-propanol, and 2-propanol aqueous mixtures and up to molar fractions 0.36 for methanol, 0.28 for ethanol, and 0.23 for propanol. The results for methanol and ethanol agree with the previous ones, while in the case of 2-propanol they observed a maximum total uptake at $x \approx 0.15$, and for 1-propanol the maximum seems to be close to $x = 0.23$.

Summarizing the results of water–methanol uptakes in Nafion membranes one can conclude that the total uptake reaches a maximum at high methanol concentration ($x = 0.5–0.8$) and the partition constant seems to be higher than one, indicating preference for methanol uptake, in the water-rich region ($x < 0.5$), approaching the unity in the methanol-rich region ($x > 0.5$). Regarding the uptake of higher alcohols, the information on partition coefficient is absent, but assuming $K_x \approx 1$ we can infer that the higher the size of the alcohol, the higher the maximum sorption of liquid (and the lower the alcohol content of the solution at the maximum), probably as a consequence of a plasticizing effect of the alcohol that increase the water uptake capacity of Nafion. On the other hand, for the pure or very concentrated alcohols the liquid uptake, measured in terms of λ_m, decreases in

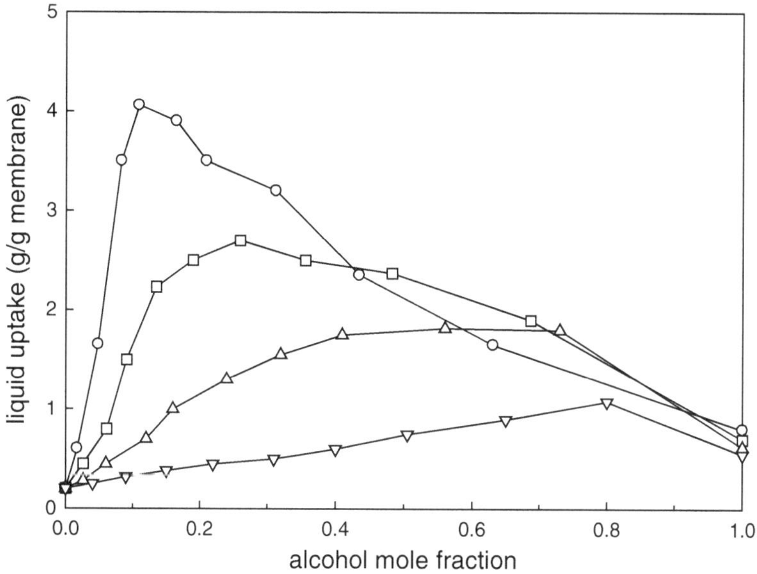

Fig. 6.16 Alcohol uptake of Nafion 117 at room temperature as a function of the alcohol concentration: (○) t-butanol: (□) 2-propanol; (△) ethanol; (▽) methanol. (Adapted from Ref. [282])

the order methanol ($\lambda_m \approx 17$) > ethanol ($\lambda_m \approx 15$) > propanol ($\lambda_m \approx 13$) > butanol ($\lambda_m \approx 12$) [287], similar to that observed by Rivin et al. [272] for the sorption of the pure alcohols from the vapor phase.

Diaz et al. [280] have recently study the behaviour of the water–methanol uptake from the vapour phase by ultra-thin (thickness < 100 nm) Nafion cast membranes and they observed similarities with the water–methanol uptake from the liquid of thick commercial membranes. As illustrated in Fig. 6.17, the total uptake also exhibits a maximum close to $x = 0.5$. Moreover, the apparent partition constant increase smoothly is $K_x = 1.20 \pm 0.06$ for methanol fraction in the vapor between 0.12 and 0.69, indicating a preference for methanol uptake.

The preference of Nafion for up taking methanol over water is certainly an undesirable property for DMFC using this proton conducting membrane. It would be worth to review the effect of inorganic or organic Nafion composite membranes on the sorption of water–methanol from the liquid phase and on the partition constant. The few reported studies include Nafion/sulfonated organosilica [47] and Nafion/zirconium phosphate [78], where a reduction of the total liquid uptake is observed for the composites in methanol solutions up to 10 M, attributed to a reduction of the free volume in the ionic clusters.

Verma and coworkers [90–92] studied the methanol and ethanol uptake in Nafion/Nd_2O_3, Nafion/Nd triflate, and Nafion/Er triflate. In all the cases they observed a slight increase of the mass total uptake with alcohol concentration (between 0.5 M and 2 M) and a very small increase of the uptake with the inorganic

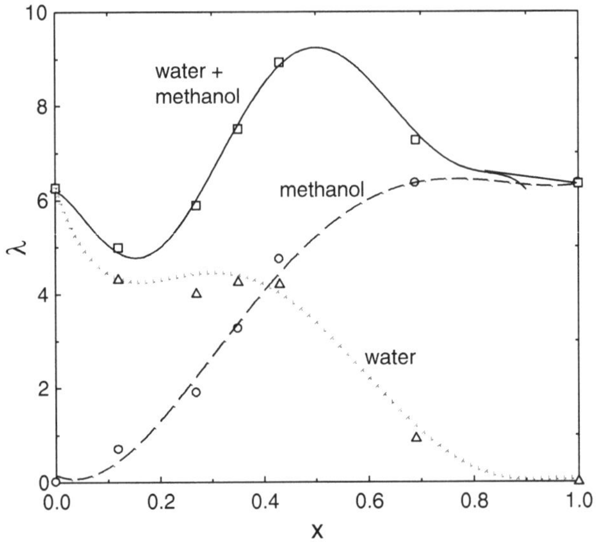

Fig. 6.17 Water–methanol sorption of ultra-thin Nafion membranes as a function of the methanol mole fraction (Adapted from Ref. [280])

phase content. Chen et al. [125] studied the total mass uptake by Nafion/polyaniline/SiO_2 membranes all over the composition range and observed a maximum sorption around 70 v/v%, along with a significant decrease of liquid uptake as compared to Nafion.

In the case of a layered Nafion/sPEEK/Nafion membrane [132], the liquid uptake from methanol solution (up to 5 M) was, as expected, in between the values for the corresponding Nafion and sPEEK membranes.

In summary, there is a lack of information on the separate uptake of methanol and water in Nafion composite membranes which preclude any conclusion about the partition of methanol, while the effect of the filler on the liquid sorption, as noted in the case of the uptake of pure methanol (Sect. 6.5.1), seems to be complex and not easy to assess.

6.5.3 Methanol Permeability

Equations 6.4 or 6.6 describe the flux of alcohol through a membrane if convective effects can be ignored. However, under normal operating conditions of DAFC, hydraulic pressure and electric potential gradients are additional driving forces that could contribute to alcohol flux. Hydraulic pressure gradient is generally low or null in a DAFC so that mechanical permeation is not important. Contrary, the ionic transport through the membrane create generates convective transport of uncharged molecules through electro-osmotic drag, that accounts for the transport of water or alcohol by coupling of the charge transport and mass transport. In that case Eq. 6.6 for the alcohol flux should be modified to incorporate convection [4],

$$J = \frac{Kv^0 \Delta \acute{c}}{1 - \exp\left(-\frac{v^0 \delta}{D}\right)} \tag{6.7}$$

where v^0 is the convective velocity, which is related to the electro-osmotic drag coefficient, ξ, by

$$v^0 = \frac{i\xi \overline{V}_w}{F\theta_w} \tag{6.8}$$

i being the current density, V_w, the molar volume of water, θ_w the volume fraction of water in the membrane, and F the Faraday's constant. Equation 6.7 simplifies to Eq. 6.6 if $v^0\delta << D$, that is, when the diffusive flux predominates over the electro-osmotic flux.

Kim and Pivovar [4] remarked that, at constant current density, δ is the parameter that determines the relative contribution of diffusion and electroosmotic to alcohol crossover because diffusion flux is depending of the membrane thickness, while electro-osmotic drag is not. Therefore, the thinner the membrane, the lower the electro-osmotic contribution, and the current density for having roughly equal diffusion and convection contribution to crossover is 4 times higher in Nafion 112 ($\delta = 50$ μm) as compared to Nafion 117 ($\delta = 178$ μm).

The most common method to measure alcohol permeability through membranes is the diffusion cell method under non-stationary conditions. In this method the membrane separates two reservoirs: the receptor reservoir containing pure water, and the donor reservoir containing the alcohol solution of known concentration. Usually the alcohol solution in the donor reservoir is refreshed during the experiment to maintain its concentration, c_d, constant along the time. The non-stationary alcohol concentration in the receptor reservoir, c_r, is followed as a function of time by in situ or ex situ sensors. By integrating Eq. 6.4 the time dependence of c_r is given by

$$\ln\left(1 - \frac{c_r(t)}{c_d}\right) = -\frac{AP}{\delta V_r} t \tag{6.9}$$

where V_r is the total closed loop volume of the receptor side, and A is the measuring membrane exposed area. This expression is simpler than that obtained when the methanol concentration in the donor compartment is not constant [288] and it facilitates the data treatment. In the case $c_r(t) << c_d$, the linear expression is recovered from Eq. 6.9,

$$\frac{c_r(t)}{c_d} = \frac{AP}{\delta\, V_r} t \tag{6.10}$$

There are several procedures to determine $c_r(t)$ in the diffusion cell method, such as gas chromatography [12, 54, 163, 188, 289–292], potentiometry or amperometry [116, 293, 294], densitometry [280, 284], and IR spectrometry [295, 296].

Other method used to determine alcohol permeability is the pervaporation method, where the membrane is embedded in a cell and one is continuously fed, by a pump, with a alcohol solution, and the other side is purged with a continuous flow of an inert gas with a fixed flow rate for carrying the permeate to a gas chromatograph [33, 34]. In this case the state of solvation of the membrane on the gas side is not well defined.

Verbrugge [297] has reported the diffusion coefficient of methanol in Nafion 117 at 25 °C using an electrodiffusion technique, and NMR has also been used for measuring the diffusion of methanol in Nafion 117 at 22 and 30 °C [270, 298].

Ren et al. [299] have studied the methanol transport through Nafion membranes from 30 °C to 130 °C by resorting to a DMFC configuration method. In this method a MEA is prepared with the membrane under study, using PtRu catalyst on one side and Pt black on the other side. A methanol solution is fed on the PtRu side and the DMFC is polarized at constant current density in such a way to oxidize the methanol that permeated through the membrane on the Pt catalyst. Thus, the methanol transport has a diffusional component in the Pt electrode direction and, because the proton migrates from the Pt electrode to the PtRu electrode (where it is reduce to hydrogen), a methanol counter flow is present due to the electro-osmotic drag effect.

It should be noted that in this method the electro-osmotic drag of methanol by the protons is opposite to the drag effects taking place in a DMFC under normal operation, where protons travel in the same direction than the methanol drived by diffusion from the anode to the cathode. The net methanol flux in this DMFC configuration is given by,

$$J = -P\left(\frac{\mathrm{d}c'}{\mathrm{d}z}\right) - \frac{\xi i}{\mathrm{F}} \tag{6.11}$$

where i is the current density through the cell, and ξ is the methanol drag coefficient, defined as the number of methanol molecules dragged by proton.

According to Eq. 6.11, the DMFC determination of the methanol permeability requires the knowledge of the methanol drag factor, because the electro-osmotic flux could afford for a considerable fraction of the methanol flow, particularly at high methanol concentrations. An important drawback of this method is that the methanol drag coefficient is not well known, so Ren et al. [299] assumed that it was similar to the water drag coefficient (ξ =2.5). However, some recent NMR [300] and electro-osmosis [301] studies would indicate that this assumption is not valid, leading to considerable uncertainties in the methanol permeability coefficients determined by this method.

Figure 6.18 summarizes the results reported in the literature for the permeability of methanol through Nafion membranes as a function of temperature. The results

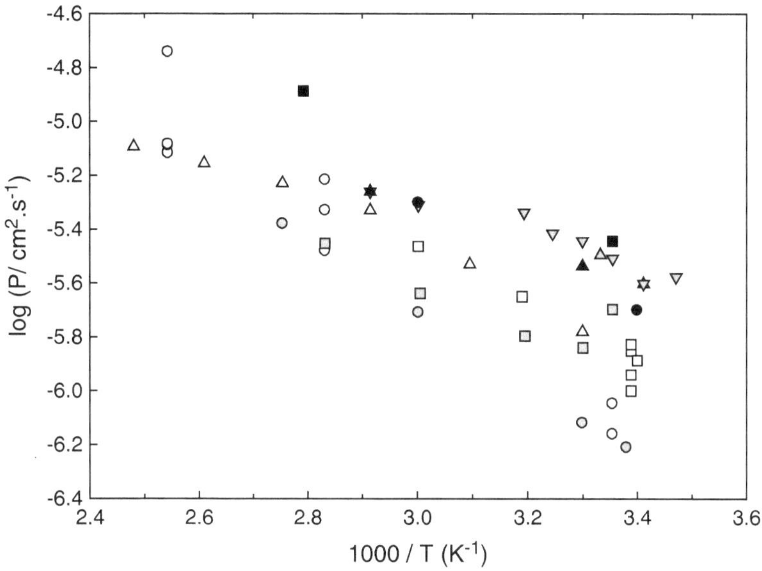

Fig. 6.18 Temperature dependence of the methanol permeability coefficient in Nafion 117 membranes: (○) [280]; (▽) [293]; (□) [289, 290]; (△) [299]; (○) [188]; (▲) [116, 288]; (▽) [284]; (□) [296], and Nafion recast membranes: (●) [51]; (▲) [58]; (■) [85]

obtained by the DMFC configuration method [299] exhibits a non-Arrhenius behaviour, probably as a consequence of the assumptions on the methanol drag coefficient mentioned above.

It can be observed that near room temperature the permeability data spreads over almost one order of magnitude, as a reflect of the different membrane treatment. In the case of expanded Nafion membranes (pretreated by boiling in H_2O_2, water and acid) [188, 280, 296, 299] the permeabilities were lower than those of the membranes hydrated in water at 20–80 °C [284, 289, 290, 293], with the exception of the data by Wu et al. [116, 288] that being expanded membrane show one of the highest methanol permeabilities.

The permeability data of Nafion recast membranes are also shown in Fig. 6.18. These membranes were studied mainly to compare its alcohol permeability behavior with composite Nafion membranes prepared by mixing Nafion with inorganic or organic fillers, followed by casting from the corresponding solution. Although there is also a high scatter of data (not shown in Fig. 6.18), the few works where the effect of temperature has been studied seems to indicate that methanol permeability through recast membrane is higher than in Nafion 117 membranes. Therefore, when comparing the methanol crossover of Nafion composite membranes one should take into account not only the relative permeability coefficient, but also the absolute permeability values.

In a very comprehensive study, Yu and coworkers [302] prepared several Nafion membranes by casting from several solvents in order to study its morphology, proton and methanol transport properties. Several solvents (methanol/water, ethanol/water, isopropyl alcohol/water, N-methyl formamide, dimethyl acetamide, and dimethyl formamide) were used in order to understand the effect of the solubility parameter and dielectric constant on the properties of the recast membranes which were annealed at 125 or 150 °C for 90 min. The results clearly show that the methanol permeability, as well as the morphology and proton conductivity of the membranes, largely depends on the solvent and annealing conditions. The membranes prepared using alcohol/water solvents exhibit the highest permeabilities, slightly lower than those reported for other recast membranes in Fig. 6.18, while the membranes cast from NMF, DMF and DMA exhibit permeabilities that are between 2 and 8 times lower than the measured for Nafion 117 over the same range of temperatures.

Several Nafion/inorganic composite membranes were prepared with the aim of reducing the alcohol permeability of pristine Nafion. Among them, one of the more studied is the Nafion/SiO_2 composite, exhibiting very modest effect on methanol permeation. The relative permeability coefficient, Pr, of these composite membranes ranges between 0.6 and 1.0, independent that the composite was prepare by mixture and recast [33, 36, 39, 40, 44] or by sol–gel formation of the silica particles inside the Nafion membrane [32, 35]. A much better reduction of the methanol permeation is achieved with sulfonated mesoporous silica [44] and organic functionalized silica [46, 47, 50] where values of P_r between 0.16 and 0.57 have been obtained, varying with the content of inorganic phase.

Silica layered Nafion membranes prepared by olasma enhanced chemical vapour deposition [45] also exhibit a moderate reduction of permeability (P_r =0.33) by those membranes fabricated by dip-coating [38] show extremely high barrier properties to methanol permeation, with $0.002 < P_r < 0.005$. However, it should be noted that the permeability reported for the Nafion 117 without silica layers is more than one order of magnitude higher than the values reported by several authors, including those shown in Fig. 6.18.

Different types of Nafion/zeolite composites are reported to have P_r between 0.4 and 1 [51, 54, 55, 57, 58, 60], with the exception of a composite of Fe-silicalite-1 with Nafion, prepared by impregnation of the colloidal zeolite or in situ zeolite synthesis in supercritical-CO_2 treated Nafion [56]. These membranes have P_r values as low as 0.002–0.007.

Montmorillonite, silicate, aluminosilicate, titanosilicate, calcium phosphate, zirconium phosphate, metal oxides, and heteropolyacids composites with Nafion exhibit in general a modest barrier effect to methanol permeation ($P_r > 0.1$), with a couple of exceptions: a Cloisite 10A composite [62] with $0.013 < P_r < 0.065$, and a chitosan-functionalized montmorillonite composite [71] with $0.028 < P_r < 0.050$, in both cases for clay contents between 1 and 10 wt%.

Organic/Nafion blends can be classified as having a small barrier effect ($P_r > 0.25$), as the case of PPy [102–104], PPO [108, 109], PFA [110], sPF [116], PPV [123], PVP [124], PE [129], sPEEK [133–135], sPAEK [136], PVI [138], PTFE [142], sPOSS [144], PEDT [146], PR [147], PM [148], PSSA [150], and AMPS [152], moderate barrier effect ($0.04 < P_r < 0.25$), as the case of PETE

[130], sPAES [137], or strong barrier effect ($P_r < 0.04$), as the case of PAES [137] and PFEP [143].

Polymers such as PVDF, PBI and PVA, for which there several studies of methanol permeability reported in the literature, are not so easy to classify because values of $P_r < 0.1$ have been measured for PBI [106], PVA [114], and PVdF [118–120], but $P_r > 0.3$ has been reported for PBI [107], PVA [112, 113], and PVdF [117].

Paradoxically, the efforts to reduce the methanol permeabilities of Nafion with inorganic or organic fillers in most cases yield composite membranes with permeabilities similar to that obtained by optimizing the cast procedure of pure Nafion [302]. Nevertheless, the reduction of methanol permeability by itself is not a criterion for improving DMFC performance because it is usually associated to a reduction of the proton conductivity. We will analyze this property in Sect. 6.5.5 as a previous step to discuss the behavior of the proton-conducting membranes in terms of alcohol selectivity defined by Eq. 6.2.

6.5.4 Ethanol and Higher Alcohols Permeability

The permeability of ethanol and higher alcohols through Nafion has not been studied so extensively as methanol permeation. Verma and coworkers reported the permeability coefficients of ethanol through Nafion composites with Nd_2O_3 [90], neodymium triflate [91], and erbium triflate [92] and the corresponding Nafion recast membranes. Unexpectedly, the measured P for ethanol differ more than a factor 20 for recast membranes (P between $1.0\ 10^{-7}$ and $2.8.10^{-6}$ $cm^2.s^{-1}$).

Song et al. [286] studied ethanol crossover through an expanded Nafion 115 membrane at 75 °C, observing a considerable increase of the crossover rate when the ethanol concentration rises from 1.5 to 7.5 M. An estimation of the permeability coefficients, neglecting contribution from the hydraulic permeability contribution yields to values from 2 to 7.10^{-7} $cm^2.s^{-1}$, which are much lower than those reported for methanol in Fig. 6.18.

The permeability of ethanol and methanol through Nafion 117 membranes was reported by Xue et al. [303] at 22 °C, founding similar values for methanol ($P = 1.77\ 10^{-6}$ $cm^2.s^{-1}$) and ethanol ($P = 1.48.10^{-6}$ $cm^2.s^{-1}$). This permeability value is much higher than the reported above, although membrane pretreatment was not indicated by the authors. Kontou et al. [304] measured the ethanol crossover through an expanded Nafion 115 using the DEFC configuration method as a function of temperature and alcohol concentration. Neglecting the electro-osmotic correction, the permeability coefficients increase from 7.10^{-7} to 3.10^{-6} $cm^2.s^{-1}$ when the temperature raised from 30 °C to 90 °C in 2 M ethanol, and from 4.10^{-7} to $1.3.10^{-6}$ $cm^2.s^{-1}$ over the same temperature interval in ethanol 12 M. These permeability values are slightly lower than those measured for methanol (Fig. 6.18) on the same range of temperature.

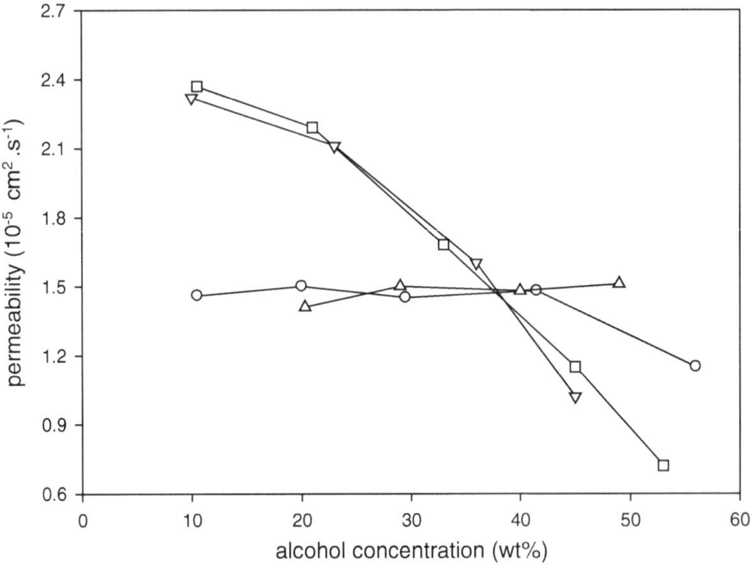

Fig. 6.19 Alcohol permeability through Nafion 117 at 25 °C as a function of the alcohol concentration: (□) 1-propanol: (▽) 2-propanol; (○) ethanol; (△) methanol. (Adapted from Ref. [287])

The permeability of methanol, ethanol, 1-propanol, and 2-propanol was measured by Godino et al. [287] at 25 °C as a function of the alcohol concentration through a non-treated Nafion 117. The results, shown in Fig. 6.19, indicate that the permeation of methanol does not depend on the alcohol concentration in the interval 10–50 wt%, but the permeation of both 1- and 2-propanol, decrease significantly and the permeation of ethanol is independent of the concentration up to approximately 40 wt%, but it decreases at higher concentrations.

The authors explained this behavior by resorting to differences in viscosity between methanol and higher alcohols, but a look to Fig. 6.16 showing the alcohol uptake of Nafion 117 membranes would indicate that the solubility of the alcohols could be also responsible for the observed dependence of permeability on alcohol concentration.

6.5.5 Proton Conductivity

The high proton conductivity of hydrated Nafion is the main reason of its wide use in PEM fuel cells. Many studies were devoted to determine this property under different conditions of temperature and water uptake (from vapor and liquid phase) on membranes subjected to different pretreatments, using AC impedance and DC techniques. Slade et al. [305] reviewed the conductivity of Nafion membranes measured till 2002, and recently Yadav and Fedkiw [306] have updated the information collected during the last decade.

Because of different measurement techniques and sample pretreatment the conductivity varies significantly even under identical measurement conditions (temperature and relative humidity). Thus, at 20 °C, the reported conductivity of Nafion 117 varies from 63 to 90 mS.cm^{-1} in liquid water, and between 51 and 96 mS.cm^{-1} in vapor phase at 100 % RH (relative humidity) [306]. Gardner and Anantaraman [307] measured the in-plane and through-plane conductivity of fully hydrated Nafion 117 and found that it was 85.6 mS.cm^{-1} in-plane, but only 24 mS.cm^{-1} through-plane. On the other hand, the pressure dependence of the through-plane conductivity depends strongly on the pressure applied on the membrane, decreasing around 30 % as the pressure increases from 3 to 7 bar. These authors also observed a dramatic reduction of the conductivity of Nafion 117 when the relative humidity is reduced from 100 % (66 mS.cm^{-1}) to 34 % (0.14 mS.cm^{-1}), although the conductivity drops to 13.6 mS.cm^{-1}, just by reducing RH to 93 % [308]. A similar pressure effect was also reported by Wintersgill and Fontanella [309], although over a higher range of pressure, out of the region of interest for PEM fuel cells. However, for $\lambda_w < 5$ they found that conductivity increases with pressure. These authors also reported de decrease of several orders of magnitude of the measured conductivity for Nafion 105, 117, and 120 as a function of λ_w.

The studies of the effect of temperature and humidity (or water uptake) are abundant, beyond those discussed in the above mentioned reviews [305, 306]. Lufrano et al. [310] studied the proton conductivity of Nafion 117 between 25 °C and 90 °C during in-situ fuel cells experiments by measuring the cell resistance by the current interruption method. Yang et al. [311] measured the conductivity of Nafion 115 membranes kept at 100 % RH between 50 °C and 100 °C. Damay and Klein extended the range of temperature of conductivity of Nafion 115 up to 130 °C, including the effect of RH [312]. More recently, Ochi et al. [313] studied the conductivity of Nafion 117 between 31 °C and 118 °C as a function of water uptake, and Wu et al. [314] measured the conductivity of Nafion 115 at 60 °C and 80 °C, as a function of RH or water activity, a_w = RH/100.

Figure 6.20 summarizes the results obtained for Nafion 115 and 117 as a function of temperature, where it can be observed that the agreement among different authors is reasonable, taking into account differences in the techniques used to measure the conductivity and the pretreatment of the membranes, which determines the water uptake.

The dependence of Nafion conductivity with temperature obeys Arrhenius law in the range of temperature 20–100 °C and, as noted by Yadav and Fedkiv [306], the activation energy depends on water activity (from 10.44 kJ.mol^{-1} at 100 % RH up to 14.11 kJ.mol^{-1} at 30 % RH). The conductivity is higher for membranes equilibrated with liquid water, where the water uptake is larger than from vapor at 100 % RH, but the activation energy is slightly lower (8.2 kJ.mol^{-1}). Dimitrova et al. [33] measured the conductivity of Nafion 117 equilibrated with liquid water and obtained lower proton conduction than Yadav and Fedkiv, probably as a consequence of a different membrane pretreatment. These authors also prepared a recast membrane whose conductivity was slightly higher than that Nafion 117.

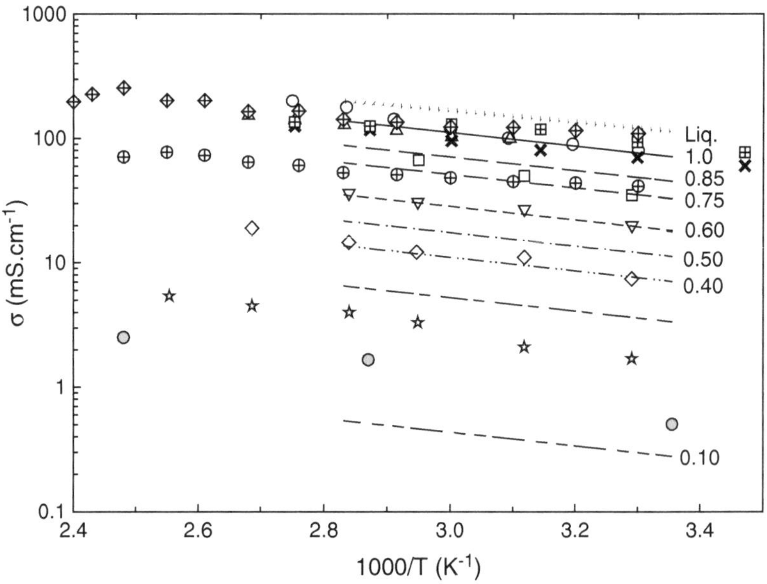

Fig. 6.20 Conductivity of Nafion membranes as a function of temperature at several water activities, a_w. Nafion 117: the lines correspond to the best Arrhenius fit of the data from Ref. [306] at the indicated a_w; Ref. [33] (✖) $a_w = 1$; Ref. [310] (○) $a_w = 1$; Ref. [313]: (□) $a_w = 0.8$, (▽) $a_w = 0.6$, (◊) $a_w = 0.4$, (★) $a_w = 0.2$. Nafion 115: Ref. [311] (△) $a_w = 1$; Ref. [312] (◎) $a_w = 0.1$. Recast Nafion: Ref. [33] (⊠) $a_w = 1$; Ref [315] (⊕) $a_w = 1$. Nafion/SiO_2/PWA Ref. [315] (⊕)

Aricó et al. [315] have also measured the conductivity of recast Nafion membrane and reported values about one half those determined by Dimitrova et al., as can be seen in Fig. 6.20. Dimitrova et al. [34] also studied the effect of thickness on the conductivity of commercial Nafion membranes (112, 115 and 117), observing an important increase of conductivity with thickness, which was modeled assuming a layered structure, with the conductivity of the surface layer lower than bulk conductivity. Similar results were found by Slade et al. [305] comparing the ex-situ and in-situ conductivities of Nafion 112, 1135, 115 and 117.

The effect of thickness on the conductivity of recast Nafion films was addressed by Siroma et al. [316] by measuring the conductivity of membranes, at 85 °C, over a wide range of thickness, from ~20 nm to 500 μm. They found, as in the above mentioned studies, that the conductivity decreases with a decrease in thickness, being the reduction a factor one order of magnitude at 85 % RH, and two order of magnitude at 60 % RH. Interestingly, the conductivity of commercial Nafion films (112,115, and 117) lies on the same line that recast Nafion membranes, while the activation energy for the proton conductivity increases from ~15 kJ.mol^{-1} for thick membranes (100 μm) up to 30 and 50 kJ.mol^{-1} for the thinnest membranes (< 100 NM) AT 85 % AND 60 % RH, respectively. The authors suggested that the lower conductivity of thin membranes is due to a reduction of water uptake, with the consequent increase in activation energy for proton, transport, a fact that has been confirmed recently [317]. These results are relevant to the description of the charge transport in the ionomer within the three phase boundary region, which will be discussed later in this Chapter.

The thermal treatment of Nafion membrane has a significant effect on conductivity as shown by Sone et al. [318], who measured the conductivity of expanded Nafion 117 without heat-treatment and after heat-treatment at 80 °C (N-form), 105 °C (S-form), and 120 °C (FS-form). The thermal treatment reduces the conductivity of the expanded Nafion from ~80 $mS.cm^{-1}$ down to ~50 $mS.cm^{-1}$ for the N-form, and down to ~30 $mS.cm^{-1}$ for the S- and FS-form. The reduction of conductivity correlates with the reduction of water uptake with increasing temperature of thermal treatment.

Figure 6.20 illustrates the strong effect of water activity on the proton conduction, which is linked to the water content of the membrane. The water sorption isotherm of Nafion (see Fig. 6.15) shows that water activity higher than 0.3 (30 % RH) is required for having more than three molecules of water per sulfonic group. This is the minimum number of water molecules need to dissociate the sulfonic acid in Nafion, according to the ab-initio calculations by Padisson [319]. Once the sulfonic group dissociates the proton conductivity increases further with the water uptake because protons separates further from the anionic fixed charges and can migrate via a Grotthuss mechanism.

The first study of the conductivity of Nafion 117 at high temperature (above 100 °C) was performed by Savinel et al. [320], in the interval 100–180 °C, including the effect of RH. They observed that the conductivity of a dried member is almost constant (~ 20 $mS.cm^{-1}$) all over the temperature range, but decrease from 90 down to 35 $mS.cm^{-1}$ between 110 and 180 °C when the membrane was humidified with water. The presence of methanol in the vapour phase in contact with the membrane increased the conductivity at 110 oC by a factor 2, but a pronounced decrease of conductivity with temperature was observed. Similar results were found by Edmondson et al. [321] studying the conductivity of Nafion 117, at room temperature, in contact with methanol and water/methanol vapor phase. In the low liquid uptake region, the presence of methanol in 1:1 mixtures with water seems to increase the conductivity in comparison with membranes in contact with pure water having the same water content, probably due to a plasticizing effect of methanol. However, proton conductivity in pure methanol is one order of magnitude lower than that observed in water.

The reduction of proton conduction of Nafion 117 in pure methanol, ethanol, and 2-propanol as compared to water was also observed by Saito et al. [322]. The measured conductivities at 25 °C were 70, 10, 2.5, and 0.6 $S.cm^{-1}$ in water, methanol, ethanol and 2-propanol, respectively. Based on NMR diffusion measurements these authors concluded that proton transport in the membrane takes place by Grotthuss and vehicular mechanisms, and the former is hindered by the presence of alcohols with larger alkyl groups. Chaabane et al. [323] have interpreted the reduction of proton conductivity with increasing methanol content in terms of a model considering the nanostructure of Nafion. In this model a core region filled with electroneutal solution and a gel phase containing the charged hydrophilic polymer domains joined to the hydrophobic polymer domains coexist, and it is assumed that the volume fraction of the joined gel phase increases with methanol content. However, the volume of the core region is almost unmodified and the lower ionization of the sulfonic group with the less polar methanol accounts for the decrease of proton conductivity, partially compensated by swelling of the gel phase.

The mechanism of proton conduction in Nafion is still in debate. Pivovar et al. [324] have discussed this mechanism in relation to electro-osmosis in Nafion 117. They argue that for proton conduction occurring entirely by the Grotthuss mechanism, the electro-osmotic drag coefficient would be zero, because proton transport proceeds without a net water (methanol) transport. A proton transport dominated by Grotthuss mechanism could suggest an independent mechanism for proton and methanol transport, leading to a highly selective DMFC membrane. From the analysis of the electro-osmitic experiments they concluded that, unfortunately, the local environments that promote the Grotthuss mechanism also lead to increased electro-osmotic drag coefficients. Thus, in Nafion the Grotthuss transport dominates at high levels of hydration, which are also the conditions for the largest electros-osmotic drag effect, suggesting that is the viscous drag instead of the proton transport mechanism that play a more important role on the observed drag coefficients.

Because the complete miscibility of water and methanol, it is unlikely that a high selectivity can be achieved in polymers with large "bulk" water domains. On contrary, materials with smaller water domains, where electrical field gradients are more intense could improve the selectivity [324]. Nevertheless, the lack of information on the relative electro-osmotic drag of water and methanol in DMFC membranes, as mentioned in Sect. 6.5.3, preclude us to establish a strong correlation between selectivity and electro-osmotic drag.

Numerous works have reported the decrease of hydrated Nafion conductivity at high temperature. Thus, Rikukawa and Sanui [325] observed a maximum in the conductivity of Nafion 115 around 80 °C, and a sudden drop at temperatures above 100 °C. Aricó et al. [315] also observed a maximum conductivity around 120 °C, as shown in Fig. 6.20, for recast Nafion membranes. This behavior is the consequence of membrane dehydration at temperatures above 100 °C, which limits the use of Nafion in high temperature PEM fuel cells.

Aricó's group in Italy have studied Nafion composite membranes and other polymers aiming to find low cost membranes with suitable proton conductivity and stability up to 150 °C for high temperature DMFC [6]. Nafion composites with SiO_2, ZrO_2, Al_2O_3, and SiO_2/PWA [315] exhibit maximum in conductivity at about 135 °C, that is, a 15 °C higher that the corresponding pure Nafion recast membrane, as shown in Fig. 6.20 for the Nafion/SiO_2/PWA composite. This behavior is related to the hygroscopic characteristics of the inorganic fillers which enhance the water retention in the membrane. However, this is not the only effect on the membrane characteristics because it was observed that the membrane conductivity increases as a function of the acid character of the filler ($SiO_2 > ZrO_2 > Al_2O_3$) probably as a consequence of the formation of strong hydrogen bonds between the surface

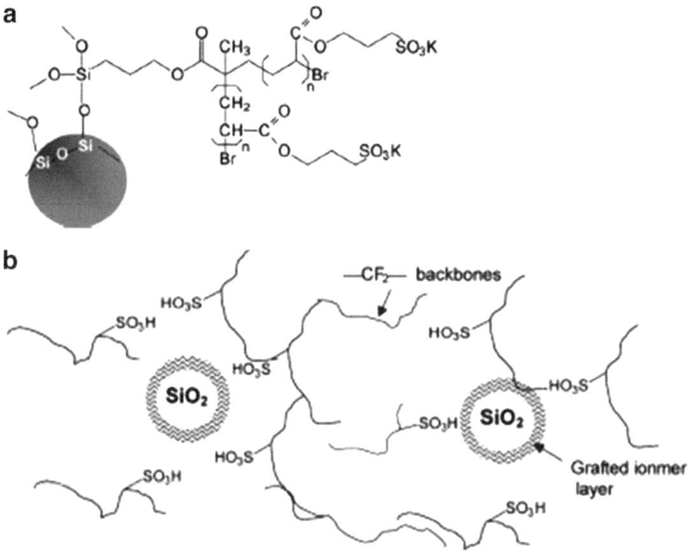

Fig. 6.21 Structure of the silica-core nanoparticles with grafted sulfonated oligomers (**a**) and physical crosslinking with Nafion (**b**) (From Ref. [50] with permission)

functionalities and the water absorbed in the filler, which could promote Grotthuss-like mechanism of proton transport.

The highest conductivity among the Nafion/inorganic fillers correspond to the ternary Nafion/SiO_2/PWA composite with conductivities about three times higher than that of recast Nafion over the entire range of temperature (see Fig. 6.20) [6, 95, 96]. Important conductivity increases were also obtained using sulfonated meso-structurated silica (MSU-F and MCM-41) [44], and silica nanoparticles with a sulfonated oligomer layer [50], as shown in Fig. 6.21a, which enhance interaction with the sulfonic groups in Nafion, illustrated in Fig. 6.21b.

Most of the zeolite and montmorillonite fillers used as methanol barrier in Nafion composites decrease the membrane conductivity, except in the case of a sulfonated montmorillonite [67] which exhibit a conductivity increase of 25 % when a composite with Nafion containing 5 wt% of organoclay is prepared. Mesoporous aluminosilicates [73] in concentration as low as 0.5 wt% are reported to increase the conductivity of Nafion by a factor higher than three between 30 °C and 80 °C.

A solid superacid (stronger than H_2SO_4 100 %) such as TiO_2-SO_4^{2-} as nano-sized particles has been successfully used as a filler yielding membranes with conductivities up to 50 % higher than the recast Nafion membrane [88]. Neodymium oxide [90] and triflate [91], along erbium triflate [92] enhance the conductivity of Nafion between 30 % and 40 % when added in concentrations between 3 and 8 wt%.

Nafion/organic composites, in most of the cases, exhibit lower proton conductivity as compared to the recast Nafion membrane. A composite of Nafion with a sulfonated β-cyclodextrin has been reported to have conductivities up to 40 %

higher than the Nafion recast membrane over the temperature range 10–80 °C [131]. Also, and inclusion complex of α-cyclodextrin and poly(ethylene glycol), call polyrotaxane, exhibit proton conductivities up to 60 % higher than the recast Nafion membrane [147].

As discussed in Sect. 6.4.3 in the case of methanol permeability through Nafion composite membranes, proton conductivity along is not a criteria for an optimal DMFC membrane. Therefore, in the next Section, membrane selectivity to methanol and other alcohols will be critically reviewed.

6.5.6 Methanol Selectivity of Nafion Composite Membranes

In Sect. 6.1 we defined the alcohol selectivity coefficient, β, of a membrane Eq. 6.2, which accounts for the ratio between the proton and alcohol transport through it, and the relative selectivity, β_r, Eq. 6.3 which compare the alcohol selectivity of the membrane to Nafion under the same experimental conditions. Here we will analyze the abundant data on the methanol selectivity in inorganic and organic Nafion composites in comparison with the selectivity of pure Nafion membranes. In order to perform this analysis we will resort to the general selectivity plot described in Fig. 6.2.

In Fig. 6.22 are plotted the relative conductivity *vs.* relative methanol permeability for Nafion/inorganic composites. The dashed lines indicate relative selectivity values of 1 and 10, as illustrative boundaries.

It can be observed that most of the studied inorganic compounds exhibit σ_r and P_r close to unity, that is, the filler does not affect significantly the conductivity neither the methanol permeability as compare to the pure Nafion membrane. Few composites lie on the undesirable C1, C2 and D region ($\beta_r < 1$), and in the B1 octant, corresponding to membranes with higher conductivity and higher permeability than pure Nafion.

A considerable number of composites exhibiting relative methanol selectivities between 1 and 10 are located in the B2 and A regions. Those in the B2 (lower conductivity but higher permeability barrier than Nafion) octant and A (higher conductivity and higher permeability barrier than Nafion) quadrant, with high relative selectivity, corresponds to Nafion composites with sulfonated mesoporous silica [44], sulfonated Fe_2O_3 [89], and sulfonated montmorillonite [66–68]. However, a number on non-sulfonated inorganic fillers shown excellent characteristics, particularly some organic silicas [46, 50], PWA [94], and aluminophosphate [72]. Remarkable high methanol selectivity coefficients were reported for aluminosilicate ($\beta_r \approx 22$) [73], and some montmorillonites ($15 < \beta_r < 27$) [62, 71].

Pd [22], SiO_2 [45], and TiO_2 [85] films deposited on the surface of Nafion membranes by sputtering, plasma vapour deposition, and dip-coating, respectively, does not show relevant characteristics for DMFC. However, SiO_2 layers deposited on both surface of Nafion 117 membranes by a simple dip-coating technique [38] leads to the highest reported values of relative selectivity ($\beta_r = 650$ at 25 °C, and $\beta_r = 220$ at 80 °C), not shown in Fig. 6.22. These membranes show an unexpected

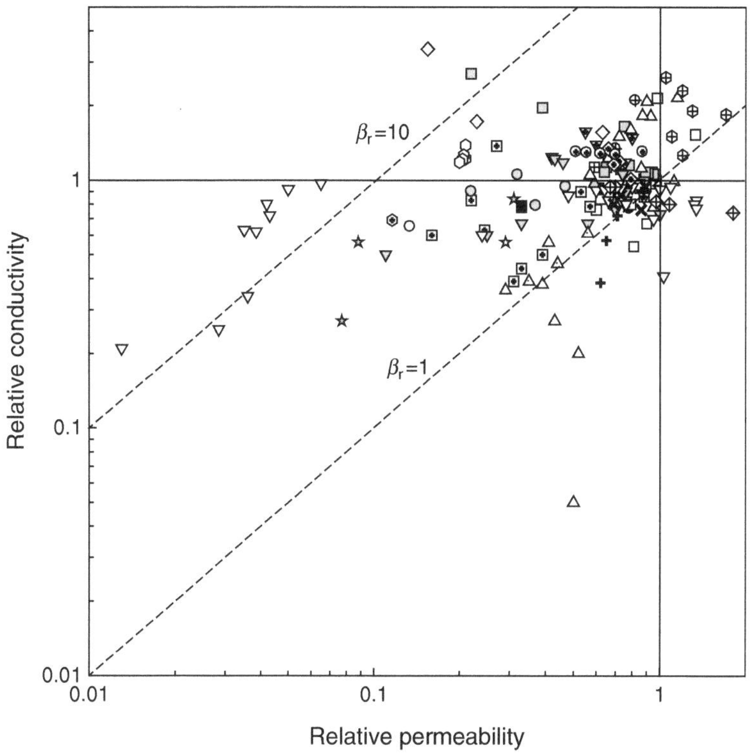

Fig. 6.22 Selectivity plot for Nafion/inorganic compounds: Pd (○) [22, 24]; SiO_2 (□) [32–36, 38–41]; organic SiO_2 (▣) [46, 47, 50], zeolite (△) [51, 54–58, 60]; montmorillonite (▽) [62–71]; silicate and aluminophosphate (☆) [72]; aluminosilicate (◇) [73]; calcium phosphate (⊕) [74]; titanosilicate (✖) [83]; sulfonated titanate (✚) [84]; titania (▲) [85]; sulphated titania (◈) [86]; Fe_2O_3 (▼) [88]; sulfonated Fe_2O_3 (○) [89]; Ne_2O_3 (◈) [90]; Nd triflate (◉) [91]; Er triflate (○) [92]; PWA (◉) [94]; PMA (⊕) [33]. Gray symbols correspond to sulfonated fillers and closed symbols to layered composites

high conductivity as compared to Nafion 117 (between 10 % and 30 %), but the high selectivity is mainly due to the high barrier toward methanol permeation ($0.002 < P_r < 0.005$) which strongly depends on the time elapsed between the deposition of the first and the second SiO_2 layers. Unfortunately, DMFC test has not been reported for these promising membranes.

Figure 6.23 shows the relative conductivity *vs.* relative methanol permeability for Nafion/organic blends. As in the case of Fig. 6.22, the dashed lines the limits of the $1 < \beta_r < 10$ region. The general aspect of the selectivity plot is similar to that shown above for Nafion/inorganic compounds, with most of the composites exhibiting σ_r and P_r values close to unity. Few composites lie on the C2 region (β_r <1), and most of them lie in the B2 octant, corresponding to membranes with lower conductivity and higher permeability than Nafion. In region A (σ_r >1, and P_r <1) only sulfonated POSS/Nafion composites [144] are clearly located,

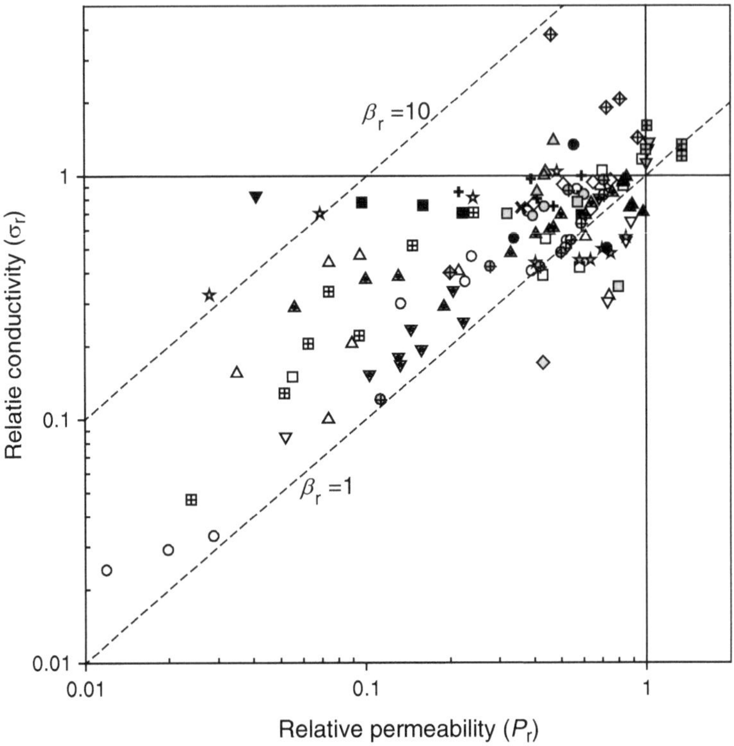

Fig. 6.23 Selectivity plot for Nafion/organic compounds: PPy (○) [102, 103]; PBI (□) [106, 107]; PPO (◊) [108, 109], PFA (✖) [110]; PVA (△) [112–114]; PF (✚) [116]; PVdF (▽) [117–120]; PPV (◎) [123];PVP (▲) [124]; PANI (▣) [126–128]; PE (◇) [129]; β-CD (▽) [131]; sPEEK (★) [133, 134]; sPAEK (⊕) [136]; sPAES (⊠) [137]; PVI (▲) [138]; PVPy (▼) [139]; PTFE (●) [142]; PFEP (▲) [143]; sPOSS (◈) [144]; PEDT (⊕) [145, 146]; PR (⊞) [147]; PSSA (▲) [150]; SEBS (◈) [151]; AMPS (☆) [152]. Closed symbols correspond to layered composites

along with some PTFE-coated Nafion membranes [142], and Nafion/PVP composites [124].

Exceptionally high relative selectivities are reported for sPOSS ($\beta_r = 8.3$) [144], PVdF ($\beta_r = 20$) [119, 120], PANI-coated Nafion ($\beta_r = 8.0$) [126], and semi-inter penetrating polymer networks of Nafion and crosslinked poly(AMPS) ($\beta_r = 16$) [152].

Other Nafion-based membranes described in Sect. 6.3.4, like organic/inorganic ternary composites with Nafion, have high methanol relative selectivity, such as the Nafion/PEG/SiO_2 ($\beta_r = 10$–20) [155] but DMFC test has not been reported for this composite. A trilayer membrane composed of one central methanol-barrier layer of PVdF and two Nafion layers was prepared [164]. Although the membranes could reduce methanol permeability between one and more than two orders of magnitude, depending on the layer thickness, proton conductivity is also reduced by a similar factor, resulting in a moderate β_r value.

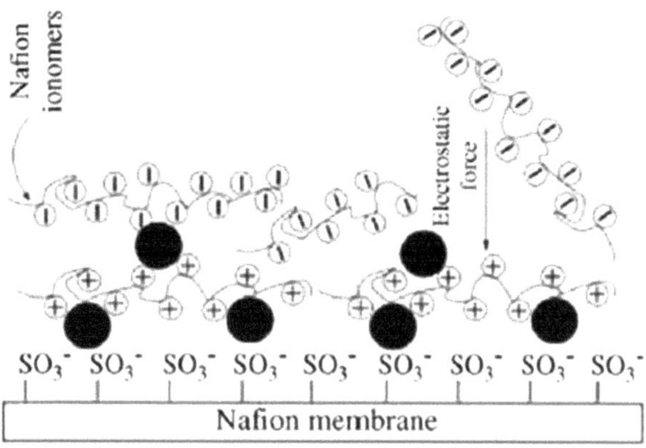

Fig. 6.24 Structure of self-assembled PDAC/Pd and Nafion layers on a Nafion membrane (From Ref. [25] with permission)

Nafion membranes treated with supercritical CO_2 become more crystalline and a moderate increase in methanol selectivity is observed [166], although DMFC performance improved significantly [165].

Special attention deserves Nafion layered membrane prepared by the LBL self assembly of polyelectrolytes [25, 167–171]. A high selectivity membrane was prepared by Tang et al. [25] by self-assembling multi-layer Pd nanoparticles onto Nafion, using poly(diallyl dimethylammonium chloride) (PDAC) for charging the Pd particles. A Nafion 112 membrane was immersed in a Pd/PDDA dispersion and then in a Nafion dispersion. The process was repeated five times to obtain a multi-layer self-assembly Nafion composite that shows a small decrease in conductivity (from 112 $mS.cm^{-1}$ for Nafion 112 to 81 $mS.cm^{-1}$ for the composite). However, the reported methanol permeability was reduced by a factor 0.0085 (out of scale in Fig. 6.22), leading to $\beta_r \approx 85$. This composite membrane, whose structure is depicted in Fig. 6.24, was not tested in a DMFC.

Hammond and coworkers [167] prepared several LBL assembled films onto Nafion using different polyanions and polycations. The best results were obtained with a membrane produced assembling poly(ethylene imine) (LPEI) and sulfonated poly(2,6-dimethyl 1,4-phenylene oxide) (sPPO), having $\beta_r \approx 47$ and a DMFC performance over 50 % compared to unmodified Nafion. Moon and Rhim [168] prepared LBL assembled membranes using poly(allylamine hydrochloride) (PAH) and poly (estyerene sulfonic acid) (PSSA), where both, methanol and conductivity decrease with the number of layers, in such a way that methanol selectivity does not improve significantly in comparison with unmodified Nafion. However, Nafion membranes coated with a layer of PAH exhibit a DMFC performance at 80 °C much better (150 $mW.cm^{-2}$) than Nafion (70 $mW.cm^{-2}$) when fed with 5 M methanol. PAH and poly(vinyl sulphate) (PVS) were LBL assembled onto Nafion achieving a reasonable methanol selectivity ($\beta_r = 3$–4) [170], without test in DMFC.

The pattern shown in Figs. 6.22 and 6.23, with most of the Nafion/composite membranes located in the B2 octant has also been observed by Pivovar and coworkers [181] for sulfonated poly(aryl ether)-type membranes. In order to analyze in detail cross-influencing parameters, these authors have defined four zones in the B2 octant, as indicated in Fig. 6.25.

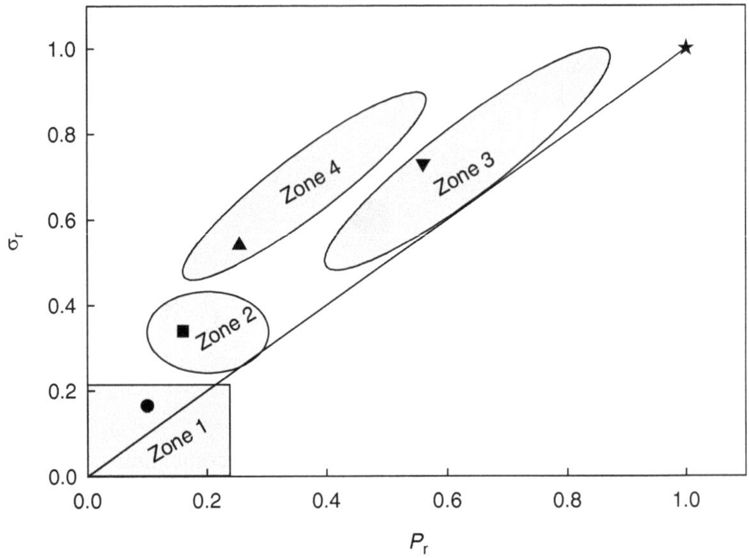

Fig. 6.25 Four zones in the relative selectivity plot as classified by Pivovar and coworkers: (●) SPEK-43; (■) Ph-SPEEKDK; (▲) m-SPAEEN-60; (▼) Ph-SPEEKK; (★) Nafion 117. (Adapted from Ref. [181])

Membranes in Zone 1 have $\sigma_r < 0.2$, and DMFC performance for these membranes are rarely reported because the reduction of membrane thickness necessary to give reasonable resistance losses is not overcome by its low methanol permeability.

Zone 2 corresponds to membranes with intermediate σ_r and P_r. Membranes in this zone should show inferior DMFC performance at low methanol feed concentration but better DMFC performance at high methanol feed concentration. Membranes in Zone 3 have relatively high σ_r and intermediate P_r, and should improve cell performance at low methanol feed concentration but should suffer high methanol crossover at high methanol concentration, like the Nafion membrane. Finally, membranes in Zone 4 exhibit better selectivity and DMFC performance should be further improved.

The agreement between the predicted behaviour of PEM in DMFC, according with this classification within the octant B2, and the more general classification shown in Fig. 6.2 is the aim of the next Section.

Kim et al. [326] have analyzed σ_r and P_r in terms of a new parameter, the percent conductive volume (PCV), defined as the ratio of the volume of the aqueous phase per acid site to the estimated volume of the wet membrane per acid site. It was found that membrane selectivity is low ($0.74 < \beta < 1.01$) for Nafion over a wide range of PCV values (form 0.21 to 0.62), but selectivity values higher than one were found for several inorganic and organic composites with lower PCV. Nevertheless, these conclusions should be taken carefully because only a limited number of examples from the vast literature on Nafion-based membranes for DMFC were analyzed.

6.5.7 DMFC Performance of Nafion Composite Membranes

If the evaluation of proton conducting membranes in relation to its potential use in DMFC is difficult due to the need of comparing methanol selectivity under similar experimental conditions, the comparison of DMFC performance of different membranes is still more complex. There is not a unique way to evaluate the cell performance because of the large number of parameters affecting the DMFC behaviour. Some of them, like membrane thickness and methanol feed concentration are related to the properties of the membrane itself, while others, such as cell operating temperature, cell humidification, cell compression, etc., influence not only the membrane but also other MEA and cell components. Therefore, a reliable comparison of DMFC performance of ionomeric membranes should be done using the same anodic and cathodic catalyst type and loading, identical GDL, procedure of MEA preparation, temperature, and alcohol concentration and flow rate.

Open circuit voltage (OCV) has been used as an indication of cell performance, although it is mainly determined by the electrochemical reversibility of the catalyst and it does not account for ohmic loss in the membrane and alcohol crossover. Thus, is not surprising to find a poor correlation between OCV of DMFC and membrane relative selectivity [4]. Other criteria used to compare DMFC performance is to compare current density at a fixed voltage [17], which is only practical if comparison is performed at the same temperature and feed methanol concentration.

Kim and Pivovar [4] suggested a comparison in terms of cell performance in terms of overall efficiency, defined as:

$$\varepsilon = \left(\frac{V}{V_t}\right)\left(\frac{i}{i + i_{cross}}\right) \tag{6.12}$$

where the first term on the right side is the voltage efficiency, that is the ratio between the cell voltage, V, and the maximum theoretical voltage, V_t, and the second term is the fuel efficiency, defined in terms of the cell current density at a given potential, i, and the crossover current density, i_{cross}, which accounts for the methanol oxidized directly in the cathode, not contribute to the cell power. Note that ε is the ratio between the power density of the cell and the theoretical power density if the voltage and current were the thermodynamic maximum and include the methanol loss by crossover, respectively.

The analysis of overall efficiency for Nafion 112, 115 and 117 membranes in DMFC at 80 °C [4] shows (see Fig. 6.26) that Nafion thicker membranes (Nafion 117) exhibits better performance at low power densities, while thinner membranes (Nafion 112) give higher performance at high power density as a consequence of a balance between methanol crossover and ohmic losses, in agreement with experimental and simulation results [327, 328]. The extension of this analysis to Nafion composites is not simple because methanol crossover current was not determined in most of the composite membranes investigated.

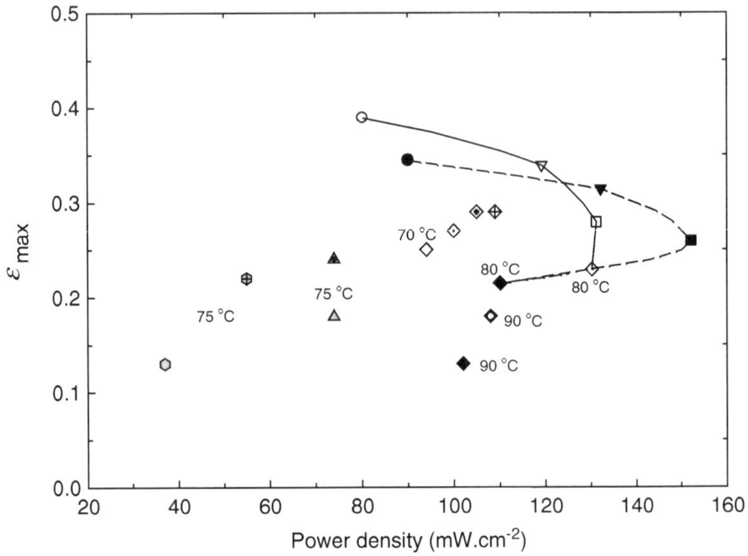

Fig. 6.26 Maximum efficiency of Nafion membranes. Open symbols: Nafion 117 at 80 °C (Ref. [4]); closed symbols: Nafion 112 at 80 °C (Ref. [4]); gray symbols: Nafion 115 at 75 °C (Ref. [78]). Methanol feed concentration: (○) = 0.3 M; (▽) 0.5 M; (□) 1 M; (◇) 2 M; (△) 5 M; (⬡) 10 M. Nafion composites are represented by crossed symbols

Figure 6.26 also shows the results of maximum overall efficiency obtained with reported results for Nafion 117/SiO_2 [41], Nafion 117 and 115/zirconium phosphate [41, 78], Nafion 117/titanium phosphate [41], and Nafion 112/PVA [111] composites at different temperatures between 75 °C and 90 °C. We have chosen these membranes because the measured power densities exceed, or are close, to 100 mW.cm^{-2}, which is a reasonable value for DMFC at temperatures between 70 °C and 90 °C.

The results indicate that the corresponding Nafion composite membranes exhibit higher maximum efficiencies and higher power densities than the corresponding pure Nafion membranes for a given thickness, temperature and methanol concentration. None of these composites were prepared by recast, but inorganic compounds are filler-impregnated Nafion membranes [41, 78], while PVA was deposited onto the Nafion 112 surface [111]. It is difficult to obtain further conclusions from Fig. 6.26 in view of differences in temperature, thickness and methanol concentration, but is clear that maximum efficiencies are not correlated to maximum power densities.

In order to extend the fuel cell performance analysis to a wider range of Nafion-based membranes the maximum power density (MPD) and the relative maximum power density (RMPD), defined as the ratio between the maximum power density of the Nafion composite and the maximum power density for the cell using pure Nafion pure will be used. Although the literature data for MPD is more abundant than for methanol crossover current, only works reporting a reasonable power density over the corresponding temperature range have been included in Table 6.3,

Table 6.3 Methanol membrane selectivity and maximum power density under different DMFC operation conditions of Nafion-based membranes

Filler (composite type)	Anode/load ($mg.cm^{-2}$)	CH_3OH flow-rate ($ml.min^{-1}$)	*c* (M)	MEA area (cm^2)	Membrane/ thick (μm)	Cathode/fuel *p* (bar)	T (°C)	β_r	PD mW. cm^{-2}	PD_r	Ref.
Pd (I)	PtRu/5	1	2	2	Nafion 117	$Pt/O_2/1$	30	4.9	65	1.1	[24]
			10						79	2.3	
PtRu (I)	PtRu/3–4	1 bar	2	7.6	Nafion 115	$Pt/O_2/1.6$	30		64	1.3	[29]
							45		105	1.5	
SiO_2(R)	PtRu (2:1)/2	4.5 bar	2	NI	80	$Pt/O_2/5.5$	145		240		[30]
SiO_2(SG)	PtRu (1:1)/2.2	1–3 bar	2	NI	NI	$Pt/O_2/4$	145		340		[31]
SiO_2(SG)	PtRu/5–10	1 bar	2	NI	Nafion 115	$Pt/O_2/1.6$	60–125	0.98–1.15	150–325		[32]
SiO_2(R)	PtRu/3	1	1–5	5	NI	$Pt/O_2/1$	90	1.0	115–165	1.11–1.15	[39]
SiO_2FSP (R)			1–5					1.5	124–165	1.11–1.42	
SiO_2-TBS (R)			1–5					1.9	137–230	1.37–2.13	
SiO_2(SG)	PtRu (1:1)/2	2	2	4	170	$Pt/O_2/1$	70		140	1.27	[41]
SiO_2(R)	PtRu (1:1)/0.8	20	2	8	NI	$Pt/O_2/1$	100		105	1.7	[42]
SiO_2-SO_3H (R)	PtRu (1:1)/2	2	2	4	170	$Pt/O_2/1$	80	2.2	68	1.5	[44]
SiO_2-MSU-F-SO_3H (R)								12.2	127	2.8	
SiO_2-MCM-41-SO_3H (R)								5.0	100	2.2	
Organic silica (SG)	PtRu (1:1)/2	2	NI	NI	Nafion 117	Pt/O_2	40	1.7–3.8	217	1.06	[46]
							80		204	1.00	
Organic silica (R)	PtRu/2	1	1–10	1	50	$Pt/O_2/1$	75		120–60	0.92–1.4	[49]
Zeolite (chabazite) (R)	PtRu (1:1)/2	2.5 (1–3.5 bar)	2	5	70	$Pt/O_2/3.5$	140		350	1.27	[52]
Zeolite (clinoptilolite) (R)							140		360	1.31	
Zeolite (mordenite) (R)							140		390	1.42	
							90–130		190–325		

(continued)

Table 6.3 (continued)

Filler (composite type)	Anode/load ($mg.cm^{-2}$)	CH_3OH flow-rate ($ml.min^{-1}$)	c (M)	MEA area (cm^2)	Membrane/ thick (μm)	Cathode/fuel p (bar)	T (°C)	β_r	PD $mW.cm^{-2}$	PD_r	Ref.
Zeolite (chabazite) (R)	PtRu (1:1)/2	2.5 (1–3.5 bar)	2	5	70	$Pt/O_2/3.5$	90–140		130–340	0.72–1.26	[53]
Zeolite (clinoptilolite) (R)			2				90–140		155–350	0.80–1.30	
Zeolite (sulfonated) (ISC)	PtRu (1:1)/2	1	1–5	5	Nafion 115	$Pt/O_2/2$	70	1.63	98–120	1.2–2.0	[54]
Zeolite (H-ZSM-5) (R)	PtRu/5	20	1	6.25	50	$Pt/O_2/2$	80	1.3–2.3	305	1.1	[57]
Montmorillonite (R)	PtRu/3	1 bar	2	7.6	130–140	$Pt/O_2/1.6$	125	1.2	181	1.1	[61]
Aluminosilicate-MSU (R)	PtRu (1:1)/2	2	2	4	170	$Pt/O_2/1$	80	2.5	156	3.25	[73]
Aluminosilicate-HMS (R)								7.5	170	3.54	
Aluminosilicate-MCM (R)								22	246	5.12	
Zirconium phosphate (SG)	PtRu (1:1)/2	2	2	4	170	$Pt/O_2/1$	70		128	1.15	[41]
Zirconium phosphate (I)	PtRu (1:1)/2.3	2.8	2	5	Nafion 115	$Pt/O_2/4$	90 145		200 375	1.7 1.3	[75]
Titanium phosphate (SG)	PtRu (1:1)/2	2	2	4	170	$Pt/O_2/1$	70		145	1.29	[41]
Titanosilicate (R)	Pt/0.5	2	2	9	NI	$Pt/O_2/1$	60		140	1.04	[82]

Titanosulfonate (R)	PtRu/8	NI	2	10	120	Pt/Air	40	0.62–1.05	73	2.0–2.3	[84]
TiO_2(sulfated) (R)	PtRu (1:1)/2	1	1–5	4	NI	$Pt/O_2/2$	75	0.41–1.39	125–118	1.04–1.2	[86]
Fe_2O_3 (sulfated) (R)	PtRu (1:1)/2.2	1	1–5	4	NI	$Pt/O_2/2$	90	0.98–2.9	175–132	1.15–1.23	[88]
Fe_2O_3_SO_3H (R)	PtRu/4	1.4 bar	1	NI	200	$Pt/O_2/1.4$	80	3.3	105	2.6	[89]
$Sr(OH)_2$(I)	PtRu/4	2	1.5	10	Nafion 115	Pt/Air	30		67	1.15	[93]
							60		123	1.1	
$SiO_2/H_4SiW_{12}O_{40}$(SG)	PtRu (1:1)/2.2	1–3 bar	2	NI	NI	$Pt/O_2/4$	145		310		[31]
$SiO_2/H_3PW_{12}0_{40}$ (SG)	PtRu (1:1)/2.2	1–3 bar	2	NI	NI	$Pt/O_2/4$	145		420		[31]
Ppy (ISP)	PtRu (1:1)/2	3	3	4	Nafion 115	$Pt/O_2/1$	30	2.6	30	1.7	[104]
			6				70		100	1.5	
PBI (R)	PtRu (1:1)/3	2	1–5	5	50–100	Pt/Air/1	60	*2.7*	100–68	1.0–1.5	[106]
PVA (I)	PtRu (1:1)/5	5	1–3	5	18–97	$Pt/O_2/1$	45	1.08	67–92	0.72–0.84	[114]
			1–3				95		147–211	073–1.0	
PVA (I)	PtRu (1:1)/5	5	2	5	47	$Pt/O_2/1$	70		150	0.94	[115]
sPF (ISP)	PtRu (1:1)/2	1	1–5	4	Nafion 115	$Pt/O_2/2$	75	1.5	107–114	1.37–1.46	[116]
PE (I)	PtRu/5	20	1–6	6.25	Nafion 117	$Pt/O_2/2$	80	1.9	280–145	1.24–1.11	[129]
PVI/(ISP)	PtRu/3	5	2	NI	Nafion 112	$Pt/O_2/1$	60	1.13	190	1.12	[138]
PTFE *(I)*	PtRu/4	5	2	NI	21	Pt/O_2	70–90		80–110	1.9–2.0	[141]
PFEP (ME)	PtRu (1:1)/3	2	1–10	5	50–100	Pt/Air/1	60	20	*96–59*	1.0–6.5	[143]
AMPS (ISP)	NI/2	1	2	2	Nafion 117	Pt/O_2	30	2.2	209	1.22	[152]
PVA/PI/TSPS (I)	PtRu/4	2.5	1.5	5	Nafion 115	Pt/Air/1	55	1.52–2.15	72–100	1.06–1.47	[154]
PVI/Pd (I)	PtRu/3	0.6	2	NI	Nafion 115	Pt/O_2	30–80	0.83–1.70	48–170	1.92–1.22	[158]
			5				30–80		74–142	2.65–1.23	

which summarizes β_r, MPD, and RMPD values for several Nafion composites, as well as the DMFC conditions used in the measurement of the power density curves.

The composites included in Table 6.3 were prepared by different methods, indicated in the first column by: impregnation (I), recast (R), sol–gel reaction (SG), in-situ crystallization (ISC), melt and extrusion (ME), and in-situ polymerization (ISP). A half of the membranes in Table 6.3 were prepared by the recast procedure, and the selectivity and maximum power density was calculated with reference to the Nafion recast membrane. For the rest of the membranes the Nafion matrix (112, 115 or 117) is indicated.

DMFC operated at temperatures above 100 °C exhibit the highest power densities, from 195 to 320 mW.cm^{-2} at 110 °C up to 240–420 mW.cm^{-2} at 145 °C, and they correspond to inorganic fillers such as SiO_2 and binary SiO_2/hetropolyacids [30–32], zeolites [52, 53], montmorillonites [61], and zirconium phosphate [75], which prevent Nafion dehydration upon . All these membranes, whose methanol selectivity were not determined in most of the cases, exhibit moderated but systematic DMFC performance improvement (1.1 < RMPD < 1.4) compared to pristine Nafion in both, passive and active cells.

Several Nafion composite membranes showing $\beta_r \geq 3$, such as Nafion/Pd [24], Nafion/PFEP [143], Nafion/Fe_2O_3-sulphated [88], are reported to have RMPD close to unity, while Nafion/PVI/Pd [158], and Nafion/titanosulfonate [88] with $\beta_r \leq 1$, exhibit significantly high RMPD. These examples clearly indicate that the correlation between membrane selectivity and DMFC performance is quite weak, as already quoted by Ahmad et al. [9], and other factors beyond the membrane characteristics should be taken into account. Figure 6.27 shows the lack of correlation, although a few system like Nafion/aluminosilicate [73], Nafion/sulfonated meso-structurated silica [44], and Nafion/sulfonated Fe_2O_3 [89], seem to exhibit a reasonable β_r *vs.* RMPD correlation.

Over the intermediate temperature range 70–95 °C it is remarkable the DMFC performance of several Nafion and Nafion composites membranes displaying rather high MPD, as it is the case of Nafion/Zeolite H-ZSM-5 (305 mW.cm^{-2} at 80 °C), [57], Nafion/organic silica (204 mW.cm^{-2} at 80 °C) [46], Nafion/aluminosilicate (156–246 mW.cm^{-2} at 80 °C) [73], Nafion/SiO_2 (140 mW.cm^{-2} at 70 °C, 137–230 mW.cm^{-2} at 90 °C) [39, 41], Nafion/Fe_2O_{3}_sulphated (132–175 mW.cm^{-2} at 90 °C) [88], Nafion/zirconium phosphate (145 mW.cm^{-2} at 70 °C) [41], Nafion/PVA (150 mW.cm^{-2} at 70 °C).

Close to room temperature the MPD reported for Nafion/AMPS (209 mW.cm^{-2} at 30 °C) [152] is outstanding, although the authors did not indicated the catalysts used in the preparation of the MEA. Also remarkable is the MPD =217 mW.cm^{-2} obtained at 40 °C using Nafion/organic silica, without indication of the methanol concentration used to feed the DMFC.

Lin et al. [8] suggested that increasing the thickness of Nafion membrane could decrease methanol crossover and improve DMFC performance. They studied several membranes prepared by stacking between 2 and 9 Nafion 112 films

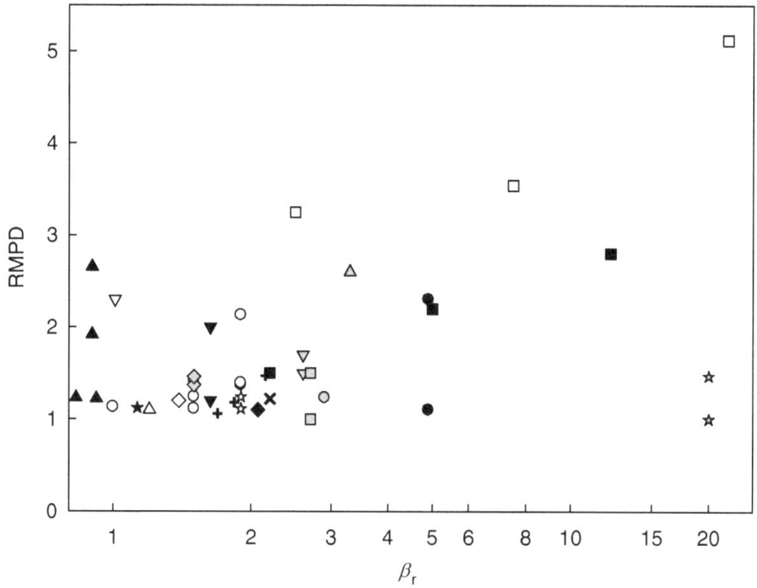

Fig. 6.27 RMPD versus membrane selectivity for Nafion-based membranes: (●) Pd [24]; (○) SiO_2 [39]; (■) sulfonated SiO_2 [44]; (▼) sulfonated zeolite [54]; (◆) zeolite [57]; (△) montmorillonite [61]; (□) aluminosilicate [73]; (▽) titanosulfonate [84]; (◊) sulfated TiO_2 [86]; (○) sulfated F_2O_3 [88]; (△) sulfonated F_2O_3 [89]; (▽) Ppy [104]; (□) PBI [106]; (◇) sPF [116]; (☆) PE [129]; (★) PVI [138]; (☆) PFEP [143]; (×) AMPS [152]; (+) PVA/PI/TSPS [154]; (▲) PVI/Pd [158]

(60 μm in thickness), obtaining membranes with thickness up to 540 μm. The results at 60 °C, with 1 M methanol show an increase in the MPD from 70 mW.cm^{-2} for the 60 μm single Nafion 112 membrane up to a maximum of 115 mW.cm^{-2} for the 240 μm (4 layers membrane). As can be observed from the data in Table 6.3, several Nafion composites, like Nafion/PBI, Nafion/PVI, and Nafion/PFEP exhibit similar or even higher MPD than thick Nafion membranes under similar operation conditions and using the same anodic catalyst. Particularly interesting is the performance of the PBI-doped Nafion [8, 106], obtained with only 3 wt% of PBI and feeding the cathode with air at ambient pressure.

It can be concluded that the MPD reported for DMFC with pristine or composite Nafion membranes at temperatures above 100 °C are similar or lower than those reported by Ren et al. [327] in 1996, using Nafion 112 (250 mW.cm^{-2} at 110 °C and 380 mW.cm^{-2} at 130 °C) meaning that, in spite of the efforts to improve the performance of Nafion-based DMFC, little progress has been achieved. The situation at room and intermediate temperatures seems to be promissory, with several Nafion composites showing very good performances in DMFC.

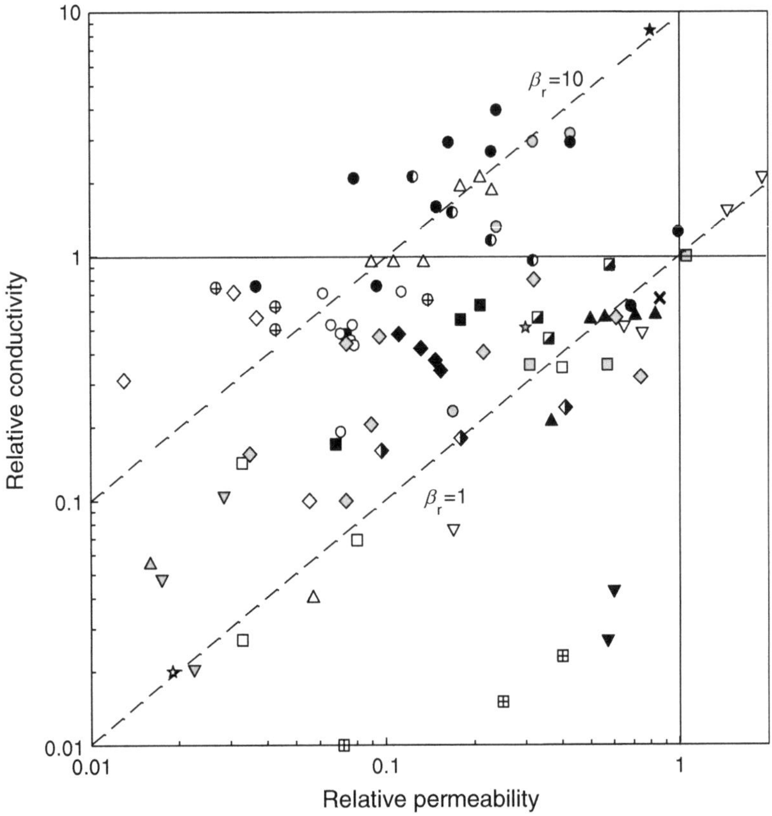

Fig. 6.28 Selectivity plot for PVA(☆) [329] and PVA-based membranes: SSA-GA (●) [330]; PSSA-MA (△) [331, 332], SSA/PSSA-MA (△) [333]; sulfonated phenolic resin (sPh) (▽) [334]; PI/3,6-pyrenetrisulfonic acid (TSGEPS) (■) [335]; SSA/PVP (◊) [336]; bis (4-γ-aminopropyldiethoxysilylphenyl)sulfone (APDSPS) (◆) [337]; Organophosphorous acid/ chitosan (○) [329]; aminopropyl triethoxysilane (APTES) (▲) [338]; sulfonated PVA (◇) [339], methylpropane sulfonic acid (MPSA) (□) [340, 341], sPOSS (◐) [342]; DSDSBA (◎) [343]; SSA/SiO_2 (▲) [344]; PAA/SiO_2 (▽) [345]; N-p-carboxy benzyl chitosan (CBC)/SiO_2 (★) [346]; benzene-silica (▣) [347]; Si-sPS/A)/PWA (⊕) [348]; sulphated β-CD (◪) [349]; Montmorillonite (★) [350, 351]; Organoclay (◆) [352]; PVA-co-Poly(vinyl acetate-co-itaconic acid/PMA (▼) [353]; poly(ether sulfone/PWA (⊞) [354]; Zirconium phosphate/Cs-salt SWA (✖) [355]

6.6 Polyvinyl Alcohol Based Membranes

As mentioned in Sect. 6.3.6., PVA-based membranes were the focus of abundant studies during the last years, and the review by Maiti et al. [180] summarized the conductivity and methanol permeability of most of the PVA-based membranes, whose preparation procedures are described in detail in that work. Figure 6.28 shows the relative selectivity plot for PVA-based membranes. Pure PVA have a low relative methanol permeability but also low proton conductivity resulting in $\beta_r \approx 1$, while a significant number of PVA-based membranes are located in the C2 octant, having selectivities lower than Nafion. However, several PVA-based membranes belong to the A-quadrant with proton conductivities and methanol barrier higher than Nafion, or are located in the B2 octant with high relative selectivity.

The highest selectivities ($7 < \beta_r < 26$) are reported for PVA membranes crosslinked with sulfo-succinic acid (SSA) [330] or poly(styrene sulfonic acid-co-maleic acid) (PSSA-MA) [332, 333], whose structures are shown in Fig. 6.6, and for a PVA/zeolite (mordenite) composite membrane, having $\beta_r \approx 120$ (out of scale in Fig. 6.28) [356]. PVA membranes prepared by crosslinking with sPOSS [342], 4-formyl-1,3-benzenedisulfonic acid disodium salt (DSDSBA) [343], and SSA and PVP [336] also exhibit β_r as high as 16.8, 9.1, and 24, respectively. High selectivities were also found for organic phosphorous acid-doped PVA/chitosan blends [329] ($2.7 < \beta_r < 11.3$), PVA blends with silicon-containing sulfonated polystyrene/acrylate (Si-sPS/A) and PWA [348] ($4.7 < \beta_r < 27$), and PVA/montmorillonite composites crosslinked with glutaraldehyde [350, 351] ($10.5 < \beta_r < 17.9$).

Unfortunately, most of the DMFC incorporating these PVA-based membranes have delivered poor power densities, ranging from 1 to 24 $mW.cm^{-2}$ at temperatures between 25 °C and 80 °C, even those exhibiting the highest β_r values, such as PVA-SSA/PVP [336], PVA/montmorillonite [350], and PVA/mordenite [356]. The values of MPD and RMPD for DMFC with PVA-based membranes are summarized in Table 6.4, where one can observe the poor correlation between β_r and RMPD. The exceptions are PVA/SSA/PSSA-MA membranes with MPD that increases from 22 $mW.cm^{-2}$ at 30 °C up to 100 $mW.cm^{-2}$ at 80 °C [333], and PVA/DSDSBA with MPD between 85 and 118 $mW.cm^{-2}$ at 60 °C [343], both having also high β_r values.

Some PVA-based membranes showing modest selectivities, like sulfonated PVA [339] and PVA/PSSA/mordenite [361] exhibit unexpected high MPD (81 and 74 $mW.cm^{-2}$, respectively) at 70 °C. PVA/poly(ether sulfone)/PWA membranes [354], with very low selectivity, show MPD in the range 95–117 $mW.cm^{-2}$ at 80 °C. In summary, the performance of DMFC with PVA-based membranes could not yet overpass the results obtained with Nafion composite membranes, although some promising results have been obtained during the last years. Indeed, the results shown in Table 6.3 for Nafion/PVA blends [114] and PVA-coated Nafion membranes are superior to those obtained with PVA-based membranes.

Few results are reported for PVA-based membranes in DEFC, and most of them correspond to alkaline DEFC, which will be discussed in Sect. 6.9. The unique study of ethanol permeability in a PVA-based membrane corresponds to PVA crosslinked with (diethylenetriamine) pentaacetic acid (DTPA) and doped with phosphotungstic acid [363]. Ethanol permeation through these membranes a factor 100 lower than in Nafion, but the proton conductivity of the membranes is lower than 10 $mS.cm^{-1}$ and DEFC performance test has not been reported.

Table 6.4 Methanol membrane selectivity and maximum power density under different DMFC operation conditions of PVA-based membranes

Filler (composite type)	Anode/load ($mg.cm^{-2}$)	CH_3OH flow-rate ($ml.min^{-1}$)	c (M)	MEA area (cm^2)	Membrane thick (μm)	Cathode/ fuel p bar)	T (°C)	β_r	PD mW. cm^{-2}	PD_r	Ref.
SSA/PSSA-MA	PtRu/3	1	2	NI	100–120	$Pt/O_2/1$	30	8.0–10.6	22		[333]
							40		34		
							50		45		
							60		67		
							70		77		
							80		100	1.03	
SSA/PVP	PtRu (1:1)/3	3	2	NI	100–120	$Pt/O_2/1$	RT	1.8–24	5	1.0	[336]
Sulfonated PVA	PtRu/4	1	3		40–110	$Pt/O_2/1$	70	1.5	81	1.33	[339]
MPSA	PtRu/2.2	Passive	2	5	100	Pt/Air/1	25	4.3	2.4	0.30	[341]
APDSPS	Pt/3	1	3	5		$Pt/O_2/1$	60	5.5–9.1	85–118		[343]
CBC/SiO_2	PtRu (2:1)/1		20 %	25		Pt/Air/1	70	1.7	7.2		[346]
			30 %						5.1		
			40 %						3.6		
Sulphated β-CD	PtRu (1:1)/2	3	2	1		Pt/Air/1	RT	1.27–1.70	14.3–18.5		[349]
Montmorillonite	PtRu (1:1)/4		2	1		Pt/	25	10.5–17.9	6.8		[350]
poly(ethersulfone/PWA)	PtRu (1:1)/4	3	1	5	100–120	$Pt/O_2/2$	80	0.14	95	2.15	[354]
			5					0.06	117	1.04	
Zirconium phosphate/ Cs-salt SWA	PtRu (1:1)/6	Passive	4	6		Pt/Air/1	100	0.78	6	1.1	[355]

Mordenite	PtRu (1:1)/2		1	9	65–80	Pt/Air/1	60	120	1.2	0.75	[356]
PVA-GA	NI/3.2		2	5	300	NI/Air	RT		2	2	[357]
PSSA/TiO_2	PtRu (1:1)/4		1		100–200	Pt/Air/1	25		24	2.8	[358]
			2						12		
			4						10		
PS SA/montmorillonite	PtRu/4		1		100–200	Pt/Air/1	25		14	2.5	[359]
			2						20		
			4						13		
PAMPS/GPTMS	PtRu (1:1)/4		1	1	120–140	Pt/Air/1	25		20	2.8	[360]
			2						23.6		
			4						19		
PSSA/mordenite	PtRu (1:1)/2	2	2	4	160	Pt/Air/1	70	1.8–2.0	35–40	0.44–0.50	[361]
								2.0–4.2	47–74	0.59–0.93	
sPEEK	Pt/0.3	1	2	4		Pt/Air/1	80		21.1		[362]
			8						23.3		

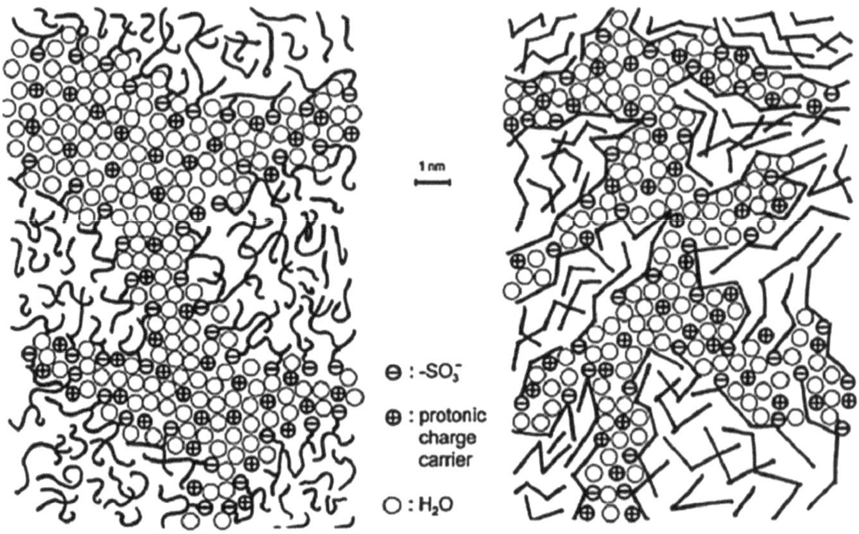

Fig. 6.29 Schematic microstructures of Nafion (*left*) and sPPEK (*right*) derived from SAXS experiments (Reproduced from Ref. [1] with permission)

6.7 Poly(arylene ether) Based Membranes

The development of polyaryls, in particular polyetherketones (PEEK), as a substitute of perfluorinated polymers was mainly based on cost and stability considerations [1]. Sulfonated polyetherketones has been found to be durable under fuel cell operation conditions over several thousand hours [185]. Victrex Company is the main producer of PEEK polymer and its sulfonation can be performed directly on the polymer backbone or by polymerization of sulfonated monomers [7]. Hoechst-Aventis and Fumatech commercialize sulfonated PEEK (sPEEK) and sulfonated poly(phtalazinone ether ketone) (sPPEK) membranes for fuel cell applications [3].

The main microstructural differences between Nafion and sPEEK, inferred from SAXS experiments, are displayed in Fig. 6.29 [1]. The water filled channels in sPEEK are narrower compared with those in Nafion, are less separated, highly branched, and with more dead-end channels.

Proton transport and electrosmotic drag in sPEEK as compared with Nafion has been analyzed by Kreuer [1], concluding that differences in the microstructure and acidity leads to disadvantageous swelling behavior and a stronger decrease of proton transport coefficients with decreasing water content in sPEEK. However, the water electro-osmotic drag and water permeation is reduced in sPEEK compared to Nafion, which an essential advantage for DMFC applications.

The sulfonation degree of PEEK membranes reported in the literature range from 10 % to 98 %, and the corresponding ion-exchange capacity (IEC) varies between 1 to more than 2.5 meq.g^{-1} [185], and the proton conductivity is strongly

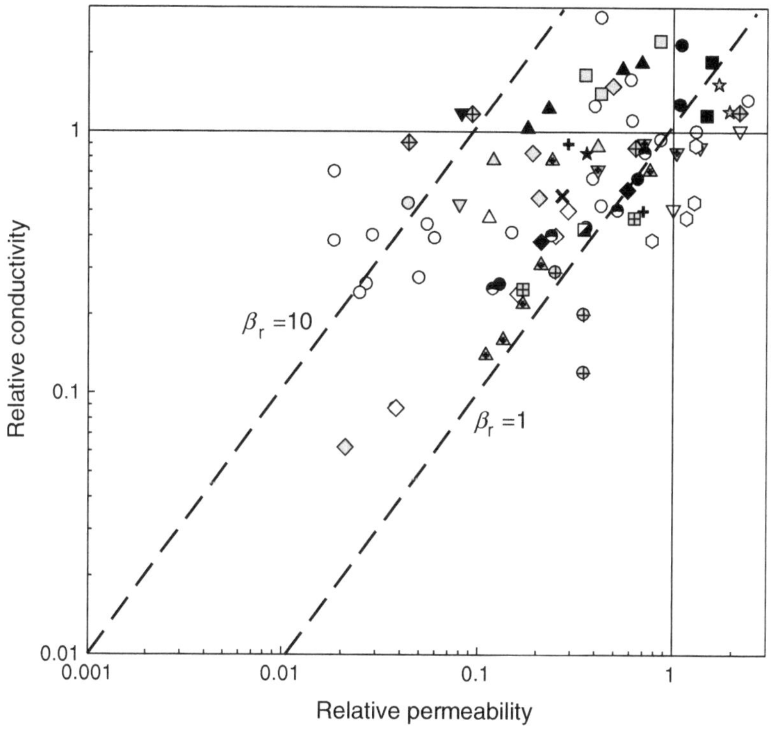

Fig. 6.30 Selectivity plot for polyaril-based membranes and their composites: sPEEK (○) [364–375]; sPEEKK (◊) [376]; sPEK (✖) [377]; sPEEK-C (△) [378]; sPEEK with pendent amino groups (Am-sPEEK) crosslinked with benzimidazole (▽) [379]; Am-sPEEK crosslinked with epoxy resin (▽) [380]; sPEEK/Nafion (+) [133, 134]; charge surface modified sPEEK (○) [368]; sPEEK/chitosan (▲) [369]; sPEEK/SiO_2/Al_2O_3/SWA (□) [370]; sPEEK/sulfonated SiO_2 (▽) [371]; sPEEK/Benzoimidazole (BI) (◇) [372, 375]; sPEEK/sulfonated CD (★) [373]; sPEEK/sulfonated MMT (⊕) [374]; sPEEK WPA/ZrO_2 (◈]; Am-sPEEK/BI (⊞) [379]; Am-sPEEK/epoxy resin (▼) [380]; sPEEK/PVDF (▲)[381, 382]; sPAEK (●) [383–385]; sPAEK with carboxylic acid groups (Carb-sPAEK) (■) [386, 387]; sPAEK/epoxy resin (▲) [383]; sPAEK/3-glycidoxy propyltrimethoxysilane (GPTMS) (★) [384]; Carb-sPAEK/PVA (▼) [386]; sPAEK/epoxy resin (◆) [387]; sPAEK/Nafion (●) [136, 385]; Carb-sPAEK/Am-PEEK (◪) [388], sPEEK/SSiGO (○) [389]

dependent of the IEC. Sulfonated PEEK membranes were prepared by casting from different solvents, such as DMA, NMP, DMSO and DMF.

The reported proton conductivity and methanol permeability of sPEEK- and sPAEK-based membranes [364–389] are summarized in Fig. 6.30 as a relative selectivity plot.

There are several sPEEK membranes with very good relative selectivity and several sPEEK composites with silica, zirconia and their mixtures with heteropolyacids [370, 371, 377] which exhibit remarkable selectivities with high proton conductivities. Composites of sPAEK with epoxy resin [383] and crosslinked with PVA [386] also show excellent selectivities and high conductivities. Unfortunately, the performances of these promising membranes in DMFC have not been reported yet.

Table 6.5 Maximum power densities of DMFC using sPEEK and sPAEK membranes and their composites

Membrane	MPD ($mW.cm^{-2}$)	MRPD	*T* (°C)	*c* (M)	Ref.
sPEEK	30	0.58	65	1	[389]
sPEEK	37	0.71	65	1	[397]
sPEEK	17.5	0.31	60	1	[390]
sPEEK	16	0.85	60	2	[391]
Sulfinated PEEK	144		110	1	[392]
MS-sPEEK	71–86	1.15–1.39	60	2	[393]
	73–116	0.85–1.35	80		
sPEEK	23		80	2	[394]
	17			8	
sPEEK/sPVA	21–23				
SPEEK	12	0.39	65	1	[395]
	27	0.46	80		
sPEEK/PSfABI	39	1.26	65		
	60	1.02	80		
	80		80	2	
sPEEK	114		70	2	[396]
sPEEK/PPy	68–86				
sPEEK	15.5	1.5	60	1.5	[374]
sPEEK/sMMT	18.5–21	1.76–2.0			
sPEEK/SiO_2/ZrP	18.7	1.0	60	2	[391]
sPEEK/sCD	15–27.5			2	[373]
	19–30			8	
sPEEK/PVDF	68	1.02	30	2	[380]
sPh-PEEK	126	0.98	80	1	[397]
sPh-PEEK/PSfBT	176	1.37			
sPEEK/SSiGO	72	1.38	65	1	[389]
	80	1.33		2	
	64	1.44		5	
sPEEK/ACrown	84	1.64	65	1	[398]
	96		80	1	
	119			2	
sPEEK/SnO_2	78	3.9	100	2	[399]
sPAEK-HQ	57	1.30	65	0.5	[400]
	81	1.14	80		
	50	1.05	65	1	
	75	1.07	80		
sPAEK-DP	54	1.23	65	0.5	
	102	1.44	80		
	57	1.20	65	1	
	96	1.37	80		

DMFC performances of sPEEK and sPAEK based membranes are summarized in Table 6.5. We have only included studies with reasonable peak power density (higher than 20 $mW.cm^{-2}$). Remarkable performances have been obtained with sPEEK [395], sulfinated PEEK [391], main-chain type sPEEK (MC-sPEEK) [392], and sPEEK composites with PPy [395] and PVDF [380] at moderate and room temperature, respectively.

A acid–base blend membrane consisting of acidic sulfophenylated poly(ether ether ketone) (sPh-PEEK) and basic polysulfone tethered with 5-amino-benzotriazole (PSfBT) recently prepared by Manthiram and coworkers [396] shows an excellent performance (176 $mW.cm^{-2}$) in DMFC at 80 °C. Recently, this group has prepared composites of sPEEK with sulfonated organosilane graphene oxide (SSiGO) [389], and with amphoteric copolymers containing sulfonic acid and basic 1-aza-18-crown groups (ACrown) [398]. Both composites exhibit very good performance and the highest relative maximum power density as compared to Nafion.

sPEEK have also be characterized in relation to its ethanol permeability [401] and maximum peak power densities above 20 $mW.cm^{-2}$ have been reported with sPEEK in DEFC at 90 °C.[402].

6.8 Polybenzoimidazole Based Membranes

Polybenzoimidazole (PBI) polymers doped with phosphoric acid have emerged in 1995 as a promising PEM for use in DMFC [403] due to its relatively low cost, low methanol permeability. Its excellent chemical and thermal stability up to 200 °C [403] has also triggered several studies aiming to test the PBI-based membrane performance in high temperature DMFC [404].

In this section the properties of PBI and modified PBI will be reviewed in relation to their use in DAFC and high-temperature DAFC.

6.8.1 Acid Doped PBI Membranes

In Sect. 6.3.6.5, Fig. 6.10, the structure of poly[2,2′-(m-phenylene)-5,5′-bibenzimidazole] (m-PBI) was shown. PBI is a neutral polymer and the doping with acid, usually phosphoric acid, takes places through the imide sites of the polymer. Bouchet and Siebert [406] analyzed the protonation of PBI by different acids and concluded that the maximum degree of protonation is 1 for H_2SO_4 and 2 for HBr and H_3PO_4. The protonation of PBI with H_3PO_4 correspond to the number of protonable imide group in PBI and it was proposed that phosphate anions are linked to the PBI structure by hydrogen bonds, forming a network favourable for proton transfer.

The doping chemistry of H_3PO_4 in PBI was analyzed in detail by He et al. [407] by resorting to the Scatchard method. Two sites of H_3PO_4 coordination to PBI, L_1

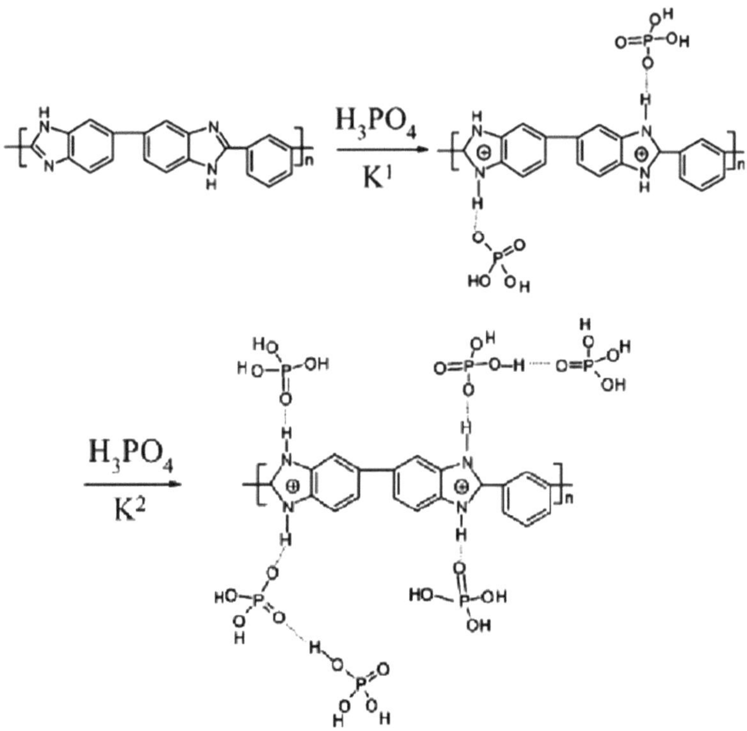

Fig. 6.31 Doping of PBI with H_3PO_4 (Reproduced from Ref. [407] with permission)

and L_2, has been considered and indicated by the coordination constants K^1 and K^2. The acid complexes dissociated according to the equilibria,

$$L_i - H_3PO_4 \leftrightharpoons L_i + H^+ + H_2PO_4^-$$

where i =1, 2, whose corresponding dissociation constants are $K^{a1}{}_{L1-H3PO4} = 5.4\ 10^{-4}$, and $K^{a1}{}_{L2-H3PO4} = 3.6\ 10^{-2}$, respectively. Thus, the dissociation constant on the L_1 sites is about 10 times smaller and the dissociation constant on the L_2 sites is 5 times higher than the first dissociation of aqueous H_3PO_4. As we will discuss later, this thermodynamics information is important to understand the proton conduction in this kind of membranes.

The degree of acid doping, λ_a, defined as the number of acid molecules per imidazole ring, could be higher than 2 as suggested by Fig. 6.31, and it increases with the concentration of the H_3PO_4 solution used for doping PBI. In fact λ_a values between 0.8 and 4.5 has been measured for PBI doped with 2 mM to 15 M H_3PO_4 [408]. PBI membranes prepared by the PPA sol–gel process (see Fig. 6.11), exhibit even larger degree of doping, in the range $\lambda_a \approx 10$–20, probably as a consequence of a more relaxed polymer structure.

The free acid content, λ_f, expressed as the fraction of leachable acid from the total acid in the membrane, is less than 10 % [409].

The water uptake from the vapor phase of PBI membranes depends not only on the water activity but also on the degree of doping. Usually the water uptake in PBI is expressed as λ_w, the number of water molecules per imidazole group, and it is lower than the corresponding λ_w (molecules of water per sulfonic group) for Nafion membranes [409], although when expressed as mass of water per gram of dry membrane the water uptake of PBI is much higher than that of Nafion.

Savinell and coworkers [410, 411] have measured the drag coefficient of water/methanol in PBI at 150 °C and 180 °C and found that it is essentially zero over a wide range of composition. The same result was reported for the drag coefficient of water in PBI at 80 °C and 150 °C by Li et al. [412]. The absence of electro-osmotic flow of water or water/methanol through PBI was interpreted as proton conduction mechanism dominated by Grotthuss proton hopping.

H_3PO_4 doped PBI membranes ($\lambda_a = 2.5$) 100 μm thick exhibit methanol crossover current densities less than 10 mA.cm^{-2} [403, 410], while the PBI/PVPA composite commercial membrane Celtec-V shows crossover current densities higher than 100 mA.cm^{-2} at 90 °C in 1 M methanol [201].

Water–methanol uptake from the vapor phase by PBI was determined in thin membranes (100 nm thick) using the quartz crystal microbalance method [280]. The results, expressed as moles of water plus methanol sorbed per imidazole ring, are shown in Fig. 6.31. It was assumed that the partition constant of methanol is $K_x = 1.29$, as determined by NMR analysis of a PBI membrane in equilibria with a methanol aqueous solution (20 wt%) [280].

The behavior observed in Fig. 6.32 is different from that observed Nafion in Fig. 6.17. In the case of PBI, and also for the modified ABPBI membrane, the pure water sorption almost doubles pure methanol sorption, that is, these membranes have a clear preference for water uptake over methanol, while no differences in the uptake of water and methanol was observed in Nafion. The water–methanol uptake from the liquid phase (methanol 20 wt%) of a thick PBI membrane (50–100 μm) shows the same behavior, that is water uptake ($\lambda_w = 3.48$) is much higher than methanol uptake ($\lambda_m = 0.63$), which is desirable behaviour for a membrane which intend to be a good barrier for methanol crossover.

Several authors have determined methanol permeability for PBI [12, 192, 197, 280, 413], methyl and ethyl-PBI [197], and fluorinated PBI [198, 414], for different acid doping degree. The reported permeability coefficients are summarized in Fig. 6.33, where the λ_a values are indicated.

Methanol permeability at room temperature spread over more than one order of magnitude, probably because the different acid doping of the membranes and methanol concentration used in the experiments. It is evident that the methanol permeability through undoped PBI membranes is lower than in doped ones, and PBI membranes with low acid doping degree, $\lambda_a = 0.9$, like those studied by Pu et al. [197, 413] exhibit lower methanol permeability than highly doped membranes. The data by Chuang et al. [198] at three different doping degrees

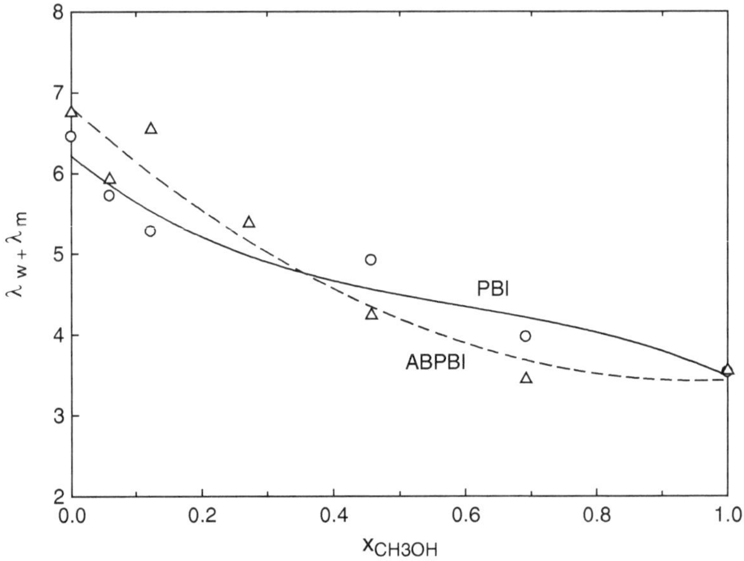

Fig. 6.32 Water–methanol uptake in PBI and ABPBI membranes at room temperature as a function of the vapor composition

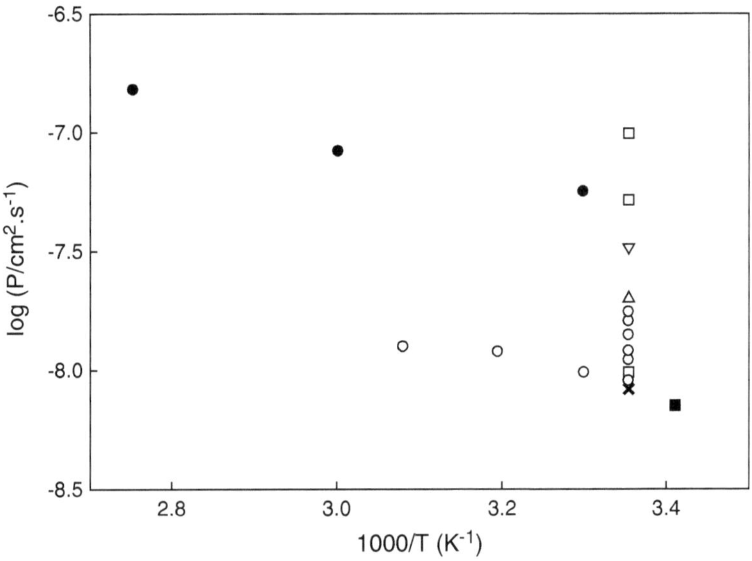

Fig. 6.33 Temperature dependence of the methanol permeability of PBI membranes: (✖) Ref. [12] ($\lambda_a = 0$); (●) Ref. [280], ($\lambda_a = 1.9$), (■) Ref. [280] ($\lambda_a = 0$); (○) Ref. [197, 413] ($\lambda_a = 0.9$); (□) Ref. [198] ($\lambda_a = 0, 0.6, 1.5$); (▽) Ref. [414] ($\lambda_a = 1.5$); (△) Ref. [192] ($\lambda_a = 2.25$)

show the same trend. On the other hand, the permeability of sulfonated PBI is two orders of magnitude higher than that of PBI [192].

The permeability coefficients of PBI membranes, even those with high doping degrees, indicate than they are more efficient methanol barrier than Nafion. For instance, $P_r = 0.01$ for undoped PBI, and $P_r = 0.040\text{–}0.036$ for PBI with $\lambda_a = 1.9$ in the range of temperature 30–90 °C. This is not only a result of the lower methanol uptake of methanol by PBI, but it is certainly due to a reduced diffusion coefficient of methanol in this polymer.

Some PBI composites membranes have also been studied in relation to methanol barrier properties. The relative permeability for PBI/Nafion composites ($0.055 < P_r < 0.70$) [106, 107] has been discussed in Sect. 6.5.6 (see Fig. 6.23), and it is clear that the blend membranes is a better methanol barrier than pure Nafion, but it is worst methanol barrier than pure PBI. Wycisk et al. [415] studied the properties of PBI/sPOP membranes and found that the relative permeability of pure sPOP membranes is reduced from $P_r = 0.56$ down to $P_r = 0.028$ by adding 12 wt% of PBI. A similar effect has been found for PBI/sPEEK composite membranes, where the permeability of pure sPEEK is reduced by a factor 50 by adding 20 wt% of PBI [416].

The proton conductivity of PBI and modified PBI membranes has been studied under different conditions of membrane preparation, acid doping degree, temperature, and water activity. The results for PBI membranes prepared by casting from dimethyl acetamide (DMA) and other solvents (NMP: N-Methyl pyrrolydone, DMSO: dimethyl sulfoxide, TFA, trifluoroacetic acid) are summarized in Table 6.6.

Dry PBI membranes ($a_w = 0$) have very low conductivities ($\sigma < 1$ mS.cm^{-1}) for doping levels $\lambda_a \leq 1.5$, and moderate conductivities ($\sigma \leq 23$ mS.cm^{-1}) for $\lambda_a \geq 3.0$ and temperatures close to 200 °C. A small increment in humidity, by exposing the membrane to the ambient [408] results in an important increase in conductivity, particularly at high doping levels [418]. Proton conductivities close to that found for Nafion membranes (around 100 mS.cm^{-1}) are found for PBI membrane with $\lambda_a > 3.0$ in high humidity conditions, but the solvent used to prepare the membrane by casting seems to have a great influence, probably due to the formation of different polymer microstructures. On the other hand, the temperatures at which PBI membranes reach such conductivity levels are above 150 °C.

Porosity in the microstructure of PBI has great influence on its properties. This has been demonstrated by a couple of works. Mecerreyes et al. [426] prepared a porous PBI membrane by leaching out a porogen, like phthalates or phosphates, from PBI/porogen films. This method allows controlling the porosity up to levels of 75 %, with pores smaller than 100 nm. Contrary to the behavior of standard membranes, the effect of temperature on the conductivity of porous PBI membranes is small, the conductivity increases from 30 mS.cm^{-1} at 24 °C up to 60 mS.cm^{-1} at 140 °C, for the material with the highest porogen content. Weber et al. [427] prepared mesoporous PBI using a hard template (silica nanoparticles 13 nm in diameter), which is removed from the membrane by treatment with NH_4HF_2. Crosslinked mesoporous PBI membranes were also prepared by incorporating a trifunctional carboxylic acid to the network. In this case the amount of H_3PO_4

Table 6.6 Conductivity of H_3PO_4-doped PBI membranes prepared by casting

Solvent	a_w	T (°C)	λ_a	σ (mS · cm^{-1})	Ref.
MSA	0.02–0.38	80	2.25	3–70	[192]
NMP	0	60–180	0.90	4.10^{-5}–0.06	[197]
DMA	0.05	130–150	1.69	3–4.5	[403]
	0.25	130–150		7.5–13	
	0.10	130–190	2.50	9.5–36	
	0.20	130–150		15–30	
DMA	0	30	0	1.10^{-9}	[406]
		30	1.5	$7 \cdot 10^{-3}$	
		60	1.5	0.19	
		110	1.5	0.26	
DMA	Ambient	110–190	1.55	0.03–0.8	[408]
			2.35	0.2–4	
			3.1	1–15	
	0.15		3.1	4–40	
DMA	0.80–0.85	30–165	2.25	4.5–46	[412]
		25–160	6.5	30–130	
		25–100	8.0	33–123	
DMSO	H_3PO_4 solution	25	0	$7 \cdot 10^{-4}$	[417]
			0.30–2.1	0.2–3	
DMA	Ambient	100–175	2.35	15–18	[418]
			3.35	18–22	
			7.25	44–78	
DMA	≈ 0	115–185	6.7	7–23	[419]
TFA/H_3PO_4	0	60–200	1.5	0.18–0.78	[420]
		60–200	3.15	3.5–4.7	
	0.10	70–200	1.5	0.3–13	
		70–200	3.15	7–77	
	0.20	70–150	1.5	0.4–8	
		60–160	3.15	8–58	
	0.30	70–140	1.5	0.4–6.5	
		50–160	3.15	8–68	
DMA	1.0–0	25–200	2.8	6–31	[421]
	0.80–0.05	80–200		21–68	
DMA	0.50–0.05	110–200	3.1	30–85	[422]
	0.50–0.05	110–200	3.3	45–95	
DMA	0.80	25–160	2.5	3–26	[423]
		25–160	5.5	22–90	
TFA	0	80–200	3.0	0.8–9.2	[424]
	0.05	80–170		32–100	
DMA	0.33	50–120	1.9	6–21	[280], [425]
	1.0	20–120		6–15	

in the membrane is given by the acid complexed to the imidazole groups plus the acid filling the pores, and it is close to 66 wt%. The conductivity of the mesoporoues membranes raise from 0.4 mS.cm^{-1} at 65 °C up to 11 mS.cm^{-1} at 180 °C, which are indeed low values for fuel cell applications.

More interesting results were obtained by Xiao et al. [200] using the PPA sol- gel process shown in Fig. 6.11. This procedure leads to very high acid doping levels ($\lambda_a \geq 10$–20), never reached by PBI membranes prepared with standard casting procedures. The conductivity of a membrane prepared with the PAA sol–gel method and $\lambda_a \approx 16$, raises from 15 mS.cm^{-1} at 20 °C up to 270 mS.cm^{-1} at 200 °C. The last value is comparable with those observed for the best proton conducting Nafion/inorganic composite membranes at temperatures above 100 °C.

Crosslinked PBI membranes have been proposed for improved mechanical strength and chemical stability. Li et al. [428] used p-xylene dibromine as cross linker agent and obtained 13 % crosslinked PBI with conductivities close to 100 mS.cm^{-1} at 180 °C and high acid doping ($\lambda_a = 7.75$). Xu et al. [429], synthesized crosslinked PBI by condensation of 1,3,5 benzenetricarboxylic acid (BTA) and 3,3′-diaminobenzidine (DAB). The membranes exhibited high acid uptake and a conductivity of 64 mS.cm^{-1} at 170 °C under dry conditions.

Other modified PBI membranes with reasonable conductivities include PBI poly(2,2′-(2,6-pyridin)-5,5′-bibenzimidazole [192], poly[2-(4′-oxyphenylene)-5-benzimidazole] (PPOBI) [430], poly[2,2′-(benzimidazole-*p*-phenylene)-5,5′-bibenzimidazole] (BI*p*PBI) [431], and hexafluoroisopropylidene- containing PBI (6 F-PBI) [198, 432, 433]. The permeability of methanol was studied only for the 6 F-PBI membrane prepared by Chuang et al. [433], which have a very small relative permeability (P_r =0.005), but a modest conductivity (2.6 mS.cm^{-1} at 90 °C and $a_w = 0.9$) over a wide range of doping levels ($\lambda_a = 1.5$–6.7). These authors reported a decrease of conductivity close to 160–180 °C, attributed to the dehydration of H_3PO_4, which would turns into $H_4P_2O_7$ [198], not observed in other PBI membranes studied up to 200 °C.

The selectivity plot for PBI membranes is shown in Fig. 6.34. Most of the membranes are in the B2 and C2 octants, with better methanol barrier properties than Nafion, but having much lower proton conductivities. The reported selectivities correspond to temperatures below 100 °C, but at higher temperatures the proton conductivities of PBI-based membranes are similar or even higher than Nafion conductivity, due to dehydration of the perfluorsulfonic membranes. Therefore, one should expect enhanced selectivities at high temperatures, where PBI-based membranes appears as a promising replace for Nafion.

In Table 6.7 are summarized the reported performance of DMFC with MEAs fabricated with of PBI and PBI composites membranes. The pioneer works by Savinell and coworkers [403, 405] demonstrated that, close to 200 °C, peak power of the order of 100 mW.cm^{-2}, could be obtained with DMFC fed with vapor methanol–water mixtures, using PBI membranes with moderate acid doping. More recently, Lobato et al. [434] reported maximum power densities above 220 mW.cm^{-2} at 200 °C, using PBI membranes with higher acid doping and pressurized oxygen in the cathode.

Several PBI blends have been investigated in relation to their DMFC performance. PBI layered Nafion and PBI/Nafion composite have been discussed in Sect. 6.5.7. PBI composites with sPOP [415], sPPO [435], and sPEEK [436] exhibit

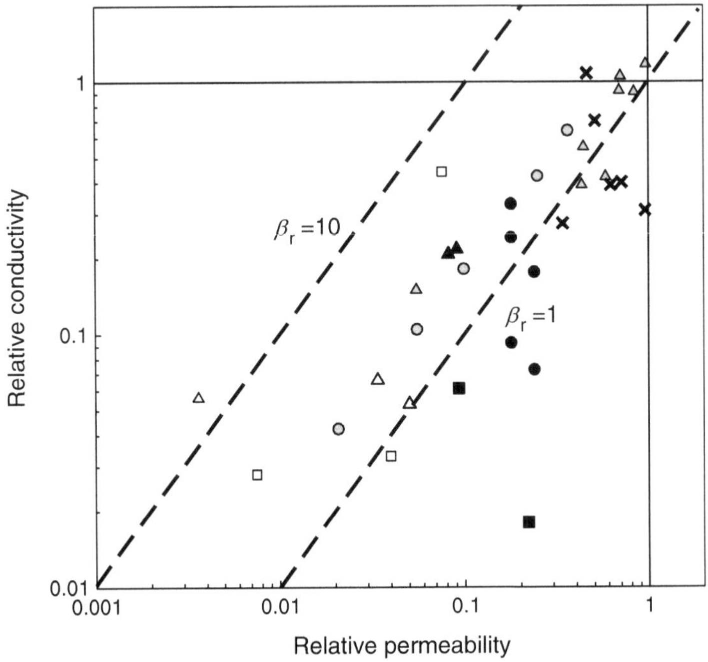

Fig. 6.34 Selectivity plot for PBI and PBI composite membranes: PBI (△) [12, 280]; 6 F-PBI (□) [198]; PBI/Nafion (▲) [106, 107]; PBI/sPOP (◎) [405]; Nafion layered PBI (✖) [105]; ABPBI (MSA cast) (▲) [280]; ABPBI (ethanol/NaOH cast) (●) [280]; commercial ABPBI (Fumatech) (■) [280]

reasonable performances at temperatures between 60 and 70 °C, while PBI/PVPA [201] shows rather good performance at 110 °C, in all cases with DMFC fed with liquid methanol aqueous mixtures.

Ternary composites of PBI with sPEK + polysulfone (PSU) [437], sPEEK + sulfonated PSU (sPSU) [438], and sPEEK + zirconium phosphate [439] were studied in DMFC. Excellent results were reported for DMFC, at 110 °C, using the first two composite membranes.

Table 6.7 also includes some results obtained for DEFC with PBI membranes, with maximum power densities in the range 55–80 mW.cm^{-2} at 200 °C, which certainly are promising results.

6.8.2 *Acid Doped ABPBI Membranes*

Poly(2,5-benzimidazole) (ABPBI) is the simplest benzimidazole polymer (see Fig. 6.10) with a higher affinity towards H_3PO_4 than PBI, and it can be polymerized from a single, inexpensive, commercial monomer (DABA) [442].

Table 6.7 Methanol membrane selectivity and maximum power density under different DMFC operation conditions of FBI-based membranes

Membrane	Anode/load (mg . cm^{-2})	CH_3OH flow-rate (ml . min^{-1})	c (M)*	MEA area (cm^2)	Membrane thick (μm)	Cathode/fuel p (bar)	T (°C)	β_r	MPD mW. cm^{-2}	RMPD	Ref.
PBI ($\lambda_a = 1.5$)	PtRu/4	Vapor	$x = 0.2$		100	$Pt/O_2/1$	200		105		[403]
PBI ($\lambda_a = 1.5$)	PtRu/4	Vapor	$x = 0.33$	1	100	$Pt/O_2/1$	190		102		[405]
						Pt/Air/1			65		
PBI ($\lambda_a = 3.3$)	PtRu (1:1)/1	Vapor	$w = 0.5–1$	4.65		$Pt/O_2/1$	125		33		[434]
							150		50		
							175		70–74		
							200		118–132		
			$w = 0.5–1$			$Pt/O_2/2$	175		126		
							200		222		
PBI/Naficn (layered)		0.6	2				60	2.3	21.7	1.04	[105]
			3.2						17.8	1.46	
PBI/Nafion	PtRu (1:1)/3	2	1	5	50–100	Pt/Air/1	60	2.7	100	1.0	[106]
			5						68	1.5	
PBI/sPOP	PtRu (1:1)/4	2	1	5	61–77	Pt/Air/1	60	1.74	89	0.93	[415]
								1.68	78	0.81	
PBI/sPPO	PtRu/4		1				70		77.3	0.89	[435]
			5						57.6	1.46	
PBI/PVPA	PtRu/1.5		2			Pt/Air/1	110		130		[201]
PBI/sPEEK	PtRu (1:1)/4	6	1		55	$Pt/O_2/1$	60		47	2.14	[436]
PBI/sPEK/PSU	PtRu/5		1			Pt/Air/43	110		250		[437]
PBI/sPEEK/sPSU	PtRu/5		1			Pt/Air/43	110		200		[438]
PBI/sPEEK/ZrP	PtRu/1					$Pt/O_2/3$	130		50.1		[439]
PBI ($\lambda_a = 3$)	PtRu (1:1)/4	2/ethanol	6.7			$Pt/O_2/1$	150		21.3–27.6		[440]
							175		47.1–48.2		
							200		55.5–79.4		
PBI	PtRu (1:1)/2	1/ethanol	$w = 0.33$			$Pt/O_2/1$	150		23		[441]
							200		55		

*x: mole fraction; w: mass fraction

Table 6.8 Conductivity of H_3PO_4-doped ABPBI membranes prepared by casting

Solvent	a_w	T (°C)	λ_a	σ (mS · cm^{-1})	Ref.
MSA	0–0.40	80	1.1	1–5	[192]
MSA	≈ 0	100–185	2.7	5–25	[194]
MSA	0.33	30–100	2.8	6.5–18	[280], [425]
	0.5	20–40		5.5–13	
	1.0	30–100		15–85	
Ethanol/NaOH	0.33	20–100	2.5	5.5–85	
	0.55	30–120		30–280	
	0.75	20–100		18–70	
	1.0	30–100		7–17	
MSA	≈ 0	100–185	2.7	4–25	[419]
MSA	0.05	100–180	3.0	9–38	[442]
	0.10	80–180		7–51	
	0.20	80–180		12–50	
	0.30	50–180		9–62	
MSA	≈ 0	100–185	2.7	4–24	[443]
MSA/H_3PO_4		112–185	3.0	2–15	
MSA	Ambient	124–185	2.7	8–24	[444]
MSA	0	100–185	2.7	5–25	[445]
		120–185	2.5	3–15	
		147–185	1.9	2–6	
MSA		75–200	1.2	9–26	[446]
MSA/H_3PO_4	Ambient	110	1.6	26	[447]
			2.4	41	
			3.7	60	
cABPBI (Fumatech)	0.33	20–80	1.55	5–6	[425]
		25–120	3.1	30–200	
cABPBI (Fumatech)	0	80–180	3.1	50–95	[448]
	0.2	60–140		80–220	
PPA (iABPBI)	0	180	24.5–36.5	195–216	[449]
MSA (sABPBI)	≈ 0	105–185	3.4	6.5–28	[194]
		80–185	4.6	7.5–34	[419]
MSA (ABPBI/PMA)	≈ 0	80–185		4–28	[419]
MSA (ABPBI/PMA)	Ambient	90–185		5–30	[445]
TFA (ABPBI/PVPA)	0	20–150		2.10^{-6}–2.10^{-3}	[450]
	0.5	20–60		0.8–3	

Methanol uptake (see Fig. 6.31) by ABPBI is similar to PBI, being water preferred over methanol. The doping chemistry of H_3PO_4 in ABPBI was analyzed using the Scatchard method [425] and the dissociation constants of the acid complexed to the sites L1 and L2 are $K^{a1}{}_{L1\text{–}H3PO4} = 9.5\ 10^{-4}$ and $K^{a1}{}_{L2\text{–}H3PO4} = 0.48$, respectively, that is, almost one order of magnitude higher than those for PBI. This result, along with that higher acid doping of ABPBI as compared with PBI for the same acid concentration of the doping solution [409, 442], indicates that ABPBI could be a better proton conducting ionomer than PBI.

In Table 6.8 are summarized the conductivity data reported in the literature for ABPBI membranes obtained by casting [192, 194, 280, 419, 425, 442–447], along with results for commercial crosslinked ABPBI membranes [423, 446], an isomer of ABPBI [449], sulfonated ABPBI [194, 419], and ABPBI composites [419, 445, 450]. Most of the membranes were prepared by casting from methanesulfonic acid (MSA) and dry ($a_w \approx 0$) ABPBI membranes reach a maximum conductivity close to 25 mS.cm^{-1}, for doping degree $\lambda_a = 2.7$ at 185 °C [194, 419, 443, 445]. A dry sulfonated ABPBI membrane [194, 419] with higher acid doping reaches slightly higher conductivities. In all cases these conductivity values are higher than that reported for dry PBI membranes (see Table 6.5) at high temperature.

By increasing humidity, the proton conductivity of the MSA-cast ABPBI membranes raise up to values of 85 mS.cm^{-1}, for doping degree $\lambda_a = 2.8$ at 100 °C [425], which is quite reasonable. Interestingly, the conductivity of ABPBI membranes prepared by low-temperature casting from alkaline ethanol, exhibit a maximum conductivity of 280 mS.cm^{-1} at 120 °C and $a_w = 0.55$, for doping degree $\lambda_a = 2.5$. The conductivity of these membranes goes through a maximum at water activities close to $a_w = 0.5-0.6$, in good agreement with the conductivity behaviour of concentrated aqueous H_3PO_4.

An isomer of ABPBI which contains head-to head and tail-to-tail benzimidazole sequences, was recently synthesized [449] and membranes were prepared by using the PPA sol- gel process (see Fig. 6.11). The membranes, like those prepared with PBI using this method PBI [200], have a much higher doping degree and their conductivities are above 200 mS.cm^{-1} at 180 °C, even without humidification. Proton conductivities above 200 mS.cm^{-1} were also reported for commercial crosslinked ABPBI membranes by Fumatech [425, 447], at 120 and 140 °C and partial humidification. ABPBI/MMA membranes exhibit only modest conductivities [419, 445], while an ABPBI/PVPA composite [450], which is the equivalent to the commercial PBI-based Celtec V by BASF Fuel Cells shows a poor conductivity.

In summary, ABPBI membranes seems to have similar or higher proton conductivity than PBI membranes, under similar conditions of acid doping, humidity and temperature, as expected from the acid uptake and acid dissociation differences discussed above.

The information on methanol permeability for ABPBI is scarce [280], and the comparison with commercial crosslinked ABPBI, PBI and Nafion is illustrated in Fig. 6.35. It can be seen that the permeability of ABPBI membranes cast from MSA is much lower than that of ABPBI cast from ethanol/NaOH all over the range of temperature studied, which is consistent with the lower water content of the former [409]. It is interesting to note that the permeability data of the commercial crosslinked ABPBI membrane lies in between the permeability of both linear polymers.

The selectivity plot of Fig. 6.34, including data for these ABPBI membranes, shows that ABPBI prepared by casting from MSA and ethanol/NaOH have rather good relative selectivity and could be good candidates for DMFC. However, studies on the performance of DMFC with MEAs prepared with these membranes have not been yet reported.

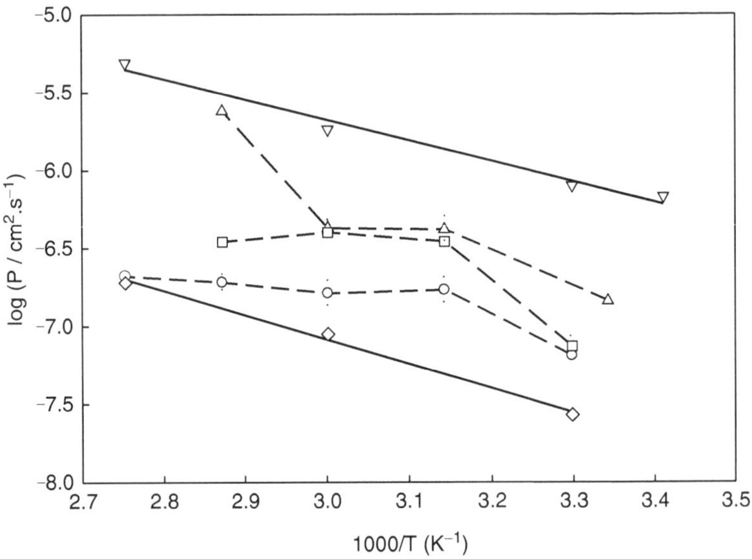

Fig. 6.35 Temperature dependence of the methanol permeability of acid doped ABPBI-MSA cast (○); ABPBI - ethanol/NaOH cast (△); commercial ABPBI (Fumatech) (□); PBI (◊), and Nafion (▽) (Reproduced from Ref. [280] with permission)

6.9 Sulfonated Polyimide Membranes

The polyimide skeleton confers to membranes for PEM fuel cells high thermal stability, high crystallinity and high mechanical strength. Mercier and coworkers [451] prepared sulfonated polyimides as described in Sect. 6.3.6.3, using BDSA, NTDA, and ODA or aromatic diamines containing phenylether bonds. The hydrolysis stability of the naphtalic copolymide BDSA/NTDA/ODA was superior to other naphtalic and phtalic copolyimides under fuel cells conditions. However, the electrical conductivities were only a few mS.cm^{-1}.

Woo et al. [452] synthesized sPI membranes for DMFC using BDSA, ODA and 3,3′,4,4′-benzophenonetetracarboxylic dianhydride (BTDA), with different sulfonation levels controlled by the BDSA/ODA molar ratio. The proton conductivity was moderate (1.7–41 mS.cm^{-1}), but the methanol permeability is much lower than Nafion 117.

McGrath and coworkers [453] prepared a sulfonated naphtalene based polyimide copolymers using disulfonated diamine 3,3′-disulfonic acid –bis [4-(3-amino phenoxy)phenyl]sulfone (SA-DADPS), NTDA and one of two nonsulfonated diamines: ODA or bis[4-(3-aminophenoxy)phenyl]sulfone (m-BAPS). Membranes prepared with ODA and mBAPS exhibit high proton

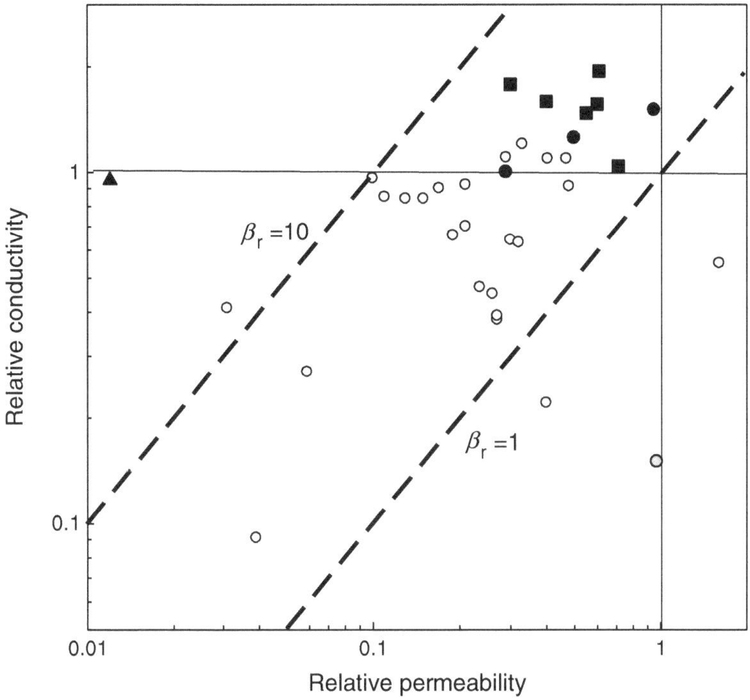

Fig. 6.36 Selectivity plot for sPI and sPI composite membranes: sPI (○) [5, 452, 453, 456, 457]; croslinked sPI (◎) [458]; sPBI/PAMPS (●) [459]; sPI/Nafion (▲) [461]; sPI/SiO_2 (■) [463, 464]

conductivity (up to 120 $mS.cm^{-1}$) and low methanol permeability. The DMFC performance of the membranes was studied at 80 °C, obtaining MPD in the order of 150 $mW.cm^{-2}$.

Watanabe and coworkers [5, 454, 455] studied the DMFC performance of sPI and crosslinked sPI membranes obtained by polycondensation of NTDA, bis (3-sulfo propoxy)benzidine (BSPB), and alkaline diamine [5]. Other studies on polyimide membranes focused on methanol selectivity [456], DMFC performance [457], and the effect of crosslinking on the methanol selectivity [458]. Hybrid sPI membranes were prepared, including composites with PAMPS [459], PTA [460], Nafion infiltrated sPI membranes [461], sPI membranes coated with crosslinkable poly(ethylene glycol) dimethylacrilate (PEGDMA) [462], and mesoporous silica [463, 464].

Figure 6.36 shows the relative selectivity plot for sPI and sPI composite or coated membranes. It can be seen that most of the sPI membranes exhibit relative selectivities higher than one, except that prepared by Miyatake and Watanabe et al. [5], which also exhibit MPD values lower than 100 $mW.cm^{-2}$ [453, 454] at 80 °C. The membranes prepared by McGrath and coworkers [453] show β_r values between 1.7 and 4.6 and have MPD as high as 150 mW.cm^{-2} at 80 °C with 2 M methanol. Similar β_r values (1.4–2.0) were reported by Hu et al. [457] for sPI membranes synthesized from NTDA, 4,4′-bis(4-aminophenoxy)biphenyl-3,3-′-disulfonic acid (BAPBDS), and 4,4′-bis(4-aminophenoxy) biphenyl (BAPB). The MPD obtained in DMFC using these membranes reached 120 mW.cm^{-2} at 60 °C. High relative selectivity ($\beta_r = 6.5$) has been reported for crosslinked sPI membranes [458], but its DMFC performance was only 55 mW.cm^{-2} at 60 °C.

All the sPI composite membranes are located in the A quadrant, that is have higher proton conductivities and lower methanol permeability than Nafion. The sPI composite with PTA [460] has a poor DMFC performance, but sPBI layered with PEGDMA reached MPD of 120 mW.cm^{-2} at 60 °C [462]. The behavior of the sPI/Nafion composite is exceptional, exhibiting β_r =79 and a MPD of 130 mW. cm^{-2} at 70 °C with 5 M methanol.

6.10 Sulfonated Polysulfone Membranes

Typical sulfonated polysulfones are synthesized by nucleophilic aromatic polycondensation of the monomer 4,4′-diclorodiphenyl sulfone (DCDPS) and its 3,3′-disulfonated derivated (SDCDPS), with bisphenol, leading to the polymer known as BPSH [188, 465–470], whose structure is shown in Fig. 6.9. Some modifications of this polysulfone include the replacement of bisphenol by hexafluoroisoproppylidene diphenol (HFPSH polysulfone) [465, 471], or bis(hidroxyphenyl) fluorene (HPFSH polysulfone) [470], and the end group modification with allyl (aBPSH) or propargyl halide (pBPSH) [472]. Many of these polysulfones exhibit proton conductivities up to 100 mS.cm^{-1}, close to the conductivity of Nafion [188, 465, 472], and in the case of BPSH/HPFSH blends the maximum conductivity reaches 130 mS.cm^{-1} [470]. The methanol permeabilities of these membranes range from 10^{-6} to 10^{-7} cm^2.s^{-1}, with relative selectivities not higher than 2. Exceptions are the allyl (aBPSH) and propargyl modified BPSH membranes, which present very low permeabilities (between 1 and 5×10^{-8} cm^2.s^{-1}) and relative selectivities in the interval 2.3–23 (see Fig. 6.37).

Other polysulfone membranes for DMFC were prepared by sulfonation of commercial polyethersulfone with Cardo group (sPES-C) [473] (Fig. 6.9), by sulfonation of a commercial poysulfone and the use of silica as a filler [474], by ¨click¨ cycloaddition of alkyne sulphonate to polysulfone containing azide moieties and crosslinked with 1,7-octadiyne [475], and by polycondensation of 4,4′-difluorodiphenyl sulfone (DFDPS) and 1,3-bis(4-fluorobenzoyl)benzene,6,7-dihydroxy-2-naphtha lene sulfonate with bisphenol [476]. Commercial poly(phtalazinone ether sulfone ketone (PPESK) [477], Udel (Solvay) polysulfone [478–480], and Lasuf Lati SPA) polysulfone [481] were sulfonated to prepare membranes for DMFC. Relative selectivity larger than 7 have been obtained with some of these membranes [477, 478], as indicated in Fig. 6.37.

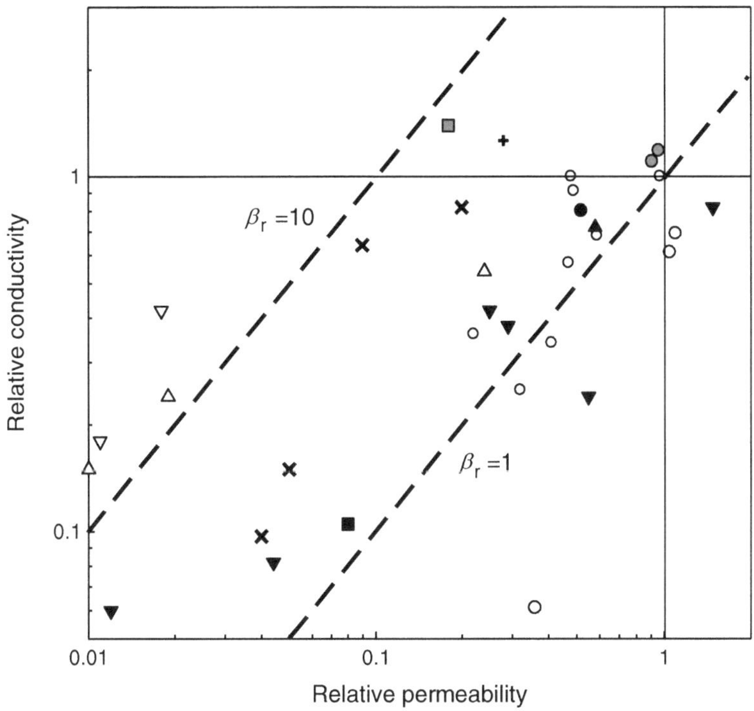

Fig. 6.37 Selectivity plot for polysulfone and polysulfone composite membranes: BPSH (○) [188, 465, 467, 469]; aBPSH (△) [472]; pBPSH (▽) [472]; HFPSH (●) [465]; HPFSH (▲) [470]; BPSH/HPFSH (◎) [470]; sPES-C (■) [473]; sPSf (▼) [475]; sPPEK (+) [477]; sPPEK/SiO_2 (▣) [477]; sPSf (Udel) (×) [478]

The performance of sulfonated polysulfones in DMFC is summarized in Table 6.9. MEAs prepared with membranes obtained by sulfonation of commercial [479–481], HFPSH [471] and aBPSH [472] membranes have moderate power even at temperatures of 80 °C. The highest powers have been obtained with BPSH and BPSH/HPFSH membranes [470] under air-passive conditions (MPD close to 140 $mW.cm^{-2}$ at 60 °C) and with BPSH/SiO_2 composites, particularly those that were surface fluorinated [468] under oxygen-active conditions (MPD close to 170 $mW.cm^{-2}$ at 90 °C). Lufrano et al. [474] have reported MPD between 124 and 180 $mW.cm^{-2}$ in the temperature range between 90 °C and 120 °C, but procedure to prepare the membrane and the catalyst characteristics are not well described.

In spite the reduce amount of information, the correlation between membrane relative selectivity and power density is poor, as already noted for other PEM.

Table 6.9 Methanol membrane selectivity and maximum power density under different DMFC operation conditions of sulfonated polysulfone membranes

Membrane	Anode/load (mg.cm^{-2})	CH_3OH flow-rate (ml.min^{-1})	c (M)	MEA area (cm^2)	Membrane thick (μm)	Cathode/fuel p (bar)	T (°C)	β_r	MPD (mW. cm^{-2})	RMPD	Ref.
BPSH			0.5		115		80	1.6	105–119	0.91–1.03	[188]
			2						128	1.07	
			5					1.9	45	2.25	
BPSH	PtRu/10		0.5	5		Pt/Air/flow	80	2.1	84	0.93	[465]
HFPSH								1.5	100	1.11	
BPSH	PtRu/3	1	1			Pt/Air/flow	90		82	1.26	[466]
BPSH/SiO_2									80–115	1.23–1.77	
BPSH	PtRu/3	1	1	5		Pt/O_2/flow	90		100	1.6	[468]
BPSH/SiO_2									160	2.5	
fBPSH/SiO_2									169–190	2.7–3.1	
BPSH	PtRu/4		2	5		Pt/Air/1	60	1.03	140	1.07	[470]
BPSH/HPFSH								1.21–1.23	126–146	0.96–1.11	
HPFSH								1.24	105	0.80	
HFPSH	PtRu/5		1	10		Pt/Air/	60		14–28		[471]
HFPSH/CNT/ PtRu									32–55		
aBPSH	PtRu/4	1	1			Pt/Air/1	40	23	11.5		[472]
							60		18.5		
							80		23.5		

$sPSf/SiO_2$			2			$/O_2/$	90		124		[474]
							100		140		
							110		155		
							120		180		
asPSf-copolymer	PtRu/5	2.5	1	5		Pt/O_2/flow	65	1.85	37	0.25	[475]
								1.33	45–130	0.30–0.87	
sDPS-DP/sPPEK	PtRu/2.5	2.5	1	5		Pt/O_2/flow	80		140	1.70	[476]
sPSf (Udel)	PtRu/1	2.5	1	5	100–120	$Pt/O_2/1.4$	65		20	1.53	[479]
							80		31	1.19	
			2				65		26	0.87	
							80		39	0.81	
sPSf (Udel)	PtRu/0.6	2.5	1	5	50	$Pt/O_2/1.4$	80		15.5	0.55	[480]
sPSf/DBImB									11.5–24.5	0.41–0.87	
sPSf (Lasulf)	PtRu/4	3	1	4	70	Pt//1	22		4.1–16		[481]
			2						3.5–19		
			5						3.7–15		
			1–5			Pt/Air/flow			31–39		

6.11 Sulfonated Polystyrene Membranes and Blends with Polyvinylidene Fluoride

Although poly(styrene sulfonic acid) (PSSA) is a low price polymer with high proton conductivity and low alcohol permeability, its poor chemical and thermal stability limit applicability to DAFC. Some efforts were devoted to increase hydrolytic and oxidative stabilities together with enhanced mechanical and thermal properties. Thus, copolymers of ethylene and diethylene glycol dimethyacrylate with styrene sulfonic acid (PEGDMA-SSA and PDEGDMA-SSA) [482], crosslinked sulfonated poly (styerene-b-butadiene-b-styrene) (sPSEBS) [483], and a grafted membrane prepared from sodium 4-styrene sulfonate (NaSS) and hydrogenated butadiene rubber (HNBR) [484] have been recently studied in relation to DMFC.

PEGDMA-SSA membranes exhibit proton conductivities similar to Nafion and methanol permeabilities slightly lower than Nafion, leading to relative selectivities in the range 1.5–2.0, while PDEGDMA-SSA membranes have $\beta_r < 1$ [482]. HNBR/SSA membranes have low methanol permeabilities but modest proton conductivities, with β_r close to unity [483]. Membranes prepared with sPSEBS, with β_r between 1.6 and 2.7, were tested in DMFC and show MPD that increases from 45 to 80 mW.cm^{-2} in the temperature range 30–60 °C [484]. On contrary, crosslinked sPSEBS membranes have poor performance in DMFC.

Poly(vinylidene fluoride) (PVDF) membranes, on the other hand, have much better chemical and thermal stabilities than PSSA, but is not a proton conducting polymer. A composite membrane of polyvinylidene fluoride –hexafluoropropyl ene (PVDF-HFP) with sulfonated silica was prepared [485] with proton conductivity up to 40 mS.cm^{-1} and good relative selectivity ($\beta_r > 3$), which delivers 43 mW.cm^{-2} in a DMFC at 80 °C. However, the most studied PVDF composites membranes, are those prepared with PSSA [486–488], poly(2-acrylamido-2-methyl propylene sulfonic acid) (PAMPS) [489].

PVDF/PSSA composite membranes were prepared by impregnating PSSA in the porous PVDF matrix. The resulting membranes exhibit high conductivities (55–93 mS.cm^{-1}) [486], which depend on the PSSA content [487] and the crosslinking degree of PSSA in the composite [488]. The highest proton conductivities were obtained for membranes impregnated in supercritical CO_2 (126 mS.cm^{-1}) [488], although the relative selectivities were modest ($\beta_r < 2$). Regarding the DMFC performances using PVDF/PSSA composite membranes, Prakash et al. [486] have reported the best results with MPD = 52 mW.cm^{-2} at 55 °C. Poor properties were reported for PVDF/PAMPS composite membranes ($\beta_r < 1$) and DMFC test were not performed [489].

A PVDF-HFP/PSSA composite membrane was tested in DMFC [490] reaching 80 mW.cm^{-2} at 60 °C. When SiO_2 was added to the composite the MPD increased up to 110 mW.cm^{-2} at the same temperature, which represents the best DMFC performance for a PVDF-based membrane. Recently, a composite of PVDF with sulfonated poly(styrene-b-ethylene butylenes-b-styrene) (sPSEBS) was reported to deliver up to 72 mW.cm^{-2} at 60 °C in a DMFC, although the characteristics of the catalysts and the MEA were not described [491].

In summary, the properties of PSSA and PVDF composites in relation to DMFC seems to be only slightly superior to Nafion in some cases, but their performances are still modest and durability test are not available yet.

6.12 Quaternized, Impregnated and Alkali-Doped Anion Exchange Membranes

AEMs containing a quaternary ammonium group as a cation exhibit higher thermal and chemical stability than other quaternary cations such as quaternary phosphonium or tertiary sulfonium groups, although they are prone to degradation in a high-pH environment by HO^- nucleophilic attack [492].

One of the reasons to use AEM in PEM fuel cells is to reduce alcohol crossover duc to thc climination of electrosmotic drag by protons moving from anode to cathode. In ADAFC the current through the AEM is conducted by HO^- ions, which are transported in the opposite direction minimizing the electrosmotic flux and, in addition, alcohol oxidation kinetics can be facilitated in alkaline media.

Although the Grotthuss conduction mechanism is also possible for HO^- ions in aqueous media, the ionic conductivity of AEM membranes are expected to be lower than the PEM for the same ion-exchange capacity. In fact, the ionic conductivity of ammonium quaternized AEM at room temperature spreads over a wide range of values, from a few $mS.cm^{-1}$ for QPEK-C [206], QSEBS [212], and QPVA [213], up to 60 and 140 $mS.cm^{-1}$ for QPAES [209] and QPPEKS [207], respectively.

Figure 6.38a shows the conductivity of QPAES membrane as a function of the KOH concentration of the solution where the membrane was equilibrated. This membrane reached a maximum conductivity of 140 $mS.cm^{-1}$ at room temperature at the 2 M KOH solution [207], but decreases at higher KOH concentrations, probably due to the loss of ammonium groups by nucleophilic displacement.

Figure 6.38b illustrates the time stability of the ionic conductivity of two QPAES membranes having different IEC. The membranes exhibit good stability at short times (few hours), and the conductivity decrease 30–40 % after testing for almost 1,500 h.

The conductivity reported for the A-201 membrane from Tokuyama is 29 $mS.cm^{-1}$, which is a modest conductivity compared with that of a Nafion proton conducting membrane. The low thickness of this membrane (28 μm) partially compensates this fact, while the alcohol transport is not probably exacerbated. Typical methanol permeabilities of quaternized membranes are lower than 10^{-7} $cm^2.s^{-1}$ and values as low as 10^{-12} $cm^2.s^{-1}$ have been reported [214], which implies that the only concern regarding the applicability in DAFC is ionic conductivity.

Crooslinked organic–inorganic membranes containing PVA and anion-exchange silica precursor with ammonium functionality [236] have been reported with conductivity up to 75 $mS.cm^{-1}$. Composite of quaternized PVA with alumina exhibited conductivities between 35 and 48 $mS.cm^{-1}$ in the temperature range 30–70 °C [242], while in the same range of temperature a composite of quaternized PVA with titania

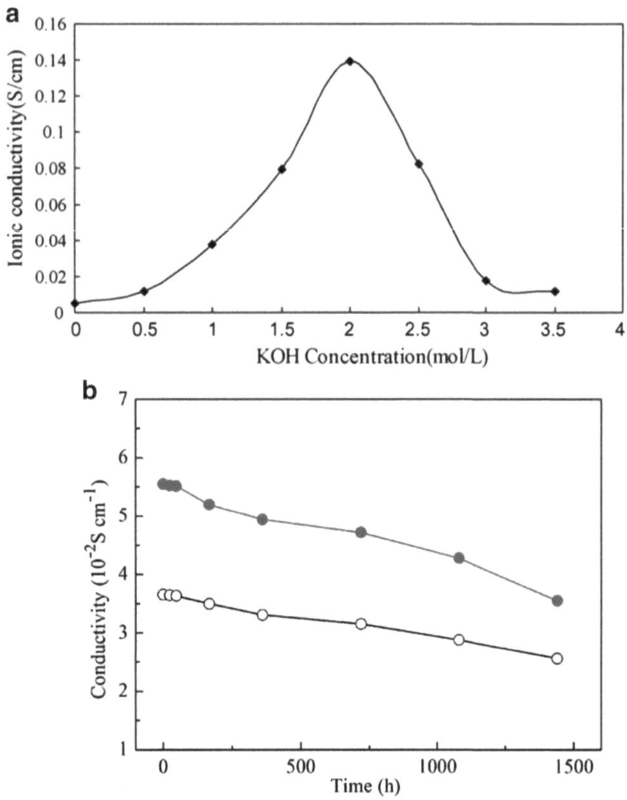

Fig. 6.38 (**a**) Dependence of the ionic conductivity of QPPESK membranes with the alkali concentration. (**b**) Evolution of the ionic conductivity with time in a QPAES membrane (Reproduced from [207] and [209] with permission)

reached conductivities between 48 and 90 mS.cm^{-1} [234], and between 56 and 74 mS.cm^{-1} with poly(epichlorohydrin) [245].

A different kind of alkaline membrane is PVA layered with double hydroxides [237, 238]. Single PVA membrane becomes conductive if impregnated with alkaline hydroxides, but a synergic effect seems to operate when composites with magnesium, nickel and aluminium are formed by casting. Probably this effect is related to the HO^- conduction capacity of the solid layer. Conductivities between 28 and 50 mS.cm^{-1} have been reported for this type of composites [237]. The methanol permeability of all these PVA organic and inorganic are in the range of 10^{-6}–10^{-7} cm^2.s^{-1}, that is, lower than Nafion membranes.

Alkali doped PBI membranes have been studied with ionic conductivities between 16 and 23 mS.cm^{-1} at room temperature [222–225], which increases up to 34 mS.cm^{-1} at 60 °C [225]. PBI has very low conductivity in deionizated water because of the lack of ionic groups, but conductivity increases almost linearly with the alkaline concentration till reach a maximum about 7 M NaOH [225], and decrease slightly at higher concentration. This behaviour is consequence of the mobility of HO^- ions complexed with the imidazole group plus free ions in the solution. The amount of charge carriers increases with concentration, but the increasing viscosity of the aqueous media with alkaline concentration results in an optimal concentration.

Curiously, most of the alkali doped PBI membranes were studied in relation to its performance in DEFC, being the methanol and ethanol permeabilities, in the range 0.86–6.5 10^{-7} $cm^2.s^{-1}$, lower than for Nafion [222, 223, 225, 493]. Recently, Yan et al. [221] introduced imidazole group in a polysulfone polymer reaching IEC in the range 0.78–2.19 $mmol.g^{-1}$, and ionic conductivities up to 53 $mS.cm^{-1}$ at 20 °C for the membrane with the highest IEC, while the methanol permeability increases with the charge content and range between 0.8 and 4.7 10^{-7} $cm^2.s^{-1}$.

A general analysis of AEM in terms of selectivity plots is not possible because Nafion membrane is not an appropriate reference membrane for ADAFC. Yu et al. [267] performed a comparative study of the performance of ADAFC with different catalysts and AEMs. Upgraded information, including the more recent studies, is reported in Table 6.10 for DMFC, in Table 6.11 for DEFC, and in Table 6.12 for cells feed with isopropanol, ethylene glycol and glycerol.

The results shown in Table 6.10 for DMFC indicate a low performance for MEAs prepared with Pt catalyst in the anode, or for cells where alkali is not added to the methanol solution. An exception is the case of the MEA prepared with crosslinked PVA with layered Ni-Al hydroxides, which reached moderate power (22 $mW.cm^{-2}$) with Pt as anodic catalyst (the Pt load is not reported) and without alkali in the methanol feeding solution.

The best performance for non-commercial AEM corresponds to active DMFC with copolymers of vinylbenzyl chloride and metacrylates [220] and alkali-doped PBI [223]. Regarding the performance of commercial AEM in DMFC assays, the results are quite modest in terms of MPD, except for the A-006 membrane from Tokuyama. Prakash et al. [248] reported MPD up to 170 $mW.cm^{-2}$ at 90 °C in active cells using a high load (8 $mg.cm^{-2}$) Pt –Ru anode, while Bianchini and coworkers [250, 251] have obtained MPD between 95 and 120 $mW.cm^{-2}$ at 80 °C in active cells using Pd supported on multiwall carbon nanotubes and on Ni-Zn/C as anodic catalyst.

The performance of AEM in DEFC, reported in Table 6.11, is quite high for non-commercial membranes such as alkali-doped PBI [222, 226], and crosslinked PVA with layered Mg-Al hydroxide [237]. MPD up to 125 $mW.cm^{-2}$ were obtained at 80 °C in active cells using Ru-V anodic catalyst [226], but around the half of power is delivered with Pt-Ru catalyst [222]. With commercial A-006 Tokuyama membranes, Bianchini and coworkers have reported MPD up to 160–165 $mW.cm^{-2}$ at 80 °C in active cells using Pd as anodic catalyst [249, 251] and methanol 10 % with 2 M KOH. Zhao and coworkers, using A201 Tokuyama membranes, obtained MPD in the range 67–130 $mW.cm^{-2}$ at temperatures between 50 °C and 80 °C in active cells using Pd as anodic catalyst [249, 251] and 3 M methanol with 5 M KOH.

Table 6.12 summarizes the results obtained with AEM for higher alcohols, most of them for commercial membranes. A-006 Tokuyama membrane exhibit the best performances for glycerol in active cells with Pd catalyst in the anode,

Table 6.10 Performance of ADMFC with different alkaline membranes

Fuel/oxidant	Alkali	Anode/load ($mg.cm^{-2}$)	Cathode/load ($mg.cm^{-2}$)	Membrane	Conditions: T (°C)/J_a/J_c ($cm^3.min^{-1}$)*	PD ($mW\ cm^{-2}$)	Ref.
2 M methanol/O_2	—	Pt/4	Pt/4	QETFE	50/10/0	1.5	[204]
					60/10/0–2.5 bar	2.4–4.2	
					80/10/2.5 bar	8.5	
1 M methanol/Air	1 M NaOH	Pt-Ru/1	Pt/1	QVBC/BMA/HFMA	65/4/100	53	[220]
2 M methanol/O_2	2 M KOH	Pt-Ru/2	Pt/1	PBI	90/1/2 bar	31	[223]
2 M methanol/Air	1 M NaOH	Pt/2	Pt/2	Nafion 112 & 117	60/?/2 bar	2.5–4.5	[228]
2 M methanol/Air	2 M KOH	Pt-Ru/4.5	MnO_2/3.63	cPVA	30–60/?/0	2.4–4.1	[230]
2 M methanol/Air	8 M KOH	Pt-Ru E-TEK	MnO_2/?	cPVA/HAP	Ambient/?/0	5.8	[232]
		Pt-Ru/4	MnO_2/?		Ambient/?/0	11.5	
2 M methanol/Air	2 M KOH	Pt-Ru/4	MnO_2/?	cPVA/TiO_2	25–60/?/0	3.9–7.5	[233]
2 M methanol/Air	4 M KOH	Pt-Ru/3.6	MnO_2/?	cPVA/TiO_2	Ambient/?/0	9.2	[234]
16 % methanol/Air	—	Pt/?	Pt/?	cPV A/layered Ni-Al hydroxide	40/100/100	22	[238]
		Pt-Ru/?	Pt/?			35	
2 M methanol/O_2	2 M KOH	Pt/0.5	Pt/0.5	cPVA/QHECE	Ambient/3/150	4.6	[239]
2 M methanol/Air	4 M KOH	Pt-Ru/4	MnO_2/4	QPVA/$QSiO_2$	25/?/0	26	[241]
	1 M KOH				50/?/0	17.5	
	2 M KOH					20.9	
	4–8 M KOH					30.1–35.1	
1-4 M methanol/Air	4 M KOH	Pt-Ru/4	MnO_2/4	QPVA/Al_2O_2	25/?/0	28.3–36.1	[242]
1-4 M methanol/Air		Pt-Ru/1.1				9.1–12.5	
	1–6 M KOH					8.5–17.7	
1-4 M methanol/Air	4 M KOH	Pt-Ru/4	MnO_2/?	QPVA	25/?/0	14.0–17.5	[245]
1-4 M methanol/Air				QPVA/PECH		17.2–22.3	
1 M methanol/O_2	1 M KOH	Pt-Ru/4	Pt/1	AHA (Tokuyama)	50/50/50	6.2	[246]

1 M methanol/O_2	1 M KOH	Pt-Ru/8	Pt/3	A-006 (Tokuyama)	30/200/1270	27	[248]
					60/200/1270	56	
					90/200/1270	63	
					90/200/200	59	
2 M methanol/O_2					90/200/1270	29	
3 M methanol/O_2						19	
1 M methanol/O_2	2 M KOH					100	
						170	
1 M methanol/O_2	3 M KOH					60	
10 % methanol/O_2	2MKOH	Pd/0.7	Hypermec	A-006 (Tokuyama)	Ambient/4/0	7.5	[250]
10 % methanol/O_2		Pd/1 On MWCNT	Hypermec		25/4/200	12	
					40/4/200	35	
					60/4/200	60	
					80/4/200	95	
10 % methanol/O_3	2 M KOH	Pd/1 On Ni-Zn/C	Hypermec	A-006 (Tokuyama)	25/4/200	19	[251]
					40/4/200	35	
					60/4/200	78	
					80/4/200	120	
3 M – 10 M methanol/Air	1 M KOH	Pt-Ru/4	Pt/1	A-010 (Tokuyama)	36–41/10/0	9.6–12.8	[252]
0.2 M-5 M methanol/O_2	.....	Pt-Ru/1	Pd/0.5	A201 (Tokuyama)	80/5/100	1.5–2.2	[256]
			Pt/0.5			1.6–2.4	
1 M methanol/?	1 M NaOH	Pt/2	Pt/2	ADP (Solvay)	20/?/?	8	[258]
2 M methanol/?	4 M NaOH					18	
1 M methanol/Air	1 M NaOH	Pt/1.24	Pt/2.05	ADP (Solvay)	60/60/0	7.8	[259]
2 M methanol/Air	1 M NaOH	Pt/2.19	Pt/2.07	ADP (Solvay)	20/60/1 bar	6.5	[260]
					40/60/1 bar	8.3	
					60/22–81/ 0–2 bar	8.3–10.6	

(continued)

Table 6.10 (continued)

Fuel/oxidant	Alkali	Anode/load ($mg.cm^{-2}$)	Cathode/load ($mg.cm^{-2}$)	Membrane	Conditions: T (°C)/J_a/J_c ($cm^3.min^{-1}$)*	PD ($mW\ cm^{-2}$)	Ref.
1 M methanol/Air					60/60/1 bar	10.0	
4 M methanol/Air					60/60/1 bar	8.5	
2 M methanol/Air	1 M NaOH	Pt/0.93–1.46	Pt/0.53–2.5	ADP (Solvay)	60/60/1 bar	5.2–7.8	[261]
		Pt/1.79	Pt/0.77			17.8	
1 M methanol/Air	1 M NaOH	Pt-Ru/1	Pt/1	ADP (Solvay)	30/2.8/400	5.6	[262]
					40/2.8/400	6.6	
					50/2.8/400	7.4	
					60/2.8/400	8.7	
1 M methanol/O_2					30/2.8/400	7.0	
					40/2.8/400	9.1	
					50/2.8/400	10.5	
					60/2.8–12.7/400	11.0–16.5	
2 M methanol/O_2					60/2.8/400	12.8	
4 M methanol/O_2					60/2.8/400	9.8	
1 M – 15 M methanol/O_2	—	Pt-Ru/1	Pt/1	FAA (Fumatech)	30/0.2/30	0.18–0.31	[265]

*In some cases the cathodic feed conditions are given in bar instead of oxygen or air flowrate

Table 6.11 Performance of ADEFC with different alkaline membranes

Fuel/oxidant	Alkali	Anode/load (mg.cm^{-2})	Cathode/load (mg.cm^{-2})	Membrane	Conditions: T (°C)/J_a/J_c (cm^3.min^{-1})*	PD (mW.cm^{-2})	Ref.
1 M ethanol/Air	—	Pt/0.4	Pt/0.2	QPVC	Ambient/1/0	1.0	[214]
2 M ethanol/Air					Ambient/1/0	1.4	
2 M ethanol/O_2	2 M KOH	Pt-Ru/2	Pt/1	PBI	75/1/2 bar	49	[222]
					90/1/2 bar	61	
2 M ethanol/Air	2MKOH	Pt-Ru/2	MnO_2/1	PBI	60/1/0	12–16	[224]
		Pd/2	MnO_2/1		60/1/0	9–30	
2 M ethanol/Air	3MKOH	Ru-V/2.1	Pt/1.24	PBI	80/5/200	90	[226]
2 M ethanol/Air		Ru-V/4.5	TMPhP/9.2			105	
2 M ethanol/O_2						125	
2 M ethanol/Air		Ru-V/0.72	TMPhP/9.2			35	
4 M ethanol/O_2	8MKOH	Ru-Ni/30	CoN4/4	PBI	60/?/?	55	[227]
2 M ethanol/O_2	2MKOH	Pd/2	MnO_2/1	Nafion 112	60/1/2 bar	11.5	[229]
		Pt-Ru/2	Pt/1		60/1/2 bar	33.6	
					75/1/2 bar	41.1	
					90/1/2 bar	58.9	
2 M ethanol/Air	8 M KOH	Pt-Ru E-TEK	MnO_2/?	cPVA/HAP	Ambient/?/?	7.5	[231]
		Pt-Ru/4	MnO_2/?		Ambient/?/?	10.7	
2 M ethanol/Air	4 M KOH	Pt-Ru/3.6	MnO_2/?	cPVA/TiO_2	Ambient/?/0	8.0	[234]
3 M ethanol/Air	5 M KOH	Hypermec/2	Hypermec/1	cPVA/$QSiO_2$	25/2/50	30	[235]
					40/2/50	35	
					50/2/50	42	
					60/2/50	48	
3 M ethanol/O_2	1 M KOH	Hypermec/1	Hypermec/1		60/1/100	32–61	[237]
					80/1/100	37–82	

(continued)

Table 6.11 (continued)

Fuel/oxidant	Alkali	Anode/load ($mg.cm^{-2}$)	Cathode/load ($mg.cm^{-2}$)	Membrane	Conditions: T (°C)/J_a/J_c ($cm^3.min^{-1}$)*	PD ($mW.cm^{-2}$)	Ref.
				cPVA/layered Mg-A1 hydroxide			
1 M ethanol/O_2	0.5 M KOH	Pt-Ru/3	Pt/3	A-006 (Tokuyama)	Ambient/4/100	58	[247]
10 % ethanol/O_2	2 M KOH	Pd/1	Hypermec	A-006 (Tokuyama)	20/4/0	55	[249]
					25/4/200	63	
					40/4/200	80	
					60/4/200	120	
					80/4/200	160	
10 % ethanol/O_2	2 M KOH	Pd/0.7	Hypermec	A-006 (Tokuyama)	Ambient/4/0	18.4	[250]
10 % ethanol/O_2		Pd/1 On MWCNT			25/4/200	25	
					40/4/200	37	
					60/4/200	51	
					80/4/200	73	
10 % ethanol/O_2	2 M KOH	Pd/1	Hypermec	A-006 (Tokuyama)	25/4/200	52	[251]
		On Ni-Zn/C			40/4/200	75	
					60/4/200	125	
					80/4/200	165	
1 M ethanol/O_2	1 M KOH	Hypermec/2	Hypermec/1	A201 (Tokuyama)	30/1/100	12	[253]
					40/1/100	17	
					50/1/100	22	
					60/1/100	30	

1 M ethanol/O_2	7 M KOH				40/2/100	60	
3 M ethanol/O_2	5 M KOH	Pd/2	Hypermec	A201 (Tokuyama)	50/1/100	87	[254]
					60/1/100	102	
					70/1/100	118	
					80/1/100	130	
		Pd/1.5			60/1/100	66	
		Pd/2.5			60/1/100	85	
		Pd/3			60/1/100	72	
1 M ethanol/O_2	1 M KOH	Pd/1	Hypermec	A201 (Tokuyama)	60/1/100	33	[257]
		Pd-Ni/1				44	
3 M ethanol/O_2	5 M KOH	Pd/1				67	
		Pd-Ni/1				90	
2 M ethanol/Air	1 M NaOH	Pt/2	Pt/2	ADP (Solvay)	60/?/2 bar	9.2	[228]
1 M ethanol/O_2	2 M KOH	Ru-V/10	CoTMPP/10	FAA (Fumatech)	45/210/5	32	[266]
					45/70/5	22	

*In some cases the cathodic feed conditions are given in bar instead of oxygen or air flowrate

Table 6.12 Performance of ADAFC fed with isopropanol, ethylene glycol and glycerol

Fuel/oxidant	Alkali	Anode/load ($mg.cm^{-2}$)	Cathode/load ($mg.cm^{-2}$)	Membrane	Conditions: T (°C)/ J_a/J_c ($cm^3.min^{-1}$)	PD ($mW.cm^{-2}$)	Ref.
2 M isopropanol/Air	4 M KOH	Pt-Ru/3.6	MnO_2/?	cPVA/TiO_2	Ambient/?/0	5.5	[234]
1 M ethylene glycol/O_2	1 M KOH	Pt-Ru/4	Pt/1	AHA (Tokuyama)	50/50/50	9.2	[246]
1 M glycerol/O_2						6.7	
1 M ethylene glycol/O_2		Pt-Ru/4	Ag/1			4.0	
3 M ethylene glycol/O_2						8.1	
5 M ethylene glycol/O_2						7.4	
5 % glycerol/O_2	2 M KOH	Pd/0.7	Hypermec	A-006 (Tokuyama)	Ambient/4/0	6.0	[250]
5 % glycerol/O_2		Pd/1			25/4/200	18	
		On MWCNT			40/4/200	30	
					60/4/200	55	
					80/4/200	79	
5 % glycerol/O_2	2 M KOH	Pd/1	Hypermec	A-006 (Tokuyama)	25/4/200	16	[251]
		On Ni-Zn/C			40/4/200	45	
					60/4/200	77	
					80/4/200	119	
0.25 M ethylene glycol/O_2	1 M KOH	Pd-Ni/2	Hypermec/1	A201 (Tokuyama)	60/2/100	27	[255]
1 M ethylene glycol/O_2						35	
3 M ethylene glycol/O_2						22	
1 M ethylene glycol/O_2	5 M KOH					53	
	7 M KOH					67	
	9 M KOH					60	

2 M ethylene glycol/?	4 M NaOH	Pt/2	Pt/2	ADP (Solvay)	20/?/?	19	[258]
2 M glycerol/?	4 M NaOH	Au/?	Pt/2	ADP (Solvay)	?	0.3	[263]
		Au-Pd/?				0.9	
		Pd/?				2.3	
		Pt/?				4.2	
2 M ethylene glycol/O_2	1 M NaOH	Pt/2	Pt/2	ADP (Solvay)	20/2/20	4.0	[264]
	2 M NaOH					12.5	
	4 M NaOH					19	
		Pd/2				14	
		Pt-Bi/2				22	
		Pt-Pd-Bi/2				28	

reaching 119 mW.cm^{-2} at 80 °C [251]. For ethylene glycol the best results (67 mW.cm^{-2} at 60 °C) were reported by Zhao and coworkers [255] for a A201 Tokuyama membranes in active cells using Pd-Ni anodic catalyst feed with 1 M ethylene glycol with 7 M KOH.

In summary, the performance of ADMFC are still behind that observed for DMFC using PEM. However, the possibility of using higher temperatures in thermically stable PBI membranes opens an interesting field of research. In the case of higher alcohols (ethanol, ethylene glycol and glycerol), which are more easily oxidized in alkaline media, the results obtained for ADAFC are promisories, particularly because of the possibility of replacing Pt for less expensive catalysts.

6.13 Perspectives

The development of new membranes for DAFC has grown exponentially during the last decade, reaching a maximum in 2009 (Fig. 6.1). The main objective was focused on reducing the alcohol crossover in relation to Nafion while maintaining high ionic conductivity.

Three approaches to reach those purpose have been reviewed and discussed in this Chapter: the improvement of Nafion with fillers that reduce alcohol permeability, the synthesis of new polymers and blends with better alcohol selectivity, and the development of AEM with good conductivities for use in alkaline fuel cells where the alcohol crossover is not important.

As expected, the intensity of such studies have decay during the last years and probably this tendency will continue during this decade, being compensate, in part, by the use of most rationale strategies for developing membranes with higher selectivities. Thus, the synthesis of tailored membranes with well controlled nanopore sizes and the functionalization of the pore walls with proton conducting and alcohol barrier moieties is one of the research areas that will deserve more attention. Some preliminary results have been already reported in the literature by Yameen et al. [494], who described the synthesis of the self-humidifying membrane with high proton conduction by surface-initiated polymerization of polyelectrolyte brushes bearing sulfonate groups on a silicon scaffold. The resulting structure, shown in Fig. 6.39, exhibits a proton conductivity that is independent of the humidity due to the humidifying effect of the polyglycol.

This approach, although it could yield membranes with tailored properties, should overcome huge challenges related to mechanical stability and low cost production.

Also promising are the layer-by-layer technique which allows the intercalation of polyelectrolytes to prepare membranes with enhanced properties for DAFC, like that described by Hammond and coworkers [167, 495] not only because it is easy to perform, but also offer a lot of alternatives to modify commercial membranes or integration with the codeposition of catalyst layers on the membrane to form integrated MEAs.

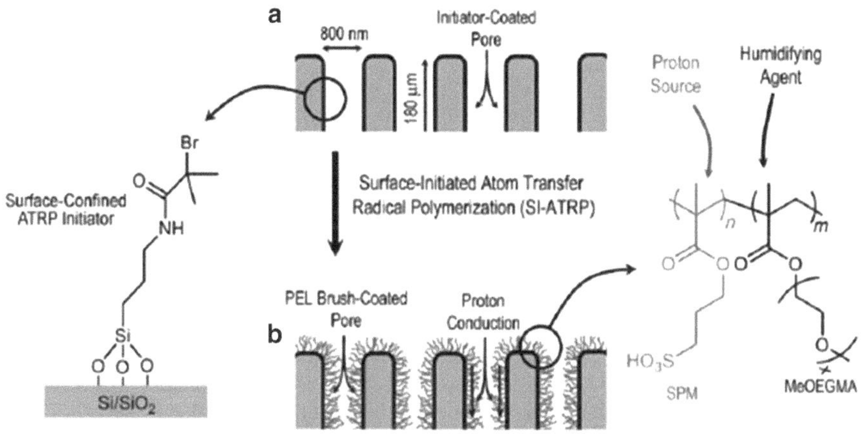

Fig. 6.39 Modification of a macroporous silicon scaffold with initiator-terminated self-assembled monolayers (**a**) and then immersed in the atom transfer radical polymerization (ATRP) solution where the surface-initiated copolymerization of sulfopropyl methacrylate (SPM) and monomethoxy oligo-(ethylene glycol) methacrylate (MeOEGMA) was carried out (**b**). Also depicted are the chemical structures of the copolymer brush (n = 0.87, m = 0.13, x = 5) and the surface-confined ATRP initiator (Reproduced from [494] with permission)

Finally, it should be emphasized that a good relative selectivity is a necessary, but not a sufficient property required for an optimal PEM or AEM in DAFC. There are multiple examples of membranes with very high β_r values, which exhibit very poor fuel cell performances. That means that the architecture of the three phases region is probably the bottleneck of the MEA performance and much attention should be paid to the properties of the ionomeric material use as a binder, forming nanometric structures whose properties could be rather different than those of the bulk materials [317]. Consequently, the properties of very thin films of ionomeric membranes on carbon and metal substrates should be studied for enhancing the performance of direct alcohol fuel cells.

References

1. Kreuer KD (2001) On the development of proton conducting polymer membranes for hydrogen and methanol fuel cells. J Membr Sci 185:29–39
2. DeLuca NW, Elabd YA (2006) Polymer electrolyte membranes for the direct methanol fuel cell: a review. J Polym Sci Pol Phys 44:2201–2225
3. Neburchilov V, Martin J, Wang H, Zhang J (2007) A review of polymer electrolyte membranes for direct methanol fuel cells. J Power Sources 169:221–238
4. Kim YS, Pivovar BS (2007) Polymer electrolyte membranes for direct methanol fuel cells. In: Zhao TS, Kreuer KD, Van Nguyen T (eds) Advances in fuel cells. Elsevier, Oxford (UK) Chap. 4, pp 187–234

5. Miyatake K, Watanabe M (2008) Polyimide ionomer membranes for PEFCs and DMFCs. In: Peinemann KV, Pereira Nunes S (eds) Membranes for energy conversion. Wiley VCH, Weinheim, pp 47–59, Chap. 2
6. Aricó AS, Baglio V, Antonucci V (2008) Composite membranes for high temperature direct methanol fuel cells. In: Peinemann KV, Pereira Nunes S (eds) Membranes for energy conversion. Wiley VCH, Weinheim, pp 123–167, Chap. 5
7. Kim DS, Guiver MD, Kim YS (2009) Proton exchange membranes for direct methanol fuel cells. In: Liu A, Zhang J (eds) Electrocatalysis of direct methanol fuel cells. Wiley-VCH, Weinheim, pp 379–416, Chap. 10
8. Lin J, Wycisk R, Pintauro PN (2009) Modified Nafion as the membrane material for direct methanol fuel cells. In: Zaidi SMJ, Matsuura T (eds) Polymer membranes for fuel cells. Springer, New York, pp 341–359, Chap. 14
9. Ahmad H, Kamarudin SK, Hasran UA, Daud WRW (2010) Overview of hybrid membranes for direct-methanol fuel-cell applications. Int J Hydrogen Energ 35:2160–2175
10. Laberty-Robert C, Vallé K, Pereira F, Sanchez C (2011) Design and properties of functional hybrid organic–inorganic membranes for fuel cells. Chem Soc Rev 40:961–1005
11. Thiam HS, Daud WRW, Kamarudin SK, Mohammad AB, Kadhum AAH, Loh KS, Majlan EH (2011) Overview on nanostructured membrane in fuel cell applications. Int J Hydrogen Energ 36:3187–3205
12. Pivovar BS, Wang Y, Cussler EL (1999) Pervaporation membranes in direct methanol fuel cells. J Membr Sci 154:155–162
13. Wee JH (2007) A feasibility study on direct methanol fuel cells for laptop computers based on a cost comparison with lithium-ion batteries. J Power Sources 173:424–436
14. Piela P, Eickes C, Brosha E, Garzon F, Zelenay P (2004) Ruthenium crossover in direct methanol fuel cell with Pt-Ru black anode. J Electrochem Soc 151:A2053–A2059
15. Choi JH, Kim YS, Bashyam R, Zelenay P (2006) Ruthenium crossover in DMFCs operating with different proton conducting membranes. ECS Trans 1:437–445
16. Mauritz KA, Moore RB (2004) State of understanding of Nafion. Chem Rev 104:4535–4585
17. Viswanathan B, Helen M (2007) Is Nafion the only choice? Bull Catal Soc India 6:50–66
18. Gierke TD, Munn GE, Wilson FC (1981) The morphology in Nafion perfluorinated membrane products, as determined by wide- and small-angle X-ray studies. J Polym Sci Pol Phys 19:1687–1704
19. Rubatat L, Gebel G, Diat O (2004) Fibrillar structure of Nafion: matching Fourier and real space studies of corresponding films and solutions. Macromolecules 37:7772–7783
20. Schmidt-Rohr K, Chen Q (2008) Parallel cylindrical water nanochannels in Nafion fuel cell membranes. Nat Mater 7:75–83
21. Mu S, Tang H, Wan Z, Pan M, Yuan R (2005) Au nanoparticles self-assembled onto Nafion membranes for use as methanol-blocking barriers. Electrochem Commun 7:1143–1147
22. Yoon SR, Hwuang GH, Cho WI, Oh IH, Hong SA, Ha HY (2002) Modification of polymer electrolyte membranes for DMFCs using Pd films formed by sputtering. J Power Sources 106:215–223
23. Ma ZQ, Cheng P, Zhao TS (2003) A palladium-alloy deposited Nafion membrane for direct methanol fuel cells. J Membr Sci 215:327–336
24. Kim YM, Park KW, Choi JH, Park IS, Sung YE (2003) A Pd-impreganted nanocomposite Nafion membrane for use in high-concentration methanol fuel in DMFC. Electrochem Commun 5:571–574
25. Tang H, Pan M, Jiang S, Wan Z, Yuan R (2005) Self-assembling multi-layer Pd nanoparticles onto Nafion membrane to reduce methanol crossover. Colloid Surface A 262:65–70
26. Brandão L, Rodrigues J, Madeira LM, Mendes A (2010) Methanol crossover reduction by Nafion modification with palladium composite nanoparticles: application to direct methanol fuel cells. Int J Hydrogen Energ 35:11561–11567

27. Liang ZX, Shi JY, Liao SJ, Zeng JH (2010) Noble metal nanowires incorporated Nafion membranes for reduction of methanol crossover in direct methanol fuel cells. Int J Hydrogen Energ 35:9182–9185
28. Jiang SP, Liu Z, Tang HL, Pan M (2006) Synthesis and characterization of PDDA-stabilized Pt nanoparticles for direct methanol fuel cells. Electrochim Acta 51:5721–5730
29. Jung EH, Jung UH, Yang TH, Peak DH, Jung DH, Kim SH (2007) Methanol crossover through PtRu/Nafion composite membrane for a direct methanol fuel cell. Int J Hydrogen Energ 32:903–907
30. Antonucci PL, Aricó AS, Creti P, Ramunni E, Antonucci V (1999) Investigation of a direct methanol fuel cell based on a composite Nafion-silica electrolyte for high temperature operation. Solid State Ion 125:431–437
31. Staiti P, Aricó AS, Baglio V, Lufrano F, Passalacqua E, Antonucci V (2001) Hybrid Nafion-silica membranes doped with heteropolyacids for application in direct methanol fuel cells. Solid State Ion 145:101–107
32. Jung DH, Cho SY, Peck DH, Shin DR, Kim JS (2002) Performance evaluation of a Nafion/silicon oxide hybrid membrane for direct methanol fuel cell. J Power Sources 106:173–177
33. Dimitrova P, Friedrich KA, Stimming U, Vogt B (2002) Modified Nafion- based membranes for use in direct methanol fuel cells. Solid State Ion 150:115–122
34. Dimitrova P, Friedrich KA, Vogt B, Stimming U (2002) Transport properties of ionomer composite membranes for direct methanol fuel cells. J Electroanal Chem 532:75–83
35. Yen CY, Lee CH, Lin YF, Lin HL, Hsiao YH, Liao SH, Chuang CY, Ma CCM (2007) Sol–gel derived sulfonated-silica/Nafion composite membrane for direct methanol fuel cell. J Power Sources 173:36–44
36. Lin Y, Yen C, Ma CM, Liao S, Lee C, Hsiao Y, Lin H (2007) High proton-conducting Nafion/-SO_3H functionalized mesoporous silica composite membranes. J Power Sources 171:388–395
37. Jin Y, Qiao S, Zhang ZP, Xu ZP, Smart S, Diniz da Costa JC, Lu GQ (2008) Novel Nafion composite membranes with mesoporous silica nanospheres as inorganic fillers. J Power Sources 185:664–669
38. Lin Y, Li H, Liu C, Xing W, Ji X (2008) Surface-modified Nafion membranes with mesoporous SiO_2 layers via a facile dip-coating approach for direct methanol fuel cells. J Power Sources 185:904–908
39. Park CH, Kim HK, Lee CH, Park HB, Lee YM (2009) Nafion nanocomposite membranes: effect of fluorosurfactants on hydrophobic silica nanoparticle dispersion and direct methanol fuel cell performance. J Power Sources 194:646–654
40. Jiang R, Kunz HR, Fenton JM (2006) Composite silica/Nafion membranes prepared by tetraethylorthosilicate sol–gel reaction and solution casting for direct methanol fuel cells. J Membr Sci 272:116–124
41. Sahu AK, Bhat SD, Pitchumani S, Sridhar P, Vimalan V, George C, Chandrakumar N, Shukla AK (2009) Novel organic–inorganic composite polymer electrolyte for DMFCs. J Membr Sci 345:305–314
42. Lei M, Wang SL, Li LH, Tang WH (2011) A Nafion-silica cathode electrolyte for durable elevated-temperature direct methanol fuel cells. J Power Sources 196:1123–1126
43. Alvarez A, Guzman C, Peza-Ledesma J, Godinez LA, Nava R, Duron-Torres SM, Ledesma Garcia J, Arriage LG (2011) Silica-based composite membranes for methanol fuel cells operating at high temperature. J New Mater Electrochem Syst 14:87–91
44. Sahu AK, Meenakshi S, Bhat SD, Shahid A, Sridhar P, Pitchumani S, Shukla AK (2012) Meso-structured silica-Nafion hybrid membranes for direct methanol fuel cells. J Electrochem Soc 159:F702–F710
45. Kim D, Scibioh MA, Kwak S, Oh IH, Ha HY (2004) Nano-silica layer composite membrane prepared by PECVD for direct methanol fuel cells. Electrochem Commun 6:1069–1074
46. Kim YJ, Choi WC, Woo SI, Hong WH (2004) Proton conductivity and methanol permeation in Nafion/ORMOSIL prepared with various organic silanes. J Membr Sci 238:213–222

47. Li C, Sun G, Ren S, Liu J, Wang Q, Wu Z, Sun H, Jin W (2006) Casting Nafion-sulfonated organosilica nanocomposite membranes used in direct methanol fuel cells. J Membr Sci 272:50–57
48. Ren S, Sun G, Li C, Liang Z, Wu Z, Jin W, Qin X, Yang X (2006) Organic silica/Nafion composite membrane for direct methanol fuel cells. J Membr Sci 282:450–455
49. Liang ZX, Zhao TS, Prabhuram J (2006) Diphenylsilicate-incorporated Nafion membranes for reduction of methanol crossover in direct methanol fuel cells. J Membr Sci 283:219–224
50. Tay SW, Zhang X, Liu Z, Hong L, Chan SH (2008) Composite Nafion membrane embedded with hybrid nanofillers for promoting direct methanol fuel cell performance. J Membr Sci 321:139–145
51. Tricoli V, Nannetti F (2003) Zeolite-Nafion composites as ion conducting membrane materials. Electrochim Acta 48:2625–2633
52. BaglioV AAS, Di Blasi A, Antonucci PL, Nannetti F, Tricoli V, Antonucci V (2005) Zeolite-based composite membranes for high temperature direct methanol fuel cells. J Appl Electrochem 35:207–212
53. Baglio V, Di Blasi A, Aricó AS, Antonucci V, Antonucci PL, Nannetti F, Tricoli V (2005) Investigation of the electrochemical behaviour in DMFCs of chabazite and clinoptilolite-based composite membranes. Electrochim Acta 50:5181–5188
54. Chen Z, Holmberg B, Li W, Wang X, Deng W, Munoz R, Yan Y (2006) Nafion/Zeolite nanocomposite membrane by in situ crystallization for a direct methanol fuel cell. Chem Mater 18:5669–5675
55. Byun SC, Jeong YJ, Park JW, Kim SD, Ha HY, Kim WJ (2006) Effect of solvent and crystal size on the selectivity of ZSM-5/Nafion composite membranes fabricated by solution-casting method. Solid State Ion 177:3233–3243
56. Gribov EN, Parkhomchuk EV, Krivobokov IM, Darr JA, Okunev AG (2007) Supercritical CO_2 assited synthesis of highly selective nafion-zeolite nanocomposite membranes for direct methanol fuel cells. J Membr Sci 297:1–4
57. Yildirim MH, Curoś AR, Motuzas J, Julbe A, Stamatialis DF, Wessling M (2009) Nafion/H-ZSM-5 composite membranes with superior performance for direct methanol fuel cells. J Membr Sci 338:75–83
58. Yoonoo C, Dawson CP, Roberts EPL, Holmes SM (2011) Nafion/mordenite composite membranes for improved direct methanol fuel cell performance. J Membr Sci 369:367–374
59. Zhang Z, Désilets F, Felice V, Mecheri B, Licoccia S, Tavares AC (2011) On the proton conductivity of Nafion-Faujasite composite membranes for low temperature direct methanol fuel cells. J Power Sources 196:9176–9187
60. Li X, Roberts EPL, Holmes SM, Zholobenko V (2007) Functionalized zeolite A-Nafion composite membranes for direct methanol fuel cells. Solid State Ion 178:1248–1255
61. Jung DH, Cho SY, Peck DH, Shin DR, Kim JS (2003) Preparation and performance of a Nafion/montmorillonite nanocomposite membrane for direct methanol fuel cell. J Power Sources 118:205–211
62. Song MK, Park SB, Kim YT, Rhee HW, Kim J (2003) Nanocomposite polymer membrane based on cation exchange polymer and nano-dispersed clay sheets. Mol Cryst Liq Crys 407:15–23
63. Thomassin JM, Pagnoulle C, Bizzari D, Caldarella G, Germain A, Jérome R (2004) Nafion-layered silicate nanocomposite membrane for fuel cell application. e-Polymers 4:182–194
64. Thomassin JM, Pagnoulle C, Caldarella G, Germain A, Jérome R (2005) Impact of acid containing montmorillonite on the properties of Nafion membranes. Polymer 46:11389–11395
65. Thomassin JM, Pagnoulle C, Bizzari D, Caldarella G, Germain A, Jérome R (2004) Improvement of the barrier properties of Nafion by fluoro-modified montmorillonite. Solid State Ion 177:1137–1144
66. Rhee CH, Kim HK, Chang H, Lee JS (2005) Nafion/sulfonated montmorillonite composite: a new concept electrolyte membrane for direct methanol fuel cells. Chem Mater 17:1691–1697

67. Lin YF, Yen CY, Hung CH, Hsiao YH, Ma CCM (2007) A novel composite membranes based on sulfonated montmorillonite modified Nafion for DMFCs. J Power Sources 168:162–166
68. Kim TK, Kang M, Choi YS, Kim HK, Lee W, Chang H, Seung D (2007) Preparation of Nafion-sulfonsted clay nanocomposite membrane for direc t methanol fuel cells via a film coating process. J Power Sources 165:1–8
69. Lee W, Kim H, Kim TK, Chang H (2007) Nafion based organic/inorganic composite membrane for air-breathing direct methanol fuel cells. J Membr Sci 292:29–34
70. Lin YF, Yen CY, Ma CCM, Liao SH, Hung CH, Hsiao YH (2007) Preparation and properties of high performance nanocomposite proton exchange membrane for fuel cell. J Power Sources 165:692–700
71. Hasani-Sadrabadi MM, Dashtimoghadam E, Majedi FS, Kabiri K (2009) Nafion/bio-functionalized montmorillonite nanohybrids as novel polyelectrolytes membranes for direct methanol fuel cells. J Power Sources 190:318–321
72. Hudiono Y, Choi S, Shu S, Koros WJ, Tsapatsis M, Nair S (2009) Porous layered oxide/Nafion nanocomposite membranes for direct methanol fuel cell applications. Micropor Mesopor Mat 118:427–434
73. Meenakshi S, Sahu AK, Bhat SD, Sridhar P, Pitchumani S, Shukla AK (2013) Mesostructured-aluminosilicate-Nafion hybrid membranes for direct methanol fuel cells. Electrochim Acta 89:35–44
74. Park YS, Yamazaki Y (2005) Low methanol permeable and high proton-conducting Nafion/calcium phosphate composite membrane for DMFC. Solid State Ion 176:1079–1089
75. Yang C, Srinivasan S, Aricó AS, Creti P, Baglio V, Antonucci V (2001) Composite Nafion/zirconium phosphate membranes for direct methanol fuel cell operation at high temperature. Electrochem Solid St 4:A31–A34
76. Bauer F, Willert-Porada M (2004) Microstructural characterization of Zr-phosphate-Nafion membranes for direct methanol fuel cell (DMFC) applications. J Membr Sci 233:141–149
77. Bauer F, Willert-Porada M (2005) Characterization of zirconium and titanium phosphates and direct methanol fuel cell (DMFC) performance of functionally graded Nafion composite membranes prepared out of them. J Power Sources 145:101–107
78. Hou YH, Sun GQ, Wu ZM, Jin W, Xin Q (2008) Zirconium phosphate/Nafion 115 composite membrane for high concentration DMFC. Int J Hydrogen Energ 33:3402–3409
79. Casciola M, Bagnasco G, Donnadio A, Micoli L, Pica M, Sganappa M, Turco M (2009) Conductivity and methanol permeability of Nafion–zirconium phosphate composite membranes containing high aspect ratio filler particles. Fuel Cells 9:394–400
80. Arbizzani C, Donadio A, Pica M, Sganappa M, Varzi A, Casciola M, Mastragostino M (2010) Methanol permeability and performance of Nafion-zirconum phosphate composite membranes in active and passive direct methanol fuel cells. J Power Sources 195:7751–7756
81. Kim YS, Cho HS, Song MK, Ghil LJ, Kang JS, Rhee HW (2008) Characterization of Nafion/zirconium sulphophenyl phosphate nanocomposite membrane for direct methanol fuel cells. J Nanosci Nanotechnol 8:4640–4643
82. Yang HN, Lee JY, Jeong JY, Na Y, Kim WJ (2011) Cell performance of DMFC fabricated with H + –ETS-10/Nafion composite membrane. Micropor Mesopor Mat 143:215–220
83. Jeon JD, Kim J, Kwak SY (2012) Nafion/microporous titanosilicate ETS-4 composite membranes for effective methanol crossover reduction in direct methanol fuel cells. J Membr Sci 415–416:353–359
84. Rhee CH, Kim Y, Lee JS, Kim HK, Chang H (2006) Nanocomposite membranes of surface-sulfonated titanate and Nafion for direct methanol fuel cells. J Power Sources 159:1015–1024
85. Liu Z, Guo B, Huang J, Hong L, Han M, Gan LM (2006) Nano-TiO_2-coated polymer electrolyte membranes for direct methanol fuel cells. J Power Sources 157:207–211
86. Wu Z, Sun G, Jin W, Hou H, Wang S, Xin Q (2008) Nafion and nano-size TiO_2-SO_4^{2-} solid superacid composite membrane for direct methanol fuel cell. J membrane Sci 313:336–343

87. Ren S, Sun G, Li C, Song S, Xin Q, Yang X (2006) Sulfated zirconia-Nafion composite membranes for higher temperature direct methanol fuel cells. J Power Sources 157:724–726
88. Sun L, Wang S, Jin W, Hou H, Jiang L, Sun G (2010) Nano-sized Fe_2O_3-SO_4^{2-} solid superacid composite Nafion membranes for direct methanol fuel cells. Int J Hydrogen Energ 35:12461–12468
89. Hasanabadi N, Ghaffarian SR, Hasani-Sadrabadi MM (2013) Nafion-based magnetically aligned nanocomposite proton exchange membranes for direct methanol fuel cells. Solid State Ion 232:58–67
90. Barbora L, Singh R, Shroti N, Verma A (2010) Synthesis and characterization of neodymium oxide modified Nafion membrane for direct alcohol fuel cells. Mater Chem Phys 122:211–216
91. Shroti N, Barbora L, Verma A (2011) Neodymium triflate modified Nafion composite membrane for reduced alcohol permeability in direct alcohol fuel cell. Int J Hydrogen Energ 36:14907–14913
92. Barbora L, Acharya S, Singh R, Scott K, Verma A (2009) A novel composite Nafion membrane for direct alcohol fuel cells. J Membr Sci 326:721–726
93. Kang S, Peck DH, Park YC, Jung DH, Jang JH, Lee HR (2008) Hydroscopic strontium hydroxide/Nafion composite membrane for a direct methanol fuel cell. J Phys Chem Solids 69:1280–1283
94. Xiang Y, Yang M, Zhang J, Lan F, Lu S (2011) Phosphotungstic acid (HPW) molecules anchored in the bulk of Nafion as methanol-blocking membrane for direct methanol fuel cells. J Membr Sci 368:241–245
95. Xu W, Lu T, Liu C, Xing W (2005) Low methanol composite Nafion/silica/PWA membranes for low temperature direct methanol fuel cells. Electrochim Acta 50:3280–3285
96. Kim YC, Jeong JY, Hwang JY, Kim SD, Yi SC, Kim WJ (2008) Incorporation of heteropoly acid, tungstophosphoric acid within MCM-41 via impregnation and direct synthesis methods for the fabrication of composite membrane of DMFC. J Membr Sci 325:252–261
97. Thiam HS, Daud WRW, Kamarudin SK, Mohamad AB, Kadhum AAH, Loh KS, Majlan EH (2013) Nafion/Pd-SiO2 nanofiber composite membrane for direct methanol fuel cell applications. Int J Hydrogen Energ 38:9474–9483
98. Easton EB, Langsdorf BL, Hughes JA, Sultan J, Qi Z, Kaufman A, Pickup PG (2003) Characteristics of polypyrrole/Nafion composite membranes in a direct methanol fuel cell. J Electrochem Soc 150:C735–C739
99. Langsdorf BL, Sultan J, Pickup PG (2003) Partitioning and polymerization of pyrrole into perfluorosulfonate (Nafion) membranes under neutral conditions. J Phys Chem B 107:8412–8415
100. Xu F, Innocenti C, Bonnet B, Jones DJ, Rozière J (2005) Chemical modification of perfluorosulfonated membranes with pyrrole for fuel cell application: preparation, characterization and methanol transport. Fuel Cells 5:398–405
101. Zhu J, Sattler RR, Garsuch YO, Pickup PG (2006) Optimisation of polypyrrole/Nafion composite membranes for direct methanol fuel cells. Electrochim Acta 51:4052–4060
102. Park HS, Kim YJ, Hong WH, Lee HK (2006) Physical and electrochemical properties of Nafion/polypyrrole composite membrane for DMFC. J Membr Sci 272:28–36
103. Li L, Zhang Y, Drillet JF, Dittmeyer R, Jüttner KM (2007) Preparation and characterization of Pt direct deposition on polypyrrole modified Nafion composite membrane for direct methanol fuel cell applications. Chem Eng J 133:113–119
104. Park HS, Kim YJ, Choi YS, Hong WH, Jung D (2008) Surface chemistry and physical properties of Nafion/polypyrrole/Pt composite membrane prepared by chemical in situ polymerization for DMFC. J Power Sources 178:610–619
105. Hobson LJ, Nakano Y, Ozu H, Hayase S (2002) Targeting improved DMFC performance. J Power Sources 104:79–84
106. Wycisk R, Chisholm J, Lee J, Lin J, Pintauro PN (2006) Direct methanol fuel cell membranes from Nafion-polybenzimidazole blends. J Power Sources 163:9–17

107. Ainla A, Brandell D (2007) Nafion-polybenzimidazole (PBI) composite membranes for DMFC applications. Solid State Ion 178:581–585
108. Sauk J, Byun J, Kim H (2005) Composite Nafion/polyphenylene oxide (PPO) membranes with phosphomolybdic acid (PMA) for direct methanol fuel cells. J Power Souces 143:136–141
109. Ma CCM, Hsiao YH, Lin YF, Yen CY, Liao SH, Weng CC, Yen MY, Hsiao MC, Weng FB (2008) Effect and properties of various molecular weights of poly(propylene oxide) oligomers/Nafion acid–base blend membranes for direct methanol fuel cells. J Power Sources 185:846–852
110. Liu J, Wang H, Cheng S, Chan KY (2005) Nafion-polyfurfuryl alcohol nanocomposite membranes for direct methanol fuel cells. J Membr Sci 246:95–101
111. Shao ZG, Wang X, Hsing IM (2002) Composite Nafion/polyvinyl alcohol membranes for the direct methanol fuel cell. J Membr Sci 210:147–153
112. DeLuca NW, Elabd YA (2006) Direct methanol fuel cell performance of Nafion/poly(vinyl alcohol) blend membranes. J Power Sources 163:386–391
113. Lin HL, Wang SH, Chiu CK, Yu TL, Chen LC, Huang CC, Cheng TH, Lin JM (2010) Preparation of Nafion/poly(vinyl alcohol) electro-spun fiber composite membranes for direct methanol fuel cells. J Membr Sci 365:114–122
114. Mollá S, Compañ V (2011) Performance of composite Nafion/PVA membranes for direct methanol fuel cells. J Power Sources 196:2699–2708
115. Mollá S, Compañ V (2011) Polyvinyl alcohol nanofiber reinforced Nafion membrane for fuel cell applications. J Membr Sci 372:191–200
116. Wu Z, Sun G, Jin W, Wang Q, Hou H, Chan KY, Xin Q (2007) Use of in situ polymerized phenol-formaldehyde resin to modify a Nafion membrane for the direct methanol fuel cell. J Power Sources 167:309–314
117. Kim HJ, Kim HJ, Shul YG, Han HS (2004) Nafion-Nafion/polyvinylidene fluoride-Nafion laminated polymer membrane for direct methanol fuel cells. J Power Sources 135:66–71
118. Cho KY, Eom JY, Jung HY, Choi NS, Lee YM, Park JK, Choi JH, Park KW, Sung YE (2004) Characteristics of PVdF copolymer/Nafion blend membrane for direct methanol fuel cells (DMFC). Electrochim Acta 50:583–588
119. Cho KY, Jung HY, Choi NS, Sung SJ, Park JK, Choi JH, Sung YE (2005) A coated Nafion membrane with PVdF copolymer/Nafion blend for direct methanol fuel cells (DMFCs). Solid State Ion 176:3027–3030
120. Cho KY, Jung HY, Sung SJ, Kim WK, Sung SJ, Park JK, Choi JH, Sung YE (2006) Preparation and characteristics of Nafion membrane coated with a PVdF copolymer/recast Nafion blend for direct methanol fuel cell. J Power Sources 159:524–528
121. Choi SW, Fu YZ, Ahn YR, Jo SM, Manthiran A (2008) Nafion-impregnated electrospun polyvinylidene fluoride composite membranes for direct methanol fuel cells. J Power Sources 180:167–171
122. Alwin S, Bhat SD, Sahu AK, Jalajakshi A, Sridhar P, Pitchumani S, Shukla AK (2011) Modified-pore-filled-PVDF-membrane electrolytes for direct methanol fuel cells. J Electrochem Soc 158:B91–B98
123. Kim HY, Kang MS, Lee DH, Won JG (2007) Proton exchange membrane with high cell performance based on Nafion/poly(p-phenylene vinylene) composite polymer electrolyte. J Membr Sci 304:60–64
124. Li T, Zhong G, Fu R, Yang Y (2010) Synthesis and characterization of Nafion/cross-linked PVP semi-interpenetrating polymer network membrane for direct methanol fuel cell. J Membr Sci 354:189–197
125. Chen CY, Garnica-Rodriguez JI, Duke MC, Dalla Costa RF, Dicks AL, Diniz da Costa JC (2007) Nafion/polyaniline/silica composite membranes for direct methanol fuel cell application. J Power Sources 166:324–330

126. Huang QM, Zhang QL, Huang HL, Li WS, Huang YJ, Luo JL (2008) Methanol permeability and proton conductivity of Nafion membranes modified electrochemically with polyaniline. J Power Sources 184:338–343
127. Choi BG, Park HS, Im HS, Kim YJ, Hong WH (2008) Influence of oxidation state on physicochemical and transport properties of nafion/polyaniline composite membrane for DMFC. J Membr Sci 324:102–110
128. Wang CH, Chen CHC, Du HY, Chen CP, Hwang JY, Chen LC, Shih HC, Stejskal J, Chen KH (2009) Low methanol-permeable polyaniline/Nafion composite membrane for direct methanol fuel cells. J Power Sources 190:279–284
129. Yildirim MH, Stamatialis D, Wessling M (2008) Dimensionally stable Nafion-polyethylene composite membranes for direct methanol fuel cell applications. J Membr Sci 321:364–372
130. Shim JH, Koo IG, Lee WM (2005) Nafion-impregnated polyethylene-terephtalate film used as the electrolyte for direct methanol fuel cells. Electrochim Acta 50:2385–2391
131. Jeon JD, Kwak SY (2008) Nafion/sulfated β-cyclodextrin composite membrane. J Power Sources 185:49–54
132. Ren S, Li C, Zhao X, Wu Z, Wang S, Sun G, Xin Q, Yang X (2005) Surface modification of sulfonated poly(ether ether ketone) membranes using Nafion solution for direct methanol fuel cells. J Membr Sci 247:59–63
133. Tsai JC, Cheng HP, Kuo JF, Huang YH, Chen CY (2009) Blended Nafion/SPEEK direct methanol fuel cell membranes for reduced methanol permeability. J Power Sources 189:958–965
134. Tsai JC, Kuo JF, Chen CY (2009) Nafion/nitrated sulfonated poly(ether ether ketone) membranes for direct methanol fuel cells. J Power Sources 194:226–233
135. Zhang N, Zhang G, Xu D, Zhao C, Ma W, Li H, Zhang Y, Xu S, Jiang H, Sun H, Na H (2011) Cross-linked membranes based on sulfonated poly (ether ether ketone) (SPEEK)/Nafion for direct methanol fuel cells (DMFCs). Int J Hydrogen Energ 36:11025–11033
136. Kim IT, Choi J, Kim SC (2007) Blend membranes of Nafion/sulfonated poly(aryl ether ketone) for direct methanol fuel cell. J Membr Sci 300:28–35
137. Choi J, Kim IT, Kim SC (2005) Nafion-sulfonated poly(arylene ether sulfone) composite membrane for direct methanol fuel cell. Macromol Res 13:514–520
138. Bae B, Ha HY, Kim D (2005) Preparation and characterization of Nafion/poly (1-vinylimidazole) composite membrane for direct methanol fuel cell application. J Electrochem Soc 152:A1366–A1372
139. Woong JC, Venkataramani SD, Kim SC (2006) Modification of Nafion membrane using poly (4-vinyl pyridine) for direct methanol fuel cell. Polym Int 55:491–499
140. Lin HL, Yu TL, Huang LN, Chen LC, Shen KS, Jung GB (2005) Nafion/PTFE composite membranes for direct methanol fuel cell applications. J Power Sources 150:11–19
141. Chen LC, Yu TL, Lin HL, Yeh SH (2008) Nafion/PTFE and zirconium phosphate modified Nafion/PTFE composite membranes for direct methanol fuel cells. J Membr Sci 307:10–20
142. Chiu KF, Chen YR, Lin HC, Ho WH (2010) PTFE coated Nafion proton conducting membranes for direct methanol fuel cells. Surf Coat Technol 205:1647–1650
143. Lin J, Lee JK, Kellner M, Wycisk R, Pintauro PN (2006) Nafion-fluorinated ethylene-propylene resin membrane blends for direct methanol fuel cells. J Electrochem Soc 153:A1325–A1331
144. Kim SH, Song K (2011) Preparation and characterization of Nafion/sPOSS polyelectrolyte nanocomposite membranes for direct methanol fuel cell applications. J Ind Eng Chem 17:170–173
145. Li L, Drillet JF, Mácová Z, Dittmeyer R, Jüttner K (2006) Poly(3,4-ethylenedioxy thiophene)-modified Nafion membrane for direct methanol fuel cells. Russ J Electrochem 42:1193–1201
146. Li L, Zhang Y (2008) Chemical modification of Nafion membrane with 3,4-ethylene dioxythiophene for direct methanol fuel cell application. J Power Sources 175:256–260

147. Cho HD, Won J, Ha HY, Kang YS (2006) Nafion composite membranes containing rod-shape polyrotaxanes for direct methanol fuel cells. Macromol Res 14:214–219
148. Zhang J, Lan F, Liang D, Xiao Y, Lu S, Xiang Y (2011) Bulk modification of Nafion with purple membrane for direct methanol fuel cell applications. J Membr Sci 382:350–354
149. Kundu PP, Kim BT, Ahn JE, Han HS, Shul YG (2007) Formation and evaluation of semi-IPN of Nafion 117 membrane for direct methanol fuel cell. 1. Crosslinked sulfonated polystyrene in the pores of Nafion 117. J Power Sources 171:86–91
150. Bae B, Ha HY, Kim D (2006) Nafion-graft-polystyrene sulfonic acid membranes for direct methanol fuel cells. J Membr Sci 276:51–58
151. Wei X, Yates MZ (2010) Nafion/polystyrene-b-poly(ethylene-ran-butylene)-b-polystyrene composite membranes with electric field-aligned domains for improved direct methanol fuel cell performance. J Power Sources 195:736–743
152. Cho KY, Jung HY, Shin SS, Choi NS, Sung SJ, Park JK, Choi JH, Park KW, Sung YE (2004) Proton conducting semi-IPN based on Nafion and crosslinked poly(AMPS) for direct methanol fuel cell. Electrochim Acta 50:589–593
153. Zhang Y, Cui Z, Liu C, Xing W, Zhang J (2009) Implantation of Nafion ionomer into polyvinyl alcohol/chitosan composites to form novel proton-conducting membranes for direct methanol fuel cells. J Power Sources 194:730–736
154. Fang Y, Wang T, Miao R, Tang L, Wang X (2010) Modification of Nafion membranes with ternary composite materials for direct methanol fuel cells. Electrochim Acta 55:2404–2408
155. Lin CW, Fan KC, Thangamuthu R (2006) Preparation and characterization of high selectivity organic–inorganic hybrid-laminated Nafion 115 membranes for DMFC. J Membr Sci 278:437–446
156. Huang LN, Chen LC, Yu TL, Lin HL (2006) Nafion/PTFE/silicate composite membranes for direct methanol fuel cells. J Power Sources 161:1096–1105
157. Cui ZM, Li NW, Zhou XC, Liu CP, Liao JH, Zhang SB, Xing W (2007) Surface modified Nafion membrane by casting proton-conducting polyelectrolyte complexes for direct methanol fuel cells. J Power Sources 173:162–165
158. Tian AH, Kim JY, Kim K (2008) Poly(1-vinylimidazole)/Pd-impregnated Nafion for direct methanol fuel cell applications. J Power Sources 183:1–7
159. Ahmad H, Kamarudin SK, Hasran UA, Daud WRW (2011) A novel hybrid Nafion-PBI-ZP membrane for direct methanol fuel cells. Int J Hydrogen Energ 36:14668–14677
160. Sauk JH, Byun J, Kim HY (2004) Grafting of styrene on to Nafion membranes using supercritical CO_2 impregnation for direct methanol fuel cells. J Power Sources 132:59–63
161. Bae B, Ha HY, Kim D (2006) Nafion-*graft*-polystyrene sulfonic acid membranes for direct methanol fuel cells. J Membr Sci 276:51–58
162. Mohy Eldin MS, Elzatahry AA, El-Khatib KM, Hassan EA, El-Sabbah MM, Abu-Saied MA (2011) Novel grafted Nafion membranes for proton-exchange membrane fuel cell applications. J Appl Polym Sci 119:120–133
163. Feichtinger J, Galm R, Walker M, Baumgärtner KM, Schulz A, Räuchle E, Schumacher U (2001) Plasma polymerized barrier films on membranes for direct methanol fuel cells. Surf Coat Technol 142–144:181–186
164. Si Y, Lin JC, Kunz HR, Fenton JM (2004) Trilayer membranes with a methanol-barrier layer for DMFCs. J Electrochem Soc 151:A463–A469
165. Cai Z, Li L, Su L, Zhang Y (2012) Supercritical carbon dioxide treated Nafion 212 commercial membranes for direct methanol fuel cells. Electrochem Commun 14:9–12
166. Li L, Su L, Zhang Y (2012) Enhanced performance of supercritical CO_2 treated Nafion 212 membranes for direct methanol fuel cells. Int J Hydrogen Energ 37:4439–4447
167. Argun AA, Ashcraft JN, Hammond PT (2008) Highly conductive, methanol resistant polyelectrolyte multilayers. Adv Mater 20:1539–1543
168. Moon GY, Rhim JW (2008) Self-assembly modification of perfluorosulfonic acid membranes for the application to direct meytahnol fuel cells. Macromol Res 16:524–531

169. Lin H, Zhao C, Ma W, Li H, Na H (2009) Low water swelling and high methanol resistant proton exchange membrane fabricated by cross-linking of multilayered polyelectrolyte complexes. J Membr Sci 345:242–248
170. Yilmaztürk S, Deligöz H, Yilmazoğlu M, Damyan H, Öksüzömer F, Koç SN, Durmuş A, Gürkaynak A (2010) Self-assembly of highly charged polyelectrolyte complexes with superior proton conductivity and methanol barrier properties for fuel cells. J Power Sources 195:703–709
171. Zhang H, Huang H, Shen PK (2012) Methanol-blocking Nafion composite membranes fabricated by layer-by-layer self-assembly for direct methanol fuel cells. Int J Hydrogen Energ 37:6875–6879
172. Scott K, Taama WM, Argyropoulos P (2000) Performance of the direct methanol fuel cell with radiation-grafted polymer membranes. J Membr Sci 171:119–130
173. Hatanaka T, Hasegawa N, Kamiya A, Kawasumi M, Morimoto Y, Kawahara K (2002) Cell performances of direct methanol fuel cells with grafted membranes. Fuel 81:2173–2176
174. Aricó AS, Baglio V, Creti P, Di Blasi A, Antonucci V, Brunea J, Chapotot A, Bozzi A, Schoemans J (2003) Investigation of grafted ETFE-based polymer membranes as alternative electrolyte for direct methanol fuel cells. J Power Sources 123:107–115
175. Shen M, Roy S, Kuhlmann JW, Scott K, Lovell K, Horsfall JA (2005) Grafted polymer electrolyte membrane for direct methanol fuel cells. J Membr Sci 251:121–130
176. Saarinen V, Kallio T, Paronen M, Tikkanen P, Rauhala E, Kontturi K (2005) New ETFE-based membrane for direct methanol fuel cell. Electrochim Acta 50:3453–3460
177. Chen J, Asano M, Yamaki T, Yoshida M (2006) Preparation and characterization of chemically stable polymer electrolyte membranes by radiation-induced graft copolymerization of four monomers into ETFE films. J Membr Sci 269:194–204
178. Chen J, Asano M, Yamaki T, Yoshida M (2006) Chemical and radiation crosslinked polymer electrolyte membranes prepared from radiation-grafted ETFE films for DMFC applications. J Power Sources 158:69–77
179. Yoshida M, Kimura Y, Chen J, Asano M, Maekawa Y (2009) Preparation of PTFE-based fuel cell membranes by combining latent track formation technology with graft polymerization. Radiat Phys Chem 78:1060–1066
180. Maiti J, Kakati N, Lee SH, Jee SH, Viswanathan B, Yoon YS (2012) Where do poly(vinyl alcohol) based membranes stand in relation to Nafion for direct methanol fuel cell applications? J Power Sources 216:48–66
181. Liu B, Kim DS, Guiver MD, Kim YS, Pivovar BS (2008) Sulfonated poly(aryl ether)-type polymers as proton exchange membranes: synthesis and performance. In: Peinemann KV, Pereira Nunes S (eds) Membranes for energy conversion, vol 2. Wiley-VCH, Weinheim, pp 1–39
182. Regina A, Fontananova E, Drioli E, Casciola M, Sganappa M, Trotta F (2006) Preparation and characterization of sulfonated PEEK-WC membranes for fuel cell applications. A comparison between polymeric and composite membranes. J Power Sources 160:139–147
183. Luo H, Vaivars G, Mathe M (2010) Covalent-ionically cross-linked polyetheretherketone proton exchange membrane for direct methanol fuel cell. J Power Sources 19:5197–5200
184. Bose S, Kuila T, Nguyen TXH, Kim NH, Lau KT, Lee JH (2001) Polymer membranes for high temperature proton exchange membrane fuel cell: recent advances and challenges. Prog Polym Sci 36:813–843
185. Iulianelli A, Basile A (2012) Sulfonated PEEK-based polymers in PEMFC and DMFC applications: a review. Int J Hydrogen Energ 37:15241–15255
186. Miyatake K, Chikashige Y, Higuchi E, Watanabe M (2007) Tuned polymer electrolyte membranes based on aromatic polyethers for fuel cell applications. J Am Chem Soc 129:3879–3887
187. Rozière J, Jones DJ (2003) Non-fluorinated polymer materials for proton exchange membrane fuel cells. Annu Rev Mater Res 33:503–555

188. Harrison WL, Hickner MA, Kim YS, McGrath JE (2005) Poly(arylene ether sulfone) copolymers and related systems from disulfonated monomer building blocks: synthesis, characterization and performance – a topical review. Fuel Cells 5:201–212
189. Vogel H, Marvel CS (1961) Polybenzimidazoles, new thermally stable polymers. J Polym Sci 50:511–539
190. Mader J, Xiao L, Schmidt TJ, Benicewicz BC (2008) Polybenimidazole/acid complexes as high-temperature membranes. Adv Polym Sci 216:63–124
191. Asensio JA, Borrós S, Gómez-Romero P (2002) Proton-conducting polymers based on benzimidazoles and sulfonated benzimidazoles. J Polym Sci Pol Chem 40:3703–3710
192. Carollo A, Quartarone E, Tomasi C, Mustarelli P, Belotti F, Magistris A, Maestroni F, Parachini M, Garlaschelli L, Righetti PP (2006) Developments of new proton conducting membranes based on different polybenzimidazole structures for fuel cells applications. J Power Sources 160:175–180
193. Uno K, Niume K, Iwata Y, Toda F, Iwakura Y (1977) Synthesis of polybenzimidazoles with sulfonic acid groups. J Polym Sci Pol Chem 15:1309–1318
194. Asencio JA, Borrós S, Gómez-Romero P (2004) Sulfonated poly(2,5-benzimidazole) (SABPBI) impregnated with phosphoric acid as proton conducting membranes for polymer electrolyte fuel cells. Electrochim Acta 49:4461–4466
195. Jones DJ, Rozière J (2001) Recent advances in the functionalization of polybenzimidazole and polyetherketone for fuel cells application. J Membr Sci 185:41–58
196. Asensio JA, Sánchez EM, Gómez-Romero P (2010) Proton-conducting membranes based on benzimidazole polymers for high-temperature PEM fuel cells. A chemical quest. Chem Soc Rev 39:3210–3239
197. Pu H, Liu Q, Liu G (2004) Methanol permeation and prόton conductivity of acid-doped poly (N-methylbenzimidazole) and poly(N-ethylbenzimidazole). J Membr Sci 241:169–175
198. Chuang SW, Hsu SL (2006) Synthesis and properties of a new fluorine-containing polybenzimidazole for high-temperature fuel cells applications. J Polym Sci Pol Chem 44:4508–4513
199. Li Q, Jensen JO, Savinell RF, Bjerrum NJ (2009) High temperature proton exchange membranes based on polybenzimidazoles for fuel cells. Prog Polym Sci 34:449–477
200. Xiao L, Zhang H, Scanlon E, Ramanathan LS, Choe EW, Rogers D, Apple T, Benicewicz BC (2005) High-temperature polybenzimidazole fuel cell membranes via a sol–gel process. Chem Mater 17:5328–5333
201. Gubler L, Kramer D, Belack J, Ünsal Ö, Schmidt TJ, Scherer GG (2007) Celtec-V. A polybenzimidazole-based membrane for the direct methanol fuel cell. J Electrochem Soc 154:B981–B987
202. Wang Y, Li L, Hu L, Zhuang L, Lu J, Xu B (2003) A feasibility analysis for alkaline membrane direct methanol fuel cell: thermodynamic disadvantages versus kinetic advantages. Electrochem Commun 5:662–666
203. Danks TN, Slade RCT, Varcoe JR (2002) Alkaline anion-exchange radiation grafted membranes for possible electrochemical application in fuel cells. J Mater Chem 13:712–721
204. Varcoe JR, Slade RCT (2006) An electron-beam-grafted ETFE alkaline anion-exchange membrane in metal-cation-free solid-state alkaline fuel cells. Electrochem Commun 8:839–843
205. Li L, Wang Y (2005) Quaternized polyethersulfone Cardo anion exchange membranes for direct methanol alkaline fuel cells. J Membr Sci 262:1–4
206. Xiong Y, Liu QL, Zeng QH (2009) Quaternized cardo polyetherketone anion exchange membrane for direct methanol alkaline fuel cells. J Power Sources 193:541–546
207. Fang J, Shen PK (2006) Quaternized poly(phthalazinon ether sulfone ketone) membrane for anion exchange membrane fuel cells. J Membr Sci 285:317–322
208. Abuin GA, Nonjola P, Franceschini EA, Izraelevitch FH, Mathe MK, Corti HR (2010) Characterization of an anionic-exchange membranes for direct methanol alkaline fuel cells. Int J Hydrogen Energ 35:5849–5854

209. Zhao CH, Gong Y, Liu QL, Zhang QG, Zhu AM (2012) Self-crosslinked anion exchange membranes by bromination of benzylmethyl-containing poly(sulfone)s for direct methanol fuel cells. Int J Hydrogen Energ 37:11383–11393
210. Zhou J, Ünlü M, Anestis-Richard I, Kohl PA (2010) Crosslinked, epoxy-based anion conductive membranes for alkaline membrane fuel cells. J Membr Sci 350:286–292
211. Liu G, Shang Y, Xie X, Wang S, Wang J, Wang Y, Mao Z (2012) Synthesis and characterization of anion exchange membranes for alkaline direct methanol fuel cells. Int J Hydrogen Energ 37:848–853
212. Zeng QH, Liu QL, Broadwell I, Zhu AM, Xiong Y, Tu XP (2010) Anion exchange membranes based on quaternized polystyrene-block-poly(ethylene-ran-butylene)-block-polystyrene for direct methanol alkaline fuel cells. J Membr Sci 349:237–243
213. Xiong Y, Fang J, Zeng QH, Liu QL (2008) Preparation and characterization of cross-linked quaternized poly(vinyl alcohol) membranes for anion exchange membrane fuel cells. J Membr Sci 311:319–325
214. Hu J, Zhang C, Cong J, Toyoda H, Nagatsu M, Meng Y (2011) Plasma-grafted alkaline anion-exchange membranes based on polyvinyl chloride for potential application in direct methanol fuel cells. J Power Sources 196:4483–4490
215. Zhang C, Hu J, Cong J, Zhao Y, Shen W, Toyoda H, Nagatsu M, Meng Y (2011) Pulsed plasma-polymerized alkaline anion-exchange membranes for potential application in direct alcohol fuel cells. J Power Sources 196:5386–5393
216. Wu L, Xu T (2008) Improving anion exchange membranes for DMAFCs by inter-crosslinking CPPO/BPPO blends. J Membr Sci 322:286–292
217. Wu L, Xu T, Wu D, Zheng X (2008) Preparation and characterization of CPPO/BPPO membranes for potential application in alkaline direct methanol fuel cell. J Membr Sci 310:577–585
218. Guo TY, Zeng QH, Zhao CH, Liu QL, Zhu AM, Broadwell I (2011) Quaternized polyepichlorohydrin/PTFE composite anion exchange membranes for direct methanol alkaline fuel cells. J Membr Sci 371:268–275
219. Xu HK, Fang J, Guo M, Lu X, Wei X, Tu S (2010) Novel anion exchange membrane based on copolymer of methyl methacrylate, vinylbenzyl chloride and ethyl acrylate for alkaline fuel cells. J Membr Sci 354:206–211
220. Zhang Y, Fang J, Wu Y, Xu H, Chi X, Li W, Yang Y, Yan G (2012) Novel fluoropolymer anion exchange membrane for alkaline direct methanol fuel cells. J Colloid Interface Sci 381:59–66
221. Yan X, He G, Gu S, Wu X, Du L, Wang Y (2012) Imidazolium-functionalized polysulfone hydroxide exchange membranes for potential applications in alkaline membrane direct alcohol fuel cells. Int J Hydrogen Energ 37:5216–5224
222. Hou H, Sun G, He R, Wu Z, Sun B (2008) Alkali doped polybenzimidazole membrane for high performance alkaline direct methanol fuel cell. J Power Sources 182:95–99
223. Hou H, Sun G, He R, Wu Z, Sun B, Jin W, Liu H, Xin Q (2008) Alkali doped polybenzimidazole membrane for alkaline direct methanol fuel cell. Int J Hydrogen Energ 33:7172–7176
224. Hou H, Wang S, Jiang Q, Jin W, Jiang L, Sun G (2011) Durability study of KOH doped polybenzimidazole membrane for air-breathing alkaline direct ethanol fuel cell. J Power Sources 196:3244–3248
225. An L, Zhao TS, Wu QX, Zeng L (2012) Comparison of different types of membrane in alkaline direct ethanol fuel cells. Int J Hydrogen Energ 37:14536–14542
226. Modestov AD, Tarasevich MR, Leykin AY, Filimonov VY (2009) MEA for alkaline direct ethanol fuel cell with alkali doped PBI membrane and non-platinum electrodes. J Power Sources 188:502–506
227. Tarasevich MR, Bogdanovskaya VA, Mazin PV (2010) Electrocatalysts and membrane for direct ethanol-oxygen fuel cell with alkaline electrolyte. Russ J Electrochem 46:542–551

228. Yu EH, Scott K, Reeve RW (2006) Application of sodium conducting membranes in direct methanol alkaline fuel cells. J Appl Electrochem 36:25–32
229. Hou H, Wang S, Jin W, Jiang Q, Sun L, Jiang L, Sun G (2011) KOH modified Nafion 112 membrane for high performance alkaline direct ethanol fuel cell. Int J Hydrogen Energ 36:5104–5109
230. Yang CC, Chiu SJ, Chien WC (2006) Development of alkaline direct methanol fuel cells based on crosslinked PVA polymer membranes. J Power Sources 162:21–29
231. Yang CC, Lee YJ, Chiu SJ, Lee KT, Chien WC, Lin CT, Huang CA (2008) Preparation of PVA/HAP composite polymer membrane for a direct ethanol fuel cell (DEFC). J Appl Electrochem 38:1329–1337
232. Yang CC, Chiu SJ, Lin CT (2008) Electrochemical performance of an air-breathing direct methanol fuel cell using poly(vinyl alcohol)/hydroxiapatite composite polymer membrane. J Power Sources 177:40–49
233. Yang CC (2007) Synthesis and characterization of the cross-linked PVA/TiO_2 composite polymer membrane for alkaline DMFC. J Membr Sci 288:51–60
234. Yang CC, Chiu SJ, Lee KT, Chien WC, Lin CT, Huang CA (2008) Study of poly(vinyl alcohol)/titanium oxide composite polymer membranes and their application on alkaline direct alcohol fuel cell. J Power Sources 184:44–51
235. Wang ED, Zhao TS, Yang WW (2010) Poly(vinyl alcohol)/3-(trimethyl ammonium)propyl-functionalized silica hybrid membranes for alkaline direct ethanol fuel cells. Int J Hydrogen Energ 35:2183–2189
236. Triphati BP, Kumar M, Shahi VK (2010) Organic–inorganic hybrid alkaline membranes by epoxide ring opening for direct methanol fuel cell applications. J Membr Sci 230:90–101
237. Zeng L, Zhao TS, Li YS (2012) Synthesis and characterization of crosslinked poly(vinyl alcohol)/layered double hydroxide composite polymer membranes for alkaline direct ethanol fuel cells. Int J Hydrogen Energ 37:18425–18432
238. Ganley JC, Karikari NK, Raghavan D (2010) Performance enhancement of alkaline direct methanol fuel cells by Ni/Al layered double hydroxides. J Fuel Cell Sci Technol 7:031019
239. Zhou T, Zhang J, Qiao J, Liu L, Jiang G, Zhang J, Liu Y (2013) High Durable poly(vinyl alcohol)/quaternized hydroxyethylcellulose ethoxylate anion exchange membranes for direct methanol alkaline fuel cells. J Power Sources 227:291–299
240. Xiong Y, Liu QL, Zhu AM, Huang SM, Zeng QH (2009) Performance of organic–inorganic hybrid anion-exchange membranes for alkaline direct methanol fuel cells. J Power Sources 186:328–333
241. Yang CC, Chiu SS, Kuo SC, Liou TH (2012) Fabrication of anion-exchange composite membranes for alkaline direct methanol fuel cells. J Power Sources 199:37–45
242. Yang CC, Chiu SJ, Chien WC, Chiu SS (2010) Quaternized poly(vinyl alcohol)/alumina composite polymer membranes for alkaline direct methanol fuel cells. J Power Sources 195:2212–2219
243. Xiong Y, Liu QL, Zhang QG, Zhu AM (2008) Synthesis and characterization of cross-linked quaternized poly (vinyl alcohol)/chitosan composite anion exchange membranes for fuel cells. J Power Sources 183:447–453
244. Yang JM, Chiu HC (2012) Preparation and characterization of polyvinyl alcohol/chitosan blended membrane for alkaline direct methanol fuel cells. J Membr Sci 419–420:65–71
245. Yang CC (2012) Alkaline direct methanol fuel cell based on a novel anion-exchange polymer membrane. J Appl Electrochem 42:305–317
246. Matsuoka K, Iriyama Y, Abe T, Matsuoka M, Ogumi Z (2005) Alkaline direct alcohol fuel cells using an anion exchange membrane. J Power Sources 150:27–31
247. Fujiwara N, Siroma Z, Yamazaki S, Ioroi T, Senoh H, Yasuda K (2008) Direct ethanol fuel cells using an anion exchange membrane. J Power Sources 185:621–626
248. Prakash GKS, Krause FC, Viva FA, Narayanan SR, Olah GA (2011) Study of operation conditions and cell design on the performance of alkaline anion exchange membrane based direct methanol fuel cells. J Power Sources 196:7967–7972

249. Bianchini C, Bambagioni V, Filippi J, Marchionni A, Vizza F, Bert P, Tampucci A (2009) Selective oxidation of ethanol to acetic acid in highly efficient polymer electrolyte membrane-direct ethanol fuel cells. Electrochem Commun 11:1077–1080
250. Bambagioni V, Bianchini C, Marchionni A, Filippi J, Vizza F, Teddy J, Serp P, Zhiani M (2009) Pd and Pt-Ru anode electrocatalysts supported on multi-walled carbon nanotubes and their use in passive and active direct methanol alcohol fuel cells with an anion-exchange membrane (alcohol = methanol, ethanol, glycerol). J Power Sources 190:241–251
251. Bianchini C, Shen PK (2009) Palladium-based electrocatalysts for alcohol oxidation in half cells and in direct alcohol fuel cells. Chem Rev 109:4183–4206
252. Kim JH, Kim HK, Hwang KT, Lee JY (2010) Performance of air-breathing direct methanol fuel cell with anion-exchange membrane. Int J Hydrogen Energ 35:768–773
253. Li YS, Zhao TS, Liang ZX (2009) Performance of alkaline electrolyte-membrane-based direct ethanol fuel cells. J Power Sources 187:387–392
254. Li YS, Zhao TS (2011) A high performance integrated electrode for anion-exchange membrane direct ethanol fuel cells. Int J Hydrogen Energ 36:7707–7713
255. An L, Zhao TS, Shen SY, Wu QX, Chen R (2010) Performance of a direct ethylene glycol fuel cell with an anion-exchange membrane. Int J Hydrogen Energ 35:4329–4335
256. Bunazawa H, Yamazaki Y (2009) Ultrasonic synthesis and evaluation of non-platinum catalysts for alkaline direct methanol fuel cells. J Power Sources 190:210–215
257. Shen SY, Zhao TS, Xu JB, Li YS (2010) Synthesis of PdNi catalysts for the oxidation of ethanol in alkaline direct ethanol fuel cells. J Power Sources 195:1001–1006
258. Coutanceau C, Demarconnay L, Lamy C, Léger JM (2006) Development of electrocatalysts for solid alkaline fuel cell (SAFC). J Power Sources 156:14–19
259. Yu EH, Scott K (2004) Direct methanol alkaline fuel cell with catalysed metal mesh anodes. Electrochem Commun 6:361–365
260. Yu EH, Scott K (2004) Development of direct methanol alkaline fuel cells using anion exchange membranes. J Power Sources 137:248–256
261. Yu EH, Scott K (2005) Direct methanol alkaline fuel cell with catalysed anion exchange membrane electrodes. J Appl Electrochem 35:91–96
262. Scott K, Yu EH, Vlachogiannopoulos G, Shivare M, Duteanu N (2008) Performance of a direct methanol alkaline membrane fuel cell. J Power Source 175:452–457
263. Lamy C, Simoes M, Coutanceau C, Léger JM (2009) Electrocatalytic oxidation of glycerol in a solid alkaline membrane fuel cell (SAMFC). In: Proceedings of ECS 216th meeting. Abstract 1046
264. Demarconnay L, Brimaud S, Coutanceau LJM (2007) Ethylene glycol oxidation in alkaline medium at multi-metallic Pt based catalysts. J Electroanal Chem 601:169–180
265. Santasalo-Aarnio A, Hietala S, Rauhala T, Kallio T (2011) In and ex situ characterization of an anion-exchange membrane for alkaline direct methanol fuel cell (ADMFC). J Power Sources 196:6153–6159
266. Mazin PV, Kapustina NA, Tarasevich MR (2011) Direct ethanol oxidation fuel cell with anionite membrane and alkaline electrolyte. Russ J Electrochem 47:275–281
267. Yu EH, Krewer U, Scott K (2010) Principles and materials aspects of direct alkaline alcohol fuel cells. Energies 3:1499–1528
268. Nandan D, Mohan H, Iyer RM (1992) Methanol and water uptake, densities, equivalental volumes and thicknesses of several uni- and divalent ionic perfluorosulphonate exchange membranes (Nafion 117) and their methanol–water fractionation behaviour at 298 K. J Membr Sci 71:69–80
269. Skou E, Kauranen P, Hentschel J (1997) Water and methanol uptake in proton conducting Nafion membranes. Solid State Ion 97:333–337
270. Hietala S, Maunu SL, Sundholm F (2000) Sorption and diffusion of methanol and water in PVDF-g-PSSA and Nafion 117 polymer electrolyte membranes. J Polym Sci Pol Phys 38:3277–3284

271. Gates CM, Newman J (2000) Equilibrium and diffusion of methanol and water in a Nafion 117 membrane. AIChE J 46:2076–2085
272. Rivin D, Kendrick CE, Gibson PW, Schneider NS (2001) Solubility and transport behavior of water and alcohols in Nafion. Polymer 42:623–635
273. Miyake N, Wainright JS, Savinell RF (2001) Evaluation of a sol–gel derived Nafion/silica hybrid membrane for polymer electrolyte membrane for polymer electrolyte membrane fuel cell applications. J Electrochem Soc 148:A905–A909
274. Jalani NH, Choi P, Datta R (2004) Phenomenological methanol sorption model for Nafion 117. Solid State Ion 175:815–817
275. Jalani NH, Choi P, Datta R (2005) TEOM: a novel technique for investigating sorption in proton-exchange membranes. J Membr Sci 254:31–38
276. Jalani NH, Datta R (2005) The effect of equivalent weight, temperature, cationic forms, sorbates and nanoinorganic additives on the sorption behaviour of Nafion. J Membr Sci 264:167–175
277. Saito M, Tsuzuki S, Hayamizu K, Okada T (2006) Alcohol and proton transport in perfluorinated ionomer membranes for fuel cells. J Phys Chem B 110:24410–24417
278. Hallinan DT Jr, Elabd YA (2007) Difusión and sorption of methanol and water in Nafion using time-resolved Fourier transform infrared-attenuated total reflectance spectroscopy. J Phys Chem B 111:13221–13230
279. Villaluenga JPG, Barragan VM, Izquierdo-Gil MA, Godino MP, Seoane B, Ruiz-Bauza C (2008) Comparative study of liquid uptake and permeation characteristics of sulfonated cation-exchange membranes in water and methanol. J Membr Sci 323:421–427
280. Diaz LA, Abuin GC, Corti HR (2012) Methanol sorption and permeability in Nafion and acid-doped PBI and ABPBI membranes. J Membr Sci 411–412:35–44
281. Zhao Q, Carro N, Ryu HY, Benziger J (2012) Sorption and transport of methanol and ethanol in H^+-Nafion. Polymer 53:1267–1276
282. Saarinen V, Kreuer KD, Schuster M, Merkle R, Maier J (2007) On the swelling properties of proton conducting membranes for direct methanol fuel cells. Solid State Ion 178:533–537
283. Ren X, Springer TE, Gottesfeld S (2000) Water and methanol uptakes in Nafion membranes and membrane effects on direct methanol cell performances. J Electrochem Soc 147:92–98
284. Godino PM, Barragán VM, Izquierdo MA, Villaluenga JPG, Seoane B, Ruiz-Bauzá C (2009) Study of the activation energy for transport of water and methanol through a Nafion membrane. Chem Eng J 152:20–25
285. Chaabane L, Dammak L, Grande D, Larchet C, Huguet P, Nikonenko SV, Nikonenko VV (2011) Swelling and permeability of Nafion 117 in water–methanol solutions: an experimental and modelling investigation. J Membr Sci 377:54–64
286. Song S, Zhou W, Tian J, Cai R, Sun G, Xin Q, Kontou S, Tsiakaras P (2005) Ethanol crossover phenomena and its influence on the performance of DEFC. J Power Sources 145:266–271
287. Godino PM, Barragán VM, Villaluenga JPG, Izquierdo-Gil MA, Ruiz-Bauzá C, Seoane B (2010) Liquid transport through sulfonated cation-exchange membranes for different water-alcohol solutions. Chem Eng J 162:643–648
288. Wu Z, Sun G, Jin W, Hou H, Wang S (2008) A model for methanol transport through Nafion membrane in diffusion cell. J Membr Sci 325:376–382
289. Tricoli V (1998) Proton and methanol transport in poly(perfluorosulfonate) membranes containing Cs^+ and H^+ cations. J Electrochem Soc 145:3798–3801
290. Tricoli V, Carretta N, Bartolozi M (2000) A comparative investigation of proton and methanol transport in fluorinated ionomeric membranes. J Electrochem Soc 147:1286–1290
291. Schaffer T, Tschinder T, Hacker V, Besenhard JO (2006) Determination of methanol diffusion and electrosmotic drag coefficients in proton-exchange-membranes for DMFC. J Power Sources 153:210–216
292. Xue S, Yin G (2006) Methanol permeability in sulfonated poly (etheretherketone) membranes: A comparison with Nafion membranes. Eur Polym J 42:776–785

293. Kauranen P, Skou E (1996) Methanol permeability in perfluoro sulfonate proton exchange membranes at elevated temperatures. J Appl Electrochem 26:909–917
294. Ramya K, Dhathathreyan KS (2008) Methanol crossover studies of heat-treated Nafion membranes. J Membr Sci 311:121–127
295. Roualdes S, Topala I, Mahdjoub H, Rouessac V, Sistat P, Durand J (2006) Sulfonated polystyrene-type plasma-polimerized membranes for miniature direct methanol fuel cells. J Power Sources 158:1270–1281
296. Elabd YA, Napadensky E, Sloan JM, Crawford DM, Walker CW (2003) Triblock copolymer ionomer membranes. Part I. Methanol and proton transport. J Membr Sci 217:227–242
297. Verbrugge MW (1989) Methanol diffusion in perfluorinated ion-exchange membranes. J Electrochem Soc 136:417–423
298. Every HA, Hickner MA, McGrath JE, Zawodzinski JTA (2005) An NMR study of methanol diffusion in polymer electrolyte fuel cell membranes. J Membr Sci 250:183–188
299. Ren X, Springer T, Zawodzinski T, Gottesfeld S (2000) Methanol transport through Nafion membranes. Electrosmotic drag effects on potential step measurements. J Electrochem Soc 147:466–474
300. Hallberg F, Vernersson T, Pettersson ET, Dvinskikh SV, Lindbergh G, Furó I (2010) Electrokinetic transport of water and methanol in nafion membranes as observed by NMR spectroscopy. Electrochim Acta 55:3542–3549
301. Tschinder T, Schaffer T, Fraser SD, Hacker V (2007) Electro-osmotic drag of methanol in proton exchange membranes. J Appl Electrochem 37:711–716
302. Ma CH, Yu TS, Lin HS, Huang YT, Chen YL, Jeng US, Lai YH, Sun YS (2009) Morphology and properties of Nafion membranes prepared by solution casting. Polymer 50:1764–1777
303. Xue S, Yin G, Cai K, Shao Y (2007) Permeabilities of methanol, ethanol and dimethyl ether in new composite membranes: a comparison with Nafion membranes. J Membr Sci 289:51–57
304. Kontou S, Stergiopoulos V, Song S, Tsiakaras P (2007) Ethanol/water mixture permeation through a Nafion based membrane electrode assembly. J Power Sources 171:1–7
305. Slade S, Campbell SA, Ralph TR, Walsh FC (2002) Ionic conductivity of an extruded Nafion 1100 EW series of membranes. J Electrochem Soc 149:A1556–A1564
306. Yadav R, Fedkiw PS (2012) Analysis of EIS technique and Nafion 117 conductivity as a function of temperature and relative humidity. J Electrochem Soc 159:B340–B346
307. Gardner CL, Anantaraman AV (1998) Studies on ion-exchange membranes. II. Measurement of the anisotropic conductance of Nafion. J Electroanal Chem 449:209–214
308. Anantaraman AV, Gardner CL (1996) Studies on ion-exchange membranes. I. Effect of humidity on the conductivity of Nafion. J Electroanal Chem 414:115–120
309. Wintersgill MC, Fontanella JJ (1998) Complex impedance measurements on Nafion. Electrochim Acta 43:1533–1538
310. Lufrano F, Gatto I, Staiti P, Antonucci V, Passalacqua E (2001) Sulfonated polysulfone ionomer membranes for fuel cells. Solid State Ion 145:47–51
311. Yang C, Costamagna P, Srinivasan S, Benziger J, Bocarsly AB (2001) Approaches and technical challenges to high temperature operation of proton exchange membrane fuel cells. J Power Sources 103:1–9
312. Damay F, Klein LC (2003) Transport properties of Nafion composite membranes for proton-exchange membranes fuel cells. Solid State Ion 162–163:261–267
313. Ochi S, Kamishima O, Mizusaki J, Kawamura J (2009) Investigation of proton diffusion in Nafion 117 membrane by electrical conductivity and NMR. Solid State Ion 180:580–584
314. Wu X, Wang X, He G, Benziger J (2011) Differences in water sorption and proton conductivity between Nafion and SPEEK. J Polym Sci Pol Phys 49:1437–1445
315. Aricó AS, Baglio V, Di Blasi A, Antonucci V (2003) FTIR spectroscopic investigation of inorganic fillers for composite DMFC membranes. Electrochem Commun 5:862–866

316. Siroma Z, Kakitsubo R, Fujiwara N, Ioroi T, Yamazaki S, Yasuda K (2009) Depression of proton conductivity in recast Nafion film measured on flat substrate. J Power Sources 189:994–998
317. Abuin GC, Fuertes MC, Corti HR (2013) Substrate effect on the swelling and water sorption of Nafion nanomembranes. J Membr Sci 428:507–515
318. Sone Y, Ekdunge P, Simonsson D (1998) Proton conductivity of Nafion 117 as measured by a four-electrode AC impedance method. J Electrochem Soc 143:1254–1259
319. Paddison SJ (2001) The modeling of molecular structure and ion transport in sulfonic acid based ionomer membranes. J New Mat Electrochem Syst 4:197–207
320. Savinell R, Yeager E, Tryk D, Landau U, Wainright J, Weng D, Lux K, Litt M, Rogers C (1994) A polymer electrolyte for operation at temperatures up to 200 °C. J Electrochem Soc 141:L46–L48
321. Edmondson CA, Stallworth PE, Wintergill MC, Fontanella JJ, Dai Y, Greenbaum SG (1998) Electrical conductivity and NMR studies of methanol/water mixtures in Nafion membranes. Electrochim Acta 43:1295–1299
322. Saito M, Ikesaka S, Kuwano J, Qiao J, Tsuzuki S, Hayamizu K, Okada T (2007) Mechanisms of proton transport in alcohol-penetrated perfluorosulfonated ionomer membranes for fuel cells. Solid State Ion 178:539–545
323. Chaabane L, Bulvestre G, Larchet C, Nikonenko V, Deslouis C, Takenouti H (2008) The influence of absorbed methanol on the swelling and conductivity properties of cation-exchange membranes. Evaluation of nanostructure parameters. J Membr Sci 323:167–175
324. Pivovar BS, Smyrl WH, Cussler EL (2005) Electro-osmosis in Nafion 117, polystyrene sulfonic acid, and polybenzimidazole. J Electrochem Soc 152:A53–A60
325. Rikukawa M, Sanui K (2000) Proton-conducting polymer electrolyte membranes based on hydrocarbon polymers. Prog Polym Sci 25:1463–1502
326. Kim YS, Kim DS, Guiver MD, Pivovar BS (2011) Interpretation of direct methanol fuel cell electrolyte properties using non-traditional length-scale parameters. J Membr Sci 374:49–58
327. Ren X, Wilson MS, Gottesfeld S (1996) High performance direct methanol polymer electrolyte fuel cells. J Electrochem Soc 143:L12–L15
328. Meyers JP, Newman J (2002) Simulation of the direct methanol fuel cell. III. Design and optimization. J Electrochem Soc 149:A729–A735
329. Jiang Z, Zheng X, Wu H, Pan F (2008) Proton conducting membranes prepared by incorporation of organophosphorous acids into alcohol barrier polymers for direct methanol fuel cells. J Power Sources 185:85–94
330. Tsai CE, Lin CW, Hwang BJ (2010) A novel crosslinking strategy for preparing poly(vinyl alcohol)-based proton-conducting membranes with high sulfonation. J Power Sources 195:2166–2173
331. Kim DS, Guiver MD, Nam SY, Yun TI, Seo MY, Kim SJ, Hwang HS, Rhim JW (2006) Preparation of ion exchange membranes for fuel cellbased on crosslinked poly(vinyl alcohol) with poly(styrene sulfonic acid-*co*-maleic acid). J Membr Sci 281:156–162
332. Kang MS, Kim JH, Won J, Moon SH, Kang YS (2005) Highly charged proton exchange membranes prepared by using water soluble polymer blends for fuel cells. J Membr Sci 247:127–135
333. Lin CW, Huang YF, Kannan AM (2007) Cross-linked poly(vinyl alcohol) and poly(styrene sulfonic acid-co-maleic anhydride)-based semi-interpene trating network as proton-conducting membranes for direct methanol fuel cells. J Power Sources 171:340–347
334. Wu CS, Lin FY, Chen CY, Chu PP (2006) A polyvinyl alcohol/p-sulfonate phenolic resin composite proton conducting membrane. J Power Sources 160:1204–1210
335. Fang Y, Miao R, Wang T, Wang X (2009) Suppresion of methanol cross-over in novel composite membranes for direct methanol fuel cells. Pure Appl Chem 81:2309–2316
336. Huang YF, Chuang LC, Kannan AM, Lin CW (2009) Proton-conducting membranes with high selectivity from cross-linked poly(vinyl alcohol) and poly(vinyl pyrrolidone) for direct methanol fuel cell applications. J Power Sources 186:22–28

337. Tripathi BP, Saxena A, Shahi VK (2008) Phosphonic acid grafted bis(4-γ-aminopropyl diethoxysilylphenyl)sulfone (APDSPS)-poly(vinyl alcohol) cross-linked polyelectrolyte membrane impervious to methanol. J Membr Sci 318:288–297
338. Binsu VV, Nagarale RK, Shahi VK (2005) Phosphonic acid functionalized aminopropyl triethoxysilane-PVA composite material: organic–inorganic hybrid proton-exchange membranes in aqueous media. J Mater Chem 15:4823–4831
339. Tseng CY, Ye YS, Kao KY, Joseph J, Shen WC, Rick J, Hwang BJ (2011) Interpenetrating network-forming sulfonated poly(vinyl alcohol) proton exchange membranes for direct methanol fuel cell applications. Int J Hydrogen Energ 36:11936–11945
340. Higa M, Sugita M, Maesowa S, Endo N (2010) Poly(vinyl alcohol)-based polymer electrolyte membranes for direct methanol fuel cells. Electrochim Acta 55:1445–1449
341. Higa M, Hatemura K, Sugita M, Maesowa S, Nishimura M, Endo N (2012) Performance of passive direct methanol fuel cell with poly(vinyl alcohol)-based polymer electrolyte membranes. Int J Hydrogen Energ 37:6292–6301
342. Chang YW, Wang E, Shin G, Han JE, Mather PT (2007) Poly(vinyl alcohol) (PVA)/ sulfonated polyhedral oligosilsesquioxane (sPOSS) hybrid membranes for direct methanol fuel cell applications. Polym Adv Technol 18:535–543
343. Hwang BJ, Joseph J, Zeng YZ, Lin CW, Cheng MY (2011) Analysis of states of water in poly (vinyl alcohol) based DMFC membranes using FTIR and DSC. J Membr Sci 369:88–95
344. Kim DS, Park HB, Rhim JW, Lee YM (2004) Preparation and characterization of crosslinked PVA/SiO_2 hybrid membranes containing sulfonic acid groups for direct methanol fuel cell applications. J Membr Sci 240:37–48
345. Kim DS, Park HB, Rhim JW, Lee YM (2005) Proton conductivity and methanol transport behavior of cross-linked PVA/PAA/silica hybrid membranes. Solid State Ion 176:117–126
346. Tripathi BP, Shahi VK (2008) Functionalized organic–inorganic nanostructurated N-p-carboxy benzyl chitosan-silica-PVA hybrid polyelectrolyte complex as proton exchange membrane for DMFC applications. J Phys Chem B 112:15678–15690
347. Cho EB, Kim H, Kim D (2009) Effect of morphology and pore size of sulfonated mesoporous benzene-silicas in the preparation of poly(vinyl alcohol)-based hybrid nanocomposite membranes for direct methanol fuel cell application. J Phys Chem B 113:9770–9778
348. Zhong S, Cui X, Dou S, Liu W, Gao Y, Hong B (2010) Improvement in silicon-containing sulfonated polystyrene/acrylate membranes by blending and crosslinking. Electrochim Acta 55:8410–8415
349. Yang T (2009) Poly(vinyl alcohol)/sulfonated β-cyclodextrin for direct methanol fuel cell applications. Int J Hydrogen Energ 34:6917–6924
350. Yang CC, Lee YJ, Yang JM (2009) Direct methanol fuel cell (DMFC) based on PVA/MMT composite polymer membranes. J Power Sources 188:30–37
351. Yang CC, Lee YJ (2009) Preparation of the acidic PVA/MMT nanocomposite polymer membrane for the direct methanol fuel cell (DMFC). Thin Solids Films 517:4735–4740
352. Kim DS, Park IC, Cho HI, Kim DH, Moon GY, Lee HK, Rhim JW (2009) Effect of organo clay content on proton conductivity and methanol transport through crosslinked PVA hybrid membrane for direct methanol fuel cell. J Ind Eng Chem 15:265–269
353. Anis A, Banthia AK, Bandyopadhyay S (2008) Synthesis and characterization of polyvinyl alcohol copolymer/phosphomolybdic acid-baed crosslinked composite polymer electrolyte membranes. J Power Sources 179:69–80
354. Madaeni SS, Amirinejad S, Aminirejad M (2011) Phosphotungstic acid doped poly(vinyl alcohol)/poly(ether sulfone) blend composite membranes for direct methanol fuel cells. J Membr Sci 380:132–137
355. Helen M, Viswanathan B, Srinivasa Murthy S (2006) Fabrication and properties of hybrid membranes based on salts of heteropolyacid, zirconium phosphate and polyvinyl alcohol. J Power Sources 163:433–439

356. Üçtuğ FG, Holmes SM (2011) Characterization and fuel cell performance analysis of polyninylalcohol-modrdenite mixed-matrix membranes for direct methanol fuel cell use. Electrochim Acta 56:8446–8456
357. Panero S, Fiorenza P, Navarra MA, Romanowska J, Scrosati B (2005) Silica-added, composite poly(vinyl alcohol) membranes for fuel cell application. J Electrochem Soc 152: A2400–A2405
358. Yang CC, Chien WC, Li YJ (2010) Direct methanol fuel cell based on poly(vinyl alcohol)/titanium oxide nanotubes/poly(styrenesulfonic acid)(PVA/nt-TiO_2/PSSA) composite polymer membrane. J Power Sources 195:3407–3415
359. Yang CC (2011) Fabrication and characterization of poly(vinyl alcohol)/montmorillonite/poly(styrene sulfonic acid) proton-conducting composite membranes for direct methanol fuel cells. Int J Hydrogen Energ 36:4419–4431
360. Yang CC, Lue SJ, Shih JY (2011) A novel organic/inorganic polymer membrane based on poly(vinyl alcohol)/poly(2-acrylamido-2-methyl-1-propane sulfonic acid/3-glycidyloxypropyl trimethoxysilane) polymer electrolyte membrane for direct methanol fuel cells. J Power Sources 196:4458–4467
361. Bhat SD, Sahu AK, George C, Pitchumani S, Sridhar P, Chandrakumar N, Singh KK, Krishna N, Shukla AK (2009) Mordenite-incorporated PVA-PSSA membranes as electrolytes for DMFCs. J Membr Sci 340:73–83
362. Yang T (2009) Composite membrane of sulfonated poly(ether ether ketone)and sulfated poly (vinyl alcohol) for use in direct methanol fuel cells. J Membr Sci 342:221–226
363. de Souza GA, Dutra Filho JC (2012) Hybrid membranes of PVA for direct ethanol fuel cell (DEFCs) applications. Int J Hydrogen Energ 37:6246–6252
364. Li L, Zhang J, Wang Y (2003) Sulfonated poly(ether ether ketone) membranes for direct methanol fuel cells. J Membr Sci 226:159–167
365. Gil M, Ji X, Li X, Na H, Hampsey JE, Lu Y (2004) Direct synthesis of sulfonated aromatic poly(ether ether ketone) proton exchange membranes for fuel cells applications. J Membr Sci 234:75–81
366. Sue S, Yin G (2006) Methanol permeability in sulfonated poly(etheretherketone) membranes: a comparison with Nafion membranes. Eur Polym J 42:776–785
367. Jaafar J, Ismail AF, Mustafa A (2007) Physicochemical study of poly (ether ether ketone) membranes sulfonated with mixtures of fuming sulfuric acid and sulfuric acid for direct methanol fuel cell application. Mater Sci Eng A 460–461:475–484
368. Mohd Norddin MNA, Ismail AF, Rana D, Matsuura T, Mustafa A, Tabe-Mohammadi A (2008) Characterization and performance of proton exchange membranes for direct methanol fuel cell: blending of sulfonated poly(ether ether ketone) with charged surface modyfying macromolecule. J Membr Sci 323:404–413
369. Zhong S, Cui X, Fu T, Na H (2008) Modification of sulfonated poly(ether ether ketone) proton exchange membrane for reducing methanol crossover. J Power Sources 180:23–28
370. Ismail AF, Othman NH, Mustafa A (2009) Sulfonated polyether ether ketone composite membrane using tungstosilic acid supported on silica-aluminium oxide for direct methanol fuel cell (DMFC). J Membr Sci 329:18–29
371. Gao Q, Wang Y, Xu L, Wei G, Wang Z (2009) Proton-exchange sulfonated poly(ether ether ketone) (SPEEK)/SiO_x-S composite membranes in direct methanol fuel cells. Chinese J Chem Eng 17:207–213
372. Li H, Zhang G, Ma W, Zhao C, Zhang Y, Han M, Zhu J, Liu Z, Wu J, Na H (2010) Composite membranes based on a novel benzimidazole grafted PEEK and SPEEK for fuel cells. Int J Hydrogen Energ 35:11172–11179
373. Yang T, Liu C (2011) SPEEK/sulfonted cyclodextrin blend membranes for direct methanol fuel cell. Int J Hydrogen Energ 36:5666–5674
374. Gosalawit R, Chirachanchai S, Shishatskiy S, Nunes SP (2008) Sulfonated montmorillonite/sulfonated poly(ether ether ketone) (SMMT/SPEEK) nanocomposite membrane for direct methanol fuel cells (DMFCs). J Membr Sci 323:337–346

375. Zhang Y, Shao K, Zhao G, Zhang G, Li H, Fu T, Na H (2009) Novel sulfonated poly(ether ether ketone) with pendant benzimidazole groups as a proton exchange membrane for direct methanol fuel cells. J Power Sources 194:175–181
376. Li X, Zhao C, Lu H, Wang Z, Na H (2005) Direct synthesis of sulfonated poly(ether ether ketone ketone)s (SPEEKKs) proton exchange membranes for fuel cells applications. Polymer 46:5820–5827
377. Ponce ML, Prado L, Ruffmann B, Richau K, Mohr R, Nunes SP (2003) Reduction of methanol permeability in polyetherketone – heteropolyacid membranes. J Membr Sci 217:5–15
378. Luo H, Vaivars G, Mathe M (2009) Cross-linked PEEK-WC proton exchange membrane for fuel cell. Int J Hydrogen Energ 34:8616–8621
379. Wang J, Zhao C, Li M, Zhang L, Ni J, Ma W, Na H (2012) Benzimidazole-cross-linked proton exchange membranes for direct methanol fuel cells. Int J Hydrogen Energ 37:9330–9339
380. Wang J, Zhao C, Zhang L, Li M, Ni J, Wang S, Ma W, Liu Z, Na H (2012) Cross-linked proton exchange membranes for direct methanol fuel cells: effects of the cross-linker structure on the performances. Int JHydrogen Energ 37:12586–12596
381. Jung HY, Park JK (2007) Blend membranes based on sulfonated poly(ether ether ketone) and poly(vinyllidene fluoride) for high performance direct methanol fuel cell. Electrochim Acta 52:7464–7468
382. Xue S, Yin G, Cai K, Shao Y (2007) Permeabilities of methanol. ethanol and fullydimethyl ether in new composite membranes: a comparison with Nafion membranes. J Membr Sci 289:51–57
383. Na T, Shao K, Zhu J, Sun H, Xu D, Zhang Z, Lew CM, Zhang G (2013) Composite membranes based on fully sulfonated poly(aryl ether ketone)/epoxy resin/different curing agents for direct methanol fuel cells. J Power Sources 230:290–297
384. Lin H, Zhao C, Jiang Y, Ma W, Na H (2011) Novel hybrid polymer electrolyte membranes with high proton conductivity prepared by a silane-crosslinking technique for direct methanol fuel cells. J Power Sources 196:1744–1749
385. Choi J, Kim DH, Kim HK, Shin C, Kim SC (2008) Polymer blend membranes of sulfonated poly(arylene ether ketone) for direct methanol fuel cells. J Membr Sci 310:384–392
386. Zhao C, Lin H, Na H (2010) Novel cross-linked sulfonated poly(arylene ether ketone) membranes for direct methanol fuel cell. Int J Hydrogen Energ 35:2176–2182
387. Zhang Y, Fei X, Zhang G, Li H, Shao K, Zhu J, Zhao C, Liu Z, Han M, Na H (2010) Preparation and properties of epoxy-based cross-linked sulfonated poly(arylene ether ketone) proton exchange membrane for direct methanol fuel cell applications. Int J Hydrogen Energ 35:6409–6417
388. Wang J, Zhao C, Lin H, Zhang G, Zhang Y, Ni J, Ma W, Na H (2011) Design of a stable and methanol resistant membrane with cross-linked multilayered polyelectrolyte complexes for direct methanol fuel cells. J Power Sources 196:5432–5437
389. Jiang Z, Zhao X, Manthiram A (2013) Sulfonated poly(ether ether ketone) membranes with sulfonated graphene oxide fillers for direct methanol fuel cells. Int J Hydrogen Energ 38:5875–5884
390. Ren S, Li C, Zhao X, Wu Z, Wang S, Sun G, Xin Q, Yang X (2005) Surface modification of sulfonated poly(ether ether ketone) membranes usin Nafion solution for direct methanol fuel cells. J Membr Sci 247:59–63
391. Zhang H, Fan X, Zhang J, Zhou Z (2008) Modification research of sulfonated PEEK membranes used in DMFC. Solid State Ion 179:1409–1412
392. Zhang W, Gogel V, Friedrich KA, Kerres J (2006) Novel covalently cross-linked poly (etheretherketone) ionomer membranes. J Power Sources 155:3–12
393. Tsai JC, Lin CK (2011) Preparation of main-chain-type and side-chain-type sulfonated poly (ether ether ketone) membranes for direct methanol fuel cell applications. J Power Sources 196:9308–9316

394. Yang T (2009) Composite membrane of sulfonated poly(ether ether ketone) and sulfated poly (vinyl alcohol) for use in direct methanol fuel cells. J Membr Sci 342:221–226
395. Fu Y, Manthiram A, Guiver MD (2007) Acid–base blend membranes based on 2-amino-benzimidazole and sulfonated poly(ether ether ketone) for direct methanol fuel cells. Electrochem Commun 9:905–910
396. Li X, Liu C, Xu D, Zhao C, Wang Z, Zhang G, Na H, Xing W (2006) Preparation and properties of sulfonated poly(ether ether ketone)s (SPEEK)/polypyrrole composite membranes for direct methanol fuel cells. J Power Sources 162:1–8
397. Li W, Manthiram A, Guiver MD, Liu B (2010) High performance direct methanol fuel cells based on acid–base blend membranes containing benzotriazole. Electrochem Commun 12:607–610
398. Zuo Z, Zhao X, Manthiram A (2013) High-performance blend membranes composed of an amphoteric copolymer containing supramolecular nanosieves for direct methanol fuel cells. RSC Adv 3:6759–6762
399. Mecheri B, D'Epifanio A, Traversa E, Licoccia S (2008) Sulfonated polyether ether ketone and hydrated tin oxide proton conducting composites for direct methanol fuel cell applications. J Power Sources 178:554–560
400. Lee JK, Li W, Manthiram A (2009) Poly(arylene ether sulfone)s containing pendant sulfonic acid groups as membrane materials for direct methanol fuel cells. J Membr Sci 330:73–79
401. Roelofs KS, Hirth T, Schiestel T (2010) Sulfonated poly(ether ether ketone)-based silica nanocomposite membranes for direct ethanol fuel cells. J Membr Sci 346:215–226
402. Maab H, Pereira Nunes S (2010) Modified SPEEK membranes for direct ethanol fuel cell. J Power Sources 195:4036–4042
403. Wainright JS, Wang J-T, Weng D, Savinell RF, Litt M (1995) Acid-doped polybenzimidazoles: a new polymer electrolyte. J Electrochem Soc 142:L121–L123
404. Samms SR, Wasmus S, Savinell RF (1996) Thermal stability of proton conducting acid doped polybenzimidazole in simulated fuel cell environments. J Electrochem Soc 143:1225–1232
405. Wang J-T, Wainright JS, Savinell RF, Litt M (1996) A direct methanol fuel cell using acid-doped polybenzimidazole as polymer electrolyte. J Appl Electrochem 26:751–756
406. Bouchet R, Siebert E (1999) Proton conduction in acid doped polybenzimidazole. Solid State Ion 118:287–299
407. He R, Li Q, Jensen JO, Bjerrum NJ (2007) Doping phosphoric acid in polybenzimidazole membranes for high temperature proton exchange membrane fuel cell. J Polym Sci Pol Chem 45:2989–2997
408. Lobato J, Cañizares P, Rodrigo MA, Linares JJ, Manjavacas G (2006) Synthesis and characterization of poly[2,2-(m-phenylene)-5,5-bibenzimidazole] as polymer electrolyte membrane for high temperature PEMFCs. J Membr Sci 280:351–362
409. Diaz LA, Abuin GC, Corti HR (2009) Water and phosphoric acid uptake of poly [2,5-benzimidazole] (ABPBI) membranes prepared by low and high temperature casting. J Power Sources 188:45–50
410. Wang J-T, Wasmus S, Savinell RF (1996) Real-time mass spectrometric study of the methanol crossover in a direct methanol fuel cell. J Electrochem Soc 143:1233–1239
411. Weng D, Wainright JS, Landau U, Savinell RF (1996) Electro-osmotic drag coefficient of water and methanol in polymer electrolytes at elevated temperatures. J Electrochem Soc 143:1260–1263
412. Li Q, Hjuller HA, Bjerrum NJ (2001) Phosphoric acid doped polybenzimidazole membranes: physicochemical characterization and fuel cell applications. J Appl Electrochem 31:773–779
413. Pu H, Liu Q (2004) Methanol permeation and proton conductivity of polybenzimidazole and sulfonated polybenzimidazole. Polym Int 53:1512–1516
414. Chuang SW, Hsu LC, Hsu CL (2007) Synthesis and properties of fluorine-containing polybenzimidazole/montmorillonite nanocomposite membranes for direct methanol fuel cell applications. J Power Sources 168:172–177

415. Wycisk R, Lee JK, Pintauro PN (2005) Sulfonated polyphosphazene-polybenzimidazole membranes for DMFCs. J Electrochem Soc 152:A892–A898
416. Zhang H, Li X, Zhao C, Fu T, Shi Y, Na H (2008) Composite membranes based on highly sulphonated PEEK and PBI: morphology characteristics and performance. J Membr Sci 308:66–74
417. Glipa X, Bonnet B, Mula B, Jones DJ, Rozière J (1999) Investigation of the conduction properties of phosphoric and sulphuric acid doped polybenzoimidazole. J Mater Chem 9:3045–3049
418. Lobato J, Cañizares P, Rodrigo MA, Linares JJ, Aguilar JÁ (2007) Improved polybenzimidazole films for H3PO4-doped PBI-based a vapour-fed polybenzoimidazole (PBI).based high temperature PEMFC. J Membr Sci 306:47–53
419. Asensio JA, Borrós S, Gómez-Romero P (2003) Enhanced conductivity in polyanion-containing polybenzimidazoles. Improved materials for próton-exchange membranes and PEM fuel cells. Electrochem Commun 5:967–972
420. Ma YL, Wainright JS, Litt M, Savinell RF (2004) Conductivity of PBI membranes for high-temperature polymer electrolyte fuel cells. J Electrochem Soc 151:A8–A16
421. He R, Li Q, Xiao G, Bjerrum NJ (2003) Próton conductivity of phosphoric acid doped polybenzimidazole and its composites with inorganic próton conductors. J Membr Sci 226:169–184
422. He R, Li Q, Bach A, Jensen JO, Bjerrum NJ (2006) Physicochemical properties of phos phoric acid doped polybenzimidazole membranes for fuel cells. J Membr Sci 277:38–45
423. Hasiotis C, Deimede V, Kontoyannis C (2001) New polymer electrolytes based on blends of sulfonated polysulfones with polybenzimidazole. Electrochim Acta 46:2401–2406
424. Schechter A, Savinell RF (2002) Imidazole and 1-methyl imidazole in phosphoric acid doped polybenzimidazole, electrolyte for fuel cells. Solid State Ion 147:181–187
425. Diaz L (2012) Thesis. study of PBI and ABPBI membranes for high temperature and methanol fed fuel cells. University of General San Martin (UNSAM)
426. Mecerreyes D, Grande H, Miguel O, Ochoteco E, Marcilla R, Cantero I (2004) Porous polybenzimidazole membranes doped with phosphoric acid: highly proton-conducting solid electrolyte. Chem Mater 16:604–607
427. Weber J, Kreuer KD, Maier J, Thomas A (2008) Proton conductivity enhancement by nanostructural control of poly(benzimidazole)-phosphoric acid adducts. Adv Mater 20:2595–2598
428. Li Q, Pan C, Jensen JO, Noye P, Bjerrum NJ (2007) Cross-linked polybenzimidazole membranes for fuel cells. Chem Mater 19:350–352
429. Xu H, Chen K, Guo X, Fang J, Yin J (2007) Synthesis of hyperbranched polybenzimidazoles and their membrane formation. J Membr Sci 288:255–260
430. Leikin AY, Rusanov AL, Begunov RS, Fomenkov AI (2009) Synthesis and properties of poly [2-(4′-oxyphenylene)-5- benzimidazole] and a proton-exchange membrane produced on its basis. Polym Sci Ser C 51:12–16
431. Kim SK, Kim TH, Jung JW, Lee JC (2009) Polybenzimidazole containing benzimidazole side groups for high-temperature fuel cells applications. Polymer 50:3495–3502
432. Qian G, Benicewicz BC (2009) Synthesis and characterization of high molecular weight hexafluoroisopropylidene-containing polybenzimidazole for high-temperature polymer electrolyte fuel cells. J Polym Sci Pol Chem 47:4064–4073
433. Chuang SW, Hsu SLC, Yang ML (2008) Preparation and characterization of fluorine-containing polybenzimidazole/imidazole hybrid membranes for proton exchange membrane fuel cells. Eur Polym J 44:2202–2206
434. Lobato J, Cañizares P, Rodrigo MA, Linares JJ, Lopez-Vizcaino R (2008) Performance of a vapor-fed polybenzimidazole (PBI)-based direct methanol fuel cell. Energy Fuel 22:3335–3345
435. Haghighi AH, Hasani-Sabradabi MM, Dashtimoghadam E, Bahlakeh G, Shakeri SE, Majedi FS, Emami SH, Moadedel H (2011) Direct methanol fuel cell performance of sulfonated poly

(2,6-dimethyl-1,4-phenylene oxide)-polybenzimi dazole blend proton exchange membrane. Int J Hydrogen Energ 36:3688–3696
436. Pasupathi S, Ji S, Bladergroen BJ, Linkov V (2008) High DMFC performance output using modified acid–base polymer blend. Int J Hydrogen Energ 33:3132–3136
437. Kerres J, Zhang W, Jörissen L, Gogel V (2002) Application of different types of polyaryl-blend-membranes in DMFC. J New Mater Electrochem Syst 5:97–107
438. Jörissen L, Gogel V, Kerres L, Garche J (2002) New membranes for direct methanol fuel cells. J Power Sources 105:267–273
439. Silva VS, Weisshaar S, Reissner R, Ruffmann B, Vetter S, Mendes A, Madeira LM, Nunes S (2005) Performance and efficiency of a DMFC using non-fluorinated composite membranes operating at low/medium temperatures. J Power Sources 145:485–494
440. Lobato J, Cañizares P, Rodrigo MA, Linares JJ (2009) Study of different bimetallic anodic catalysts supported on carbon for a high temperature polybenzimidazole-based direct ethanol fuel cell. Appl Catal B-Environ 91:269–274
441. Linares JJ, Rocha TA, Zignani S, Paganin VA, Gonzalez ER (2013) Different anode catalyst for high temperature polybenzimidazole-based direct ethanol fuel cells. Int J Hydrogen Energ 38:620–630
442. Asensio JA, Borrós S, Gómez-Romero P (2004) Polymer electrolyte fuel cells based on phosphoric acid-impregnated poly(2,5-benzimidazole) membranes. J Electrochem Soc 151: A304–A310
443. Asensio JA, Borrós S, Gómez-Romero P (2004) Proton-conducting membranes based on poly (2,5-benzimidazole) (ABPBI) and phosphoric acid prepared by direct acid casting. J Membr Sci 241:89–93
444. Gómez-Romero P, Asensio JA, Borrós S (2005) Hybrid proton-conducting membranes for polymer electrolyte fuel cells. Phosphomolybdic acid doped poly(2,5-benzimidazole) – (ABPBI-$H_3PMo_{12}O_{40}$). Electrochim Acta 50:4715–4720
445. Asensio JA, Gómez-Romero P (2005) Recent developments on proton conducting poly (2,5-benzimidazole) (ABPBI) membranes for high temperature polymer electrolyte membrane fuel cells. Fuel Cells 5:336–343
446. Krishnan P, Park JS, Kim CS (2006) Performance of a poly(2,5-benzimidazole) membrane based high temperature PEM fuel cell in the presence of carbon monoxide. J Power Sources 159:817–823
447. Kim HJ, Cho SY, An SJ, Eun YC, Kim JY, Yoon HK, Kweon HJ, Yew KH (2004) Synthesis of poly(2,5-benzimidazole) for use as a fuel-cell membrane. Macromol Rapid Commun 25:894–897
448. Wannek C, Lehnert W, Mergel J (2009) Membrane electrode assemblies for high-temperature polymer electrolyte fuel cells based on poly (2,5-benzimidazole) membranes with phosphoric acid impregnation via the catalyst layer. J Power Sources 192:258–266
449. Gulledge AL, Gu B, Benicewicz BC (2012) A new sequence isomer of AB-polybenzimidazole for high temperature PEM fuel cells. J Polym Sci Pol Chem 50:306–313
450. Acar O, Sen U, Bozkurt A, Ata A (2009) Proton conducting membranes based on poly (2,5-benzimidazole) (ABPBI)-poly(vinylphosphonic acid) blends for fuel cells. Int J Hydrogen Energ 34:2724–2730
451. Genies C, Mercier R, Sillion B, Cornet N, Gebel G, Pineri M (2001) Soluble sulfonated naphtalenic polyimides as materiasl for proton exchange membranes. Polymer 42:359–373
452. Woo Y, Oh SY, Kang YS, Jung B (2003) Synthesis and characterization of sulfonated polyimide membranes for direct methanol fuel cell. J Membr Sci 220:31–45
453. Einsla BR, Kim YS, Hickner MA, Hong YT, Hill ML, Pivovar BS, McGrath JE (2005) Sulphonated naphtalene dianhydride based polyimide copolymers for proton-exchange-membrane fuel cells. II. Membrane properties and fuel cell performance. J Membr Sci 255:141–148

454. Song JM, Miyatake K, Uchida H, Watanabe M (2006) Investigation of direct methanol fuel cell performance of sulfonated polyimide membrane. Electrochim Acta 51:4497–4504
455. Higuchi E, Asano N, Miyatake K, Uchida H, Watanabe M (2007) Distribution profile of water and supression of methanol crossover in sulfonated polyimide electrolyte membrane for direct methanol fuel cells. Electrochim Acta 52:5272–5280
456. Zhai F, Guo X, Fang J, Xu H (2007) Synthesis and properties of novel sulfonated polyimide membranes for direct methanol fuel cell application. J Membr Sci 296:102–109
457. Hu Z, Ogou T, Yoshino M, Yamada O, Kita H, Okamoto KI (2009) Direct methanol fuel cell performance of sulfonated polymide membranes. J Power Sources 194:674–682
458. Park HB, Lee CH, Sohn JY, Lee YM, Freeman BD, Kim HJ (2006) Effect of crosslinked chain lenght in sulfonated polyimide membranes on water sorption, proton conduction, and methanol permeation properties. J Membr Sci 285:423–443
459. Munukata H, Yamamoto D, Kanamura K (2008) Three-dimensionally ordered macroporous polyimide composite membrane with controlled pore size for direct methanol fuel cells. J Power Sources 178:596–602
460. Alcaide F, Alvarez G, Ganborena L, Iruin JJ, Miguel O, Blazquez JA (2009) Proton-conducting membranes from phosphotungstic acid-doped sulfonated polyimide for direct methanol fuel cell applications. Polym Bull 62:813–827
461. Nguyen T, Wang X (2010) Multifunctional composite membranes based on a highly porous polyimide matrix for direct methanol fuel cells. J Power Sources 195:1024–1030
462. Sung KA, Cho KY, Kim WK, Park JK (2010) Sulfonated polyimide coated with crosslinkable layer for direct methanol fuel cell. Electrochim Acta 55:995–1000
463. Liu D, Geng L, Fu Y, Dai X, Lü C (2011) Novel nanocomposite membranes based on sulfonated mesoporous silica nanoparticles modified sulfonated polyimides for direct methanol fuel cells. J Membr Sci 366:251–257
464. Geng L, He Y, Liu D, Dai X, Lü C (2012) Facile in situ template synthesis of sulfonated polyimide/mesoporous silica hybrid proton exchange membrane for direct methanol fuel cells. Micropor Mesopor Mat 148:8–14
465. Kim YS, Sumner MJ, Harrison WL, Riffle JS, McGrath JE, Pivovar BS (2004) Direct methanol fuel cell performance of disulfonated poly(arylene ether benzonitrile) copolymers. J Electrochem Soc 151:A2150–A2156
466. Lee CH, Min KA, Park HB, Hong YT, Jung BO, Lee YM (2007) Sulfonated poly(arylene ether sulfone)-silica nanocomposite membrane for direct methanol fuel cell (DMFC). J Membr Sci 303:258–266
467. Kim DH, Kim SC (2008) Transport properties of polymer blend membranes of sulfonated and nonsulfonated polysulfones for direct methanol fuel cell application. Macromol Res 16:457–466
468. Lee CH, Lee SY, Lee YM, Lee SY, Rhim JW, Lane O, McGrath JE (2009) Surface-fluorinated proton exchange membrane with high electrochemical durability for direct methanol fuel cells. ACS Appl Mater Interface 1:1113–1121
469. Kim HK, Kim DH, Choi J, Kim SC (2010) Compositional effect on the properties of sulfonated and nonsulfonated polymer blends for direct methanol fuel cell. Macromol Res 19:928–942
470. Oh SY, Park JY, Yu DM, Hong SK, Hong YT (2012) Preparation and characterization of acid-acid blend membranes for direct methanol fuel cell applications. Macromol Res 20:121–127
471. Joo SH, Pak C, Kim EA, Lee YH, Chang H, Seung D, Choi YS, Park JB, Kim TK (2008) Functionalized carbon nanotube-poly(arylene sulfone) composite membranes for direct methanol fuel cells with enhanced performance. J Power Sources 180:63–70
472. Gil SC, Kim JC, Ahn D, Jang JS, Kim H, Jung JC, Lim S, Jung DH, Lee W (2012) Thermally crosslinked sulfonated polyethersulfone proton exchange membranes for direct methanol fuel cells. J Membr Sci 417–418:2–9

473. Li L, Wang Y (2005) Sulfonated polyethersulfone Cardo membranes for direct methanol fuel cell. J Membr Sci 246:167–172
474. Lufrano F, Baglio V, Staiti P, Aricó AS, Antonucci V (2006) Development and characterization of sulfonated polysulfone membranes for direct methanol fuel cells. Desalination 199:283–285
475. Norris BC, Li W, Lee E, Manthiram A, Bielawski CW (2010) Click-functonalization of poly (sulfone)s and a study of their utilities as proton conductive membranes in direct methanol fuel cells. Polymer 51:5352–5358
476. Li W, Manthiram A (2010) Sulfonated poly(arylene ether sulfone) as a methanol-barrier layer in multilayer membranes for direct methanol fuel cells. J Power Sources 195:962–968
477. Kim DS, Shin KH, Park HB, Lee YM (2004) Preparation and characterization of sulfonated poly(phthalazinone ether sulfone ketone) (SPPESK)/silica hybrid membranes for direct methanol fuel cell applications. Macromol Res 12:413–421
478. Park HB, Shin HS, Lee YM, Rhim JW (2005) Annealing effect of sulfonated polysulfone ionomer membranes on proton conductivity and methanol transport. J Membr Sci 247:103–110
479. Fu YZ, Manthiram A (2006) Synthesis and characterization of sulfonated polysulfone membranes for direct methanol fuel cells. J Power Sources 157:222–225
480. Fu Y, Li W, Manthiram A (2008) Sulfonated polysulfone with 1,3-1H-dibenzimidazole-benzene additive as a membrane for direct methanol fuel cells. J Membr Sci 310:262–267
481. Lufrano F, Baglio V, Staiti P, Stassi A, Aricó AS, Antonucci V (2010) Investigation of sulfonated polysulfone membranes as electrolyte in a passive-mode direct methanol fuel cell ministack. J Power Sources 195:7727–7733
482. Roh SC, Hong JH, Kim CK (2012) Polymer electrolyte membranes fabricated from poly (ethylene glycol dimethylmethacrylate-co-styrene sulfonic acid) copolymers for direct methanol fuel cell application. Marcromol Res 20:197–204
483. Kim S, Lee H, Ahn D, Park HW, Chang T, Lee W (2013) Direct sulfonation and photocrosslinking of unsaturated poly(styrene-b-butadiene-b-styrene) for proton exchange membrane of direct methanol fuel cell. J Membr Sci 427:85–91
484. Wei Z, He S, Liu X, Qiao J, Lin J, Zhang L (2013) A novel environment-friendly route to prepare proton exchange membranes for direct methanol fuel cells. Polymer 54:1243–1250
485. Kumar GG, Kim AR, Nahm KS, Yoo DJ (2011) High proton conductivity and low crossover of polyvinylidene fluoride-hexafluoro propylene-silica sulphuric acid composite membranes for direct methanol fuel cells. Curr Appl Phys 11:896–902
486. Surya Prakash GK, Smart MC, Wang QJ, Atti A, Pleynet V, Yang B, McGrath K, Olah GA, Narayanan SR, Chun W, Valdez T, Surampudi S (2004) High efficiency direct methanol fuel cell based on poly(styrenesulfonic) acid (PSSA) – poly(vinylidene fluoride) (PVDF) composite membranes. J Fluorine Chem 125:1217–1230
487. Nasef M, Zubir NA, Ismail AF, Khayet M (2006) Sulfonated radiation grafted polystyrene pore-filled poly(vinylidene fluoride) membranes for direct methanol fuel cells: structure—property correlations. Desalination 200:642–644
488. Byun J, Sauk J, Kim H (2009) Preparation of PVdF/PSSA composite membranes using supercritical carbon dioxide for direct methanol fuel cells. Int J Hydrogen Energ 34:6437–6442
489. Shen J, Xi J, Zhu W, Chen L, Qiu X (2006) A nanocomposite proton exchange membrane based on PVDF, poly(2-acrylamido-2-methyl propylene sulfonic acid), and nano-Al_2O_3 for direct methanol fuel cells. J Power Sources 159:894–899
490. Kumar GG, Shin J, Nho YC, Hwang IS, Fei G, Kim AR, Nahm KS (2010) Irradiated PVdF-HFP-tin oxide composite membranes for the applications of direct methanol fuel cells. J Membr Sci 350:92–100
491. Bhavani P, Sangeetha D (2012) Blend membranes for direct methanol and proton exchange membrane fuel cells. Chinese J Polym Sci 30:548–560

492. Merle G, Wessling M, Nijmeijer K (2011) Anion exchange membranes for alkaline fuel cells: a review. J Membr Sci 377:1–35
493. Leykin AY, Ahkrebko OA, Tarasevich MR (2009) Ethanol crossover through alkali-doped polybenzoimidazole membrane. J Membr Sci 328:86–89
494. Yameen B, Kaltbeitzel A, Langer A, Müller F, Gösele U, Knoll W, Azzaroni O (2009) Highly proton-conducting self-humidifying microchannels generated by copolymer brushes on a scaffold. Angew Chem Int Ed 48:3124–3228
495. Ashcraft JN, Avni A, Argun AA, Hammond PT (2010) Structure–property studies of highly conductive layer-by-layer assembled membranes for fuel cell PEM applications. J Mater Chem 20:6250–6257

Chapter 7
Carbon Materials for Fuel Cells

Mariano M. Bruno and Federico A. Viva

Abstract Carbon materials are fundamental for the manufacturing of fuel cells. Several fuel cell components are made entirely of carbon in a graphitic form. In the present chapter, an overview of the different fuel cell carbon components and the materials used in their preparation will be presented. Novel approaches in the synthetic method, in order to impart desired properties, and in the manufacturing of the components will be shown. Also, relevant results on the latest research conducted will be discussed.

7.1 Introduction

Fuel cells (FC) are electrochemical devices that convert chemical energy directly into electrical energy and are considered to be the best choice to replace batteries for delivering power for a wide range of applications. The advances made in the basic understanding of fuel cells as well as improvements in fuel cell technology allow us to realize the benefits of switching to fuel cells as reliable power sources.

The main component of a Polymer Electrolyte Membrane Fuel Cell (PEMFC) is the Membrane Electrode Assembly (MEA) [1, 2]. The MEA is formed by a polymer membrane flanked by two electrodes. The membrane acts as the ionic conductor between the two electrodes; the anode, where the fuel is oxidized, and the cathode, where the oxidant is reduced. The electrodes are formed by a porous material named Gas Diffusion Layer (GDL) with a thin layer of an electrocatalyst denominated the Catalyst Layer (CL), The electrocatalyst, responsible of driving

M.M. Bruno (✉) • F.A. Viva
Grupo Celdas de Combustible, Departamento de Física de la Materia Condensada, Centro Atómico Constituyentes, Comisión Nacional de Energía Atómica (CNEA), Av General Paz 1499 (1650), San Martín, Buenos Aires, Argentina
e-mail: mbruno@tandar.cnea.gov.ar; viva@tandar.cnea.gov.ar

H.R. Corti and E.R. Gonzalez (eds.), *Direct Alcohol Fuel Cells: Materials, Performance, Durability and Applications*, DOI 10.1007/978-94-007-7708-8_7,

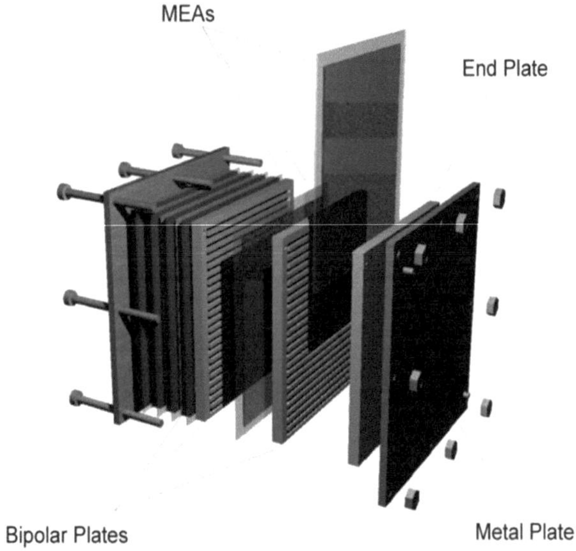

Fig. 7.1 Fuel cell stack components

the electrochemical reactions that takes place at each electrode, is composed by metal nanoparticles (3–5 nm) dispersed on carbon nanoparticles of bigger size (20–40 nm). The GDL–CL combination is commonly referred as gas diffusion electrode (GDE). In a single cell the MEA is placed between a pair of Current Collector Plates (CCP) with channels machined on one of its faces that allows the reactants to flow through the MEA surface. A FC stack is formed by intercalating MEAs with plates machined on both faces (bipolar plates-BP) and contained between two end plates. Figure 7.1 shows a schematic representation of a fuel cell with its different components.

The voltage of a FC is given by [3]

$$E = E_{rev} - \eta_{act} - \eta_{ohm} - \eta_{mass \cdot trsp}$$

Where E_{rev} is the reversible potential, which is the theoretical cell voltage at open circuit conditions. The other three terms represent losses to the voltage known as polarization (η). These are the activation polarization (η_{act}), resistance polarization (η_{ohmic}) and mass transport or concentration polarization ($\eta_{mass\ trsp}$), each of which predominates on a different zone of the polarization plot as shown in Fig. 7.2 and are a function of the current density (j). Under experimental conditions, the thermodynamic E_{rev} is never observed due to irreversible losses such as parasitic electrochemical processes, mixed potentials and even corrosion process [4]. Only the experimental open circuit voltage (OCV) is observed under zero current flow, indicated as E_o in the graph. Apart from the ohmic polarization, related with an energy loss due to the internal resistance of the cell which plays a predominant role on the central part of the plot, the other two polarizations can be divided into contributions from the anode and the cathode. The activation polarization,

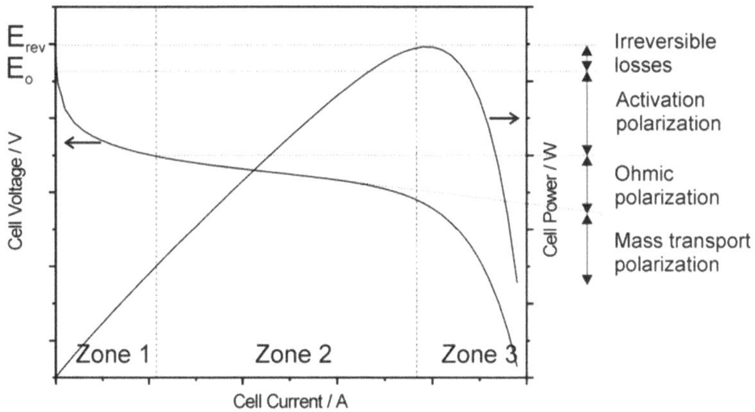

Fig. 7.2 Polarization and power plots and the different polarization contributions

dominant at low current densities, depends on both the chemical nature of the reacting compound and the electrode composition in terms of chemical and physical nature. The mass transport polarization is related to the rate at which the compounds reach the surface of the electrode, therefore it has a dominant contribution at high current densities.

In between the GDL and the membrane, i.e. where the CL lies, emerges a zone denominated the triple phase boundary where the electrocatalyst, the ionic conductor (membrane) and the reactants (liquid or gases) meet. Figure 7.3 shows a simplified representation of the triple phase boundary. In that zone takes place the electrochemical processes that generate the FC voltage an electric current. On the electrocatalyst surface the fuel is oxidized or the oxygen is reduced, the ionic conductor (polymer membrane) allows the H^+ to move from one electrode to the other while the electrons leave or arrive the electrocatalyst through the carbon based conductive material (carbon support, GDL, current collector plate). The properties of the materials involved and the way in which the components are arranged in the triple phase boundary plays a major role in the correct functioning of the FC.

Carbon constitutes the most abundant element of the different FC components. Setting aside the membrane, which is a polymer with a carbon backbone, all the other components, i.e. the CL, GDL and current collector plates (bipolar plates) are made almost entirely of graphitic carbon. The electrocatalyst support of the CL is commonly carbon black in the form of fine powder. GDLs are thin porous layers formed by carbon fibers interconnected as a web or fabric, while current collector plates are carbon monoliths with low bulk porosity. As explained above each of these components has a particular function within the fuel cell and in particular in the triple phase boundary. The structure and properties of the carbon in each of the different FC components will determine the whole performance of the cell.

Cost is a major factor for the implementation of fuel cells in the everyday use [5, 6]. There is some variation in the cost percentages assigned to each component

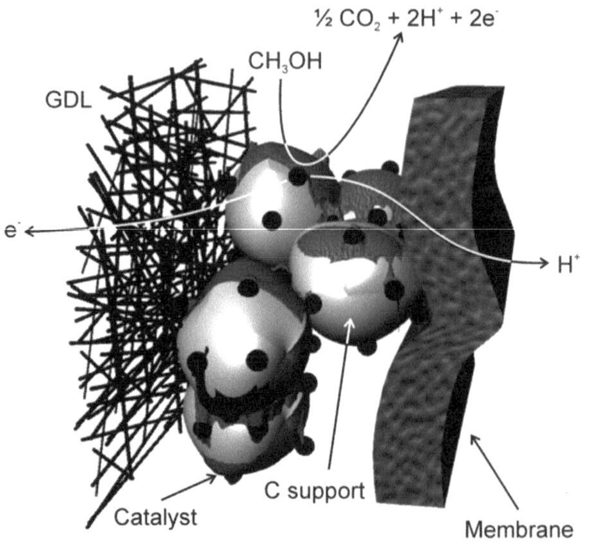

Fig. 7.3 Schematic representation of the triple phase boundary in the MEA

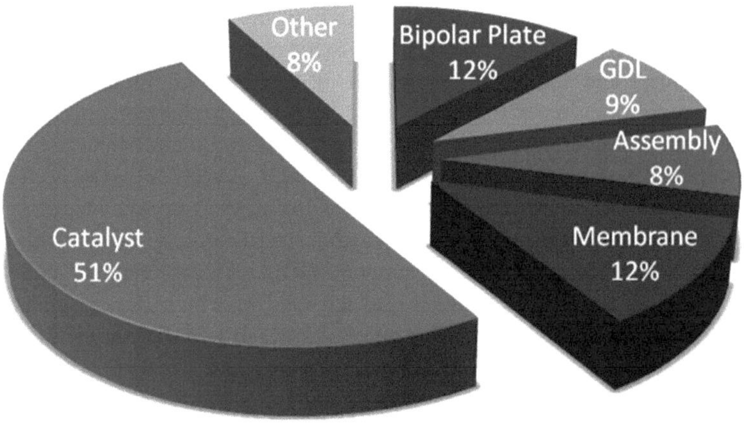

Fig. 7.4 Cost distribution of a fuel cell stack

[5, 7–9] which is related to the kind and size of system considered and to the analysis model used [8, 10]. Figure 7.4 shows the different cost contribution for a FC stack compiled from several sources [5, 9–11]. The bipolar plates and GDL contributes together to 21 % of the stack cost making those components an important factor in the final cost. Reducing the cost of the carbon based components is as important as improving their individual properties.

7.2 Carbon as Catalyst Support

Carbonous materials have excellent properties as support for electrocatalysts in fuel cells. Besides being an electronic conductor with low cost, they possess a good corrosion resistance to acidic and basic environments and low density. In the last decades, the syntheses of structured carbonous materials with controlled and designed properties have taken major attention in the scientist community, where their application can generate a technological advance on energy devices i.e., nowadays the production of carbon materials with tailored physical and chemical properties can help to the developing of power sources devices.

Direct methanol fuel cell (DMFC) suffers some drawbacks which are currently investigated in order to improve their performance and durability, and reducing their cost. In this regard, several approaches are being used to overcome these issues. Among those, can be mentioned [12]:

- Reducing Pt content by developing metallic alloy or Pt-free electrocatalysts.
- Reducing the electrocatalyst loading in fuel cell electrodes by decreasing particle size and avoiding their sintering.
- Increasing mass-transport at the fuel cell.
- Increasing the durability of the fuel cell components.

Porous carbons with tuneable properties have great potential to achieve the goals listed above. An adequate pore size of the carbon support would provide high catalytic dispersion and would enhance the mass transport, which implies a significant cost diminution of the fuel cell by reducing the catalyst loading.

The pore size distribution in the carbon support is an important factor for a well performance of the catalyst. Pores in the nanometric scale are classified by IUPAC in three groups: the micropores are those with diameters lower than 2 nm, the mesopores with diameters between 2 and 50 nm and the macropores with diameters larger than 50 nm. Each pores size offers different benefits, the micropores produce materials with high surfaces area but could be inaccessible to liquid solutions or have slow mass transport. A material with mesopores has a lower surface area but better accessibility than those with micropores. Finally, materials with macropores show the lowest surface area, but they are easily accessible to liquid fuel. For this reason, the structured carbons, principally mesoporous carbon, have attracted considerable attention due to their potential application in the catalyst area, where the challenging is to favour the dispersion of catalyst and allow the accessibility of liquids that feed the anode side of a DMFC. In the following sections a description of different carbons support will be discussed stressing on the effect of the porous structure.

7.2.1 *Carbon Blacks*

Carbon blacks are produced by pyrolysis of hydrocarbons from petroleum derivatives. Generally, the carbon blacks are composed of spherical nanoparticles that have crystalline domains composed by stacked layers parallel to the surface.

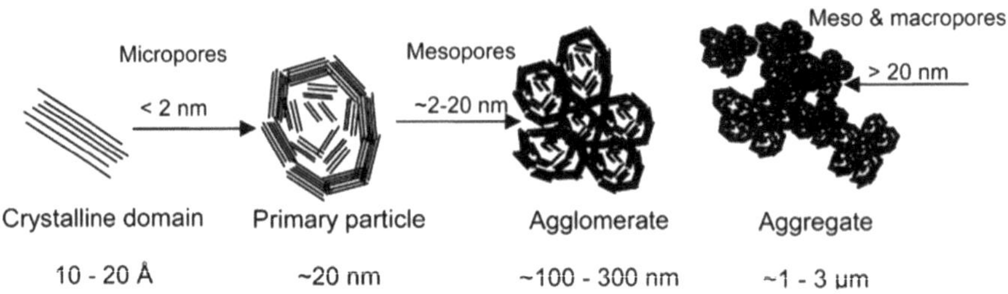

Fig. 7.5 Schematic representation of the structures of carbon black supports: crystalline domain, primary carbon particle, agglomerate and aggregate [13] (Reprinted by permission of the publisher)

This order ensures some degree of graphitization for the materials. The particles are found building aggregates in the form of chains or clusters forming meso and macropores, as show in the Fig. 7.5.

Carbons black includes several types of carbons, such as acetylene black, channel black, furnace black, lamp black. Commonly, their names are referred to the process or the source material from which they are made. Among those, the production of furnace black is the most important. Its production process consists in feeding a furnace with natural gas and aromatics oils as feedstock, where is vaporized and then pyrolyzed. Vulcan XC-72 (a furnace black from Cabot Corporation) is the most widely used catalyst support for low-temperature fuel cells due to their low cost and high availability, being this material used as standard to compare other types of carbons. Vulcan XC-72, formed by nanoparticles of 20–40 nm, has an electrical conductivity of 4 S cm^{-1}, a sulphur content of 0.05 %, and a negligible oxygen content [13]. Within the textural properties Vulcan carbon has a superficial area of 252 m^2 g^{-1}, with a total pore volume of 0.63 cm^3 g^{-1} and a pore size distribution around 15 nm [14].

Like most of the materials (GDL, membranes) used in DMFC, the application of carbon black as catalyst support is a direct extrapolation of their use in hydrogen fuel cell. However, the processes occurring in the anode side are more complex. The carbonous support besides promoting a high dispersion of the catalyst (catalytic area), should allows also the free entrance of the liquid alcohol to the catalyst and avoid the occluding of the gas close to the catalytic zone. Nevertheless the pore structure should not hinder the triple phase boundary formation.

Active sites on the carbon surface, which are unsaturated carbon atoms at the edges planes, defects of the graphitic hexagonal crystallites and oxygen surface groups, provides anchoring centres for catalyst particles favouring the nanoparticle dispersion and inhibiting their sintering [15]. The proportion and nature of these active sites have a strong effect on the preparation, morphology, particle size and catalytic activity of supported catalysts. Fortunately, several authors have been devoted to study the effect of carbon properties on the preparation, dispersion and catalytic activity of supported nanoparticles [16–24]. Takasu et al. [21] analyzed the effect of the surface area of the carbon support on the characteristics of PtRu nanoparticles which were prepared by

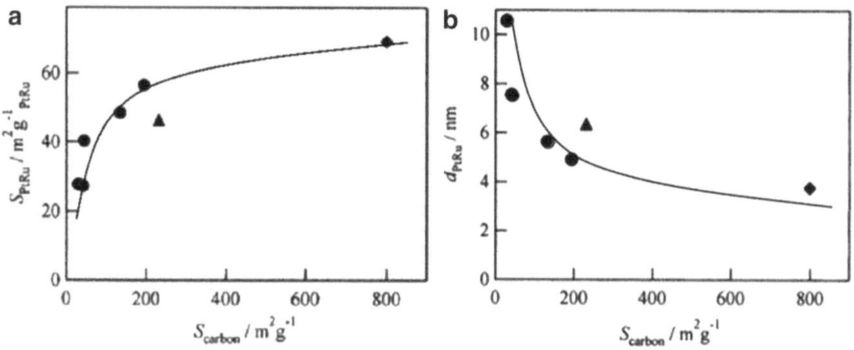

Fig. 7.6 The dependence of (**a**) the specific surface area, SPtRu, and (**b**) particle size, dPtRu, for Pt50Ru50(30 mass%)/C catalysts on the specific surface area, Scarbon, of the carbon black support: ● Mitsubishi, ▲ Vulcan XC-72R, ♦ Ketjen EC [21] (Reprinted by permission of the publisher)

conventional impregnation method. Figure 7.6 a, b show the dependence of the specific surface area of carbon blacks supports, with specific surface area in the range of 29–800 m^2 g^{-1}, on the electrochemical surface area and the particle size of $Pt_{50}Ru_{50}$, determined by CO_{ad} stripping voltammetry and from HRSEM micrographs of the catalysts, respectively. It is important to remark that the increasing of superficial area indicates that a larger portion of the pores is in the size range of micropores.

Carbon blacks as support lead to an increase of electrochemical surface area and a decrease of the nanoparticle diameter of the catalyst when increasing the carbon surface area, at equal synthesis conditions. However, the carbons with higher micropores content did not show a positive effect on the methanol oxidation reaction. Uchida et al. [25, 26] showed for Pt/C that the catalyst structure on the carbon support has a strong effect on the performance of hydrogen fed fuel cells, focusing their analysis in the performance of triple phase boundary. In order to get the optimal microstructure of the catalyst layer they analyzed its components and preparation, reporting that the colloids particle size distribution of the perfluorosulfonate ionomer (Flemion®) in butyl acetate had a mean diameter *ca.* 43 nm. This result suggested that the ionomer colloids could not penetrate into pores $<$ 40 nm and the Pt nanoparticles inside of these pores did not take part of the triple phase boundary. Therefore, a diameter of 40 nm could be considered as the minimum diameter of pore to take advantage of all the catalyst deposited.

A similar study was carried by Rao et al. [23] in Sibunit carbons. They analyzed the pore effect on the performance of DMFC with carbon supports with a range of surface areas from 6 to 415 m^2 g^{-1}. Figure 7.7 shows the observed specific activity for methanol oxidation at an anode potential of 0.5 V as a function of the surface area.

A downward trend of specific catalytic activity when carbon surface area is increased is observed. This behaviour can be attributed to many factors, but the authors emphasized that the particle size (indicated in the labels of Fig. 7.7) and

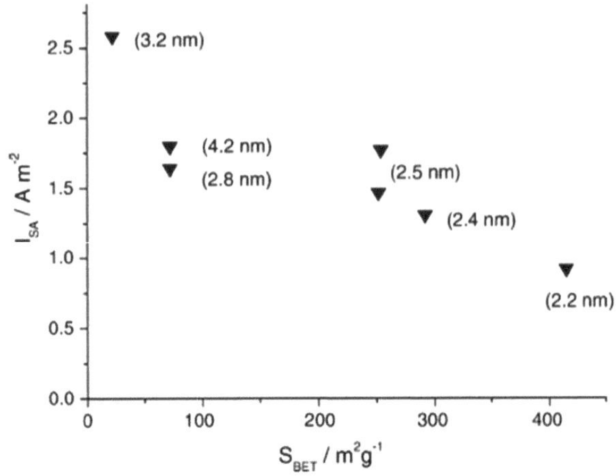

Fig. 7.7 Specific activities (A m^{-2}) at 0.5V RHE plotted versus surface area (SBET) of carbon supports [23] (Reprinted by permission of the publisher)

dispersion of catalyst were maintained almost constant. Moreover, the authors reported a similar trend for catalyst utilisation versus surface area. This trend agrees with those observed by Uchida et al. [26], indicating that the perfluorosulfonate ionomer colloids do not penetrate into the carbon pores of small diameter. However, Rao et al. suggested that with pores size larger than 20 nm optimal results were achieved.

As stated above, high surface area of a carbon support has a positive effect on the dispersion of the catalyst. However, catalyst particles within small pores may not contribute to the triple phase boundary. Therefore, microporous materials should be discarded as catalyst support, while carbons with large mesopores would promote a high electroactive area of the PtRu catalyst and would facilitate the mass transport.

7.2.2 Carbon Nanotubes and Nanofibers

The extraordinary mechanical, electronic and thermal properties of carbon nanofibers (CNFs) and carbon nanotubes (CNTs) make them suitable in several fields of materials technology, including supported catalysts for energy conversion. Graphite nanofibers, carbon filaments and carbon nanotubes, are terms employed to refer to nanofilamentous carbon. These materials can be classified into two categories: fibers and tubes. A schematic representation of their structural features is shown in the Fig. 7.8.

Although, the first report about the production of graphite nanofibers is older than a century [27], the discovery, in 1991, of multi-wall carbon nanotubes (MWNT) by Iijima [28], caused a renewed interest in this kind of carbons. There are several types of carbon nanofibers which differ in the disposition of the graphene layers. In Fig. 7.8 are shown the ribbon-like graphite nanofiber, where the graphene layers are parallel to the growth axis, the herringbone nanofibers having their layers stacked obliquely

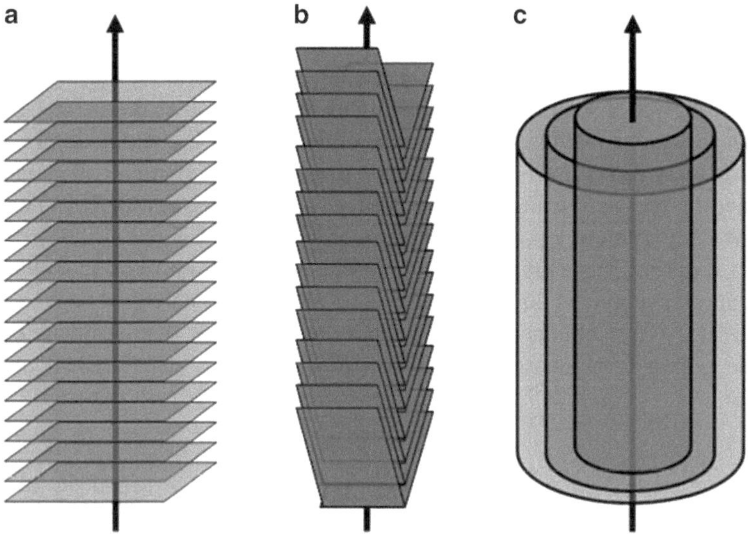

Fig. 7.8 Structure of different types of carbon nanofibers and nanotubes: (**a**) graphitic platelets, (**b**) graphitic herringbone and (**c**) multi-wall nanotube

respect to the growth axis, and the carbon nanotubes composed by graphene layers wrapped around a hollow cavity. Among them, carbon nanotubes have become one of the most active materials under study [29]. The most suitable method to produce them is by chemical vapor deposition. In this versatile process, a gas-phase rich of hydrocarbons are catalytically decomposed at high temperatures.

Carbon nanotubes can be formed by one or several wrapped graphene layers, accordingly they are classified as single-wall carbon nanotubes (SWCNT) or multi-wall carbon nanotubes (MWCNT). Usually, the internal diameter of the tube varies between 2 and 50 nm, the interlayer spacing is *ca.* 0.34 nm and the length ranges from few microns to several millimeters. Pores in MWCNT can be mainly divided into the inner hollow cavities of small diameter and pores (widely distributed, 20–40 nm) formed inside the aggregates of MWCNT. The surface area of MWCNT varies from about 50 to 400 $m^2\ g^{-1}$. The rolled up graphene layer forming the nanotube has a sp^2 hybridization, with structural defects that helps to anchor the catalyst support. In order to increase anchoring sites, many authors suggest different types of activation processes to promote superficial groups or structural defects necessary to have a good fixation of the nanoparticles [17].

Most of the alternative carbons support proposed to replace carbon black still have a lack of test analysis in a fuel cell. However, among those, the carbon nanofibers were the most studied in DMFC. Drillet et al. [30] analyzed the performance of SWCNTs, produced by the arc-discharge method, as catalyst support in the anode side of fuel cell. They used hot air (300 °C) or concentrated nitric acid as activation treatments to promote the catalyst anchoring. The performance of DMFC using SWCNT as PtRu support was 10–15 % higher than those using Vulcan carbon.

In the case of commercial MWCNTs as PtRu support, Jeng et al. [31] used this kind of support previously activated by chemical treatment. Well dispersed PtRu 1:1 nanoparticles of 3.5–4 nm were obtained by a polyol synthesis method. The fuel cell test showed a performance 50 % higher than that of a commercial PtRu on Vulcan support (E-TEK). Similar results were found by Prabhuram et al. [32] for PtRu on oxidized MWCNT, where well dispersed nanoparticles of 4 nm were obtained by the $NaBH_4$ method. The DMFC performance test of PtRu supported on MWCNTs showed a power density *ca.* 35 % higher than that using the Vulcan carbon support. Outstanding results were obtained by Tsuji et al. [33] with PtRu nanoparticles supported on carbon nanofibers prepared by polyol method and tested in a DMFC. They obtained a performance 200 % higher than standard PtRu on Vulcan carbon from Johnson Matthey.

7.2.3 Structured Porous Carbon Materials

The syntheses of materials with tailored properties have been proposed to improve the performance of the fuel cells [12, 34, 35] by increasing the dispersion of the metal, decreasing the formation of agglomerates of catalyst nanoparticles, and raising the diffusion of species from/to the electroactive area. Several synthesis routes produce a carbon with tailored porous structure and composition by controlling the properties of the carbon precursor. The fabrication method to obtain a porous material, in general, can be enclosed in three steps:

SYNTHESIS → DRYING → CARBONIZATION

This sequence of steps offers numerous benefits to obtain a carbon material with tuneable properties because is possible to introduce modifications, during the procedure of fabrication, depending on the conditions used or the customized steps (one o more) during the fabrication route. It is possibility to get desired characteristics of the final product, such as:

- Controllable carbon composition.
- Controllable graphitization degree.
- Variable surface functionalities.
- Tailored pore size and surface area.
- Tuneable morphology.
- Moldability of monolithic precursors.

In the following sections the most known fabrication routes will be discussed, including:

1. Modifying the drying conditions. The drying condition applied has a strong effect on the final properties of the polymer gel:

- Xerogels: the wet polymeric gel is dried under ambient pressure.
- Cryogels: the gel is frozen and then the solvent is removed by sublimation under low pressure.
- Aerogels: The solvent occluded into the polymer is exchanged with liquid CO_2, which is then removed by using supercritical conditions.

2. Nanocasting-hard template: a rigid structured inorganic template is infiltrated by a carbon precursor to synthesize carbons with uniform pore structure.
3. Mesoporous carbon by soft template: Carbons with mesostructures have been synthesized by using amphiphilic block copolymer as direct template, self-assembly surfactants or polyelectrolytes in the polymerization media of the carbon precursor.

All the above mentioned routes introduce modifications during the synthesis and drying stages. Once the dry structured polymer is obtained, the carbon material is formed by carbonization. During this process the polymer precursor is converted into carbonous materials by heating in an inert atmosphere up to final temperature which is generally above 800 °C. Phenolic polymers are usually used as carbon precursors due to their easy availability, low cost and simple synthesis. These materials have a carbon yield around of 52 % wt at 1,000 °C. An activation process can be performed, during or after carbonization stage, in order to increase the pore volume and the number of superficial groups of carbon materials. However, the activation process is more common in natural carbon precursors than for structured materials.

7.2.3.1 Mesoporous Carbons by Sol-Gel Process: Aerogels, Xerogels and Cryogels

Organics sol-gel refers to the product obtained by a hydrolysis-condensation reaction mechanism. This mechanism is based on a route analogous to sol-gel inorganic oxides. Sol-gel route involves the conversion of a colloidal suspension of a solid in a liquid (sol stage) to a semi-rigid colloidal dispersion (gel stage). The gel is composed by solid and liquid phases, which are independent of each other. If the liquid phase is removed from the gel through a non-destructive manner, a highly porous solid will be left with approximately the same shape and volume as the original gel. The first carbon aerogel, obtained from the carbonization of an organic gel, was produced by Pekala et al. [36] in 1989. The gel was synthesized by polymerization of resorcinol (R) and formaldehyde (F) and dried using supercritical conditions. Briefly, sol-gel reaction in a RF polycondensation occurs through the formation of a gel composed by nanosized pores and polymeric particles highly crosslinked. The polymerization begins with the addition of formaldehyde to the benzene ring, generally activated by basic catalyzers (C). During the second stage, successive condensation reactions produce a polymer highly interlinked. The formation process is schematized in Fig. 7.9.

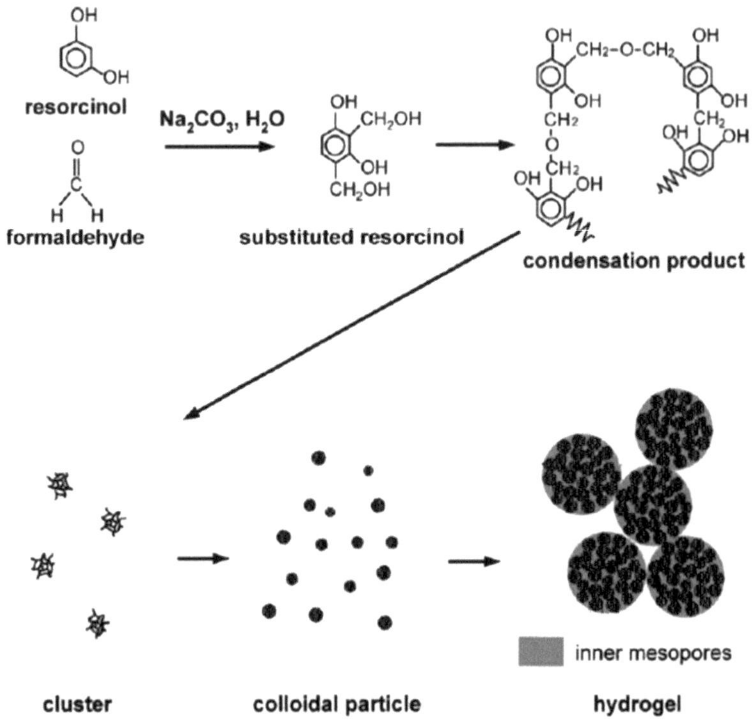

Fig. 7.9 Formation of an organic gel [37] (Reprinted by permission of the publisher)

Although the most common precursors are resorcinol and formaldehyde, the synthesis of polymers by sol-gel method have been extent to others hydroxylated benzene compounds (phenol, catechol, hydroquinone, cresols, phloroglucinol) and aldehyde compounds (e.g. paraformaldehyde, acetaldehyde, furfural) [38]. Moreover, water (W) as solvent and Na_2CO_3 as catalyst have been replaced by other compounds, opening the possibilities to a great variety of materials with different properties [39]. The sol-gel method is a very versatile route and the properties of the gel are strongly affected for the conditions used during its preparation e.g. pH, ratio R/C, reactant concentration, curing time, solvent [37, 40–42]. Since the work by Pekala the organic gels have been extensively studied, analyzing the synthesis route in order to control and tailor the gel properties. In Table 7.1 are summarized some synthesis conditions and the effect produced on the gel properties.

After obtaining the wet gel, it follows a solvent extraction process to dry it. When a gel with low mechanical strength is dried by simple evaporation, capillary forces created at the curved liquid–vapour interface can produce the shrinkage of the pore structure. Therefore, a carbon without appreciable surface area (closed pores) is obtained. For this reason, the drying method has a decisive effect in the structure of material. There are several routes to dry the carbon precursor and each of them produces different change in the properties of material.

Table 7.1 Effect of the synthesis conditions on the properties of resorcinol-formaldehyde carbon gels [43]

Factor	Effect
Decreasing reactant concentration (equivalent to reducing R/F, R/W or R/C)	Smaller particles and pore sizes Increases the surface area of xerogels Decreases the elastic modulus of carbon aerogels
Acidic catalytic solutions	Low concentrations: small, smooth, fractal aggregates of particles with wide pore size distribution High concentrations: no fractal aggregates very narrow pore size distribution May reduce gelation time
Alkaline catalyst solution	High concentrations: polymer gel (small polymer particles interconnected with fibrous appearance, high surface area, high mechanical strengths), reduces gelation time Low concentrations: colloidal gels (large particles interconnected with narrow necks, low surface areas, low mechanical strengths)
Increasing gel pH	Increases surface area and pore volumes of carbon aerogels Insignificant effect on surface area of carbon xerogels Increases the pore volume of carbon xerogels at high density of reactants Gelation time decreases
Gelation and curing period	Required to improving the crosslinking of the polymer particles
Temperature	High temperatures causes a shrinkage of porosity

If the wet gel is dried at ambient pressure by solvent evaporation as it was describe above, the resulting organic gel is called xerogel. However, in order to reduce the capillary forces which cause collapse of the structure, previously drying, the water can be exchanged by solvents of lower surface tension (e.g. ethanol acetone, etc). Some authors have subclassified these materials in xerogels and ambigels, according to the last solvent evaporated. Ambigels are those materials obtained from hexane or ciclohexane solvent evaporation.

A supercritical condition is other method to suppress the liquid–vapour interface or reduce the capillary forces. This route, used commonly to produce inorganic aerogels, was applied by Pekala et al. [36] to obtain organic aerogels, where the water in the wet gel is exchanged by a solvent soluble in liquid CO_2 (typically acetone), which is removed using supercritical conditions. These materials are called carbon aerogels.

One of the three best known drying methods is sublimation. The initial solvent in the wet gel is exchanged by one with lower density change on freezing and larger vapour pressure than water. After freezing, the solvent is removed by sublimation under low pressure. The so obtained dried gel is called cryogel.

All these methods have advantages and disadvantages. Concerning to the gel structure, supercritical drying is the best method to obtain an unaffected gel structure, because no appreciable contraction affects the pore structure. However, economical reasons reduce the possibility of this method to be applied at industrial scale.

Few reports were found dealing with the fuel cell performance of PtRu supported on this type of materials as anode in methanol oxidation. A mesoporous carbon dried under ambient conditions was analyzed as anode support in DMFC by Du et al. [44]. PtRu was deposited on a carbon with a pore size diameter *ca.* 11 nm. The same power density of commercial catalyst in single DMFC test was reached using only two third of noble metal used. Calderón et al. [45] synthesized PtRu catalysts supported on carbon xerogels. The carbon supports have high surface area and a pore size distribution centred to *ca.* 16 nm [46]. Well dispersed PtRu nanoparticles of 3.5–4.6 nm were obtained. Although the catalyst formed showed a higher current densities for the MOR than the commercial E-TEK catalyst, its power density in a DMFC single cell test was poorer.

Following the analysis of carbon pore size effect on fuel cell performance, Arbizzani et al. [47] reported DMFC tests for PtRu on mesoporous carbon cryogels. They achieved the best results for the catalyst supported on mesoporous carbon with pore size distribution around 20 nm in diameter. Again, the high surface area is sacrificed to improve the access of the perfluorosulfonate ionomer through the carbon support having bigger pores sizes.

It is important to remark the difference between the performance of catalyst support obtained by using electrochemical cell and fuel cell test. Such a differences were not only found for mesoporous carbons but also for commercial catalysts, indicating that the processing conditions used to form a catalyst layer, such as ink formulation, catalyst deposition method, pressing conditions, etc., have a strong effect in the good performance of fuel cell. Unfortunately, to the best of our knowledge, there is not experimental report analyzing the best conditions for processing the anodic side of a DMFC with catalyst supported on mesoporous carbon.

7.2.3.2 Mesoporous Carbon by Nanocasting

Although the replication method is one of the oldest ones in the human culture, Knox et al. [48] were the first that proposed to synthesize mesoporous carbon by the hard template method. In 1989 they reported the fabrication of porous carbon by using amorphous silica gels as template. However, Ryoo et al. [49] broke new ground for the fabrication of mesoporous carbons with ordered structures from replication of mesostructured silica. After that, an explosive grow in the structured carbons fabrication from replication method started. Ordered Mesoporous Carbon (OMC) is usually synthesized by nanocasting method, which consists in the infiltration of an inorganic hard template, that was produced by a soft template method (*see section below*), with the carbon precursor. The final material is obtained by carbonization of the precursor and the removing of the template. A scheme of the fabrication procedure of mesoporous carbon is shown in Fig. 7.10.

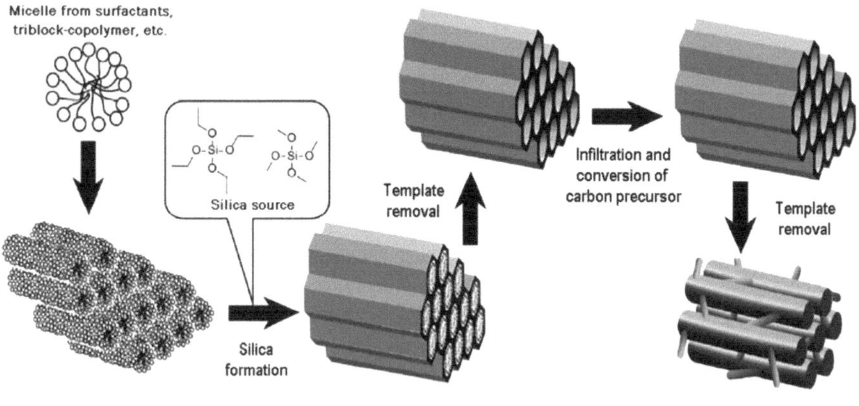

Fig. 7.10 Schematic representation of the synthesis procedures for mesoporous carbon (CMK- · 3) from a mesostructured silica (SBA-15) template [50] (Reprinted by permission of the publisher)

OMCs have been synthesized using various templates, including porous zeolites, and anodic alumina membranes, porous silica materials and silica nanoparticles [51, 52], as well as different carbon precursor: phenolic resins, furfuryl alcohol, mesophase pitch, sucrose, etc. In Table 7.2 are summarized the characteristic of OMC reported in the literature.

A wide variety of carbon with different physical and morphological properties (pore shapes, pore wall thickness, etc.) are reached depending on the template and the carbon precursor used [58]. Moreover, desired properties of the material can be introduced by carbon functionalization [59]. Therefore, a tunable carbon obtained through replication method has a broad spectrum of possibilities that can be used to suit the demand of any system. Among these materials, the structured carbon CMK-3 reported by Ryoo et al. [60] was the most studied as catalyst support in DMFC.

Nanocasting can be considered as the best method to obtain a narrow pore size distribution. This is a very important feature because carbon supports with tunable properties allow the modeling of the system, in order to optimize the processes of methanol oxidation (adsorption and re-dissolution) and mass transport in fuel cells.

For anode in DMFC, Chai et al. [61] used ordered porous carbons with pore sizes in the range of 10–1,000 nm as catalyst support, the material were synthesized using colloidal silica crystalline templates and phenol-formaldehyde as a carbon precursor. They reported a high dispersion of PtRu 1:1 nanoparticles of 2–3 nm on the carbonous materials. Among the catalysts tested, mesoporous carbon with pore size distribution centered on 25 nm showed the best performance in DMFC. Qi et al. [62] also showed that a better performance in DMFC test is achieved by using carbons with higher percentage of pores above 20 nm. Graphitic mesoporous carbons with different surface area, used as PtRu support, were synthesized using resorcinol and formaldehyde and particles of SiO_2 as a template. A power density 24 % higher than that obtained with the commercial PtRu-Vulcan XC-72 was achieved in a DMFC.

Table 7.2 Mesoporous carbons (OMCs) generated by nanocasting

OMC	Space group	Surface area and pore size	Precursor	Template	Space group	Ref.
CMK-1 (SNU-U)	*I41/a or lower*	~1,500 m^2g^{-1} ~3 nm	Sucrose, phenol resin	MCM-48	*Ia3d*	[49]
CMK-2	*Unknow cubic*	~1,400 m^2g^{-1} ~3 nm	Sucrose	SBA-1	*Pm3n*	[53]
CMK-3	*p6mm*	~1,000 m^2g^{-1}	Sucrose	SBA-15	*p6mm*	[54]
CMK-3 analogue	*p6mm*	~3.8 nm	Furfuryl alcohol, phenol resin	HMS SBA-3 MCM-41		
CMK-4	*Ia3d*	~3 nm	Acetylene	MCM-48	*Ia3d*	[55]
NCC-1	*p6mm*	~1,800 m^2g^{-1} ~5 nm	Furfuryl alcohol	SBA-15	*p6mm*	[56]
OMC (cubic)	*Ia3d*	~710 m^2g^{-1} ~6–9 nm	Sucrose, Furfuryl alcohol	KIT-6, FDU-5	*Ia3d*	[57]

OMC carbons obtained from replication of structured mesoporous silicas show a narrow pore size distribution, but with mesopores smaller (< 15 nm) than those obtained from colloidal replication. PtRu nanoparticles supported on ordered mesoporous carbon CMK-3 were analyzed by Din et al. [63] Although a good nanoparticle dispersion was obtained, the catalyst synthesized showed a worse performance than nanoparticles supported on Vulcan XC-72 for methanol oxidation.

Kim et al. [64] analyzed porous carbon by using colloidal silica particles as templates. Carbon with micro, meso and macropores were obtained modifying the initial pH of the carbon precursor solutions. This fabrication method produces materials with narrow pore size distribution in a broad range of pore size. The fuel cell test showed better DMFC performance for carbons with high meso-macropore area with large pore than that with micropores. Again, this effect was attributed to the fact that meso and macropores produce a favorable dispersion of PtRu metal species and allow the access of perfluorosulfonate ionomer for the formation of the triple phase boundary.

7.2.3.3 Mesoporous Carbon by Soft Template

A soft template route involves an organic compound, such as polymers or surfactants, which is used as a direct mold in order to obtain a structured carbon precursor. Then, the soft template is eliminated during the carbonization stage. Nowadays, there are a large number of preparation methods reporting materials with a wide variety of structures and pore size distributions. In general, these methods are more versatile and cheaper than nanocasting ones. Moreover, the soft template route eliminates steps on the carbon fabrication, because avoids the use of hard template and consequently the chemicals (HF or NaOH) to remove it.

The soft templates are organic compounds able to produce supramolecular arrangements by self-aggregation. The most commonly used are block copolymers and surfactant. The organic template self-assembles forming ordered structures which, through phase separation (nanoscale) and structure stabilization, induce the pore formation in the carbon precursor. Soft template pathways are very dependent of the synthesis conditions (template, solvent, temperature ionic strength, drying conditions, etc.), and this makes them less predictable than the nanocast pathways [65]. The family of phenolic resins are the most used common carbon precursors [66].

Bell et al. [67] were one of the pioneers in proposing the fabrication of mesoporous carbon by using surfactants as template. This novel route produces mesoporous carbon monolithic using directly cationic micelles as nanomolds in the polymerization media. More recently, Fujikawa et al. showed a possible route to control the morphology of the polymer and carbon nanostructures [68, 69].

All authors conclude that carbons with adjusted pore size distribution in the entire range of the nanopores can be obtained, depending the synthesis conditions and cationic surfactants used as templates. Despite this, it is an effective pathway to obtain porous carbons, even though the pore formation mechanism is not well understood. Hence, different mechanisms are used to explain its effect emerged, such as liquid crystal templating mechanism, cooperative self-assembly, electrostatic interaction between cationic surfactant molecules and the anionic RF polymer chain and micelles as nanoreactors to produce RF nanoparticles [51, 70]. In these cases, the simple mold effect from the globular form and the RF polymerization around it is insufficient to explain the structuring of the material by the template, where spherical closed pores would be expected.

Earlier reports showed that with a low surfactant concentration, enough for micelles formation but without reaching the liquid crystalline phase, a porous material with open structure may be obtained [71]. The authors suggested that the cationic micelles around polymeric clusters produce their stabilization and avoid the collapse during the drying step.

Soft template method by using block copolymers was reported for first time by Liang et al. [72] in 2004. After that, a significant progress on the fabrication of carbon with a well ordered mesopores was achieved [32, 73–77]. Zhao and coworkers performed a widespread study of soft template via the triblock poly (ethylene oxide) (PEO) and poly(propylene oxide) (PPO) based systems, PEO-PPO-PEO [65, 78]. Several ordered pore structures corresponding to various surfactant liquid crystal phases were synthesized by liquid crystal template pathway, a schematized synthesis procedure is shown in the Fig. 7.11.

The procedure was divide in five steps: (1) the preparation of resol precursors, (2) the mixed of carbon precursor with triblock copolymer and formation of ordered structure by self-assembly of the block copolymer, (3) the thermopolymerization of the carbon precursor around the template, (4) the removal of the template by calcination, and (5) the carbonization of the carbon precursor.

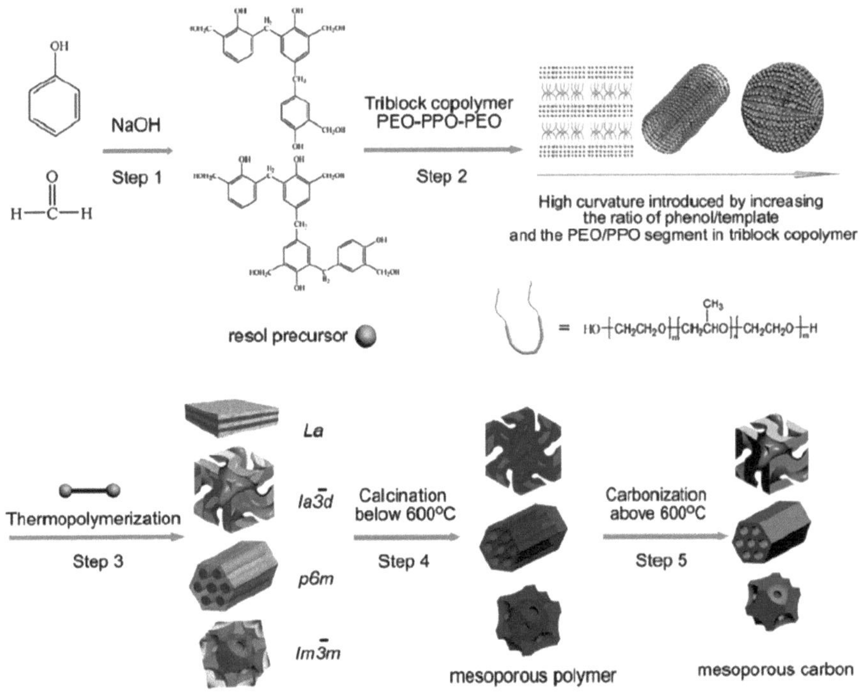

Fig. 7.11 Scheme for the preparations of the ordered mesoporous polymer resins and carbon frameworks [78] (Reprinted by permission of the publisher)

In general, carbons with narrower pore size distribution are obtained by using block copolimers than the ones from cationic surfactant. However, in general their pore sizes are smaller than 12 nm using commercial block copolymers. Deng et al. [79] reported the synthesis of ordered mesoporous carbons with large mesopores by using laboratory-made poly(ethylene oxide)-b-polystyrene (PEO-b-PS) diblock copolymers as templates. Bruno et al. [71, 80, 81] show alternative preparation method to form large mesopores. They found that polyelectrolytes also act as mesopores former, reporting carbons with mesopores size from 8 to 60 nm without ordered structure. Furthermore, these materials have a broader pore size distribution than those obtained by using block copolymers.

As it was discussed above, the carbon with small mesopores used as support produce high dispersion of the PtRu nanoparticles. However, in the fuel cell test they can show a poorer performance than the catalysts supported on Vulcan. Catalyst nanoparticles deposited in a tight pore might be unconnected to the perfluorosulfonate ionomer and inaccessible to the methanol. However, other factors affect the triple phase boundary and in consequence the performance of the cell using small mesopores carbon supports. For example, the method of catalyst layer formation, including ink formation and dispersion of the catalyst over the anode side, is a key factor in order to increase the fraction of catalyst contributing to the triple phase boundary region.

Viva et al. [82] synthesized PtRu 1:1 nanoparticles on mesoporous carbon and Vulcan carbon by using an impregnation method. The mesoporous carbon, obtained by using polyelectrolyte as structuring agent, had a surface are of 580 $m^2\ g^{-1}$ and a broad mesopores size distribution with a maximum centred at 20 nm. TEM images show the formation of highly dispersed nanoparticles of *ca.* 4 nm. PtRu nanoparticles exhibited a better methanol efficiency conversion when supported on mesoporous carbon as compared to Vulcan, and also showed an improved fuel cell performance. These results are in agreement with those showed by Chai et al. [61] where carbons with the larger mesopores exhibit the best cell performance.

7.3 Gas Diffusion Layers

As it was described in the introduction one of the many components of the MEA is the gas diffusion layer. The roles of the GDL are multiple and described below:

- Provides an even distribution of the reactants throughout the CL and the elimination of water and products (CO_2) out of the cell.
- Provides a substrate for the catalyst, present in the form of a powder. Also avoids deformation of the membrane-CL from the channels machined on the CCP.
- Provides the electrical connection between the CL and the CCP allowing electrons flow to/from the load.
- Transfers the heat generated on the catalyst away from the MEA.

The tasks mentioned above demands a particular set of properties for the compounds conforming the GDL: good mechanical properties, good porosity for the inlet and outlet of the reactant and products, good electrical conductivity and good thermal conductivity. The most widely used materials for GDLs is carbon in the form of carbon fibers, which is commercially available in the form of paper or fabric. It is worth to mention, even though they will not be discussed in this chapter, that metal meshes and porous metal foils are also used as GDL [83].

7.3.1 Carbon Fibers

Carbon fibers (CF) are known since the late 1800s. Thomas Edison produced the first carbon fibers from the pyrolyzation of bamboo and cotton fibers for the light bulbs filament. The fibers sustained high temperatures although they lacked tensile strength. The development of modern carbon fibers began in 1960 when Shindo produced the first carbon fiber from the pyrolysis of polyacrylonitrile (PAN) [84, 85]. Other materials also used for CF fabrication are rayon, phenolic resins and pitch. The CF properties like; high modulus, high tensile strength and low weight produced the widespread in the utilization of carbon fibers for a vast range of applications. In particular CF based composites possess excellent mechanic properties which are ideal for a wide range of applications particularly ones where a light and strong material is needed such as in airplanes, cars, boats, aerospace, sporting goods, etc.

The CF fabrication starts with the formation of PAN fibers by a spun process, from a solution or suspension of acrylonitrile, a co-monomer such as methyl acrylate or methyl methacrylate and a catalyst in a solvent like dymethylformamide. The most common spinning processes to form the fiber are: dry spinning, where the fibers are form in drying chamber while the solvent evaporates, and wet spinning where the fibers are form in a coagulating bath. Another method used to spun fibers is electrospinning which allows the formation of fibers a few nanometers in diameter [86, 87]. In the electrospinning method a potential difference is applied between a syringe needle and a surface where the fibers will be collected.

Once the fibers are formed they go through a stretching process where the polymer chain aligns and orientates parallel to the fiber longitudinal axis. An elongation of 500–1,500 % occurs contributing to the high strength of the final CF. The fiber might be subjected to a modification process previous to the pyrolysis which might include resin coating, chemical impregnation and stretching with plasticizer. Modification of the fibers affects the pyrolysis process by reducing the activation energy of cyclization reactions, decreasing the stabilization energy, and also improving the orientation of molecular chains in the fibers [88, 89]. The fibers are then subjected to the pyrolysis processes which has different stages depending on the type of CF desired [89]. Figure 7.12 shows a schematic representation of the pyrolysis process and the types of fiber obtained. The first stage is termed stabilization where the fibers are heated between 180 °C and 300 °C in the presence of air for 30–120 min. The working temperature depends on the manufacturer. In this stage there is a change from the linear atomic bonding to a more thermally stable ladder bonding [90] (i.e., thermoset material), by a reorganization of the bonding patter through the uptake of oxygen molecules from the air [91]. The next stage is either carbonization or graphitization, depending on the temperature, by heating the stabilized fibers in an inert atmosphere. The temperature used will define the properties of the final fiber. Carbonization up to 1,300 °C will produce Type III CF, which presents low modulus and low tensile strength. For temperatures of 1,500 °C Type II CF are obtained with high tensile strength and medium modulus. For temperature above 2,500 °C (graphitization) Type I carbon fibers, with high modulus and low tensile strength, are obtained [90, 92].

7.3.2 Carbon Paper

Carbon paper (CP) preparation begins by chopping carbon fibers obtained between 1,000 °C and 1,400 °C. Typical fiber diameter is between 5 and 15 μm. The porosity of the final product will depend on the length of the chopped fibers: short fibers will render a small pore size. The chopped fibers are soaked in water and a binder,

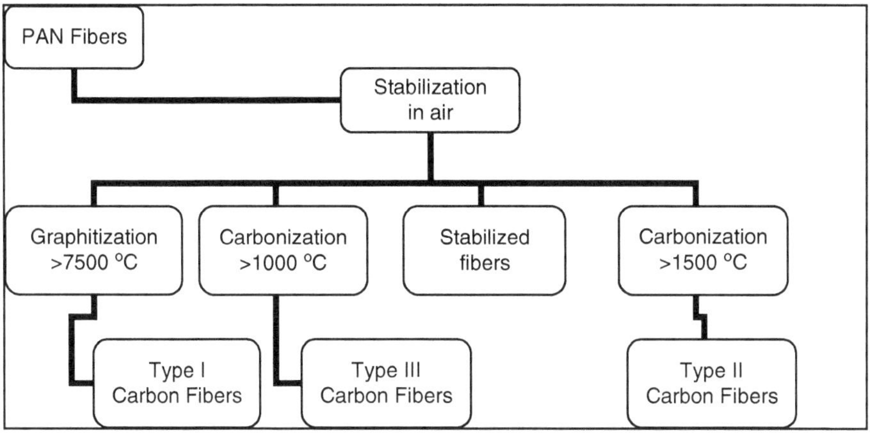

Fig. 7.12 Stages in the carbon fiber pyrolysis process from PAN fibers

typically polyvinyl alcohol [93], followed by a drying step. The coated fibers are impregnated with a resin and subjected to a pre-curing step at 150–175 °C [94]. A number of resins can be used, being the phenolic resin the most common because of its carbon structure, good adherence, and its low cost [94, 95]. The final properties of the CF will depend on the amount of the phenolic resin used [95, 96]. The material with the resin coating is molded into sheets by applying pressure and several sheets can be piled and re-pressed for a thicker material. The sheets are then cured in an oven at 150–200 °C for at least 2 h [96]. After curing, the carbonization and graphitization steps take place. The temperatures and the heating rate are different depending of the manufacturer, therefore modifying the CP final characteristics. As an option, carbon black or graphite powder can be added to the resin before impregnation of the CF improving the electrical conductivity without the need of the carbonization and graphitization steps [95].

In 1970 Toray Inc. was awarded the license of Shindo's patent [85] and started production of CF in 1971. Nowadays Toray is the world largest producer of CF and the Toray CP is one of the most referenced in the literature. Table 7.3 shows the properties of Toray CP. Figure 7.13 shows SEM micrograph of Toray CP TGP-H-060 where the morphology of the CP can be clearly observed.

7.3.3 Carbon Cloth

Carbon cloth (CC) is another kind of GDL used for PEM fuel cell application. Typically carbon cloth is thicker than the average carbon paper but has the important advantage of being flexible. Regarding the through plane electrical

Table 7.3 Physical properties of different types of Toray CP [97]

Product name	Thickness (mm)	Electrical resistivity (mΩ cm) Through plane	In plane	Thermal conductivity (W/m K) Through plane (room temp)	In plane (room temp)	In plane (100 °C)	Gas permeability (ml·mm/(cm^2·hr·mmAq)	Porosity (%)	Bulk density (g/cm^3)	Surface roughness (μm)	Coefficient of thermal expansion (in plane) (25~100 °C) ($\times 10^{-6}$/°C)	Flexural strength (MPa)	Flexural modulus (GPa)	Tensile strength (kgf/cm)
TGP-H-030	0.11	80	–	–	–	–	2,500	80	0.40	8	−0.8	40	8	–
TGP-H-060	0.19	80	5.8	(1.7)	21	23	1,900	78	0.44	8	−0.8	40	10	5
TGP-H-090	0.28	80	5.6	(1.7)	21	23	1,700	78	0.44	8	−0.8	40	10	7
TGP-H-120	0.37	80	4.7	(1.7)	21	23	1,500	78	0.45	8	−0.8	40	10	9

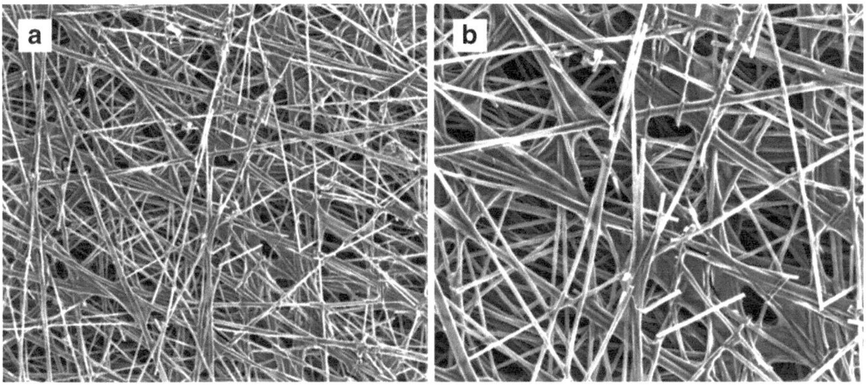

Fig. 7.13 SEM images of Toray TGP-H-60 carbon paper at 200× (**a**) and 400× (**b**)

Table 7.4 Physical properties of different carbon cloth GDL [98]

Product name	Thickness (mm)	Weight (g/m^2)	Bulk density (g/cm^3)	Porosity (%)	Tensile strength (MPa)	Through-plane air permeability (sec/100 cc)	Through-plane resistivity ($m\Omega\ cm^2$)
1071HCB Ballar	0.356	123	0.35	–	–	1.3	7.7
2002HD Ballar	0.229	96	0.42	–	–	2.6	7.2
Elat LT1400w NuVant	0.454	250	0.8	31	–	10	2 (bulk)

conductivity, in average the CC has higher values than CP with the exception of the ones manufactured by Ballard [98] as shown in Table 7.4.

Fabrication of CC begins after the PAN fiber stabilization step. With the stabilized fiber a spun yarn is formed. The yarn is put in a stretch-breaking machine, of common use in the fabric industry, which produces a fiber tow of certain lengths in a continuous way. The yarn is then homogenized with the use of different machines and subsequently braided or directly weaved. Once the cloths are ready, they are heated between 1,000 °C and 2,500 °C (carbonization or graphitization) under an inert atmosphere. As for the CP, the final temperature and heating rate will depend on the desired properties [99, 100].

Figure 7.14 shows SEM micrograph of Ballard carbon cloth while Table 7.4 shows the properties of some commercial CC.

Fig. 7.14 SEM images of Ballard carbon cloth at 200× (**a**) and 400× (**b**)

7.3.4 Carbon Felt

Carbon felt is another kind of CF arrange, generally used as thermal insulation in inert and vacuum furnaces, and that has seen its use in the fuel cell area only in microbial fuel cells (MFC) [101].

7.3.5 Diffusion Layer Coatings

In order to improve the properties of the aforementioned GDLs, they can be further modified by different coatings. These are mainly: treatment with polytetrafluourethylene (PTFE) and/or the deposition of a microporous layer (MPL) on the face at which the catalyst will be added [102].

7.3.5.1 PTFE Treatment

PTFE treatment of the GDL is one method to control internally the water content. Water management in PEM fuel cells is a key factor for the correct functioning of the device. Water management refers to the control of the water content inside the fuel cell. Low amounts of water within the boundaries of the MEA causes the membrane to dry, consequently reducing the ion transport properties. On the other hand, excessive amount of water particularly on the cathode side, known as cathode flooding, hinders the reaction on the catalyst surface. Both high and low water content may produce the shutdown of the cell. In direct alcohols fuel cells (DAFC), the water management is a key factor because as the fuel is introduced as an aqueous solution the amount of water inside the cell could be excessive.

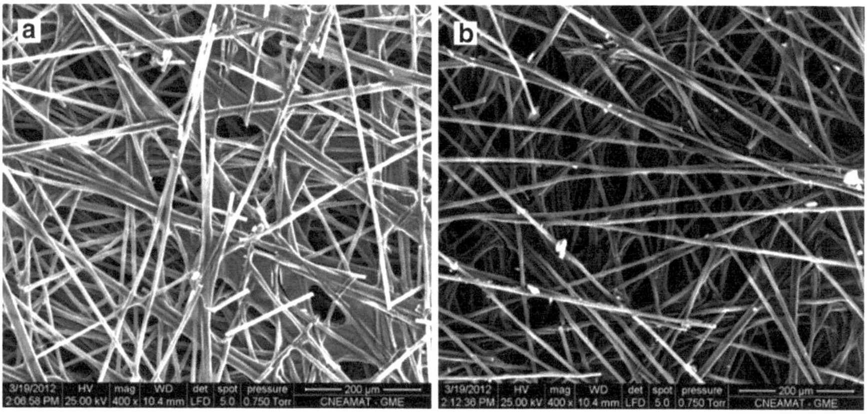

Fig. 7.15 SEM images of Toray TGP-H60 10 wt% PTFE treated (**a**) and untreated (**b**)

The GDL is simply coated with a thin layer of PTFE which renders the material hydrophobic. The coating procedure uses an aqueous suspension of PTFE with a nonionic surfactant such as polyoxyethylene alkyl phenyl ether or polyoxyethylene alkyl ether. The GDL material is soaked in the suspension or it is applied by spraying, this step is repeated according to the desired PTFE coverage. The GDL is then dried in air at temperatures below 200 °C and finally sintered at *ca.* 350 °C for 30–45 min [103, 104]. GDL with different amounts of PTFE are available ranging from 5 % to 30 % where an increased content means higher hydrophobicity. The hydrophobic surface expels the water coming to the anode or through the membrane and prevents condensation of water coming in the gases stream [105]. However, a higher amount of PTFE increases the contact resistance [106]. Moreover, the sintering temperature has been seen to adversely affect the conductivity [107]. Figure 7.15 shows the SEM micrograph of Toray TGP-H60 with 10 % of PTFE treatment and without treatment. The treated paper shows a lighter image due to the decreased electrical conductivity.

Several authors have shown the fuel cell performance with varying PTFE content [103, 106, 108, 109]. Figure 7.16 shows the polarizations curves of DMFC with different contents of PTFE on the anode and cathode GDL respectively [82]. An excess of PTFE has an adverse effect on the fuel cell polarization as observed in both graphs. The better performance is attained with different PTFE amounts at the anode and the cathode, indicating that an optimum amount has to be chosen according to the particular material and application. This is related to the fact that different amounts of water are present in the anode and the cathode. While an aqueous solution is fed to the anode, water reaches the cathode mainly by transport from the anode through the membrane, because dried O_2 or air is typically supply to the cathode.

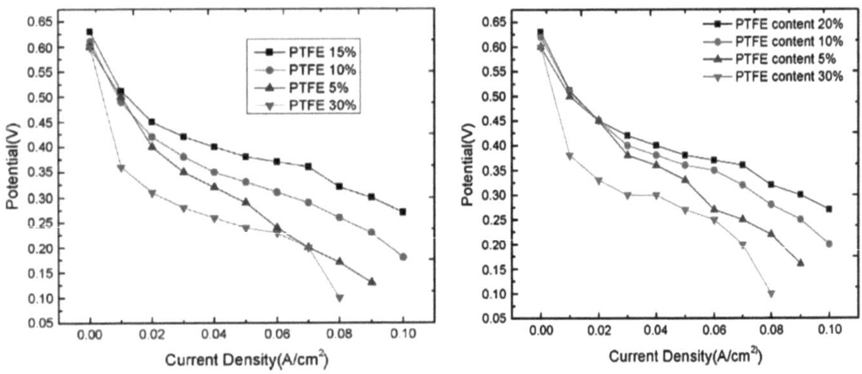

Fig. 7.16 Effect of PTFE content on the anode and cathode backing layer in the performance of DMFC [82] (Reprinted by permission of the publisher)

7.3.5.2 MPL Coverage

The coverage of the GDL by a carbon MPL is another common treatment in order to modify the layer properties. The MPL consist of a thin layer of carbon black powder with PTFE as binder, which is applied on one face of the GDL. Fabrication starts by preparing a slurry of the carbon powder and PTFE in an alcohol solution, typically isopropanol. The GDL is then coated with the mixture either by brushing, spraying, screen printing or with a doctor-blade system. The material is then dried at ca. 80–120 °C for 1 h followed by another heating step at 280 °C, in order to remove the PTFE dispersion agent, and finally sintered at 350 °C for 1 h. MPL thickness varies from manufacturer to manufacturer but typically lay in the order of 5–10 nm. The MPL coverage reduces the surface porosity of the plain CP or CC. Covering of the GDL by the MPL renders pores in the order of 20–200 nm, when in comparison without coverage the GDL pores range between 0.05 and 100 μm [110]. As the catalyst is applied on the MPL side, the pore size reduction effect is quite important because it prevents penetration of the catalyst particles deep into the GDL since the support catalyst particle sizes are between 40 and 80 nm. Thus, the MPL keeps the catalyst layer on the surface of the electrode improving the contact with the membrane and avoiding clogging of the GDL. As PTFE is used for the MPL preparation, this is another option for water management in the MEA. Figure 7.17 shows SEM images corresponding to both sides of Ballard's AVcarb 2240, a flexible carbon paper with a MPL, recommended by Ballard for DMFC applications. The image clearly shows the low porosity of the MPL covered face.

Wang et al. [111] presented a thorough study of MEAs for DMFC prepared with different loadings of Vulcan XC-72 as MPL. Polarization and power plots were obtained with 1.5 M methanol as fuel and humid air (50 % RH) as oxidant. The results (Fig. 7.18) show that 1 mg cm^{-2} of carbon as MPL at the anode or the cathode presents the highest power density when compared to 0.5 and 1.5 mg cm^{-2}

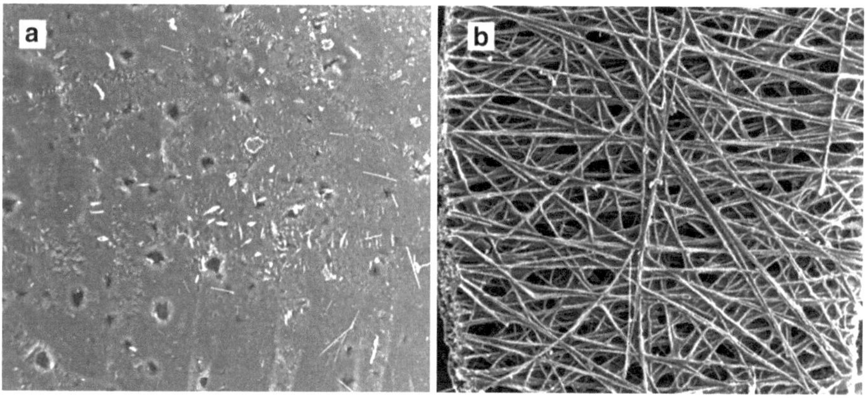

Fig. 7.17 SEM images of both faces of Ballard AVCarb 2240. (**a**) MPL covered face. (**b**) Non covered face

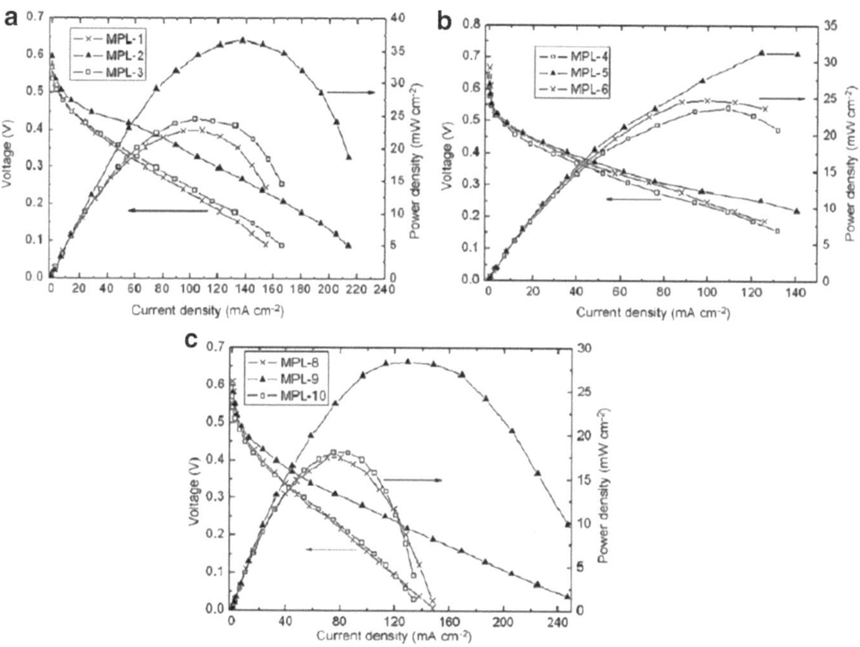

Fig. 7.18 Influence of different carbon loadings on DMFC performance at 35 °C and atmospheric pressure. (**a**) Anode: without the microporous layer, cathode: 0.5 (MPL-1), 1.0 (MPL-2) and 1.5 mgcm^{-2} (MPL-3) Vulcan XC-72 carbon with 20 wt% Teflon; (**b**) anode: 0.5 (MPL-4), 1.0 (MPL-5) and 1.5 mg cm^{-2} (MPL-6) Vulcan XC-72 carbon with 20 wt% Teflon, cathode: without the microporous layer; (**c**) anode and cathode with: 0.5 (MPL-8), 1.0 (MPL-9) and 1.5 mg cm^{-2} (MPL-10) Vulcan XC-72 carbon with 20 wt% Teflon [111] (Reprinted by permission of the publisher)

loadings. On a different MEA the same loading of 1 mg cm^{-2}, both at the anode and cathode, also presents the highest power density. The authors claim that an intermediate loading of Vulcan gives the best electron conduction and mass transfer while the lowest and highest loading had a diminished electron conduction and mass transfer respectively.

7.3.6 Non Commercial GDLs

In the previous section, an overview of the most common commercially available GDLs has been discussed. However, a great deal of research has been conducted on further modifications of commercial GDL or the preparation, at laboratory scale, of GDLs with particular properties. CP and CC processes produces material with low control of the porosity. Mainly, the preparation of non commercial GDLs aims to improve the porosity control in order to obtain a better diffusion of fuel and oxygen and for water management.

Yang et al. [112] prepared GDL with carbon black and PTFE. The layers were like a standalone MPL with high gas permeability. However, the cell performance was half of the commercial materials although the authors stressed the low cost and simplicity of preparation.

Carbon nanotubes (CNT) were used by several researchers to modify commercial GDLs [113–116]. Wang et al. [116] grew CNTs directly over CC by microwave plasma-enhanced chemical vapor deposition using $CH_4/H_2/N_2$ as precursors for DMFC. They deposited PtRu nanoparticles by a sputtering method, obtaining a loading of 0.4 mg cm^{-2}. The anode high performance was attributed to the high conductivity and low interfacial resistance of CNT.

Gao et al. [117] went further and prepared GDL made entirely with CNT. They mixed PAN CF, CNT and PTFE and sintered the mixture at 340 °C for 30 min. The prepared GDL was compared against Toray TGP-H-60 and they observed better electrical conductivity and mass transport properties by electrochemical impedance spectroscopy (EIS). Polarization plot also exhibited a better performance as show in Fig. 7.19.

Gerteisen et al. [118, 119] modified the GDL by drilling holes across it. The holes having 80 μm in diameter were drilled with a laser along the channels of the current plates. As the in-plane water saturation is higher than the through plane, it is expected that holes through the GDL will help in the fast evacuation of water from the GDL/membrane interface. The authors showed an improvement in the fuel cell polarization. In a similar work, Mench and coworkers [120] drilled 300 μm holes on GDL with MPL using also a laser. However, the holes were drilled in an array along the whole GDL surface equally spaced in the XY direction. The authors observed a lower performance compared to non-drilled material with low and very high humidity. Figure 7.20 shows a schematic representation of the differences in liquid water transport within a GDL with and without macro holes. Without water transport channels (WTC) the O_2 path through the GDL to the catalyst might be

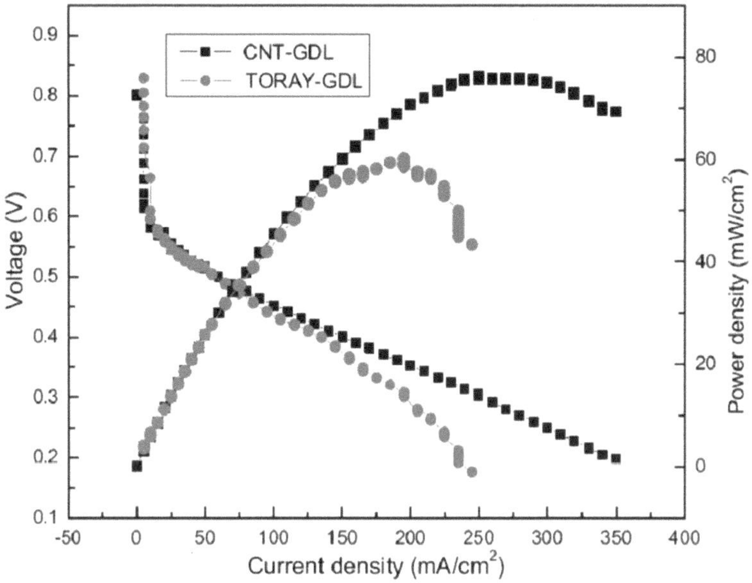

Fig. 7.19 Fuel cell performance comparison at 60 °C for a CNT GDL against a Toray GDL. Anode operation conditions: 1.0 M methanol at 1 ml min^{-1}. Cathode operation conditions: ambient pressure air at 80 ml min^{-1} [117] (Reprinted by permission of the publisher)

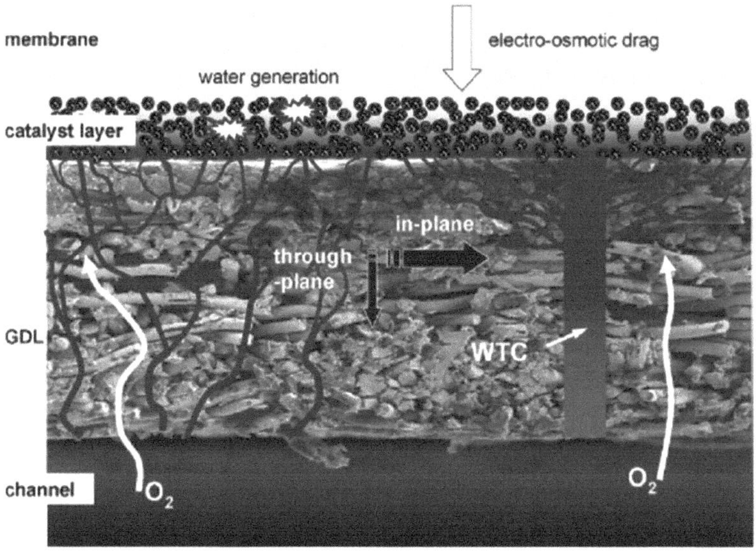

Fig. 7.20 Schematic of the liquid water transport with and without water transport channel (WTC) [119] (Reprinted by permission of the publisher)

Fig. 7.21 SEM images of a mesoporous carbon monolith. (**a**) Transversal cut showing the through plane channels. (**b**) Surface close up

hindered by water moving in the opposite direction. On the other hand, with WTC the water exits through the transport channels leaving a more open path for the O_2 transport.

The aforementioned results were very promising in order to improve the fuel cell performance. Working in a similar idea, Bruno et al. [81, 121] presented a novel method to produce mesoporous carbon monolith, obtained by a soft template method, with channels in the through plane direction of the materials using polypropylene fibers as a hard template. This simple method produces channels, 15 μm in diameter, together with meso and micropores in a single carbonization step of the resorcinol-formaldehyde resin used as precursor. Figure 7.21 shows SEM images of a transversal cut of the monolith exhibiting the macroscopic capillaries and the material surface showing the carbon particles forming the monolith. This material was later used as GDL and support of a electrodeposited mesoporous Pt catalyst for the anode in an H_2 fed fuel cell [122]. The observed results showed a better performance of the GDL/catalyst combination as compared to the commercial GDLs and catalyst.

7.4 Current Collector Plates and Bipolar Plates

The final component of a fuel cell to be discussed in this chapter is the current collector plates (CCP) or more commonly called the bipolar plates (BP) or flow field plate (FFP). As observed in Fig. 7.1 the fuel cell stack is composed by a number of BP, each of which will separate a pair of MEA and with two end plates completing the stack. At either side of the BP, an arrange of channels provide the flow paths for the fuel and oxygen. One of those faces is in contact with the anode of one MEA and the other face with the cathode of the other MEA, hence, the name bipolar plate. The end plates have channels only on one face, and the stack is typically completed by two metal plates with a series of bolts that holds the stack

Table 7.5 DOE technical targets for bipolar plate [123]

Property	Value
Flexural strength[a]	≥25 MPa
Contact resistance (at 140 N cm^{-2})	<20 mΩ cm^2
In-plane electrical conductivity	>100 S cm^{-1}
Thermal conductivity	>10 W $(m\ K)^{-1}$
Gas permeability	$<2 \times 10^{-6}$ cm^3 cm^{-2} s^{-1} at 80 °C and 3 atm.
Corrosion resistance	<1 μA cm^{-2}

[a]Using ASTM C651 – 11 Standard Test Method

tight. The shape and size of the channel's pattern will depend mainly on the application and final size of the stack. The BPs are the major contributor of a fuel cell stack in weight and volume, accounting for about 60–80 % of the total weight [9].

The BP functions are:

- Distribute the fuel and oxidant uniformly across the GDL outer surface
- Conduct the electrical current from the MEA to the load
- Remove the heat from the MEA
- Remove the reactants in excess and products formed
- Provide the structural support for the MEAs and sustain the compression
- Prevent, along with the seals, reactants or products leaks

Some of these functions are similar to the ones of the GDL and although the main component is also carbon material, the processes to obtain GDL and BP do differ.

Table 7.5 shows the target values of the BP properties set by the US department of energy (DOE) and taken as the goal by researcher and manufacturers.

Carbon is also the main material used for BP. As in the case of GDLs, metals were also used in the fabrication of BP perhaps with more extent than with the GDL. Carbon as material for BP is used mainly as graphite block (electronic grade graphite) or as a carbon–polymer composite, although some other forms have been tested [124].

7.4.1 Synthetic Graphite

Graphite was the first material used as BP, still is the most used at lab scale, mainly because it has an excellent corrosion resistance, high chemical stability and good electrical conductivity.

To produce graphite blocks, a filler is mixed with pitch as binder. Coke is the most used filler although natural graphite, carbon blacks and recycled graphite particles are also used. The coke is usually obtained from petroleum or coal tar pitches. The filler, particularly coke, is first pyrolyzed to release volatiles to avoid the formation of crack or distortions in the final product. Regarding the binder, petroleum and coal tar pitches are commonly used. These pitches are used due to their high carbon content, ca. 60 % wt, and the fact that they produce a carbon similar to the filler coke after it is carbonized. Other binders used are phenol and epoxy resins. The binder is heated, in order to decrease the viscosity for allowing its mixture with the filler. The filler-pitch mixture, often called carbon paste, is shaped in blocks by extrusion, molding, or cold isostatic pressing (CIP). Controlling the filler particle size and the filler-to-binder ratio will produce the desired material. The carbon blocks are then carbonized at a temperature between 700 °C and 1,000 °C (often called calcination) and graphitized at temperatures above 2,500 °C. The material treated at high temperature is called polycrystalline graphite or synthetic graphite. In order to produce graphite blocks, a two step liquid-phase carbonization is involved, first during the filler coke preparation and then during the carbonization of the binder pitch. The process of forming the graphite block is key to determine the prefered orientation of the filler particles. Extrusion gives flaky or needle like structure along the extrusion direction. By a molding process, the particles are statistically aligned perpendicular to the compression direction. While in CIP, the particles are randomly oriented producing high density isotropic graphite blocks.

The most available graphite plate is made by POCO (Pure Oil Company), a division of Entegris Inc. A variety of high-strength isotropic graphite is produced by POCO where PyroCell is the most known product with very low porosity.

The main drawback of graphite as BP is that flow channels have to be machined on, increasing the cost and the time in mass production. Moreover, a resin impregnation step has to be added post machining of the plate to avoid gas permeation. Such process also increases the cost of the plate. Another problem is the low flexural strength and brittle nature, making the material prone to fracture and forbidding the BPs to be thinner that 5 mm, hence increasing the stack size and weight.

7.4.2 Carbon Composites

Composite is a term that refers to a material prepared by introducing a filler, typically as powder, flakes or fibers, in a continuous matrix in order to obtain a material with better mechanical and corrosion properties. Several composite materials have been assessed as BP material for FC, including carbon–carbon composites, graphite/polymer mixtures (thermoplastic based composites), carbon fiber/epoxy resin (thermosetting based composites), and others conducting fillers with different polymers [8, 125–129]. The most common composites employed are prepared with carbon or graphite filler and a thermoplastic or thermosetting as matrix [130, 131]. The use of the later implies that the plates can be fabricated by injection molding or by hot pressing [126, 131]. The filler is the main responsible for the electrical conductivity of the material, therefore loadings of carbon between 60 % and 90 % by volume are commonly used. However, so high loadings tend to make the plates brittle and therefore suffering similar problems as with synthetic graphite.

The three most common composites used for BP are: carbon/carbon composite (CCC), thermoplastic based composites (TpC) and thermosetting based composites (TsC). The filler in all cases is carbon with a given degree of graphitization. Increasing the graphitization degree increases the electrical conductivity of the composite. However, carbon with low graphitization degree is less expensive, reducing the plate cost. The graphite can be natural or synthetic, the former having a higher graphitization degree but might include some impurities. On the other hand, the synthetic graphite has a higher purity but with a lower degree of graphitization. In all cases the filler is a mixture of graphite and carbon in order to have good conductivity and low cost. The percolation model can give an idea of the maximum conductivity with the lower amount of graphite.

In the CCC the filler is mixed with a resin, commonly phenolic resin, and sintered in a carbonization or graphitization step similar to the fabrication of synthetic graphite. The material obtained has a lower density, typically 20–40 % than synthetic graphite. The CCC allows the use of molding or vacuum molding processes. Lower plate thickness than synthetic graphite can be obtained.

In the TsC the filler is also inserted in a resin matrix, typically phenolic although epoxy and vinyl ester are also used. However, there is not a carbonization step like in the CCC. Due to the similarity in composition, CCC is sometimes confused with TsC. Channel manufacturing in TsC is done by compression molding. Entegris Inc. commercializes TsC BPs based on vinyl ester and phenolic resin.

In the TpC the filler is introduced in a more common plastic such as polypropylene. Even though the use of that kind of plastic reduces the production cost they have a low working temperature (70–80 °C) and mechanical properties that limit the range of applications. Other thermoplastic used as matrix are polyphenylene sulfide (PPS), polyvinylidenefluoride (PVDF), polyarylene disulfide and liquid crystalline polymer (LCP) [130, 132] which might increase the range of applications. TpC allows the use of injection molding process, a very well known industrial process, for channel fabrication. Typical conductivity values reported for TpC materials are 100 S cm^{-1} for the in-plane direction and 20 S cm^{-1} for the through-plane direction [130]. The British company Bac2 produces a TsC composite for BP using a proprietary conducting polymer that cures at room temperature called Electrophen®.

7.5 Conclusions

Carbon compounds are present in almost all the fuel cell components as was described along the chapter.

The catalyst carbon support plays an important role in the catalytic activity and the fuel cell performance. Different carbon supports were reviewed, analyzing their pore size distribution and fuel cell performance. An overview of the fabrication routes used to synthesize them was included. The carbon porosity affects the catalytic activity by modifying the catalyst dispersion, particle size distribution, alloying degree and mass transport. Several authors concluded that a better FC performance could be achieved when the carbon pore size distribution is above 20 nm improving the contact between the PtRu catalyst, methanol, and perfluorsulfonate ionomer (triple phase boundary) in the anodic layer of a DMFC. These experimental results show that the fabrication of materials with negligible microporosity and mesoporosity larger than 20 nm is a challenge for the synthetic chemist, being very few the materials showing such characteristics. Therefore, the developing of new routes to fabricate carbon with the mentioned tailored structure can be promising for improving the performance of fuel cells fed with methanol. Likewise, these materials could also be considered for other systems that use similar fuels like ethanol or formic acid (liquid fuels), although research with such fuels has just began.

Similarly, research conducted on materials and fabrication methods for GDL and bipolar plates aim to tune their properties in order to improve the fuel cell performance. It is clear that the current trend is the integration of the MEA components in order to improve the architecture of the triple phase boundary region and, consequently, the mass and charge transport.

References

1. Litster S, McLean G (2004) PEM fuel cell electrodes. J Power Sources 130:61–76
2. Mehta V, Cooper JS (2003) Review and analysis of PEM fuel cell design and manufacturing. J Power Sources 114:32–53
3. Zhang J (2008) PEM fuel cell electrocatalysts and catalyst layers: fundamentals and applications. Springer, New York
4. Chen E (2003) Thermodynamics and electrochemical kinetics. In: Hooger G (ed) Fuel cell technology handbook. CRC Press, New York
5. DOE Hydrogen and Fuel Cells Program. 2009 Annual Progress Report (2009) No DOE/GO-102009-2950. U.S. Department of Energy, Washington, DC
6. Marcinkoski J, James BD, Kalinoski JA, Podolski W, Benjamin T, Kopasz J (2011) Manufacturing process assumptions used in fuel cell system cost analyses. J Power Sources 196:5282–5292
7. Baker AA (1975) Carbon fibre reinforced metals – a review of the current technology. Mater Sci Eng 17:177–208
8. Dhakate S, Mathur R, Kakati B, Dhami T (2007) Properties of graphite-composite bipolar plate prepared by compression molding technique for PEM fuel cell. Int J Hydrogen Energy 32:4537–4543
9. Kamarudin SK, Daud WRW, Som AM, Takriff MS, Mohammad AW (2006) Technical design and economic evaluation of a PEM fuel cell system. J Power Sources 157:641–649
10. Bar-On I, Kirchain R, Roth R (2002) Technical cost analysis for PEM fuel cells. J Power Sources 109:71–75
11. Guy R, Lancaster T, Thornton J, Hart A, Sun J, Wilde J (2012) Polymer fuel cells – cost reduction and market potential. Carbon Trust, London
12. Sharma S, Pollet BG (2012) Support materials for PEMFC and DMFC electrocatalysts – a review. J Power Sources 208:96–119

13. Soboleva T, Zhao X, Malek K, Xie Z, Navessin T, Holdcroft S (2010) On the micro-, meso-, and macroporous structures of polymer electrolyte membrane fuel cell catalyst layers. ACS Appl Mater Interfaces 2:375–384
14. Kinoshita K (1988) Carbon: electrochemical and physicochemical properties. Wiley, New Jersey
15. Rodríguez-Reinoso F (1998) The role of carbon materials in heterogeneous catalysis. Carbon 36:159–175
16. Antolini E (2009) Carbon supports for low-temperature fuel cell catalysts. Appl Catal, B 88:1–24
17. Figueiredo JL, Pereira MFR, Freitas MMA, Órfão JJM (2006) Characterization of active sites on carbon catalysts. Ind Eng Chem Res 46:4110–4115
18. Calvillo L, Gangeri M, Perathoner S, Centi G, Moliner R, Lázaro MJ (2011) Synthesis and performance of platinum supported on ordered mesoporous carbons as catalyst for PEM fuel cells: effect of the surface chemistry of the support. Int J Hydrogen Energy 36:9805–9814
19. Uchida M, Aoyama Y, Tanabe M, Yanagihara N, Eda N, Ohta A (1995) Influences of both carbon supports and heat-treatment of supported catalyst on electrochemical oxidation of methanol. J Electrochem Soc 142:2572–2576
20. Liu H, Song C, Zhang L, Zhang J, Wang H, Wilkinson DP (2006) A review of anode catalysis in the direct methanol fuel cell. J Power Sources 155:95–110
21. Takasu Y, Kawaguchi T, Sugimoto W, Murakami Y (2003) Effects of the surface area of carbon support on the characteristics of highly-dispersed PtRu particles as catalysts for methanol oxidation. Electrochim Acta 48:3861–3868
22. Watanabe M, Sei H, Stonehart P (1989) The influence of platinum crystallite size on the electroreduction of oxygen. J Electroanal Chem Interf Electrochem 261:375–387
23. Rao V, Simonov PA, Savinova ER, Plaksin GV, Cherepanova SV, Kryukova GN, Stimming U (2005) The influence of carbon support porosity on the activity of PtRu/Sibunit anode catalysts for methanol oxidation. J Power Sources 145:178–187
24. Aricò AS, Srinivasan S, Antonucci V (2001) DMFCs: from fundamental aspects to technology development. Fuel Cells 1:133–161
25. Uchida M, Fukuoka Y, Sugawara Y, Ohara H, Ohta A (1998) Improved preparation process of very-low-platinum-loading electrodes for polymer electrolyte fuel cells. J Electrochem Soc 145:3708–3713
26. Uchida M, Fukuoka Y, Sugawara Y, Eda N, Ohta A (1996) Effects of microstructure of carbon support in the catalyst layer on the performance of polymer-electrolyte fuel cells. J Electrochem Soc 143:2245–2252
27. Hughes TV, Chambers CR (1889) Manufacture of carbon filaments. US Patent 405,480, 18 June 1889
28. Iijima S (1991) Helical microtubules of graphitic carbon. Nature 354:56–58
29. Serp P, Corrias M, Kalck P (2003) Carbon nanotubes and nanofibers in catalysis. Appl Catal A 253:337–358
30. Drillet J-F, Bueb H, Dittmeyer R, Dettlaff-Weglikowska U, Roth S (2009) Efficient SWCNT-based anode for DMFC applications. J Electrochem Soc 156:F137–F144
31. Jeng K-T, Chien C-C, Hsu N-Y, Yen S-C, Chiou S-D, Lin S-H, Huang W-M (2006) Performance of direct methanol fuel cell using carbon nanotube-supported Pt–Ru anode catalyst with controlled composition. J Power Sources 160:97–104
32. Prabhuram J, Zhao TS, Tang ZK, Chen R, Liang ZX (2006) Multiwalled carbon nanotube supported PtRu for the anode of direct methanol fuel cells. J Phys Chem B 110:5245–5252
33. Tsuji M, Kubokawa M, Yano R, Miyamae N, Tsuji T, Jun M-S, Hong S, Lim S, Yoon S-H, Mochida I (2006) Fast preparation of PtRu catalysts supported on carbon nanofibers by the microwave-polyol method and their application to fuel cells. Langmuir 23:387–390
34. Joo SH, Pak C, You DJ, Lee S-A, Lee HI, Kim JM, Chang H, Seung D (2006) Ordered mesoporous carbons (OMC) as supports of electrocatalysts for direct methanol fuel cells

(DMFC): effect of carbon precursors of OMC on DMFC performances. Electrochim Acta 52:1618–1626
35. Chang H, Joo SH, Pak C (2007) Synthesis and characterization of mesoporous carbon for fuel cell applications. J Mater Chem 17:3078–3088
36. Pekala RW (1989) Low density, resorcinol-formaldehyde aerogels. US Patent 4,873,218, 10 Oct 1989
37. Yamamoto T, Mukai SR, Endo A, Nakaiwa M, Tamon H (2003) Interpretation of structure formation during the sol-gel transition of a resorcinol-formaldehyde solution by population balance. J Colloid Interf Sci 264:532–537
38. Aegerter MA, Leventis N, Koebel MM (2011) Aerogels handbook. Springer, New York
39. Job N, Théry A, Pirard R, Marien J, Kocon L, Rouzaud J-N, Béguin F, Pirard J-P (2005) Carbon aerogels, cryogels and xerogels: influence of the drying method on the textural properties of porous carbon materials. Carbon 43:2481–2494
40. Takashi K (2000) Control of pore structure in carbon. Carbon 38:269–286
41. Lu AH, Spliethoff B, Schüth F (2008) Aqueous synthesis of ordered mesoporous carbon via self-assembly catalyzed by amino acid. Chem Mater 20:5314–5319
42. Zhang F, Meng Y, Gu D, Yan Y, Yu C, Tu B, Zhao D (2005) A facile aqueous route to synthesize highly ordered mesoporous polymers and carbon frameworks with Ia3d bicontinuous cubic structure. J Am Chem Soc 127:13508–13509
43. Al-Muhtaseb SA, Ritter JA (2003) Preparation and properties of resorcinol–formaldehyde organic and carbon gels. Adv Mater 15:101–114
44. Du H, Li B, Kang F, Fu R, Zeng Y (2007) Carbon aerogel supported Pt–Ru catalysts for using as the anode of direct methanol fuel cells. Carbon 45:429–435
45. Calderón JC, Mahata N, Pereira MFR, Figueiredo JL, Fernandes VR, Rangel CM, Calvillo L, Lázaro MJ, Pastor E (2012) Pt-Ru catalysts supported on carbon xerogels for PEM fuel cells. Int J Hydrogen Energy 37:7200–7211
46. Mahata N, Silva AR, Pereira MFR, Freire C, de Castro B, Figueiredo JL (2007) Anchoring of a [Mn(salen)Cl] complex onto mesoporous carbon xerogels. J Colloid Interf Sci 311:152–158
47. Arbizzani C, Beninati S, Soavi F, Varzi A, Mastragostino M (2008) Supported PtRu on mesoporous carbons for direct methanol fuel cells. J Power Sources 185:615–620
48. Knox JH, Kaur B, Millward GR (1986) Structure and performance of porous graphitic carbon in liquid chromatography. J Chromatogr A 352:3–25
49. Ryoo R, Joo SH, Jun S (1999) Synthesis of highly ordered carbon molecular sieves via template-mediated structural transformation. J Phys Chem B 103:7743–7746
50. Vinu A, Mori T, Ariga K (2006) New families of mesoporous materials. Sci and Technol Adv Mat 7:753–771
51. Lee KT, Oh SM (2002) Novel synthesis of porous carbons with tunable pore size by surfactant-templated sol-gel process and carbonisation. Chem Commun 22:2722–2723
52. Lu AH, Schüth F (2006) Nanocasting: a versatile strategy for creating nanostructured porous materials. Adv Mater 18:1793–1805
53. Ryoo R, Joo SH, Jun S, Tsubakiyama T, Terasaki O (2001) Ordered mesoporous carbon molecular, sieves by templated synthesis: the structural varieties. In: Galarneau A, Fajula F, Renzo FD, Vedrine J (eds) Studies in surface science and catalysis. Elsevier, Amsterdam
54. Shin HJ, Ryoo R, Kruk M, Jaroniec M (2001) Modification of SBA-15 pore connectivity by high-temperature calcination investigated by carbon inverse replication. Chem Commun 4:349–350
55. Liu X, Tian B, Yu C, Gao F, Xie S, Tu B, Che R, Peng L-M, Zhao D (2002) Room-temperature synthesis in acidic media of large-pore three-dimensional bicontinuous mesoporous silica with Ia3d symmetry. Angew Chem Int Ed 41:3876–3878
56. Lu AH, Schmidt W, Spliethoff B, Schüth F (2003) Synthesis of ordered mesoporous carbon with bimodal pore system and high pore volume. Adv Mater 15:1602–1606
57. Fuertes AB (2003) Template synthesis of mesoporous carbons with a controlled particle size. J Mater Chem 13:3085–3088

58. Lu AH, Zhao D, Wan Y (2009) Nanocasting: a versatile strategy for creating nanostructured porous materials. Royal Society of Chemistry, London
59. Stein A, Wang Z, Fierke MA (2009) Functionalization of porous carbon materials with designed pore architecture. Adv Mater 21:265–293
60. Jun S, Joo SH, Ryoo R, Kruk M, Jaroniec M, Liu Z, Ohsuna T, Terasaki O (2000) Synthesis of new, nanoporous carbon with hexagonally ordered mesostructure. J Am Chem Soc 122:10712–10713
61. Chai GS, Yoon SB, Yu J-S, Choi J-H, Sung Y-E (2004) Ordered porous carbons with tunable pore sizes as catalyst supports in direct methanol fuel cell. J Phys Chem B 108:7074–7079
62. Qi J, Jiang L, Tang Q, Zhu S, Wang S, Yi B, Sun G (2012) Synthesis of graphitic mesoporous carbons with different surface areas and their use in direct methanol fuel cells. Carbon 50:2824–2831
63. Ding J, Chan K-Y, Ren J, F-s X (2005) Platinum and platinum–ruthenium nanoparticles supported on ordered mesoporous carbon and their electrocatalytic performance for fuel cell reactions. Electrochim Acta 50:3131–3141
64. Kim P, Kim H, Joo JB, Kim W, Song IK, Yi J (2005) Preparation and application of nanoporous carbon templated by silica particle for use as a catalyst support for direct methanol fuel cell. J Power Sources 145:139–146
65. Zhao D (2012) Ordered mesoporous materials. Wiley, New Jersey
66. Liang C, Dai S (2006) Synthesis of mesoporous carbon materials via enhanced hydrogen-bonding interaction. J Am Chem Soc 128:5316–5317
67. Bell W, Dietz S (2001) Mesoporous carbons and polymers. US Patent 6,297,293, 2 Oct 2001
68. Fujikawa D, Uota M, Sakai G, Kijima T (2007) Shape-controlled synthesis of nanocarbons from resorcinol–formaldehyde nanopolymers using surfactant-templated vesicular assemblies. Carbon 45:1289–1295
69. Fujikawa D, Uota M, Yoshimura T, Sakai G, Kijima T (2006) Surfactant-templated synthesis of resorcinol-formaldehyde polymer and carbon nanostructures: nanospheres and nanowires. Chem Lett 35:432–433
70. Nishiyama N, Zheng T, Yamane Y, Egashira Y, Ueyama K (2005) Microporous carbons prepared from cationic surfactant–resorcinol/formaldehyde composites. Carbon 43:269–274
71. Bruno MM, Cotella NG, Miras MC, Barbero CA (2010) A novel way to maintain resorcinol–-formaldehyde porosity during drying: stabilization of the sol–gel nanostructure using a cationic polyelectrolyte. Colloids Surface A 362:28–32
72. Liang C, Hong K, Guiochon GA, Mays JW, Dai S (2004) Synthesis of a large-scale highly ordered porous carbon film by self-assembly of block copolymers. Angew Chem Int Ed 43:5785–5789
73. Pantea D, Darmstadt H, Kaliaguine S, Sümmchen L, Roy C (2001) Electrical conductivity of thermal carbon blacks: influence of surface chemistry. Carbon 39:1147–1158
74. Tanaka S, Nishiyama N, Egashira Y, Ueyama K (2005) Synthesis of ordered mesoporous carbons with channel structure from an organic-organic nanocomposite. Chem Commun 16:2125–2127
75. Xu J, Wang A, Zhang T (2012) A two-step synthesis of ordered mesoporous resorcinol–formaldehyde polymer and carbon. Carbon 50:1807–1816
76. Wang X, Liang C, Dai S (2008) Facile synthesis of ordered mesoporous carbons with high thermal stability by self-assembly of resorcinol – formaldehyde and block copolymers under highly acidic conditions. Langmuir 24:7500–7505
77. Meng Y, Gu D, Zhang F, Shi Y, Yang H, Li Z, Yu C, Tu B, Zhao D (2005) Ordered mesoporous polymers and homologous carbon frameworks: amphiphilic surfactant templating and direct transformation. Angew Chem Int Ed 44:7053–7059
78. Meng Y, Gu D, Zhang F, Shi Y, Cheng L, Feng D, Wu Z, Chen Z, Wan Y, Stein A, Zhao D (2006) A family of highly ordered mesoporous polymer resin and carbon structures from organic – organic self-assembly. Chem Mater 18:4447–4464

79. Deng Y, Yu T, Wan Y, Shi Y, Meng Y, Gu D, Zhang L, Huang Y, Liu C, Wu X, Zhao D (2007) Ordered mesoporous silicas and carbons with large accessible pores templated from amphiphilic diblock copolymer poly(ethylene oxide)-b-polystyrene. J Am Chem Soc 129:1690–1697
80. Atiyeh H, Karan K, Peppley B, Phoenix A, Halliop E, Pharoah J (2007) Experimental investigation of the role of a microporous layer on the water transport and performance of a PEM fuel cell. J Power Sources 170:111–121
81. Thomas YRJ, Bruno MM, Corti HR (2012) Characterization of a monolithic mesoporous carbon as diffusion layer for micro fuel cells application. Micropor Mesopor Mat 155:47–55
82. Krishnamurthy B, Deepalochani S (2009) Effect of PTFE content on the performance of a Direct Methanol fuel cell. Int J Hydrogen Energy 34:446–452
83. Wilkinson DP, Zhang J, Hui R, Fergus J, Li X (2009) Proton exchange membrane fuel cells: materials, properties and performance. CRC Press, London
84. Akio S (1964) On the carbonization of polyacrylonitrile fiber. Carbon 1:391–392
85. Shindo A, Fujii R, Sengoku T (1959) Method for manufacturing carbon product from acrylonitrile synthetic macromolecular substance. Japan Patent 37–4405
86. Greiner A, Wendorff JH (2007) Electrospinning: a fascinating method for the preparation of ultrathin fibers. Angew Chem Int Ed 46:5670–5703
87. Sutasinpromprae J, Jitjaicham S, Nithitanakul M, Meechaisue C, Supaphol P (2006) Preparation and characterization of ultrafine electrospun polyacrylonitrile fibers and their subsequent pyrolysis to carbon fibers. Polym Int 55:825–833
88. Mittal J, Mathur RB, Bahl OP (1997) Post spinning modification of PAN fibres – a review. Carbon 35:1713–1721
89. Yusof N, Ismail AF (2012) Post spinning and pyrolysis processes of polyacrylonitrile (PAN)-based carbon fiber and activated carbon fiber: a review. J Anal Appl Pyrol 93:1–13
90. Pierson HO (1995) Handbook of carbon, graphite, diamonds and fullerenes: processing, properties and applications. Noyes Publications, New Jersey
91. Ko T-H, Chiranairadul P, Lin C-H (1991) The influence of continuous stabilization on the properties of stabilized fibers and the final activated carbon fibers. Part I. Polym Eng Sci 31:1618–1626
92. Donnet JB (1998) Carbon fibers. CRC Press, New York
93. Zhang X, Shen Z (2002) Carbon fiber paper for fuel cell electrode. Fuel 81:2199–2201
94. Mathur RB, Maheshwari PH, Dhami TL, Tandon RP (2007) Characteristics of the carbon paper heat-treated to different temperatures and its influence on the performance of PEM fuel cell. Electrochim Acta 52:4809–4817
95. Liu C-H, Ko T-H, Liao Y-K (2008) Effect of carbon black concentration in carbon fiber paper on the performance of low-temperature proton exchange membrane fuel cells. J Power Sources 178:80–85
96. Mathur RB, Maheshwari PH, Dhami TL, Sharma RK, Sharma CP (2006) Processing of carbon composite paper as electrode for fuel cell. J Power Sources 161:790–798
97. Toray carbon paper specification (2005) http://www.torayca.com/en/index.html. Accessed 18 July 2012
98. Ballard gas difussion layer (2012) http://www.ballard.com/material-products/gas-diffusion.aspx. Accessed 18 July 2012
99. Liu C-H, Ko T-H, Kuo W-S, Chou H-K, Chang H-W, Liao Y-K (2009) Effect of carbon fiber cloth with different structure on the performance of low temperature proton exchange membrane fuel cells. J Power Sources 186:450–454
100. Ko T-H, Liao Y-K, Liu C-H (2007) Effects of graphitization of PAN-based carbon fiber cloth on its use as gas diffusion layers in proton exchange membrane fuel cells. New Carbon Mater 22:97–101
101. Wei J, Liang P, Huang X (2011) Recent progress in electrodes for microbial fuel cells. Bioresource Technol 102:9335–9344
102. Dicks AL (2006) The role of carbon in fuel cells. J Power Sources 156:128–141

103. Qi Z, Kaufman A (2002) Improvement of water management by a microporous sublayer for PEM fuel cells. J Power Sources 109:1–9
104. Tucker MC, Odgaard M, Lund PB, Yde-Andersen S, Thomas JO (2005) The pore structure of direct methanol fuel cell electrodes. J Electrochem Soc 152:A1844–A1844
105. Park G-G, Sohn Y-J, Yang T-H, Yoon Y-G, Lee W-Y, Kim C-S (2004) Effect of PTFE contents in the gas diffusion media on the performance of PEMFC. J Power Sources 131:182–187
106. Velayutham G (2011) Effect of micro-layer PTFE on the performance of PEM fuel cell electrodes. Int J Hydrogen Energy 36:14845–14850
107. Sun X, Saha MS (2008) Nanotubes, nanofibers and nanowires as supports for catalysts. In: Zhang J (ed) PEM fuel cell electrocatalysts and catalyst layers: fundamentals and applications. Springer, New York
108. Lin G, Nguyen TV (2005) Effect of thickness and hydrophobic polymer content of the gas diffusion layer on electrode flooding level in a PEMFC. J Electrochem Soc 152: A1942–A1948
109. Oedegaard A, Hebling C, Schmitz A, Møller-Holst S, Tunold R (2004) Influence of diffusion layer properties on low temperature DMFC. J Power Sources 127:187–196
110. Wang X, Zhang H, Zhang J, Xu H, Tian Z, Chen J, Zhong H, Liang Y, Yi B (2006) Micro-porous layer with composite carbon black for PEM fuel cells. Electrochim Acta 51:4909–4915
111. Wang T, Lin C, Fang Y, Ye F, Miao R, Wang X (2008) A study on the dissymmetrical microporous layer structure of a direct methanol fuel cell. Electrochim Acta 54:781–785
112. Chen-Yang YW, Hung TF, Huang J, Yang FL (2007) Novel single-layer gas diffusion layer based on PTFE/carbon black composite for proton exchange membrane fuel cell. J Power Sources 173:183–188
113. Du H-Y, Wang C-H, Hsu H-C, Chang S-T, Yen S-C, Chen L-C, Viswanathan B, Chen K-H (2011) High performance of catalysts supported by directly grown PTFE-free micro-porous CNT layer in a proton exchange membrane fuel cell. J Mater Chem 21:2512–2516
114. Tang Z, Poh CK, Tian Z, Lin J, Ng HY, Chua DHC (2011) In situ grown carbon nanotubes on carbon paper as integrated gas diffusion and catalyst layer for proton exchange membrane fuel cells. Electrochim Acta 56:4327–4334
115. Jeng K-T, Chien C-C, Hsu N-Y, Huang W-M, Chiou S-D, Lin S-H (2007) Fabrication and impedance studies of DMFC anode incorporated with CNT-supported high-metal-content electrocatalyst. J Power Sources 164:33–41
116. Wang CH, Du HY, Tsai YT, Chen CP, Huang CJ, Chen LC, Chen KH, Shih HC (2007) High performance of low electrocatalysts loading on CNT directly grown on carbon cloth for DMFC. J Power Sources 171:55–62
117. Gao Y, Sun GQ, Wang SL, Zhu S (2010) Carbon nanotubes based gas diffusion layers in direct methanol fuel cells. Energy 35:1455–1459
118. Gerteisen D, Heilmann T, Ziegler C (2008) Enhancing liquid water transport by laser perforation of a GDL in a PEM fuel cell. J Power Sources 177:348–354
119. Gerteisen D, Sadeler C (2010) Stability and performance improvement of a polymer electrolyte membrane fuel cell stack by laser perforation of gas diffusion layers. J Power Sources 195:5252–5257
120. Manahan MP, Hatzell MC, Kumbur EC, Mench MM (2011) Laser perforated fuel cell diffusion media. Part I: Related changes in performance and water content. J Power Sources 196:5573–5582
121. Bruno MM, Corti HR, Balach J, Cotella NG, Barbero CA (2009) Hierarchical porous materials: capillaries in nanoporous carbon. Funct Mat Lett 2:135–138
122. Bruno MM, Franceschini EA, Viva FA, Thomas YRJ, Corti HR (2012) Electrodeposited mesoporous platinum catalysts over hierarchical carbon monolithic support as anode in small PEM fuel cells. Int J Hydrogen Energy 37:14911–14919

123. DOE Hydrogen and Fuel Cells Program. 2011 Annual Progress Report (2011) No DOE/GO-102011-3422. U.S. Department of Energy, Washington, DC
124. Middelman E, Kout W, Vogelaar B, Lenssen J, de Waal E (2003) Bipolar plates for PEM fuel cells. J Power Sources 118:44–46
125. Blunk R, Abd Elhamid MH, Lisi D, Mikhail Y (2006) Polymeric composite bipolar plates for vehicle applications. J Power Sources 156:151–157
126. Heinzel A, Mahlendorf F, Niemzig O, Kreuz C (2004) Injection moulded low cost bipolar plates for PEM fuel cells. J Power Sources 131:35–40
127. Kim M, Yu HN, Lim JW, Lee DG (2012) Bipolar plates made of plain weave carbon/epoxy composite for proton exchange membrane fuel cell. Int J Hydrogen Energy 37:4300–4308
128. Lee Y-B, Lee C-H, Kim K-M, Lim D-S (2011) Preparation and properties on the graphite/polypropylene composite bipolar plates with a 304 stainless steel by compression molding for PEM fuel cell. Int J Hydrogen Energy 36:7621–7627
129. Mathur RB, Dhakate SR, Gupta DK, Dhami TL, Aggarwal RK (2008) Effect of different carbon fillers on the properties of graphite composite bipolar plate. J Mater Process Tech 203:184–192
130. Huang J, Baird DG, McGrath JE (2005) Development of fuel cell bipolar plates from graphite filled wet-lay thermoplastic composite materials. J Power Sources 150:110–119
131. Chen W, Liu Y, Xin Q (2010) Evaluation of a compression molded composite bipolar plate for direct methanol fuel cell. Int J Hydrogen Energy 35:3783–3788
132. de Oliveira MCL, Ett G, Antunes RA (2012) Materials selection for bipolar plates for polymer electrolyte membrane fuel cells using the Ashby approach. J Power Sources 206:3–13

Chapter 8
Physical Modeling and Numerical Simulation of Direct Alcohol Fuel Cells

Scales, Mechanisms and Approaches

Alejandro A. Franco

Abstract This chapter focuses on the role of physical theory, atomistic/molecular simulation and computational electrochemistry for fundamental understanding, diagnostics and design of new electrochemical materials and operation conditions for Direct Alcohol Fuel Cells (DAFCs). Development of stable and inexpensive materials and components is among the most important technological challenges that DAFCs nowadays are facing. Deep insight based on physical modeling of the materials behavior and aging will advise how these components with optimal specifications could be made and how they can be integrated into operating devices. Ongoing efforts within the community to understand from physical modeling and numerical simulation electrochemical mechanisms and degradation processes in DAFCs are critically reviewed. The capabilities of such approaches to propose innovative procedures (operation strategies and electrodes formulation) to enhance the DAFCs performance and durability are also illustrated through several examples. Finally, emerging multiscale simulation techniques allowing bridging the gap between processes simulated at different scales as well as major challenges and perspectives for DAFC modeling are presented.

8.1 Introduction: Why Physical Modeling?

8.1.1 Direct Alcohol Fuel Cells: General Context

From the second half of twentieth century, acidic (proton exchange) Polymer Electrolyte Membrane Fuel Cells (PEMFC) have attracted much attention due to their potential as a clean power source for portable applications (alcohol feed),

A.A. Franco (✉)
Laboratoire de Réactivité et de Chimie des Solides (LRCS), Université de Picardie Jules Verne & CNRS, UMR 7314-33, rue St. Leu, 80039 Amiens Cedex, France

Réseau sur le Stockage Electrochimique de l'Energie (RS2E), FR CNRS 3459, France
e-mail: a.a.franco.electrochemistry@gmail.com

H.R. Corti and E.R. Gonzalez (eds.), *Direct Alcohol Fuel Cells: Materials, Performance, Durability and Applications*, DOI 10.1007/978-94-007-7708-8_8,

vehicles traction (hydrogen feed) and stationary power generation (alcohol or hydrogen feed). Since the first PEMFCs proposed by Grubb and Niedrach in 1959 [1] there has been a remarkable technological progress towards the increase of their efficiency and the reduction of platinum loading, through the development of new membranes [2] and electro-catalytic nanoparticles [3, 4] or the improvement of the electrode structure thanks to the growing understanding in modern materials and porous media science. On the other hand, the platinum loading reduction resulted on increasing the components structural complexity, especially of the Catalyst Layers (CLs): from this, even if the overall operation principles of a single cell remains relatively simple, complex mechanisms at different spatial scales strongly interplay limiting the effectiveness of the catalyst activity. In fact, processes at the smaller scales (e.g. Oxygen Reduction Reaction -ORR- on the cathode platinum nanoparticles) dominate the processes at the larger scales (e.g. liquid water transport through the cathode carbon support secondary pores) which in turn affect the processes at the smaller ones (e.g. through the water flooding limiting O_2 transport in the cathode). PEMFC technologies have not yet reached all the required characteristics to be competitive, in particular regarding their high cost and their low durability [5, 6].

Among PEMFCs, Direct Alcohol Fuel Cells (DAFCs) present several advantages compared to other types of fuel cells, including their low working temperature (and thus they do not have material problems due to, for example, thermal expansion), the easy handling of storage and supply of fuel -in particular in comparison to hydrogen- and the options to create alcohols from renewable sources (e.g. biogas, syngas from biomass, etc.).

Among the alcohols, methanol has been considered as one of the most suitable fuels for DAFCs: methanol releases six protons and electrons per molecule during its electrocatalytic oxidation in acidic media. For instance Direct Methanol Fuel Cells (DMFCs) are being investigated since many years as complementary or alternative power source to batteries for portable applications because they can offer high energy densities. DMFC works at low and intermediate temperatures (up to 150 °C) and can be fed with a diluted aqueous solution of methanol in water. Alternatively, DMFC operation with methanol in gas phase also gives good performance. Actually, the higher temperature enhances kinetics and some problems such as methanol crossover (see below) are lowered with a gas phase feed.

Some other types of DAFCs feed by other alcohols such as ethanol or ethylene-glycol are also being studied [7, 8]. Compared with other direct liquid fuel cells, DAFCs with ethanol as fuel (or Direct Ethanol Fuel Cells -DEFC-) have several advantages: ethanol is non-toxic, less volatile, safer and has a greater energy density than methanol [9, 10]. Another advantage is that ethanol can be extracted in significant amount from sugar containing biomass resources. However, C–C bond of ethanol molecule is difficult to break to accomplish complete electrooxidation at low temperature with the most current electrocatalysts, for instance, Pt and Pt based bi-metallic and tri-metallic (Pt–Ru, Pt–Sn) alloys. High anodic overpotential, unavailability of stable Polymer Electrolyte Membranes (PEM) at high temperatures and crossover of ethanol are some of the issues limiting the DEFC development.

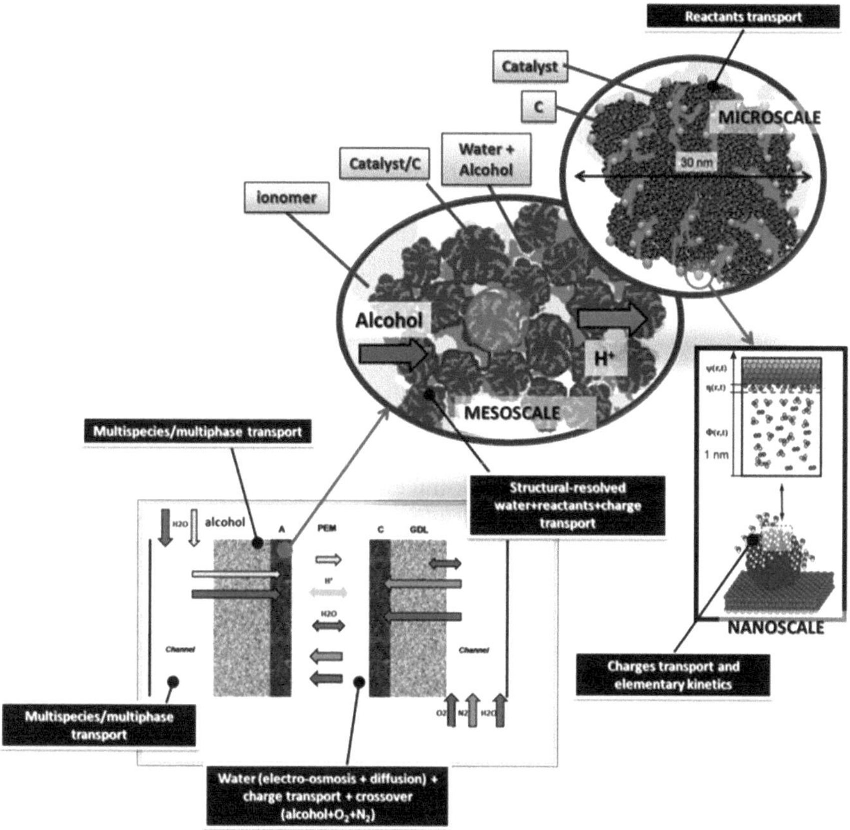

Fig. 8.1 Main mechanisms and scales in an operating DAFC

Figure 8.1 summarizes the operation principle and the main mechanisms occurring at multiple scales in a DAFC: alcohol and water transport in the anode (air and water in the cathode) at the macroscale (within the distributor), mesoscale (within the secondary pores formed by C in the electrodes) and microscale (within the primary pores of the C), nanoscale electrochemical double layer formation around the catalyst nanoparticles, alcohol electrochemical oxidation in the anode and ORR in the cathode.

8.1.2 Technical Issues: Overview

Despite the promising characteristics of DAFCs, several factors limit their performance, including the slow kinetics of the electrochemical reactions in the electrodes (oxidation of the alcohol at the anode and ORR in the cathode) and the

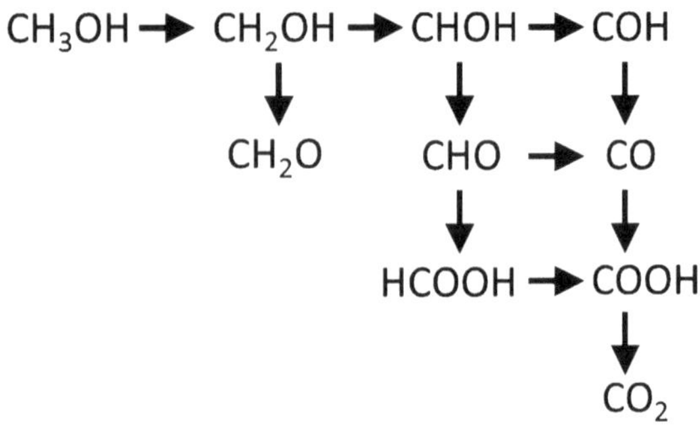

Fig. 8.2 A possible reaction pathway for the methanol oxidation on Pt surfaces

crossover of the alcohol from the anode to the cathode, both issues leading to high voltage losses, high noble metal catalyst loadings and thus high costs.

Electrocatalysis in DAFC anodes is complex because the reaction mechanism involves adsorption of alcohol and several elementary reaction steps including the CO oxidation. Figure 8.2 shows a possible network of reaction pathways by which the electrochemical oxidation of methanol occurs. Since more than 50 years detailed catalysis studies have attempted to analyze possible reaction pathways to find the main pathway of methanol oxidation [11, 12] (see next Section). Most studies conclude that the reaction can proceed according to multiple mechanisms and that the most significant reactions are the adsorption of the alcohol and the oxidation of CO.

Various surface intermediates are formed during methanol electrooxidation. Methanol is mainly decomposed to CO which is then further oxidized to CO_2. Other CO-like species are also formed such as COH_{ads}, HCO_{ads}, $HCOO_{ads}$ [13]. Main by-products are formaldehyde and formic acid. Some of these intermediates are not easily oxidizable and remain strongly adsorbed to the catalyst surface. Consequently, they prevent methanol molecules adsorbing and undergoing further reaction. Thus the electrooxidation of the reaction intermediates reveals to be the rate limiting step.

This catalyst poisoning seriously slows down the oxidation reaction. Besides, a small percentage of the intermediates desorbs before being oxidized to CO_2 and hence reduce fuel efficiency but undergoing in complete oxidation. Thus, a very important challenge is to develop new electrocatalysts that inhibit the poisoning and increase the rate of the reaction i.e. that provide a better activity toward carbon dioxide formation.

For instance, DMFC uses frequently PtRu alloys as the anode catalyst to prevent fast poisoning of the catalyst surface by CO molecules produced during methanol oxidation.

The large amount of gaseous CO_2 produced in the anode can also reduce the cell performance. Methanol oxidation gives one CO_2 per each methanol molecule. Already at medium current densities (100 $mA.cm^{-2}$) the amount of dissolved CO_2 becomes negligible and the molar flux of gaseous CO_2 in the anode channel is comparable to the flux of liquid methanol. Thus, most of the CO_2 formed during methanol oxidation is in the gas phase and has to be removed via the feed channel through the gas diffusion layer. There, CO_2 bubbles form at the outlet of the pores.

Fig. 8.3 Anode flow field with bubbles (Source: [16] reproduced with permission of Elsevier)

Their size depends on the size of the pores and the wettability of the electrode surface. After they reach a certain size, these bubbles are released into the methanol solution [14]. There is thus a two phase flow inside the feed channels. The high concentration of bubbles makes them coalesce and form slugs: large gaseous bubble in the anode channel blocks transport of liquid methanol to the adjacent electrocatalytic sites. This results in a reduction of the methanol diffusion area. Consequently, the supply of methanol may not be sufficient to maintain a desired current density. The current density becomes limited by methanol diffusion. This phenomenon is in particular an important issue at high current densities where a large amount of CO_2 is formed (Fig. 8.3). On the other hand, mass conservation prescribes that the flow with bubbles must move faster, thus providing faster transport of reactants along the channel. Which effect dominates in a cell with the conventional serpentine flow field is not obvious and is strongly dependent on the channel geometry (Fig. 8.4). Experimental studies indicate degradation of cell performance under intensive formation of bubbles in the flow fields of various designs [14, 15]. Two techniques are usually considered to solve the problem of the bubbles formation. First, the patterns of the feed channels can be changed to help a quick removal of CO_2. Second, Lu and Wang [14] investigated the structure of the diffusion layer itself and formation of bubbles on the top. According to these authors, small and homogeneously spread pores favours the creation of small diameter bubbles rather than slugs.

The high rate of alcohol crossover through currently available PEM materials (e.g. PFSA polymers) also lowers the system efficiency and decreases the cell potential due to the formation of a "mixed potential" in the cathode. Crossoving methanol reacts directly with oxygen at the cathode [17, 18]. Electrons are brought directly from the anode to the cathode along with methanol resulting in an electrolyte short circuiting and consequently a loss of current (Fig. 8.5). Besides, the cathode catalyst, which is pure platinum, is fouled by methanol oxidation intermediates similarly to the anode.

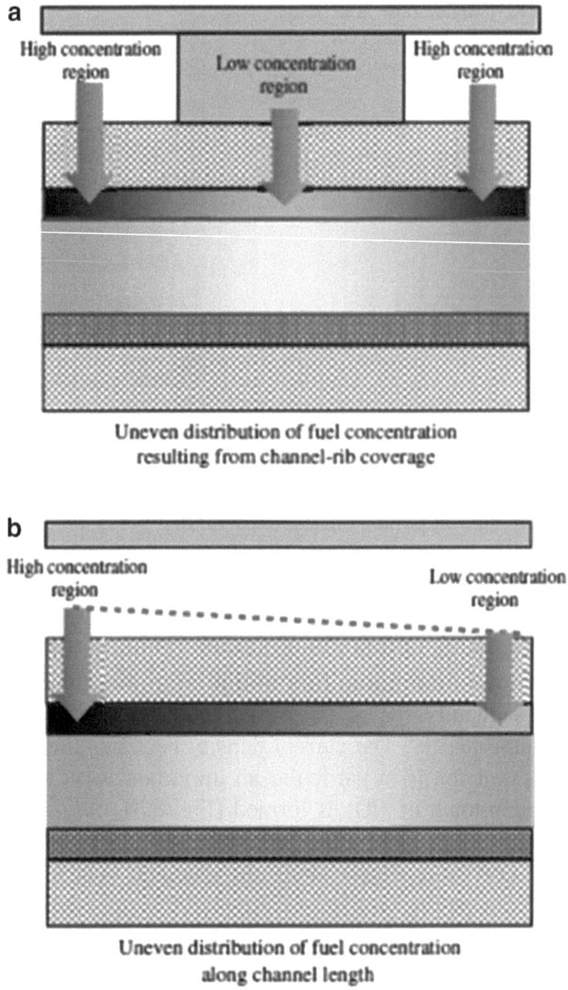

Fig. 8.4 Schematic illustration of the uneven fuel distribution in the in-plane direction of a DAFC (Source: [23] reproduced with permission of Elsevier)

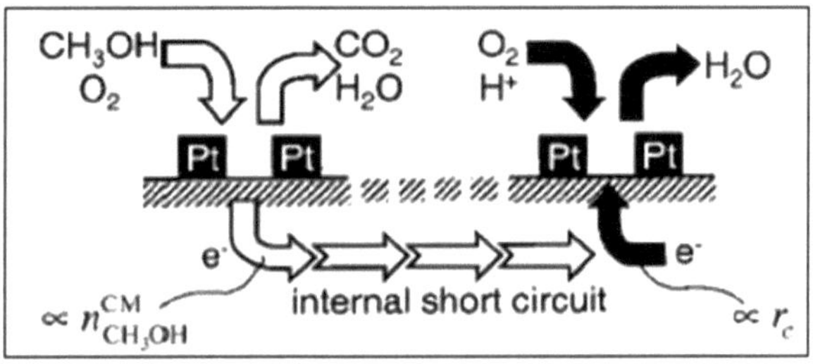

Fig. 8.5 Internal short circuit created by alcohol crossover (Source: [19] reproduced with permission of the royal society of chemistry)

Several studies have characterized methanol crossover in DMFCs [20, 21]. Due to the hydroxyl group and its hydrophilic properties, methanol interacts with the ion exchange sites and is dragged by hydronium ions in addition to diffusion as a result of concentration gradient between anode and cathode.

Another problem is the flooding of the cathode CL due to the high electro-osmotic flux of water through the PEM, largely studied in the context of hydrogen-feed PEMFCs [22].

Zhao et al. reported an interesting review on the transport phenomena issues in DMFCs [23].

In addition to the electrochemical reactions, reactants and biphasic water transport, other mechanisms limiting optimal catalyst utilization are charge transfer, thermo-mechanical stresses and irreversible materials degradation. Typical experiments show that the state–of–the-art DMFC in a stack environment loses 10–20 μV per hour of operation at a current density in the order of 100 mA.cm^{-2} [24]. This limits DMFC lifetime by several thousand hours, which still is not sufficient for widespread commercialization.

For instance, microstructural degradation leading to the DAFC components aging can occur in a similar way that it has been observed with hydrogen-feed PEMFC, and can be attributed to several complex physicochemical phenomena not yet completely understood:

- Dissolution and redistribution of the catalyst due to the high potentials of the cathode electrode [25]. This phenomenon reduces the specific catalyst surface area leading to the loss of the electrochemical activity [26–29]. Although Pt nanoparticles are very stable in acidic electrolytes at low electrode potentials, considerable dissolution of cathodic Pt can take place at high potentials representative of typical cathode operating conditions (potential between 0.7 and 1.1 V vs. NHE) [30]. Both steady-state and drive-cycle conditions produce extensive catalytic oxidation, dissolution and redistribution of the Pt atoms [31–34]. Extensive nanoparticle coarsening and redistribution is also observed, leading to the reduction of the ECSA and to the loss of the electrochemical activity [35]. Alternatively, it has been shown that the specific activity and stability of hydrogen-fed PEMFC electrodes can be improved by alloying Pt with some transition metals (e.g. Co, Ni, V. . .): these electrodes also achieve lower global Pt loadings compared to pure Pt-based electrodes. Among these Pt-alloys, Pt-Co nanomaterials have been widely studied as promising alternative catalysts for PEMFC applications, showing enhanced ORR activity and better stability than pure Pt materials [36–38]. Because composition combinatorial methods for improving Pt_xCo_y activity and stability have had limited success, several studies have been dedicated to the understanding of the Pt-Co nanostructure impact on its activity and stability properties [39, 40]. The CL elaboration method and coordination of Co with Pt atoms could also affect the performance and durability [41]. However detailed Pt_xCo_y degradation mechanisms are still very poorly understood. Only few papers are devoted to the experimental study of the long term (> 300 h) durability of these materials [42].
- Another of the most detrimental processes for DMFC lifetime is the catalyst dissolution in the anode side [43]. The present DMFC anode material of choice is an alloy of Pt, Ru and sometimes a third or fourth metal [44–47]. These alloys

have higher activity and less severe CO poisoning (parasitically produced by the MOR) than monometallic catalysts because of the bifunctionality of their component metals [48]. Some experimental studies (e.g. by combining electrochemical and X-ray fluorescence characterizations, [43]) evidenced Ru crossover using state-of-the-art Pt-Ru alloy catalyst at the anode. Ru susceptible to leaching out from the highly active Pt-Ru catalyst crosses the PEM and redeposits at the Pt cathode The degree of cathode contamination by Ru-based species (of chemical form not known yet) depends on, among other factors, the DMFC anode potential and the cell operating time. This process lowers the number of active sites available for methanol oxidation on the anode and reduces ORR activity of the cathode catalyst [49]. In addition, Ru_{y+} ions block sulfonic acid groups in the PEM thereby reducing its proton conductivity.

- Corrosion of the catalyst carbon-support: carbon is thermodynamically unstable at typical cathode operating conditions (potential >0.2 V vs. NHE at 25 °C) [50]. Despite that carbon corrosion reaction reveals to be quite slow at common DAFC operating temperatures (<120 °C) with a steady non-zero current demand, severe CB structure damage can be observed after power-cycled and startup/shut down operation in PEMFCs [51–53]. In-line direct gas mass spectroscopy analysis has successfully detected intermittent peaks of CO_2 emission in exhausted gas from the cathode, indicating that carbon corrosion occurs during the load cycle in the PEMFC [54, 55]. Thus, carbon degrades more rapidly during transient start-up and shut-down conditions and high humidification levels. Indeed, oxygen permeation to the anode induces a reverse proton current between the electrodes increasing the cathode carbon oxidation, which can manifest as dramatic structural changes in the cathode (e.g. porosity increase, electrode thickness decrease) [56]. Furthermore, it is believed that mal-distribution or anode local interruptions of fuel supply (the so-called fuel starvation phenomenon) enhances oxygen permeation from the cathode into the anode in the DAFCs. Following Fuller and Gray [57] as well as Darling and Jayne studies for hydrogen-feed PEMFCs [58], formation of CO_2 bubbles in the DAFC anode could also locally block the hydrogen access to the CL. When carbon corrosion occurs on DMFCs, the carbon support in the cathode CL is reduced in volume and connectivity, which causes also catalyst particles to become isolated or less effective to ORR (carbon corrosion-driven catalyst coarsening [59]). The CL then suffers from reduction of ECSA, which leads to irreversible performance degradation. Furthermore, Pt-based catalysts are believed to bond with the carbon support through the interaction between Pt and π electrons of the graphene sheets in the carbon support: the weaker the π bond (sp^2-hybridized carbon), the weaker the interaction, thus carbon corrosion can also induce Pt-based particles detachment, but this depends clearly on the chemistry of the carbon and of the Pt-carbon bond [60, 61]. Moreover, CB corrosion can also contribute onto the delamination between the CL and the PEM, and/or between the CL and the GDL.
- Loss or decrease of the hydrophobicity: caused by an alteration of the PTFE [62], which is used to give hydrophobic properties to the CLs as well as to the

Gas Diffusion Layers (GDLs) and the Micro Porous Layers (MPLs). This can affect the alcohol and water management in the cell and thus the electrochemical performance;

- The PFSA PEM aging is a very complex issue as it can result from combined effects between electrochemical, physicochemical and mechanical mechanisms. Regarding the chemical and electrochemical degradation, hydrogen peroxide (H_2O_2) can be formed at the surface of the Pt-based particles, if H_2 and O_2 coexist (from crossover) in the CLs in an operating PEMFC [63]. Being the H_2O_2 a highly oxidative reagent with a high reduction potential (1.77 V vs. RHE) it may deteriorate the ionomer in the PEM (maybe even inside the CLs), through the formation of radical species, such as OH° and OOH°, according to a mechanism not yet fully understood [64]. The decomposition of H_2O_2 to give the peroxide radical species is definitely increased by the presence of traces of metal ions (the so-called Fenton cations) such as iron [65, 66], copper [67] or titanium [68], but the PFSA PEM aging is described even in the absence of these pollutants [69]. Some researchers suggest the direct formation of peroxide radicals on Pt catalyst, from the molecular H_2 and O_2 reaction on the surface, may be the main cause of degradation, although H_2O_2 can also be formed during the cathodic ORR, in particular if Pt-M catalysts (with M being a transition metal element) are used [70, 71]. However, the study of these mechanisms is very difficult because H_2O_2 decomposition rate is extremely high on Pt and Pt-M catalysts [72]. Chemical decomposition of the ionomer produces fluoride ions (measured in hydrogen-feed PEMFC effluent water), probably in the form of HF, which in turn may deteriorate other materials used in MEA and cell stacks. Furthermore, PEM degradation facilitates reactants crossover between the CLs, and hence the performance drop ("mixed-potential phenomenon"). By using XPS methods in multi-layered MEA, Chen and Fuller demonstrated that H_2O_2 emission rates are strongly dependent on the PEM relative humidity, oxygen partial pressure and PEM thickness [73]. Shim et al. recently reported accelerated experiments showing that the anode side of the PEM was more deteriorated under H_2-O_2 supply conditions, and that the cathode side of the PEM was more deteriorated under H_2-air supply conditions [74]. In other paper, Hatanaka et al. carried out an OCC durability test with a fluoride emission rate (FER) measurement for intentionally Pt-precipitated PEM with and without CLs [75]. The FER of the PEM with Pt-band was clearly higher than that of the PEM without Pt band. The results showed that precipitated Pt contributes to the PEM degradation. On the other hand, the PEM with both Pt-band and CLs showed much higher FER, clearly showing that the Pt-band might not be the main cause of PEM degradation. Furthermore, it has also been also claimed that crossover oxygen promotes the degradation of the PFSA PEM in the presence of Pt. For the case of DMFCs, crossover Ru could also impact the PEM chemical degradation, but very few experimental studies have been reported in this sense until the time this chapter has been written [76].

These spatiotemporal nano/microstructural changes translate into irreversible long-term cell power degradation. Moreover, the ways of how aging mechanisms occur are expected to be strongly sensitive to the PEMFC operation mode. Understanding the relationship between operation mode and degradation mode remains a challenging task. The competition and cooperation between all these "non-aging" and "aging" mechanisms determine the effective instantaneous electrochemical performance and durability of the cell.

It is extremely important for the DAFC applications to accurately predict their performance, state-of-health and remaining lifetime. For that purpose, it is necessary to develop diagnostic schemes that can evaluate DAFC performance and state-of-health adequately. In order to achieve this, several steps are required:

- To develop a better understanding of the several individual mechanisms in the cell components;
- To understand the interplay between individual scales over the spatiotemporal hierarchies with their possible competitive or synergetic behavior;
- To identify the contribution of each mechanism into the global cell response under dynamic conditions;
- To design separated controllers for an online control of the DAFC behavior to enhance its durability under specific operation conditions (e.g. by controlling the dynamics of the alcohol concentration, the temperature, etc.).

Because of the structural complexity and multi-physics character of modern DAFCs, an analysis through a consistent multiscale physical modeling approach is required to elucidate the efficiency limitations and their location, the degradation and failure mechanisms.

In the following, a review on some modeling efforts for the study of DAFCs or for hydrogen feed PEMFC with approaches that can be extended for DAFCs is provided. This review does not intend to be exhaustive as numerous DAFC models are available in published literature, but instead, to provide a critical overview on the most recent approaches that have been developed. Finally the goal is highlighting the remaining methodological challenges.

8.2 Modeling of Mechanisms and Processes: A Critical Review

8.2.1 Atomistic Level Modeling Studies

The growing use of nanosciences and nanotechnologies are encouraging to understand and thus to explore ways to control the fundamental structure and behavior of the DAFC materials at the atomic and molecular level. First-principles or ab initio calculations (e.g. on the basis of the Density Functional Theory -DFT- method) can be used to predict important quantities such as the adsorbate atomic structures and bonding energies, and provide key information on the reaction mechanisms and pathways. This includes the determination of the controlling elementary reaction pathways and intrinsic kinetics involved in the alcohol oxidation and ORR over Pt and Pt based alloys and their potential dependent behavior. This is also related to the understanding of the influence of the reaction environment including the surface coverage, alloy composition, electrolyte and electric field. New functionalities or lower Pt loadings (e.g. via the development of multi-metallic catalysts with lower Pt loading, e.g. [77]) are made available by manipulation of matter at this scale or through specificities of the nanodimensions, where the physicochemical properties of materials differ from those of the bulk matter. In this context DFT has been largely used to explore different reactions in the absence of interfacial electric field (e.g. cathode ORR steps in [78–90]). Generally these studies were performed using a few atoms/small clusters or extended surfaces to simulate the catalyst.

Significant efforts combining experiments with theory have been focused on the understanding on the mechanisms underlying the total electrooxidation of methanol, ethanol, glycerol and other polyols [91–99]. The oxidation of alcohols occurs in acidic media but usually requires significantly higher overpotentials than reactions carried out in alkaline media [100–103]. The oxidation of methanol proceeds over most transition metals through a sequence of elementary C-H and O-H bond activation steps which occur on the metal, and at the end results in the formation of CO. Through the combination of cyclic voltammetry, chronoamperometry and DFT studies in gas phase (neglecting solvent effects) it is believed that the methanol decomposition occurs via a dual path mechanism [104]. This mechanism for the oxidation of methanol to CO over Pt(111) along with their corresponding reaction energies calculated at 0.5 V RHE are shown in Fig. 8.6 [104].

In order to capture the possible effects of the solvent on the kinetics, Hartnig and Spohr studied by DFT and Ab Initio Molecular Dynamics (AIMD) the oxidation of adsorbed methanol at a water/platinum interface [105]. The authors concluded that water does not only impact the reaction pathway as compared to the corresponding gas phase reaction, but also it modifies the sequence of reactive events by directly, i.e., without intermediate adsorption, incorporating hydrogen ions into its hydrogen bond network, as illustrated in Fig. 8.7.

While CO is an intermediate to CO_2, it readily builds up on the surface and poisons most of the active metals such as Pt. As such, Pt is typically alloyed with a more oxophilic metal such as Ru to promote the adsorption and dissociation of water thus creating bifunctional sites on the surface. The OH groups that result interrupt the CO adlayer and readily oxidize CO to CO_2. The addition of Ru also helps to weaken the Pt-CO bond thus enhancing CO desorption [106].

The oxidation of ethanol, glycerol and other larger alcohols in acidic media result in dehydrogenation which forms the corresponding aldehyde [107, 108]. The

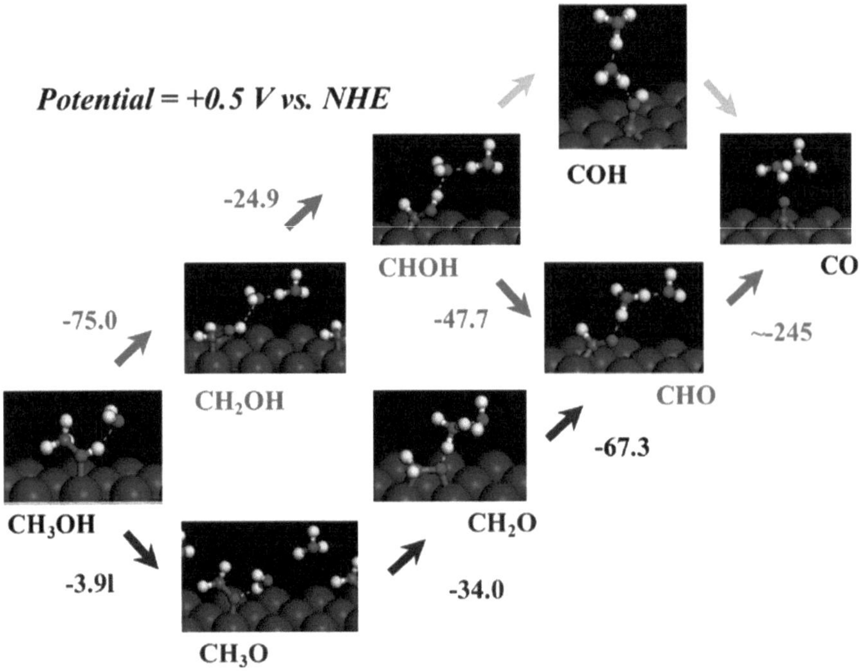

Fig. 8.6 Dual pathway as found by DFT on the oxidation of methanol on Pt(111) (Source: [104]. Copyright 2005, the American Chemical Society)

subsequent activation of the C-C bond, however, is very difficult over typical metals such as Pt, Pd or Au, and as such, very limited CO_2 is formed [109].

In particular, the electrooxidation of ethanol over Pt in acidic media has two major limitations which prevent its viability as it was discussed by Koper [109]. The first relates to the fact that reaction predominantly produces acetate and acetic acid intermediates thus resulting in only 2 and 4 electrons respectively which are only very minor contributions to the possible current. Both are thus unwanted side products for fuel cell applications. The second limitation is that the path to CO_2 is rather difficult in that it requires the activation of C-C bond as well as the oxidation of both the CH_x and CO intermediates that form. Both of these intermediates tend to inhibit or poison metal surfaces at lower potentials [110].

The ethanol oxidation on Pd electrocatalysts is dramatically affected by the pH of the aqueous ethanol solution: no reaction occurs in acidic solutions, while the reaction is fast in alkaline solutions. Some rationale for the origin of this pH effect on the ethanol oxidation to acetaldehyde has been provided by DFT calculations [111] (Fig. 8.8). DFT calculations show that in acidic media continued dehydrogenation of ethanol is difficult due to the lack of OH species to instantly remove hydrogen, which inhibits the ethanol electrooxidation. Conversely, both ethanol and sufficient OH can adsorb on Pd in alkaline media, leading to continuous ethanol electrooxidation. DFT calculations show that in acidic media continued

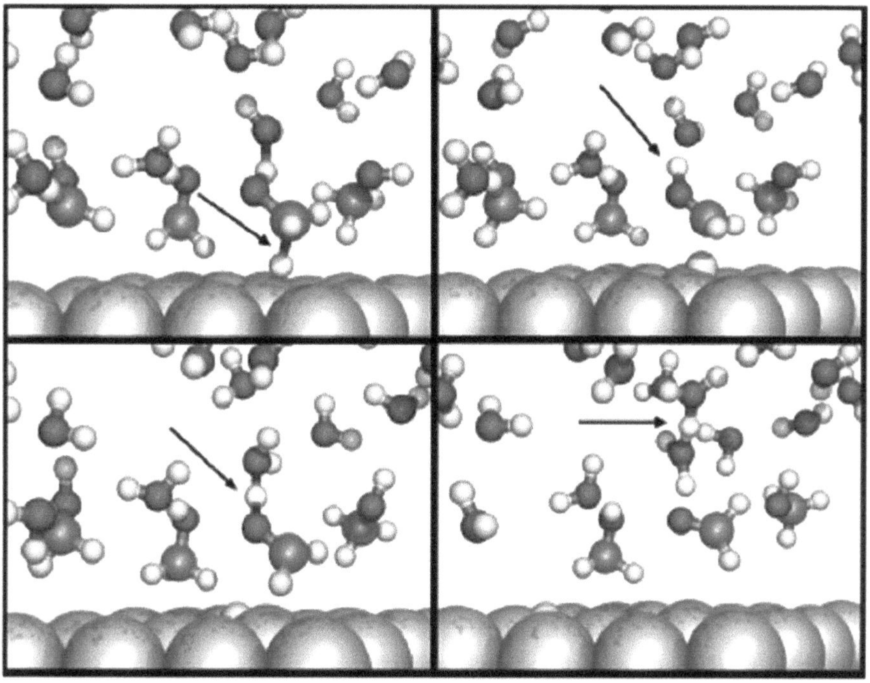

Fig. 8.7 Chemical reaction dynamics of methanol oxidation to formaldehyde as calculated from ab initio Molecular Dynamics. One of the C-H bonds of the methyl group becomes elongated (*top left*) and eventually breaks (*top right*). The adsorbed hydrox-methyl group stabilized by forming a hydrogen bonded complex to a water molecule (*bottom left*) and dissociates rapidly into adsorbed formaldehyde and a hydronium ion (*bottom right*) (Source: [105] with permission)

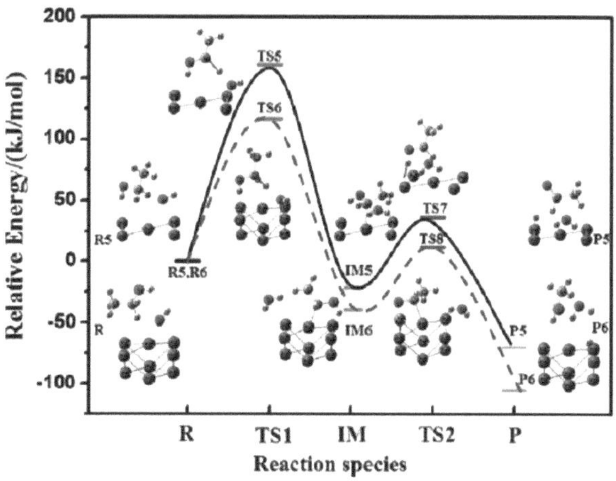

Fig. 8.8 Potential energy profile of the initial reaction processes of ethanol oxidation on Pd5 and Pd9 in alkaline solution. *Blue*, *gray*, *red* and *white* spheres represent the Pd, carbon, oxygen, and hydrogen atoms, respectively (Source: [111], Copyright 2009, the American Chemical Society)

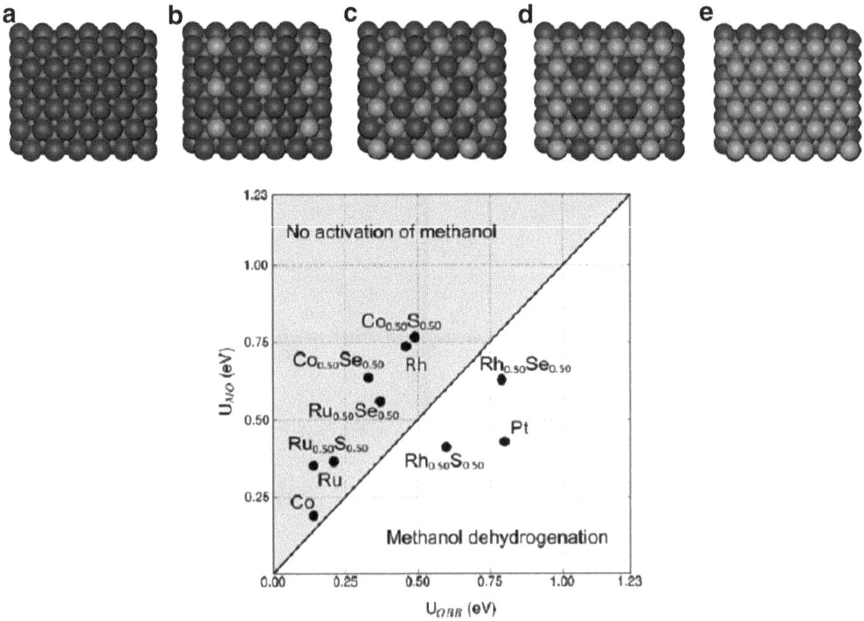

Fig. 8.9 (*Up*) Model of a (**a**) pure fcc(111) surface (x = 0), Se-containing surface with (**b**) small (x = 0.25), (**c**) moderate (x = 0.50), (**d**) high (x = 0.75) content in Se, and a fcc(111) surface covered with a monolayer of Se (x = 1); (*Bottom*) Change in free energy associated with the initial activation of methanol (UMO) plotted against the ORR activity (UORR). For structures lying on the *upper left* part of the plot methanol activation is thermodynamically unfavorable, with UMO > UORR. For structures lying on the *lower right* part such as Pt, methaonol readily dehydrogenates since UMO < UORR (Source: [126] reproduced with permission of Elsevier)

dehydrogenation of ethanol is difficult due to the lack of OH species to instantly remove hydrogen, which inhibits the ethanol electrooxidation. Conversely, both ethanol and sufficient OH can adsorb on Pd in alkaline media, leading to continuous ethanol electrooxidation.

Transition metal surfaces enriched with S, Se and Te, have been considered as candidates for DAFC cathode catalysts [112–115]. For example, ruthenium selenium (RuSe) is a well-studied electro-catalyst for the ORR [116, 117]. The ORR catalysis on pure Ru surfaces depends on the formation of a Ru oxide-like phase [118]. Ru is also an active catalyst for methanol oxidation. On the other hand, the activity of the ORR on RuSe is found not be affected by methanol [116]. RuS, has also been reported insensitive to methanol [119–122]. DFT studies of model transition metal surfaces have provided with atomistic insights into different classes of reactions relevant to fuel cells operation, such as the hydrogen evolution [123], the oxygen reduction [124], and the methanol oxidation [125] reaction. Tritsaris, et al. [126] recently used DFT calculations to study the ORR and methanol activation on selenium and sulfur-containing transition metal surfaces of Ru, Rh, Ir, Pd, Co and W (Fig. 8.9). With RuSe as a starting point, the authors studied the effect of the Se on the activity, selectivity and stability of the catalyst. Ruthenium surfaces with moderate content of Se are found to be active for the ORR and insensitive to methanol.

The AIMD method, based on the Carr and Parrinello approach [127], has also been applied in the study of electrochemistry [128]. Reactive Force Field approaches are now being used to study the ionomer/water/catalyst interfaces during an electrochemical reaction [129]. Neurock et al. developed a detailed first-principles approach that employs a double-reference method to simulate the influence of the electrochemical potential on the chemistry at the metal/solution interface [130]. In this method the aqueous solution metal interface and the interfacial potential drop are explicitly treated. However the choice of an appropriate water surface structure is critical for establishing the appropriate electrochemical behavior at the atomistic scale. This method has been applied to study some electrochemical steps involved in the ORR and methanol oxidation on Pt (e.g. [131, 132]).

Kawakami et al. performed a series of MD simulations on Nafion® PEM which contain various quantities of water and methanol [133]. The simulations gave a variety of nanoscale phase separated structures, such as cluster structures, channel structures, and cluster-channel structures. This was maybe the first time that an atomistic MD simulation reproduced such nanoscale phase separations from random-phased amphipathic polymer solutions. The calculated cluster-channel structure is consistent with experimental results of X-Ray Diffraction (XRD). The authors also investigated the diffusion coefficients of water, methanol and proton in these PEM. Estimated diffusion coefficients of water in the membranes depends very significantly on the water contents in qualitatively good agreement with NMR results. The simulation indicated that water was rarely found in the hydrophobic domains of the Nafion® membranes, whereas methanol penetrated into the hydrophilic/hydrophobic interfacial regions. The authors found that the methanol permeability around sulfonic groups is one-third of that in the water clusters. It means that methanol molecules are trapped by the hydrophilic groups of the PEM and the majority of methanol passes through the water clusters.

The development and application of carbonate as well as anion-exchange PEM electrolytes have significantly renewed interest in the development of alkaline-based DAFC [134]. Many of the reaction intermediates, products and paths discussed above for the catalytic oxidation of alcohols in alkaline media have also been identified or speculated to take part in the electrocatalytic oxidation of these same alcohols.

Despite all this significant theoretical progress on the understanding the DAFC reactions at the atomistic and molecular level, there exist very few bottom-up efforts concerning the use of ab initio data to build a elementary kinetic model allowing to predict electrochemical observables (e.g. i-V curves at the single cell level): we are only aware of the use of Mean Field (MF) or Kinetic Monte Carlo (KMC) approaches, simulating potentiostatic operation modes and using simple Tafel-type or Butler-Volmer–type expressions to simulate the electrode kinetics [135–138].

Application of atomistically-resolved methods for modeling the electrochemical processes in a PEMFC-like environment remains however a major challenge as the description of the coupling between chemistry and the electric field at the atomic scale is a difficult task. Some tentative steps in this direction have been already reported, but under restricted (low Butler-Volmer overpotential, or system close to equilibrium) DAFC operating conditions. In general, no realistic descriptions of the interface are provided: e.g. the interfacial morphology of Nafion® and the catalyst-support electronic structure probably have an impact on the specific electrocatalytic properties [139, 140].

8.2.2 Modeling of Processes at the Single Cell Level

Since the pioneering work of Scott et al. [21], a considerable number of theoretical works have also aimed at the simulation of the processes occurring in DAFC single cells [141]. The majority of the models have been developed for acidic DMFCs, but the mathematical equations can be transposed to other types of DAFCs, actually considerably less studied from a modeling perspective.

8.2.2.1 Analogies with Passive Electric Circuits

Electrochemical impedance was used by Hsu et al. to analyze the reaction kinetics and interfacial characteristics of an anode in a DMFC [142]. The authors proposed an empirical "equivalent"-circuit model incorporating Constant Phase Elements (CPEs) in analogy-circuits taking into account the porous structure of the anode CL. It effectively simulated the electrochemical characteristics of a DMFC porous anode. The impedance model incorporates the interface factor, resulting in very good fitting of the simulation results with experimental data in the Nyquist and the Bode plots over a wide range of frequencies. In addition, the differences among methanol electrooxidation reaction kinetics at various operating potentials are clearly observed and satisfactorily explained using electrochemical impedance spectroscopy and the CPE-based analogy-circuit model.

A model-based EIS approach that combines an analogy electrical circuit method and a mathematical model derived from the reaction kinetics has been proposed to investigate the simultaneous ORR and methanol oxidation reaction at the cathode of a DMFC [143]. Good agreements between the calculated results and the experimental data validated the proposed method. Detailed kinetic parameters and state variables of the cathode were extracted and the concerned reaction processes were further analyzed, which demonstrated the comprehensive applicability of this method (Fig. 8.10). The results showed a significant poisoning effect on the ORR by the presence of methanol at the cathode.

Still in Hsu et al. work, electrochemical impedance was used to analyze the reaction kinetics and interfacial characteristics of an anode in DMFC [142]. Several analogy-circuit models are proposed (Fig. 8.11). The new model incorporates CPEs rather than conventional capacitors in the equivalent-circuits taking into account the porous structure of the anode, particularly that in the CL and at the anode/

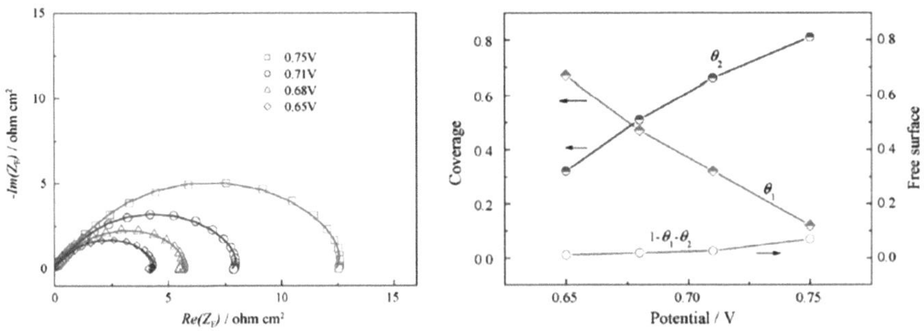

Fig. 8.10 (*Left*) experimental complex phase impedance patterns of the electrode at various potentials fitted by an electric circuit model; (*right*) calculated CO and O_2H coverage and the free surface as a function of the electrode potential (Source: [143] reproduced with permission of Elsevier)

membrane interface. It simulated the electrochemical characteristics of a DMFC porous anode. The differences among methanol electro-oxidation reaction kinetics at various operating potentials are clearly observed and explained using EIS and the CPE-based equivalent-circuit model.

Indeed, many of these equivalent electric circuit models have little physical basis, and therefore no larger or no predictive capability. They are often based on ex situ experiments which are not representative enough of real operating conditions. Analogies with electrical circuit models are often used in order to give some interpretation of DAFC transient operation [144]. Under specific operating conditions, this methodology can be used to determine and separate the effect of the various degradation processes. However, these models are not predictive because their impedance parameters have to be fitted at each DAFC operation point. From the fact that their parameters are temporally constant, these models cannot be used in order to predict the evolution of performances of an aging DAFC.

8.2.2.2 Mechanistic Modeling

Mechanistic models account for the detailed fundamental phenomena such as heat, momentum, multi-component mass transport and electrochemical processes [145]. Since the pioneering works of Scott et al. and Kulikovsky et al. considerable effort has been devoted to the development of mechanistic models for DMFCs [146, 147]. In the following, some of the key models developed are discussed on the basis of their increasing complexity.

Empirical performance models combine theory-based differential and algebraic equations with empirical determined correlations. One characteristic of these models is their simple structure and the small computational effort to perform calculations. However the estimated parameters from the experimental data are

Fig. 8.11 Different electric circuit models proposed by Hsu et al. matching experimental EIS patterns in a DMFC anode (Source: [142] reproduced with permission of Elsevier)

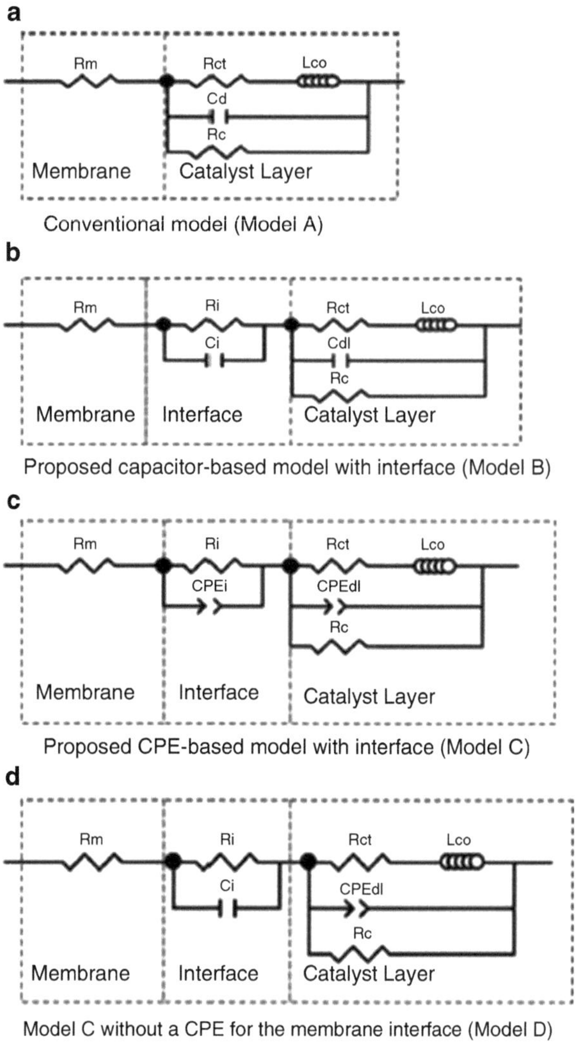

normally specific to certain types of cells and valid for a limited range of operating conditions. Thus, these models are useful to carry out fast predictions for existing designs but fail to predict innovative ones. The number of semi-empirical models developed for DMFCs is limited. Kauranen and Skou reported a model describing both the ORR and the methanol oxidation in the cathode of a DMFC and concluded that the oxygen reduction current is reduced in the presence of methanol oxidation due to surface poisoning [148]. Simouglou et al. developed an empirical model using statistical methods to predict the dynamic voltage response by extrapolation of experimental cell voltage vs. current density databases [149, 150].

Argyropoulos et al. presented a model to predict the cell voltage versus current density for a liquid feed DMFC [151]. The model is based on a semi-empirical approach in which methanol oxidation and ORR kinetics are combined with effective mass transport coefficients for the electrodes. However, the calculated effect of temperature in the polarization curve is in disagreement with experimental knowledge since an increase in temperature leads to a decrease in the cell voltage at a given current density. Dohle and Wipperman developed a model to predict polarization curves and the permeability of methanol through a DMFC based on a set of parameters adjusted from experimental data obtained in a wide range of operating conditions [152].

Baxter et al. developed a mathematical model for a liquid-feed DMFC, mainly focused on the anode CL following an isothermal one-dimensional approach (i.e. across the CL thickness) [153]. A major assumption of their study was that the CO_2 is only dissolved in the liquid and hence their model of transport and electrochemical processes in the anode CL is single-phase only (no bubbles considered). Using a macro-homogeneous model to describe the reaction and transport in the CL of a vapor-feed anode, Wang and Savinell discussed the effects of the anode CL structure on the cell performance [154]. Kulikovsky presented an analytical model of a DMFC anode [155]. Based on this model, Kulikovsky et al. simulated a vapor-feed DMFC with a two-dimensional model and compared the detailed current density distributions in the backing, CL and PEM between two designs of current collectors [156]. In another paper, Kulikovsky numerically studied a liquid feed DMFC considering methanol transport through the liquid phase and inside the hydrophilic pores of the anode backing by using also a 2D approach [157]. Kulikovsky concluded that, near the fuel channel, transport of methanol is determined mainly by a pressure gradient. In constrast to this, he found that diffusion dominates in the CLs and in the PEM. Zones where there is a lack of methanol appear in front of the current collectors. However, in both publications of Kulikovsky, methanol crossover was ignored.

Dohle et al. presented a one-dimensional model for the vapor-feed DMFC, including a description of the methanol crossover [158]. The effects of methanol concentration on the cell performance were studied. Scott et al. also developed several simplified single-phase models to study transport and electrochemical processes in liquid-feed DMFC and showed that the cell performance is limited by the slow diffusion of methanol in the liquid [13, 159–171]. Siebke et al. presented a 1D mathematical model and a numerical simulation to explore the influence of different physical and electrochemical phenomena in the MEA of the liquid feed DMFC [162]. Dohle et al. presented a model to describe the heat and the power management of a DMFC system [163].

Scott et al. also presented an isothermal 1D mathematical model that describes mass transport in a detailed way in the CL [164]. Nordlund and Lindbergh proposed an isothermal agglomerate model based on the reaction mechanism for the electrochemical oxidation of methanol to study the influence of the porous structure on the DMFC performance [165]. Murgia et al. developed a one-dimensional, isothermal,

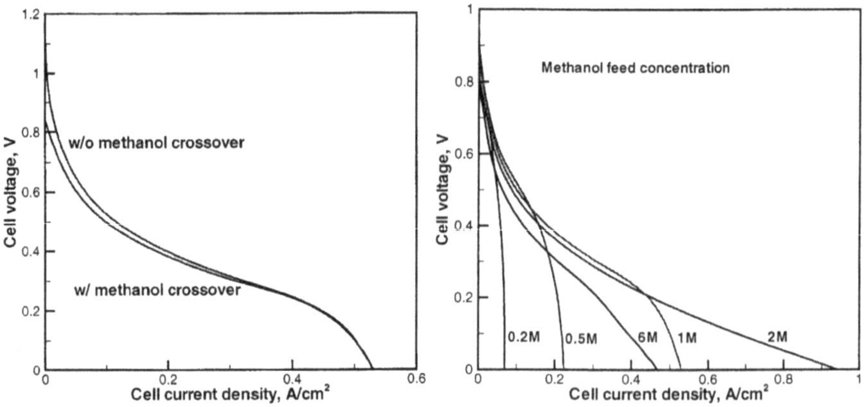

Fig. 8.12 (*Left*) calculated methanol crossover effect on the i-V curve; (*right*) calculated i-V curves for different methanol concentrations (Source: [167], Copyright 2003, The Electrochemical Society)

two-phase, multi-component steady-state model based on the phenomenological transport equations. In their model, the influences of the methanol concentration, the pressure gradient in the CL, flooding in the cathode GDL, and methanol crossover on the cell performance were investigated [166].

Wang and Wang proposed a very complete one-dimensional two-phase, multi-component model for an isothermal liquid-feed DMFC [167]. In addition to the anode and cathode electrochemical reactions, the model considers diffusion and convection of both gas and liquid phases in the backing layers and flow channels. In particular, the model was probably one of the first to account for the mixed potential effect of methanol oxidation at the cathode as a result of methanol crossover caused by diffusion, convection, and electro-osmosis. This model was solved numerically using a Computational Fluid Dynamics (CFD) code. The effects of the methanol feed concentration on cell performance were explored (Fig. 8.12). Among the several results shown by the authors, it was found that for the parameters assumed, when the methanol feed concentration is larger than 2 M, the cell voltage is greatly reduced by excessive methanol crossover and the maximum current density begins to be limited by the oxygen supply at the cathode. The oxygen depletion results from excessive parasitic oxygen consumption by methanol crossing over.

White et al. proposed an one-dimensional, isothermal model for a DMFC [168]. This model accounts for the kinetics of the multi-step methanol oxidation reaction at the anode. Diffusion and crossover of methanol are modeled and the mixed potential of the oxygen cathode due to methanol crossover is included. Kinetic and diffusional parameters are estimated by comparing the model to experimental data. The authors claim that their semi-analytical model can be solved rapidly so that it could be suitable for inclusion in real-time system level DMFC simulations.

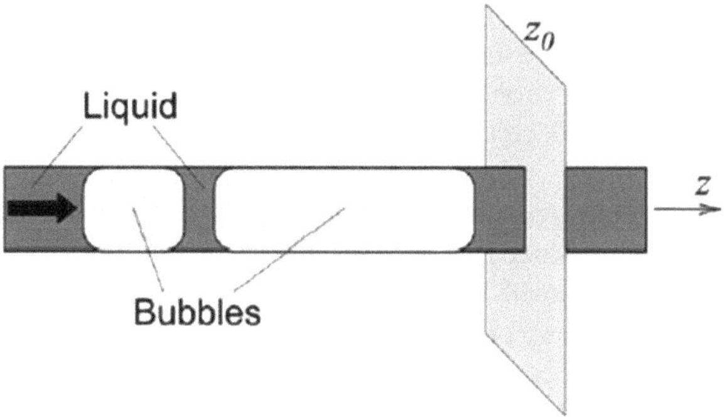

Fig. 8.13 Scheme of Kulikovsky's model of bubbles transport in DMFCs (Source: [171] reproduced with permission of Elsevier)

Guo and Ma reported an isothermal two-dimensional analytical model describing electrochemical reactions on the anode and cathode and main transport phenomena in a liquid-feed DMFC including methanol crossover, diffusion of reactants in porous media layers and fluid flow in the reactants distributor [169]. In fact, the model is one-dimensional since the authors neglect the transport in the channel direction. Garcia et al. provided a model accounting for the kinetics of the multi-step methanol oxidation reaction at the anode, the diffusion and crossover of methanol and the mixed potential of the oxygen on the cathode due to methanol crossover [168]. Some of the kinetic and diffusional parameters are estimated by adjusting the model to experimental data. The authors claim that this model has also the merit of being suitable for inclusion in real-time system level DMFC calculations.

Scharfer et al. set up a multi-component transport model to describe the diffusion driven mass transport of water and methanol in PEM [170]. For a PEM in contact with liquid methanol and water on one side and conditioned air on the other, the corresponding differential equations and boundary conditions were derived taking into account the polymers three-dimensional swelling. Phase equilibrium parameters and unknown diffusion coefficients for Nafion®117 were obtained by comparing the simulation results to water and methanol concentration profiles measured with confocal Raman spectroscopy. The influence of methanol concentration, temperature and air flow rate was predicted by the model. Although there are indications for an influence of convective fluxes, the measured profiles are ascribed to a Fickean diffusion. Furthermore, the assumption to describe the thermodynamic phase equilibrium as liquid-type equilibrium also at the lower surface of the membrane, which is in contact with a gas phase, can be confirmed by their results.

A simple isothermal model of the flow with gaseous bubbles in the anode channel of a DMFC has been developed by Kulikovsky [171, 172] (Fig. 8.13). The model is based on continuity equations for time-average concentrations in the anode channel. The model shows that formation of bubbles (i) accelerates the flow and (ii) increases the gradient of time-average concentration of methanol along the channel. The latter effect is found to dramatically reduce the limiting current density of the cell. According to the author, the optimization of the flow field in order to reduce the rate of bubbles formation or their size ("bubbles management") may have a strong impact on the overall DMFC performance. The model predicts significant decrease of cell performance in the regime with intensive formation of bubbles.

The electrochemical reactions taking place at a DMFC are exothermic. Heat can be also produced by irreversibilities in the cell (ohmic or activation losses) and thus its removal is a critical issue for fuel cells. Argyropoulos et al. developed a thermal energy one-dimensional mechanistic model for a DMFC stack based on the differential thermal energy equation [173, 174]. The model allows capturing the effect of operating parameters such as fuel and oxidant inlet temperature, flowrate, pressure and current density and also of some design parameters (active area, material properties and geometry) in the cell temperature along the stack. Shultz and Sundmacher developed a non-isothermal, one-dimensional, dynamic model using the multi-domain approach [175]. In this model, mass transport within the different porous structures of the DMFC was described using the generalized Stefan–Maxwell equations. For the membrane, an activity model based on the Flory–Huggins approach is used accounting for swelling phenomena, non-idealities and phase equilibria between the pore liquid inside the PEM and the fluids inside the CL. The model showed good agreement to experimental data obtained by the authors, in particular regarding methanol crossover and steady state current–voltage characteristics.

Meyers and Newman developed a comprehensive model which describes the thermodynamics, transport phenomena, and electrode kinetics of a DMFCs [176]. On the basis of their model, Kauranen and Skou indicated that the saturation by the OH on the surface of Pt leads to limiting current when oxidation of methanol was catalyzed by pure platinum [177].

An intrinsic time-dependent one-dimensional (1D) model and a macro two-dimensional (2D) model for the anode of the DMFC are presented in [178]. The two models are based on the dual-site mechanism, which includes the coverage of intermediate species of methanol, OH, and CO on the surface of Pt and Ru. The intrinsic 1D model focused on the analysis of the effects of operating temperature, methanol concentration, and overpotential on the transient response. The macro 2D model emphasizes the dimensionless distributions of methanol concentration, overpotential and current density in the CL which were affected by physical parameters such as thickness, specific area, and operating conditions such as temperature, bulk methanol concentration, and overpotential. The models were developed and solved in the PDEs module of COMSOL Multiphysics, giving good agreement with experimental data. The dimensionless distributions of methanol concentration, overpotential, and current density and the efficiency factor were calculated quantitatively. The models can be used to give accurate simulations for the polarizations of methanol fuel cell.

Krewer et al. presented a three-dimensional (3D) model but only concerning the anode of a DMFC [179]. The authors studied the residence time behavior and

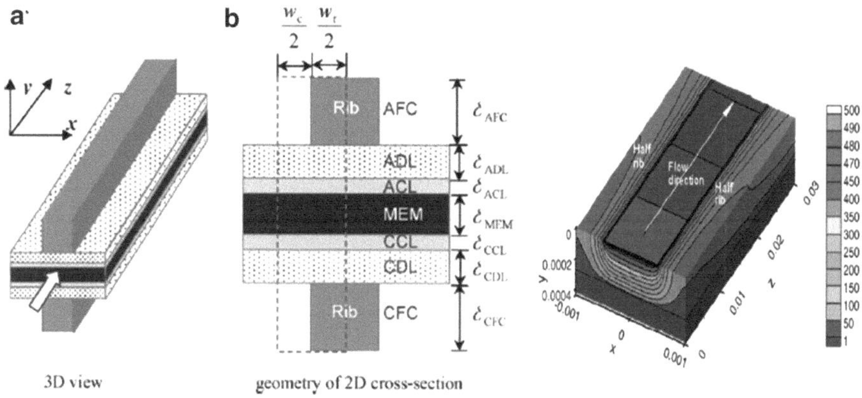

Fig. 8.14 (*Left*) Computational domain in the model by Yang et al.; (*right*) calculated distribution of methanol concentration in the anode porous region and membrane (unit: m^3/s) (Source: [180] reproduced with permission of Elsevier)

concentration distribution in a simplified rhomboidal anode flow bed in comparison with experimental data.

A 3D steady-state model for liquid feed DMFCs is presented in the paper by Yang et al. [180] (Fig. 8.14). This 3D mass transport model is formed by integrating five sub-models, including a modified drift-flux model for the anode flow field, a two-phase mass transport model for the porous anode, a single-phase model for the PEM, a two-phase mass transport model for the porous cathode, and a homogeneous mist-flow model for the cathode flow field. The two-phase mass transport models take account the effect of non-equilibrium evaporation/condensation at the gas-liquid interface. A 3D computer code is then developed based on the integrated model. After being validated against the experimental data reported in the literature, the code was used to investigate numerically transport behaviors at the DMFC anode and their effects on cell performance.

Omran et al. have proposed a 3D, single phase steady-state model of a liquid feed DMFC [181]. Their model is implemented into the commercial computational fluid dynamics (CFD) software package FLUENT®. The continuity, momentum, and species conservation equations are coupled with mathematical descriptions of the electrochemical kinetics in the anode and cathode channel and MEA. For electrochemical kinetics, the Tafel equation is used at both the anode and cathode sides. Results are validated against DMFC experimental data with reasonable agreement and used to explore the effects of cell temperature, channel depth, and channel width on polarization curve, power density and crossover rate. The results show that the power density peak and crossover increase as the operational temperature increases. It is also shown that the increasing of the channel width improves the cell performance at a methanol concentration below 1 M.

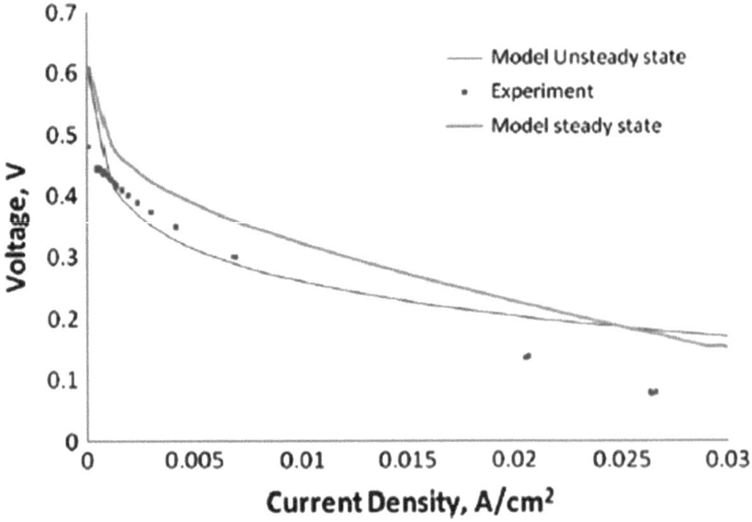

Fig. 8.15 Calculated and experimental polarization curves (Source: [183] reproduced with permission of Elsevier)

Furthermore, an one-dimensional model has been developed by Lam et al. to explore multiple 3D anode configurations of a DMFC [182]. The model was fitted to experimental data for a single electrode structure having 1, 2, or 4 mg cm^{-2}, and the resulting parameters were used in the model to qualitatively predict the performance, current distribution, and methanol concentration distribution of a multilayered anode structures having one, two, or four layers and a total catalyst loading of 4.0 mg cm^{-2} carbon-supported (Vulcan XC-72) 40 wt% Pt–Ru. This model has been designed to provide insights into the conditions for which crossover can be completely eliminated.

The importance of considering transient behavior in the DMFC models is highlighted in a more recent work by Basri et al. [183]. The authors proposed an unsteady-state model for a liquid-feed DMFC considering two-phase transport and heat management. The results were then compared by the authors with the experimental data from an in-house fabricated DMFC and a steady-state model. The authors found that the unsteady-state model fits better the experimental data than the steady-state model (Fig. 8.15). The effects of feed methanol concentration in the reservoir and current density on mass transport to the CL were in particular explored.

Other types of DMFCs have been also addressed by modeling, as for example a passive liquid-feed DMFC with neither external liquid pumps nor gas blowers by Chen and Zhao [184] (Fig. 8.16). An one-dimensional model is developed by considering coupled heat and mass transport, along with the electrochemical reactions occurring in passive DMFCs. The analytical solutions predicting the performance of the cell operating with different methanol concentrations are

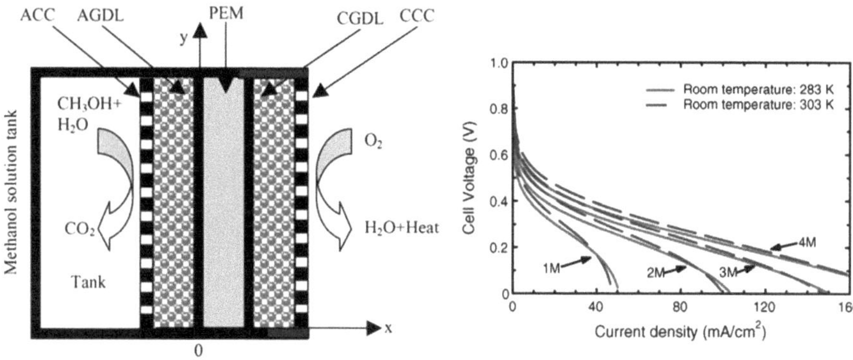

Fig. 8.16 (*Left*) modeled passive DMFC configuration (in the cathode, the air is transported by natural convection); (*right*) calculated i-V curves for different ambient temperatures (Source: [184] reproduced with permission of Elsevier)

obtained. It is shown that the performance of passive DMFCs increases with methanol concentration. It is further revealed that the improved performance with higher methanol concentrations is due primarily to the increased operating temperature resulted from the exothermic reaction between permeated methanol and oxygen in the cathode.

Moreover, a new structure of passive DMFC with two methanol reservoirs separated by a porous medium layer was designed and a corresponding mathematical model was presented by Cai et al. [185]. This type of DMFC can be directly fed with highly concentrated methanol solution or neat methanol. Modeling and experiments are used by the authors to optimize the porosity of the components. It was found that the new designed DMFC can be continuously operated for about 4.5 times longer than a conventional passive DMFC with the optimum parameters.

A two-dimensional, two-phase, non-isothermal model was developed by Xu and Faghri to investigate the water transport characteristics in a passive liquid-feed DMFC [186]. The liquid–gas two-phase mass transport in the porous anode and cathode was formulated by the authors based on multi-fluid model in porous media, and water and methanol crossover through the membrane were considered with the effect of diffusion, electro-osmotic drag and convection. The authors investigated the effects of various operating parameters, such as current density, methanol concentration, and air humidity, as well as the effect of the cathode hydrophobic air filter layer, on the water transport and cell performance. The results showed that for the free-breathing cathode, gas species (Fig. 8.17) concentration and temperature showed significant differences between the cell and the ambient air. The use of a hydrophobic air filter layer at the cathode helped to achieve water recovery from the cathode to the anode, even if oxygen transport resistance was increased. Furthermore, the authors highlight that the water transport can be influenced by the ambient relative humidity.

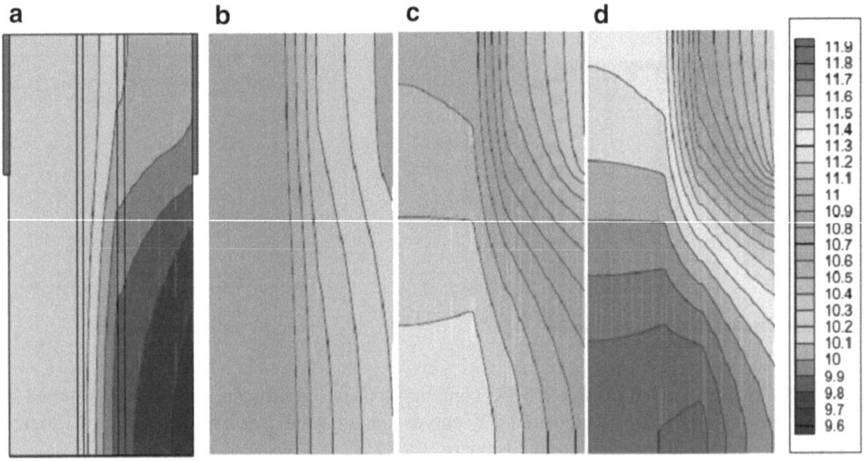

Fig. 8.17 Calculated distribution of temperature rise (°C) across the MEA at a current density of 100 mA.cm^{-2} for different methanol concentrations: (**a**) 1 M, (**b**) 2 M, (**c**) 3 M, (**d**) 4 M (Source: [186] reproduced with permission of Elsevier)

DAFCs with ethanol as fuel have been also studied by modeling. Andreadis et al. proposed a model for a DEFC anode [187]. The model accounts for the influence of different parameters, e.g., temperature, electrode thickness and ethanol concentration on anode overpotential.

CFD models of DEFC have been also proposed [188]. Suresh and Jayanti developed an one-dimensional, single phase, isothermal mathematical model for a liquid-feed DEFC, taking into account mass transport and electrochemical phenomena on both the anode side and the cathode side [189]. Tafel kinetics expressions have been used to describe the electrochemical oxidation of ethanol at the anode and the simultaneous ethanol oxidation and ORR at the cathode. The model in particular accounts for the mixed potential effect caused by ethanol cross-over at the cathode and is validated using the data from the literature. Model results show that ethanol cross-over can cause a significant loss of cell performance.

Andreadis and Tsiakaras reported an one-dimension, steady-state and single phase model with the purpose of describing the mass transport within a PtRu/Nafion® 115/Pt MEA and the performance of a DEFC [190]. The effect of important cell operating parameters on the ethanol crossover rate and the fuel cell performance is investigated. According to their results, in the case of low current density values and high concentrations of ethanol, ethanol crossover could pose serious problems to the DEFC operation (Fig. 8.18). Moreover, it was pointed out that the ethanol crossover rate dependence on the ethanol feed concentration is an almost linear function presenting a maximum at about 10.0 M. A further increase of the ethanol feed concentration leads to a steep decrease of ethanol crossover rate. This behavior is attributed by the authors to the PEM swelling which is responsible for the PEM volume fraction decrement.

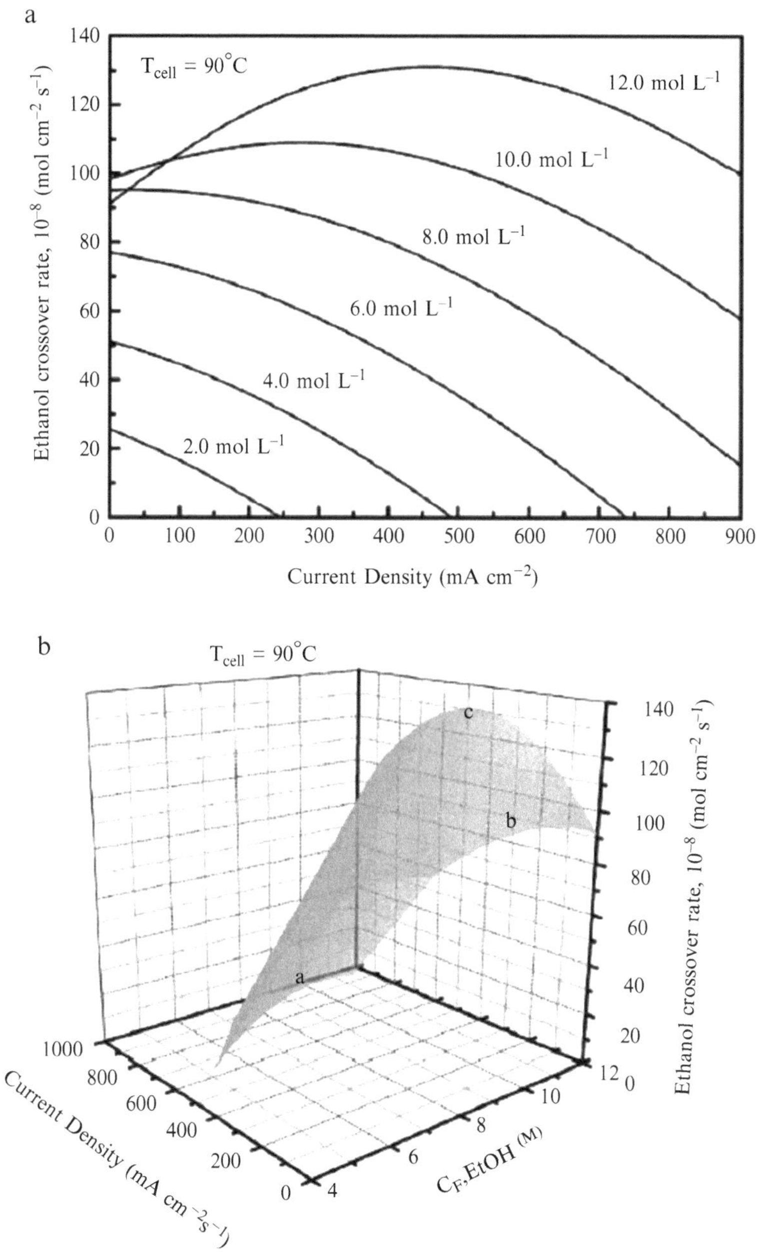

Fig. 8.18 (**a**) Effect of current density on ethanol crossover rate at different ethanol concentrations at 90 °C. (**b**) Effect of both current density and ethanol feed concentration on ethanol crossover rate at 90 °C (Source: [190] reproduced with permission of Elsevier)

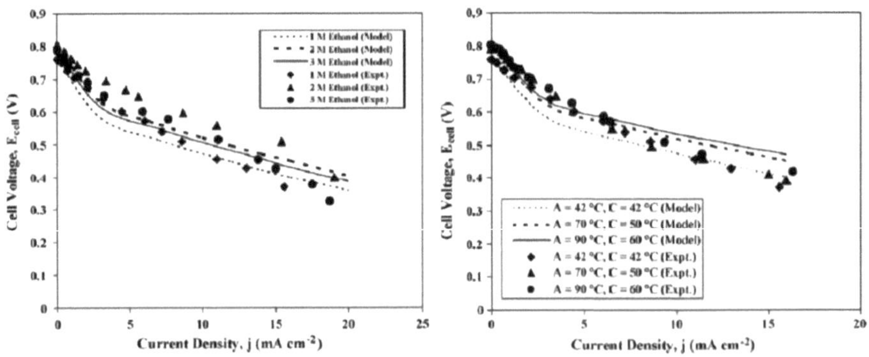

Fig. 8.19 (*Left*) calculated and experimental i-V curves for different ethanol concentrations at 42°; (*right*) calculated and experimental i-V curves at different cell temperatures for 1 M ethanol concentration (Source: [191] reproduced with permission of Elsevier)

A mathematical model for DEFC was proposed by Pramanik and Basu describing different overpotentials [191]. The assumptions of their model are: (i) the anode compartment considered as a well-mixed reactor, (ii) 1 bar pressure maintained both at the anode and cathode compartments, (iii) the transport processes are modelled in one dimension. The model accounts for Butler-Volmer-based descriptions of the ethanol electrooxidation mechanisms, diffusive reactants transport and ohmic losses at the electrode, current collector and electrode-current collector interfaces. The experiment data on current–voltage characteristics is predicted by the model with reasonable agreement and the influence of ethanol concentration and temperature on the performance of DEFC is studied by the authors (Fig. 8.19).

A detailed one-dimensional mathematical model of a DEFC has been very recently proposed by Meyer et al. [192]. The electrochemical oxidation of ethanol in the CL is described by several reaction steps leading to surface coverage with adsorbed intermediates (CH_3CO, CO, CH_3 and OH) and to the final products acetaldehyde, acetic acid and CO_2. A bifunctional reaction mechanism is assumed for the activation of water on a binary catalyst favouring the further oxidation of adsorbates blocking active catalyst sites. The mathematical descriptions of the chemical reactions are coupled with the models describing the charge and reactant transport. The model accounts for crossover of the reactants through the membrane leading to the phenomenon of cathode and anode mixed potentials due to the parasitic oxidation and reduction of ethanol and oxygen, respectively. Polarisation curves of a DEFC were experimentally measured for various ethanol feed concentrations and were used as reference data for the simulation.

Tang et al. proposed a two-dimensional mathematical model of a tubular-based DEFC cathode, describing electrochemical kinetics and multi-component transport [193]. A spherical agglomerate model is used in the CL, and the effect of ethanol penetration on the ORR of the tubular cathode is also considered. The model, coded

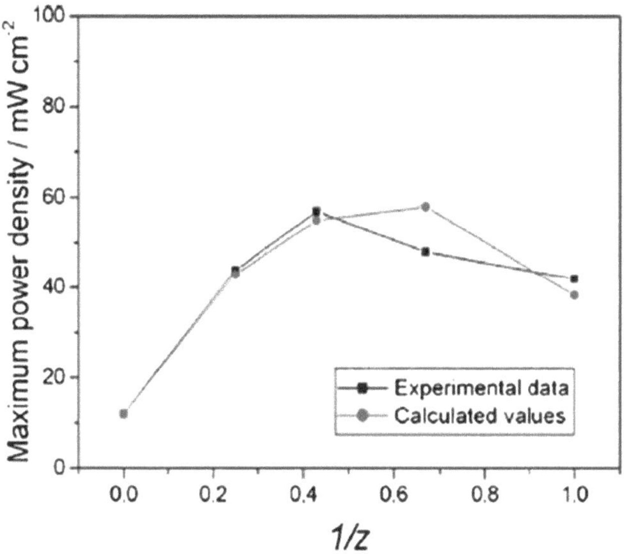

Fig. 8.20 Dependence of experimental and calculated maximum power density of a DEFC on Sn content in non-alloyes Ptz-Sn/C catalysts (Source: [194] reproduced with permission of Elsevier)

in Comsol, is used to simulate the concentration distribution of components and the current density distribution.

Antolini and Gonzalez recently proposed a simple empirical model to evaluate the contribution of alloyed and non-alloyed platinum and tin to the ethanol oxidation reaction on Pt–Sn/C catalysts for DEFC [194]. On the basis of the model, the ethanol oxidation on partially alloyed catalysts occurs through a dual pathway mechanism, separately involving the Pt_3Sn and Pt–SnOx phases. The model, validated by experimental data, can predict the performance of a DEFC by varying the Sn content and/or the degree of alloying of Pt–Sn/C catalysts used as the anode material (Fig. 8.20).

As the best of the author's knowledge, almost no work has been published on the performance modeling and numerical simulation at the single cell level of DAFCs with other types of alcohols.

Furthermore, surprisingly, very few works have been reported on the study of materials degradation phenomena in DAFCs. For instance, a model for carbon and Ru corrosion in a DMFC anode under strong methanol depletion has been very recently proposed by Kulikovsky [195]. The model is based on the mathematical description of the current conservation in the membrane. In the methanol–depleted domain, methanol oxidation reaction is substituted by the carbon oxidation (corrosion). This is supposed to dramatically lowering the membrane potential and to create an environment for electrochemical oxidation of Ru. His calculations show that 50–100 mV loss in the cell potential manifests quite a significant (above 50 %) methanol–depleted fraction of the cell active area (Fig. 8.21).

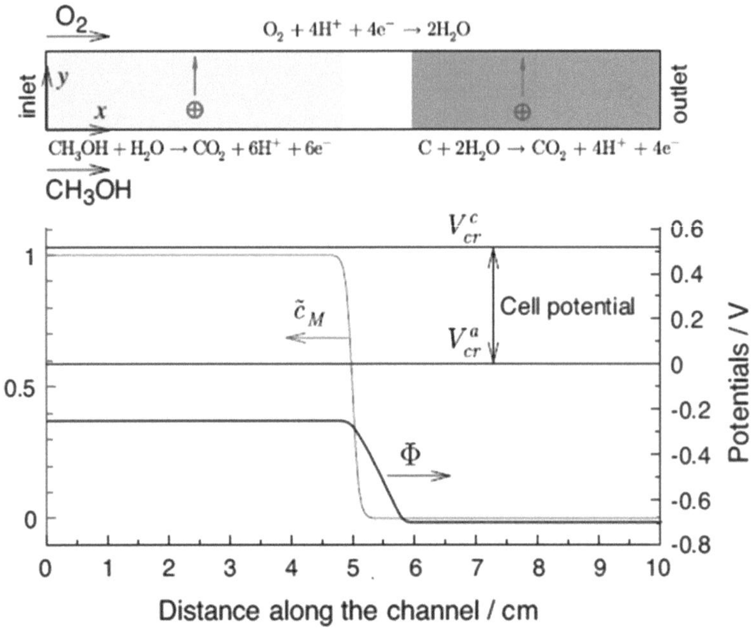

Fig. 8.21 (*Left*) schematic of a DMFC with strong methanol depletion in the second half of the channel (the normalized methanol concentration, membrane potential and carbon phase potentials are shown in the *bottom*); (*right*) calculated cell potential and carbon and Ru overpotentials as function of the fraction of methanol-depleted domain (Source: [195] reproduced with permission of Elsevier)

Numerous modeling works addressing degradation of H_2-feed PEMFC materials have been reported [196, 197], and several conclusions from them can be certainly applied to the DAFC. Some examples on this are discussed in the next Section.

8.2.3 Theoretical Challenges

Despite that the atomistic and processes models available in the literature provide very useful information about the mechanisms in DAFCs, they still present several drawbacks that make them not appropriate enough to predict the cell performance durability under realistic operation conditions [196, 197]:

- No prediction of cell potential evolution and MEA durability. The kinetic models use to represent steady-state regimes with time-independent local operating conditions (e.g. local water content within the CL: assumption of liquid water fully-saturated CL) and they show for instance poor predictive capability. One reason is that many of these models are parameterized by using ex situ experiments which are not representative enough of real operating conditions (e.g. potential holding between 1.2 and 1.6 V in [202]). Another reason is that they do not take into account the impact of the modeled aging phenomena on the variation of the structural and physicochemical properties of the MEA materials (e.g. impact of the aging mechanisms on the variation of charge conductivity, variation of porosity, variation of contact resistances...). In other words, the instantaneous feedback between performance and aging, occurring in real DAFC operation, is not described: they describe which operating conditions enhances a given degradation process (e.g. carbon corrosion) but do not describe the impact of this degradation process on the instantaneous performance (any prediction of the transient behavior of the DAFC MEA electrochemical observables, such as the cell potential degradation or durability, is provided).

- Aging mechanisms addressed in a separated (uncoupled) way. Most of the available kinetic models describing degradation phenomena focus on Pt-based MEA, where Pt and carbon materials are treated as a single phase (no distinction between Pt and carbon phases). As in the ex situ experimental approaches, each material aging phenomenon has been the subject of individual modeling studies. However, in realistic DAFC environments, aging mechanisms of the different individual materials are expected to compete or interact.
- Potentiostatic-potentiodynamic simulations. In most of the available kinetic degradation models, the Butler-Volmer electrode potential is the input variable, the output being a material corrosion rate and the cell current. Implicitly, it is assumed that the potential of the nanomaterials is equal to the external/macroscopic applied potential. Again, the cell potential evolution and the associated MEA durability cannot be predicted in this way. In fact, the majority of the single-cell tests available in the literature are made with the current being the input variable.
- Use of the classical Butler-Volmer theory. This empirical theory, largely used in the PEMFC (and in particular, DAFC) modeling community [198], describes electrochemical (electron-transfer) reactions on ideal planar electrodes. By using empirical Butler-Volmer equations written in terms of the electrode potential (as done e.g. in [185–187]), oxide formation and corrosion reactions (on Pt and carbon) are implicitly supposed to take place in the bulk, just outside the catalyst/Nafion® interfacial electrochemical double layer (i.e. far from the electrified substrates, catalyst and/or carbon support). Thus the possible interplaying of these aging mechanisms with ORR intermediate reaction species and the parasite water adsorption on the catalyst or carbon surface, expected in realistic PEMFC environments [199], is not taken into account. In fact, the electrochemical double layer capacity is usually assumed to be constant (i.e. the electrochemical double layer structure is assumed to be uncoupled from the elementary reactions). In my opinion this could be an important assumption as the electrochemical interface is expected to evolve under transient conditions (such as in the case of aging nanoparticles), and the structure of the electrochemical double layer influences in turn the electron transfer rate and thus the effective electroactivity properties of the catalyst surface. More generally, experimental evidence has been reported by Adzic et al. that Butler-Volmer equations are inappropriate for describing HOR and ORR reactions on Pt surfaces [200]. There is also some experimental evidence for nano-sized electrodes, where

pronounced nanoscale non-linear effects of charge transfer in the surrounding electrolytic environment are important and cannot be explained using conventional theories [201]. Furthermore, the use of such a macroscopic Butler-Volmer theory cannot be really justified for describing electron transfer reactions on nanomaterials with an evolving structure (in fact, standard transition-state theory assumes that the "catalyst" properties, by definition of "catalyst", are time-invariant or recovering its morphology after reaction). For example, water splitting and hydroxyl adsorption are speculated to modify the surface Pt-alloys dissolution thermodynamics [202]: thus, the stability properties of nanostructured catalysts (e.g. facility to oxidize) can be strongly affected by the electrochemical reactions of interest in PEMFC environments, such as the HOR or the ORR. Structure and interatomic distance can also change during reactant (alcohol or O_2) exposure [203, 204].

8.3 Towards a Multiscale Modeling Framework for DAFCs

8.3.1 General Overview

In 2002 I have invented a transient, multiscale and multiphysics single fuel cell model, called MEMEPhys. This model, that I have continuously developed since then, accounts for the coupling between self-consistent physical-based mechanistic descriptions of the PEM and the CL phenomena (e.g. reactants, water and charge transport and detailed electrochemistry) and different materials aging mechanisms [59, 205–215]. The model is designed for simulating hydrogen-feed PEMFC, PEM Water Electrolyzers and Li Ion batteries, but could be easily extended to simulate DAFCs.

The model was designed to connect atomistic phenomena (elementary kinetic processes) with macroscopic electrochemical observables (e.g. polarization curves-i-V characteristics -, EIS…) with reasonable computational efforts by using ab initio and surface science databases. Such a model is a multiscale one in the sense that it is made of a set of interconnected sub-models describing the phenomena occurring at different scales in the fuel cell. However, this description remains macroscopic in the sense that it is based on irreversible (non-equilibrium) thermodynamics concepts particularly adapted for the description of non-equilibrium physicochemical systems, as they are extensively used in chemical engineering: use of conservation laws coupled to closure equations (e.g. flux expressions, chemical rate models, thermodynamic models…). A good way to derive such a model is to use the infinite-dimensional Bond Graph approach: this approach allows building the model as a modular collection of reusable sub-models through the concept of port power-conjugate variables, in particular for distributed-parameter systems [216]. Such an approach allows to easily modify the sub-models

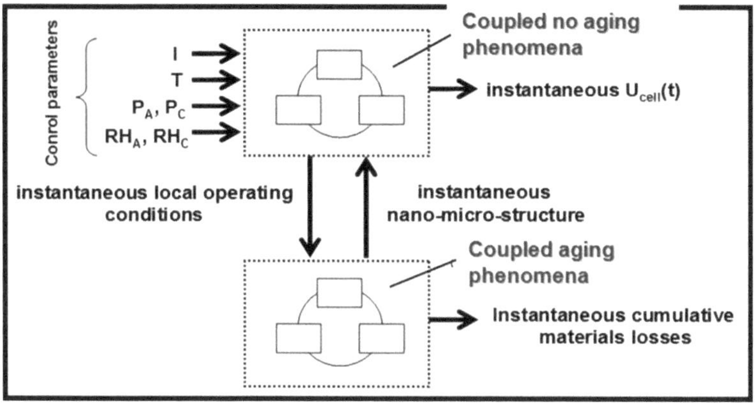

Fig. 8.22 Algorithmic principles behind MEMEPhys simulation package

and to test new assumptions as it is computationally causal-independent. The advantages of the Bond Graph approach utilization in chemical engineering and fuel cells modelling have been already demonstrated [217, 218].

In contrast to the standard Butler-Volmer-based fuel cell models, MEMEPhys describes (mechanistically) the feedback between the instantaneous performance and the intrinsic MEA aging processes (Fig. 8.22): that means that the model takes into account, at each simulated time step, both the effect of performance on aging and the effect of aging on performance. This powerful characteristic arises naturally from the coupled structure of this model, and thus, it allows the prediction of material properties evolution (e.g. the temporal evolution of the interlinked activity, selectivity and stability properties of the electrocatalyst). The prediction of the cell potential degradation over operation time, such as MEA durability (defined here as the operation time where the cell potential collapses to zero) as a function of the cell operating conditions and the initial MEA design parameters, also arises from the model. Furthermore, this model allows carry out temporal multiscale analysis (e.g. calculation of the EIS shape evolution during the MEA aging) and provides insights on the competitions and synergies between the different degradation mechanisms.

More generally, in contrast to the classical theories used to describe the PEMFC electrochemical behavior, this multiscale modeling approaches the fuel cell environment as a full non-equilibrium physicochemical system. The nano/microstructural properties of the PEMFC electrodes (e.g. catalyst active surface area, cathode thickness. . .) are not time-invariant: they are defined as instantaneous state-variables. The characteristics of this multiscale modeling approach represent important prerequisites for engineering practice to understand the synergies and interactions between the aging phenomena, to interpret fuel cell behavior under realistic operating conditions, to predict the MEA degradation and durability as a function of its initial nanomaterials composition and nano/microstructure, and to find new operation strategies to mitigate the degradation.

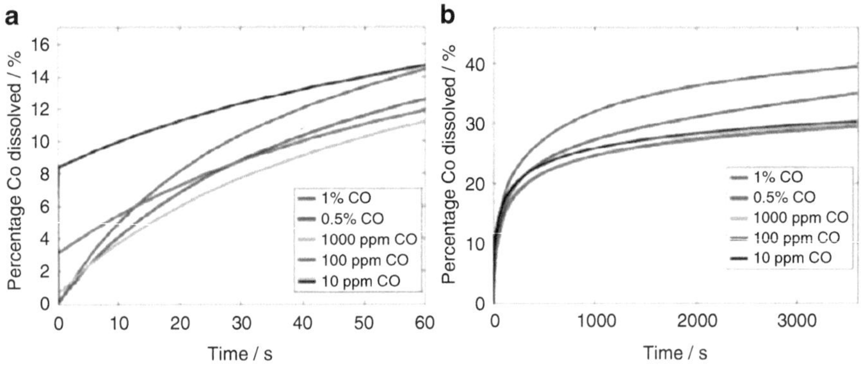

Fig. 8.23 Percentage of cobalt dissolved from PtCo-catalyst versus operation time with H_2 + CO mixtures, under 0.5 A.cm^{-2} constant current. (**a**) simulated results for the first 60 s, (**b**) simulated results for the long run (Source: [208], Copyright 2011, The Electrochemical Society)

8.3.2 Modeling Competitive Degradation Phenomena

The model includes novel descriptions of coupled electrochemical aging processes in hydrogen-feed PEMFCs (e.g. Pt and Pt_xCo_y oxidation/dissolution/ripening, carbon catalyst-support corrosion, PEM degradation…) but could be certainly adapted for the simulation of DAFCs. For instance, within the context of PEMFCs, the tool has provided very interesting information on the competition of aging phenomena. Some experimental data suggests that external anode and cathode contaminants (e.g. CO in the anode, SO_2 in the cathode) can enhance the damage of the PEMFC materials. But according to my modeling work, the injection of these contaminants can mitigate, under appropriate current-cycled conditions, the intrinsic materials aging mechanisms as demonstrated based on an approach combining experiments and this multiscale numerical model [208, 213]: this work clearly illustrated the interest of treating the complex mechanisms interacting between them towards engineering optimization of the PEMFC operation.

Figure 8.23 reports one of the simulation results which illustrates the importance of accounting for the interaction between mechanisms when modeling PEMFCs: this concerns a simulated PEMFC anode based on PtCo catalyst, operating with H_2 + CO mixtures. In this case, CO presence enhances Co dissolution and thus decreases the fuel cell durability.

Figure 8.24 shows the model capabilities to address competitive degradation phenomena: membrane chemical degradation makes the proton conductivity decrease and thus potential cathode decreases which in turn decreases the kinetics of the cathode Pt degradation. In other words, membrane degradation can mitigate Pt degradation in the cathode, under specific operation conditions.

Coarse Grained Molecular Dynamics (CGMD) models have been developed to predict the self-organization of the electrodes and PEM and to understand its impact on the effective transport and electrochemical properties [209, 220]. CGMD is essentially a multiscale technique (parameters are directly extracted from classical

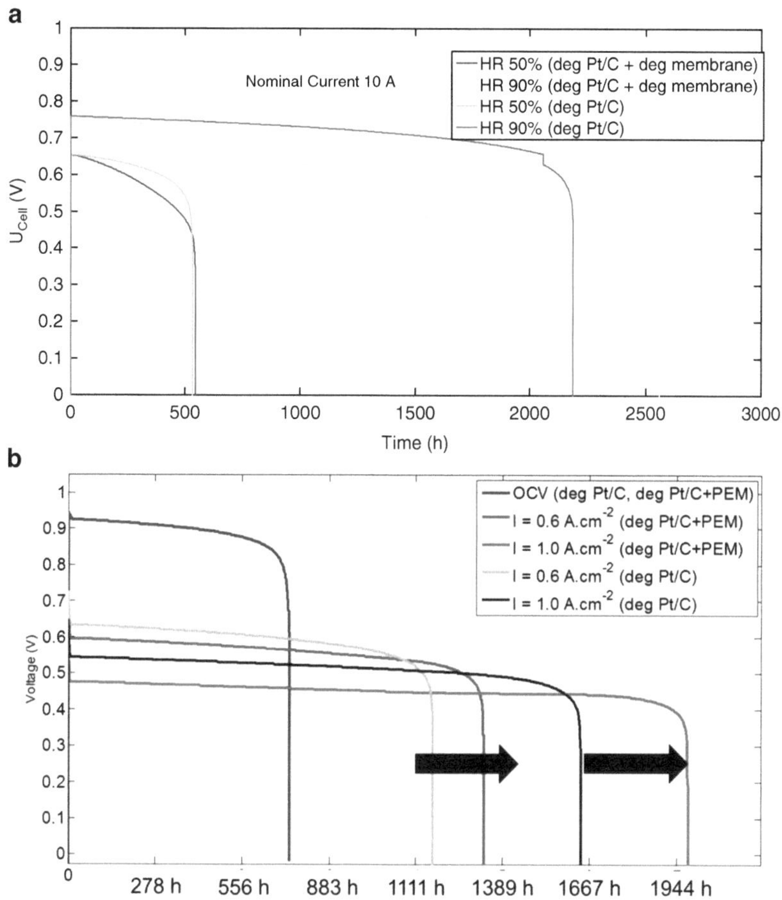

Fig. 8.24 Example of calculated competitive materials aging impact on the global cell performance decay for several operation condition. (**a**) constant current impact on the performance decay for two cathode relative humidities and three current values with the description of Pt/C degradation only vs. the description of both Pt/C and membrane degradation, (**b**) cycled vs. constant current impact on the PEM degradation (Source: [219])

atomistic MD) and account for the conformational flexibility of ionomer molecules appropriately. Only few teams worldwide have employed the CGMD simulations for characterising microstructure of the electrodes in view of effect of solvent, ionomer, and Pt particles. Moreover, the CGMD simulations were employed to develop a microstructure based modeling scheme of carbon corrosion in the PEMFC cathode and integrated in MEMEPhys (Fig. 8.25). This illustrates the capabilities of the approach to capture materials degradation features in three dimensions, something helpful for the analysis of experimental characterization. Again, this work could be also applied for the modeling and numerical simulation of DAFCs.

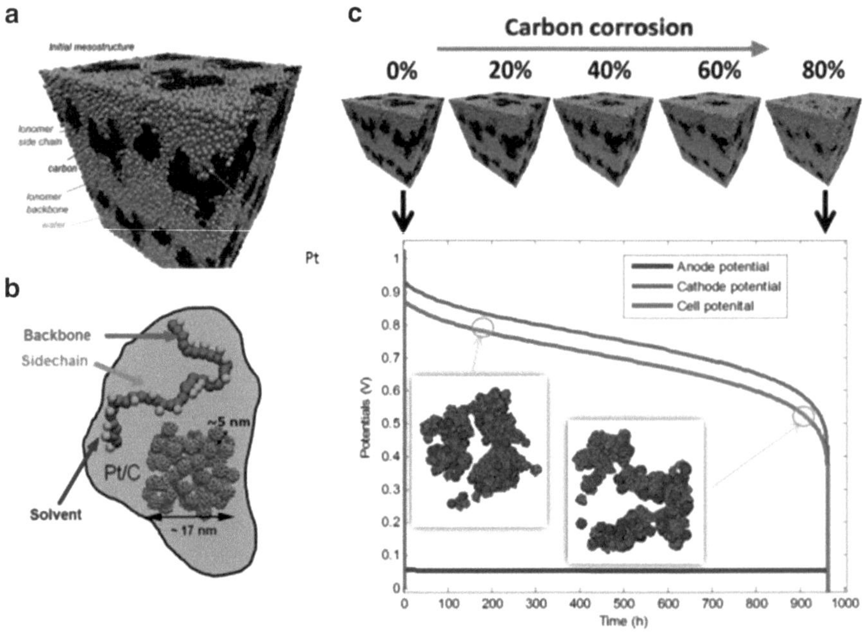

Fig. 8.25 (**a**) CGMD model of the CL; (**b**) coarse-grained constituents accounted in the simulations; (**c**) example of result from the combination of the CGMD database into a multiscale simulation package: multiscale simulation of cathode CL carbon corrosion. Reprinted from Ref. [209] with permission

Finally, since recently, I am proposing a new multiscale computational framework of electrochemical devices for energy storage and conversion [221]. This new model, called MS LIBER-T (*Multiscale Simulator of Lithium Ion Batteries and Electrochemical Reactor Technologies*), constitutes a breakthrough compared to the previously developed MEMEPhys simulation package penalized by its dependence on commercial software toolboxes and solvers such as Simulink. MS LIBER-T is coded on an independent C/Python language basis, highly flexible and portable (it can eventually be coupled to commercial software such as Matlab/Simulink), and supports direct multiparadigm calculations, for instance, simulations coupling on the fly the numerical resolution of continuum models (e.g. describing reactants transport in a bulk) with the numerical resolution of discrete models (e.g. Kinetic Monte Carlo codes resolving detailed electrocatalytic reactions) [222, 223].

8.4 Summary and Perspectives

DAFCs have attracted much attention due to their potential as a clean power source for many applications. For example, stationary systems providers have now demonstrated, all over the world, the ability of low temperature PEMFCs to answer their customers' needs as far as performance, efficiency, environmental goals are concerned. Cost and durability are now the main priorities towards the large scale commercialisation of these systems. Demonstrating a durability of several thousands of hours, up to 60,000 h for some applications, is thus now the prime requirement. The longer the durability, the higher the price customers will be ready to pay for their investment. Demonstrating such a long life-time in real operating conditions within an autonomous system is a real challenge. System failures often impair satisfactory demonstrations of the stack reliability. Statistical proof of a repeated success in achieving a long life time asks for a large number of units operating at customers facilities, which is very costly. Before launching such large scale demonstrations, necessary when approaching market maturity, fuel cell system providers need a reliable prediction of their products lifetime.

The main bottleneck now, if one wants to shorten the "time to market" for new PEMFCs, is that one needs an efficient method to take into account durability targets in all R&D actions on components and unit operating management strategies. A simple "try and error" method is manageable to get system durability from a few hundreds to a few thousands of hours. It is practically impossible when one wants to get from a few thousands of hours up to several tens of thousands of hours.

In this Chapter, I aimed to demonstrate that modeling at multiple scales (from atomistic level to the simulation of processes at the cell and stack level) combined with experiments can constitute a reliable method to predict the DAFC system performance and lifetime and to benchmark components and improve operating strategies with respect to a durability target. Predictive modeling is a requisite to establish this methodology for DAFC. Performance and durability of a DAFC is the result of a very complex set of interrelated events, with competitive effects but also synergies between performance degradation processes. By accelerating one phenomenon, one usually creates conditions that are no more representative of the subtle balance between reactions in the real operating conditions. One can thus either overestimate performance losses degradation rates (leading to developing very highly resistant membranes, support carbons or catalysts for conditions never encountered in a real system), or underestimate degradation because some negative feedback loops (cancellation or synergetic effects) are not taken into account. Only a physically based, multiscale and multitemporal model can provide the tool to combine all possible degradation phenomena and analyze their global impact on durability in a given set of operating conditions.

As an input for this model, one has to understand the fundamentals of performance losses and degradation. Specific experimental data is needed, to correlate degradation and deterioration phenomena to operating conditions for stationary applications, and identify the paths leading to failure phenomena. This experimental part is an effort to obtain data on degradation quantitatively and reproducibly, since the understanding of kinetics of various degradation processes are the key of final performance and lifetime prediction.

Ab initio-thermodynamics models have been largely developed to screen the activity, selectivity and stability of catalyst candidates and have allowed the

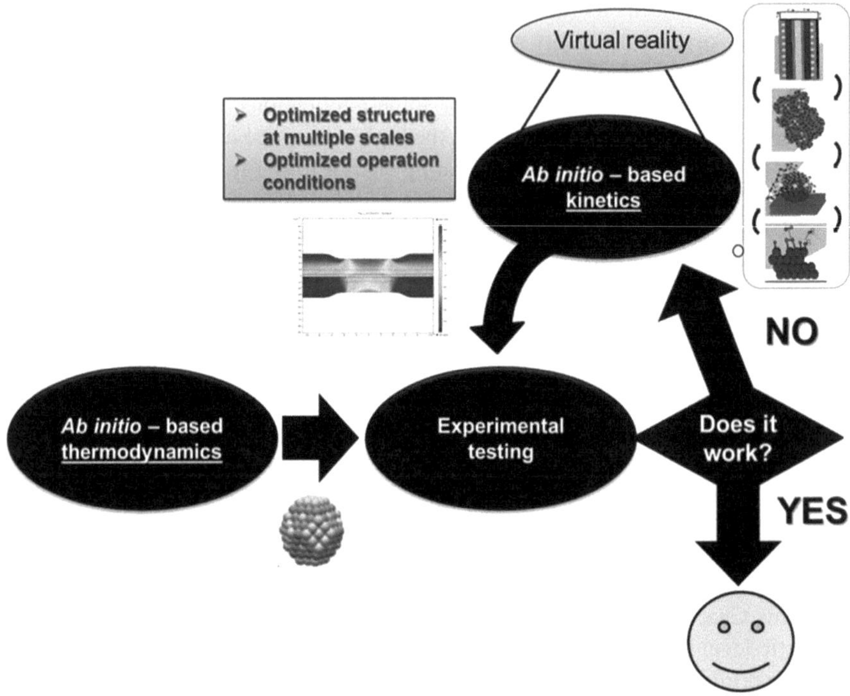

Fig. 8.26 Towards a physical modelling assisted materials development through ab initio based kinetics and the virtual reactor approach

selection of the most interesting ones. Those predicted catalysts can be then tested experimentally, and sometimes they really work and sometimes not. The catalysts that do not work in real conditions probably do not work because of the lack of consideration of electrochemical environment (electric field, solvent, etc.) in classical ab initio thermodynamics approach. Then complementary ab initio kinetic models can provide a "virtual simulator framework" to achieve the optimization of the operation conditions for which the predicted catalyst should work in the real environment (Fig. 8.26). Thus, more efforts should be developed within this sense, and even for other electrochemical devices for energy conversion and storage such as electrolyzers, batteries and supercapacitors.

The development of such multiscale modelling approach for DAFC must have several properties:

- Predictive capabilities of the relative contributions of the different scales and mechanisms into the macroscopic DAFC efficiency and durability;
- High flexibility towards its application to any type of chemical and structural properties of the used materials and components;
- Easily adaptable to any type of operation condition and system.

The modeling approach, consisting in building up a diagnostic and control-dedicated physical model with large prediction capabilities, enables:

- Reduction of the amount of experiments (and thus the cost) currently needed to build up classical empirical models with limited prediction capabilities;
- A better targeting of experimental characterizations in representative conditions of the end-user application;
- New operation strategies reducing the performance degradation and also strategies to improve the stability of the materials and components;
- The integration at the international level of modelling efforts usually developed separately. This will be done with the development of a modelling platform for more efficient communication and coordination for higher impact of the use of modelling on the DAFC optimization in engineering practice.

All this will contribute on placing the modeling community at the forefront of fuel cell technologies worldwide devoted to drive the substantial potential public benefits.

References

1. Grubb WT (1959) Batteries with solid ion exchange electrolyte I. Secondary cells employing metal electrodes. J Electrochem Soc 106:275–277
2. Kolde JA, Bahar B, Wilson MS, Zawodzinski TA, Gottesfeld S (1995) Proton conducting membrane fuel cells, PV95-23. Electrochemical Society, Pennington
3. Wilson MS, Gottesfeld S (1992) Thin-film catalyst layers for polymer electrolyte fuel cell electrodes. J Appl Electrochem 22:1–7
4. Perry ML, Fuller TF (2002) A historical perspective of fuel cell technology in the 20th century. J Electrochem Soc 149:S59–S67
5. Jollie D (2005) Fuel cell market survey: portable applications. www.fuelcelltoday.com, 6 Sept 2005
6. Mathias M, Gasteiger HA (2002) Electrochemical society proceedings, vol PV 2002-31. In: Murthy M, Fuller TF, Van Zee JW, Gottesfeld S (eds) Third international symposium on PEM fuel cells, Salt Lake City
7. Wang J, Wasmus S, Savinell RF (1995) Evaluation of ethanol, 1-propanol, and 2-propanol in a direct oxidation polymer-electrolyte fuel cell. J Electrochem Soc 142:4218–4224
8. An L, Zhao TS, Shen SY, Wu QX, Chen R (2010) Performance of a direct ethylene glycol fuel cell with an anion-exchange membrane. Int J Hydrogen Energy 35:4329–4335
9. Lamy C, Belgsir EM, Léger J-M (2001) Electrocatalytic oxidation of aliphatic alcohols: application to the direct alcohol fuel cell (DAFC). J Appl Electrochem 31:799–809
10. Léger J-M, Rousseau S, Coutanceau C, Hahn F, Lamy C (2005) How bimetallic electrocatalysts does work for reactions involved in fuel cells ?: example of ethanol oxidation and comparison to methanol. Electrochim Acta 50:5118–5125
11. Petry OA, Podlovchenko BI, Frumkin AN, Lal H (1965) The behavior of platinized-platinum and platinum-ruthenium electrodes in methanol solutions. J Electroanal Chem 10:253–269
12. Biegler T, Koch DA (1967) Adsorption and oxidation of methanol on a platinum electrode. J Electrochem Soc 114:904–909
13. Sundmacher K, Schultz T, Zhou S, Scott K, Ginkel M, Gilles ED (2001) Dynamics of the direct methanol fuel cell (DMFC): experiments and model-based analysis. Chem Eng Sci 56:333–341

14. Lu GQ, Wang CY (2004) Electrochemical and flow characterization of a direct methanol fuel cell. J Power Sources 134:33–40
15. Bewer T, Beckmann T, Dohle H, Mergel J, Stolten D (2004) Novel method for investigation of two-phase flow in liquid feed direct methanol fuel cells using an aqueous H_2O_2 solution. J Power Sources 125:1–9
16. Yang H, Zhao TS, Ye Q (2005) In situ visualization study of CO_2 gas bubble behavior in DMFC anode flow fields. J Power Sources 139:79–90
17. Aricò AS, Srinivasan S, Antonucci V (2001) DMFCs: from fundamental aspects to technology development. Fuel Cells 1:133–161
18. Hamnett A (1997) Mechanism and electrocatalysis in the direct methanol fuel cell. Catal Today 38:445
19. Hoster H, Iwasita T, Baumgärtner H, Vielstich W (2001) Pt–Ru model catalysts for anodic methanol oxidation: Influence of structure and composition on the reactivity. Phys Chem Chem Phys 3:337–346
20. Scott K, Taama WM, Argyropoulos P, Sundmacher K (1999) The impact of mass transport and methanol crossover on the direct methanol fuel cell. J Power Sources 83:204
21. Scott K, Taama WM, Cruickshank J (1997) Performance and modelling of a direct methanol solid polymer electrolyte fuel cell. J Power Sources 65:159–171
22. Tüber K, Pócza D, Hebling C (2003) Visualization of water buildup in the cathode of a transparent PEM fuel cell. J Power Sources 124:403–414
23. Zhao TS, Xu C, Chen R, Yang WW (2009) Mass transport phenomena in direct methanol fuel cells. Prog Energy Combus Sci 35:275–292
24. Vielstich W, Yokokawa H, Gasteiger HA (eds) Handbook of fuel cells: fundamentals technology and applications. Advances in electrocatalysis, materials, diagnostics and durability: part 1, volume 5.
25. Wang X, Kumar R, Myers DJ (2006) Effect of voltage on platinum dissolution: relevance to polymer electrolyte fuel cells. Electrochem Solid-State Lett 9:A225–A227
26. Swider-Lyons KE, Teliska ME, Baker WS, Bouwman PJ, Pietron JJ (2006) Leveraging metal-support interactions to improve the activity of PEMFC cathode catalysts. ECS Trans 1(6):97–105
27. Xu H, Kunz R, Fenton JM (2007) Investigation of platinum oxidation in PEM fuel cells at various relative humidities. Electrochem Solid-State Lett 10:B1–B5
28. Yazuda K, Taniguchi A, Akita T, Ioroi T, Siroma Z (2006) Platinum dissolution and deposition in the polymer electrolyte membrane of a PEM fuel cell as studied by potential cycling. Phys Chem Chem Phys 8:746–752
29. Ferreira PJ, Shao-Horn Y (2007) Formation mechanism of Pt single-crystal nanoparticles in proton exchangemembrane fuel cells. Electrochem Solid-State Lett 10:B60–B63
30. Franco AA, Coulon R, Ferreira de Morais R, Cheah S-K, Kachmar A, Gabriel MA (2009) Muli-scale modelin-based prediction of PEM fuel cells MEA durability under automotive operating conditions. ECS Trans 25:65–79
31. Ferreira PJ, la O' GJ, Shao-Horn Y, Morgan D, Makharia R, Kocha S, Gasteiger HA (2005) Instability of Pt/C electrocatalysts in proton exchange membrane fuel cells: a mechanistic investigation. J Electrochem Soc 152:A2256–A2271
32. Borup RL, Davey JR, Garzon FH, Wood DL, Inbody MA (2006) PEM fuel cell electrocatalyst durability measurements. J Power Sources 163:76–81
33. Bett JAS, Kinoshita K, Stonehart P (1976) Crystallite growth of platinum dispersed on graphitized carbon black II. Effect of liquid environment. J Catalysis 41:124–133
34. Park JY, Scibioh MA, Kim SK, Kim HJ, Oh IH, Lee TG, Ha HY (2009) Investigations of performance degradation and mitigation strategies in direct methanol fuel cells. Int J Hydrogen Energy 34:2043–2051
35. Franco AA (2007) Transient multi-scale modeling of aging mechanisms in a polymer electrolyte fuel cell: an irreversible thermodynamic approach. ECS Trans 6(10):1–23
36. Ball SC, Hudson S, Theobald B, Thompsett D (2006) Carbon supported Pt and PtCo alloys with improved corrosion resistance for PEMFC. 210th ECS meeting, abstract # 552

37. Ball SC, Hudson SL, Theobald BR, Thompsett D (2007) PtCo, a durable catalyst for automotive PEMFC? ECS Trans 11:1267–1278
38. Ball SC, Hudson SL, Hei Leung J, Russell AE, Thompsett D, Theobald BR (2007) Mechanisms of activity loss in PtCo alloy systems. ECS Trans 11:1247–1257
39. Stamenkovic VR, Mun BS, Arenz M, Mayrhofer KJJ, Lucas CA, Wang G, Ross PN, Markovic NM (2007) Trends in electrocatalysis on extended and nanoscale Pt-bimetallic alloy surfaces. Nat Mater 6:241–247
40. Watanabe M, Tsurumi K, Mizukami K, Nakamura T, Stoneharf P (1994) Activity and stability of ordered and disordered Co/Pt alloys for phosphoric acid fuel cells. J Electrochem Soc 141:2659–2668
41. Antolini E, Salgado JRC, Gonzalez ER (2006) The stability of Pt-M alloy catalysts and its effect on the activity in low temperature fuel cells. J Power Sources 160:957–968
42. Colon-Mercado HR, Popov BN (2006) Stability of platinum based alloy cathode catalysts in PEM fuel cells. J Power Sources 155:253–263
43. Piela P, Eickes C, Brosha E, Garzon F, Zelenay P (2004) Ruthenium crossover in direct metanol fuel cell with Pt-Ru black anode. J Electrochem Soc 151:A2053–A2059
44. Surampudi S, Narayanan SR, Vamos E, Frank H, Halpert G, LaConti A, Kosek J, Prakash GKS, Olah GA (1994) Advances in direct oxidation methanol fuel cells. J Power Sources 47:377–385
45. Brankovic SR, Wang JX, Adzic RR (2001) Pt submonolayers on Ru nanoparticles: a novel low Pt loading, high CO tolrerance fuel cell electrocatalyst. Electrochem Solid State Lett 4: A217–A220
46. Waszczuk P, Solla-Gullon J, Kim HS, Tong YY, Montiel V, Aldaz A, Wieckowski A (2001) Methanol electrooxidation on platinum/ruthenium nanoparticle catalysts. J Catal 203:1–6
47. Strasser P (2008) Combinatorial optimization of ternary Pt alloy catalysts for the electrooxidation of methanol. J Combin Chem 10:216–224
48. Markovic NM, Gasteiger HA, Ross PN, Jiang XD, Villegas I, Weaver MJ (1995) Electrooxidation mechanisms of methanol and formic acid on Pt-Ru alloy surfaces. Electrochim Acta 40:91–98
49. Gancs L, Hakim N, Hult BN, Mukerjee S (2006) Dissolution of Ru from PtRu electrocatalysts and its consequences in DMFCs. ECS Trans 3:607–618
50. Ball SC, Hudson SL, Thompsett D, Theobald B (2007) An investigation into factors affecting the stability of carbons and carbon supported platinum and platinum/cobal alloy catalysts during 1.2 V potentiostatic hold. J Power Sources 171:18–25
51. Teranishi K, Kawata K, Tsushima S, Hirai S (2006) Degradation mechanism of PEMFC under open circuit operation. Electrochem Solid-State Lett 9:A475–A477
52. Perry ML, Patterson TW, Reiser C (2006) Systems strategies to mitigate carbon corrosion in fuel cells durability—fuel starvation and start/stop degradation. ECS Trans 3:783–795
53. Li J, He P, Wang K, Davis M, Ye S (2006) Characterization of catalyst layer structural changes in PEMFC as a function of durability testing. ECS Trans 3:743–751
54. Fujii Y, Tsushima S, Teranishi K, Kawata K, Nanjo T, Hirai S (2006) Degradation investigation of PEMFC by scanning electron microscopy and direct gas mass spectroscopy. ECS Trans 3:735–741
55. Baumgartner WR, Wallnöfer E, Schaffer T, Besenhard JO, Hacker V, Peinecke V, Prenninger P (2006) Electrocatalytic corrosion of carbon support in PEMFC at fuel starvation. ECS Trans 3:811–825
56. Tang H, Qi Z, Ramani M, Elter JF (2006) PEM fuel cell cathode carbón corrosión due to the formation of air/fuel boundary at the anode. J Power Sources 158:1306–1312
57. Fuller TF, Gray G (2006) Carbon corrosion induced by partial hydrogen coverage PEMFC stack and system membrane degradation/reliability. ECS Trans 1:345–353
58. Darling RM, Jayne D (2007) Corrosion of polymer-electrolyte fuel cells caused by water at the fuel inlet. ECS Trans 11:975–980
59. Franco AA, Gerard M (2008) Multiscale model of carbon corrosion in a PEFC: coupling with electrocatalysis and impact on performance degradation. J Electrochem Soc 155:B367–B384

60. Shao YY, Yin GP, Gao YZ, Shi PF (2006) Durability study of Pt/C and Pt/CNTs catalysts under simulated PEM fuel cell conditions. J Electrochem Soc 153:A1093–A1097
61. Coloma F, Sepulveda Escribano A, Rodriguez Reinoso F (1995) Heat-treated carbon-blacks as supports for platinum catalysts. J Catal 154:299–305
62. Schulze M, Christenn C (2005) XPS investigation of the PTFE induced hydrophobic properties of electrodes for low temperature fuel cells. Appl Surf Sci 252:148–153
63. Liu W, Zuckerbrod D (2005) In situ detection of hydrogen peroxide in PEM fuel cells. J Electrochem Soc 152:A1165–A1170
64. Merlo L, Ghielmi A, Cirillo L, Gebert M, Arcella V (2007) Resistance to peroxide degradation of Hyflon® ion membranes. J Power Sources 171:140–147
65. Pozio A, Silva RF, De Francesco M, Giorgi L (2003) Nafion degradation in PEFCs from end plate iron contamination. Electrochim Acta 48:1543–1549
66. Inaba M, Kinumoto T, Kiriake M, Umebayashi R, Tasaka A, Ogumi Z (2006) Gas crossover and membrane degradation in polymer electrolyte fuel cells. Electrochim Acta 51:5746–5753
67. Kadirov MK, Bosnjakovic A, Schlick S (2005) Membrane derived fluorinated radicals detected by electron spin resonance in UV-irradiated nafion and dow ionomers. Effect of counterions and H_2O_2. J Phys Chem B 109:7664–7670
68. Bosnjakovic A, Schlick S (2004) Nafion perfluorinated membranes treated in Fenton media: radical species detected by ESR spectroscopy. J Phys Chem B 108:4332–4337
69. Aoki M, Uchida H, Watanabe M (2005) Novel evaluation method for degradation rate of polymer electrolytes in fuel cells. Electrochem Commun 7:1434–1438
70. Mittal VO, Kunz HR, Fenton JM (2007) Membrane degradation mechanisms in PEMFCs. J Electrochem Soc 154:B652–B656
71. Madden TH, Atrazhev VV, Sultanov VI, Timokhina EN, Burlatsky SF, Gummalla M (2007) Direct mechanism of OH radicals formation in PEM fuel cells. 211th ECS meeting, Abstract 107
72. Mittal VO, Kunz HR, Fenton JM (2006) Is H_2O_2 involved in the membrane degradation mechanism in PEMFC? Electrochem Solid-State Lett 9:A299–A302
73. Chen C, Fuller TF (2007) H_2O_2 formation under fuel cell conditions. ECS Trans 11:1127–1137
74. Shim JY, Tsushima S, Hirai S (2008) Characterization and modeling of gas crossover and its impact on the membrane degradation of PEMFCs. ECS Trans 16:1705–1712
75. Hatanaka T, Takeshita T, Murata H, Hasegawa N, Asano T, Kawasumi M, Morimoto Y (2008) Electrode and membrane durability issues of PEFCs. ECS Trans 16:1961–1965
76. Chung Y, Pak C, Park GS, Jeon WS, Kim JR, Lee Y, Chang H, Seung D (2008) Understanding a degradation mechanism of direct methanol fuel cell using TOF-SIMS and XPS. J Phys Chem C 112:313–318
77. Fang B, Luo J, Njoki PN, Loukrakpam R, Wanjala B, Hong J, Yin J, Hu X, Last J, Zhong CJ (2010) Nano-engineered PtVFe catalysts in proton exchange membrane fuel cells: electrocatalytic performance. Electrochim Acta 55:8230–8236
78. Norskov JK, Rossmeisl J, Logadottir A, Lindqvist L, Kitchin JR, Bligaard T, Jonsson HJ (2004) Origin of the overpotential for oxygen reduction at a fuel cell cathode. J Phys Chem B 108:17886–17892
79. Greeley J, Rossmeisl J, Hellmann A, Norskov JK (2007) Theoretical trends in particle size effects for the oxygen reduction reaction. Z Phys Chem 221:1209–1220
80. Jacob T, Goddard WA (2006) Water formation on Pt and Pt-based alloys: a theoretical description of a catalytic reaction. Chem Phys Chem 7:992–1005
81. Jacob T (2006) The mechanism of forming H_2O from H_2 and O_2 over a Pt catalyst via direct oxygen reduction. Fuel Cells 6:159–181
82. Jacob T, Goddard WA III (2004) Agostic interactions and dissociation in the first layer of water on Pt(111). J Am Chem Soc 126:9360–9368
83. Eichler A, Mittendorfer F, Hafner J (2000) Precursor-mediated adsorption of oxygen on the (111) surfaces of platinum-group metals. Phys Rev B 62:4744–4755

84. Eichler A, Hafner J (1997) Molecular precursors in the dissociative adsorption of O_2 on Pt (111). Phys Rev Lett 79:4481–4484
85. Kandoi S, Gokhale AA, Grabow LC, Dumesic JA, Mavrikakis M (2004) Why Au and Cu are more selective than Pt for preferential oxidation of Co at low temperature? Catal Lett 93:93–100
86. Jacob T, Goddard WA III (2004) Adsorption of atomic H and O on the (111) surface of Pt_3Ni alloys. J Phys Chem B 108:8311–8323
87. Xu Y, Ruban AV, Mavrikakis M (2004) Adsorption and dissociation of O_2 on Pt-Co and Pt-Fe alloys. J Am Chem Soc 126:4717–4725
88. Ma Y, Balbuena PB (2007) OOH dissociation on Pt clusters. Chem Phys Lett 447:289–294
89. Wang Y, Balbuena PB (2005) Potential energy surface profile of the oxygen reduction reaction on a Pt cluster: adsorption and decomposition of OOH and H_2O_2. J Chem Theory Comput 1:935–943
90. Wang Y, Balbuena PB (2005) Design of oxygen reduction bimetallic catalysts: ab-initio-derived thermodynamic guidelines. J Phys Chem B 109:18902–18906
91. Spendelow JS, Babu PK, Wieckowski A (2005) Electrocatalytic oxidation of carbon monoxide and methanol on platinum surfaces decorated with ruthenium. Curr Opin Solid State Mat Sci 9:37–48
92. Spendelow JS, Wieckowski A (2007) Electrocatalysis of oxygen reduction and small alcohol oxidation in alkaline media. Phys Chem Chem Phys 9:2654–2675
93. Borkowska Z, Tymosiak-Zielinska A, Shul G (2004) Electrooxidation of methanol on polycrystalline and single crystal gold electrodes. Electrochim Acta 49:1209–1220
94. Parpot P, Bettencourt AP, Carvalho AM, Belgsir EM (2000) Biomass conversion: attempted electrooxidation of lignin for vanillin production. J Appl Electrochem 30:727–731
95. Parpot P, Bettencourt AP, Chamoulaud G, Kokoh KB, Beigsir EM (2004) Electrochemical investigations of the oxidation–reduction of furfural in aqueous medium: application to electrosynthesis. Electrochim Acta 49:397–403
96. Parpot P, Nunes N, Bettencourt AP (2006) Electrocatalytic oxidation of monosaccharides on gold electrode in alkaline medium. Structure-reactivity relationship. J Electroanal Chem 596:65–73
97. Parpot P, Pires SG, Bettencourt AP (2004) Electrocatalytic oxidation of D-galactose in alkaline medium. J Electroanal Chem 566:401–408
98. Parpot P, Santos PRB, Bettencourt AP (2007) Electro-oxidation of d-mannose on platinum, gold and nickel electrodes in aqueous medium. J Electroanal Chem 610:154–162
99. Yan SH, Zhang SC, Lin Y, Liu GR (2011) Electrocatalytic performance of gold nanoparticles supported on activated carbon for methanol oxidation in alkaline solution. J Phys Chem C 115:6986–6993
100. Markovic NM, Ross PN (2002) Surface science studies of model fuel cell electrocatalysts. Surf Sci Rep 45:121–229
101. Yu E, Krewer U, Scott K (2010) Principles and materials aspects of direct alkaline fuel cells. Energies 3:1499–1528
102. Tripkovic AV, Popovic KD, Grgur BN, Blizanac B, Ross PN, Markovic NM (2002) Methanol electrooxidation on supported Pt and PtRu catalysts in acid and alkaline solutions. Electrochim Acta 47:3707–3714
103. Colmati F, Tremiliosi G, Gonzalez ER, Berna A, Herrero E, Feliu JM (2009) The role of the steps in the cleavage of the C-C bond during ethanol oxidation on platinum electrodes. Phys Chem Chem Phys 11:9114–9123
104. Lu GQ, Wieckowski A, Wasileski S, Neurock M (2005) Mechanisms of methanol decomposition on platinum: a combined experimental and ab initio approach. J Phys Chem B 109:11622–11633
105. Hartnig C, Spohr E (2005) The role of water in the initial steps of methanol oxidation on Pt (111). Chem Phys 319:185–191

106. Brankovic SR, Wang JX, Zhu Y, Sabatini R, McBreen J, Adzic RR (2002) Electrosorption and catalytic properties of bare and Pt modified single crystal and nanostructured Ru surfaces. J Electroanal Chem 524:231–241
107. Lai SCS, Kleyn SEF, Rosca V, Koper MTM (2008) Mechanism of the dissociation and electrooxidation of ethanol and acetaldehyde on platinum as studied by SERS. J Phys Chem C 112:19080–19087
108. Lai SCS, Koper MTM (2008) Electro-oxidation of ethanol and acetaldehyde on platinum single-crystal electrodes. Faraday Discuss 140:399–416
109. Lai SCS, Kleijn SEF, Ozturk FTZ, Vellinga VCV, Koning J, Rodriguez P, Koper MTM (2010) Catalysis Today 154:92–104
110. Neurock M, Vielstich HA, Yokokawa GH (2009) First principles modeling for the electrooxidation of small molecules. In: Vielstich W, Gasteiger HA, Yokokawa H (eds) Handbook of fuel cells. Wiley, Hoboken
111. Cui G, Song S, Kang Shen P, Kowal A, Bianchini C (2009) First-principles considerations on catalytic activity of Pd toward ethanol oxidation. J Phys Chem C 113:15639–15645
112. Zhang L, Zhang J, Wilkinson DP, Wang H (2006) Progress in preparation of non-noble electrocatalysts for PEM fuel cell reactions. J Power Sources 156:171–182
113. Cao D, Wieckowski A, Inukai J, Alonso-Vante N (2006) Oxygen reduction reaction on ruthenium and rhodium nanoparticles modified with selenium and sulfur. J Electrochem Soc 153:A869–A874
114. Trapp V, Christensen P, Hamnett A (1996) New catalysts for oxygen reduction based on transition-metal sulfides. Faraday Trans 92:4311–4319
115. Vayner E, Sidik RA, Anderson AB, Popov BN (2007) Experimental and theoretical study of cobalt selenide as a catalyst for O_2 electroreduction. J Phys Chem C 111:10508–10513
116. Alonso-Vante N, Bogdanoff P, Tributsch H (2000) On the origin of the selectivity of oxygen reduction of ruthenium-containing electrocatalysts in methanol-containing electrolyte. J Catal 190:240–246
117. Racz A, Bele P, Cremers C, Stimming U (2007) Ruthenium selenide catalysts for cathodic oxygen reduction in direct methanol fuel cells. J Appl Electrochem 37:1455–1462
118. Le Rhun V, Garnier E, Pronier S, Alonso-Vante N (2000) Electrocatalysis on nanoscale ruthenium-based material manufactured by carbonyl decomposition. Electrochem Commun 2:475–479
119. Lewera A, Inukai J, Zhou WP, Cao D, Duong HT, Alonso-Vante N et al (2007) Chalcogenide oxygen reduction reaction catalysis: X-ray photoelectron spectroscopy with Ru, Ru/Se and Ru/S samples emersed from aqueous media. Electrochim Acta 52:5759–5765
120. Bron M, Bogdanoff P, Fiechter S, Dorbandt I, Hilgendorff M, Schulenburg H et al (2001) Carbon supported catalysts for oxygen reduction in acidic media prepared by thermolysis of Ru_3C_{12}. J Electroanal Chem 500:510–517
121. Dassenoy F, Vogel W, Alonso-Vante N (2002) Structural studies and stability of cluster-like Ru_xSe_y electrocatalysts. J Phys Chem B 106:12152–12157
122. Zaikovskii VI, Nagabhushana KS, Kriventsov VV, Loponov KN, Cherepanova SV, Kvon RI et al (2006) Synthesis and structural characterization of Se-modified carbon-supported Ru nanoparticles for the oxygen reduction reaction. J Phys Chem B 110:6881–6890
123. Skúlason E, Karlberg GS, Rossmeisl J, Bligaard T, Greeley J, Jónsson H, Nørskov JK (2007) Density functional theory calculations for the hydrogen evolution reaction in an electrochemical double layer on the Pt(111) electrode. Phys Chem Chem Phys 9:3241–3250
124. Nørskov JK, Rossmeisl J, Logadottir A, Lindqvist L, Kitchin JR, Bligaard T, Jónsson H (2004) Origin of the overpotential for oxygen reduction at a fuel-cell cathode. J Phys Chem B 108:17886–17892
125. Ferrin P, Nilekar AU, Greeley J, Mavrikakis M, Rossmeisl J (2008) Reactivity descriptors for direct methanol fuel cell anode catalysts. Surf Sci 602:3424–3431
126. Tritsaris GA, Greeley J, Rossmeisl J, Norskov JK (2011) Trends in oxygen reduction and methanol activation on transition metal chalcogenides. Electrochim Acta 56:9783–9788

127. Carr R, Parrinello M (1985) Unified approach for molecular dynamics and density functional theory. Phys Rev Lett 55:2471–2474
128. Grob A (2003) Theoretical surface science: a microscopic perspective. Springer, New York
129. Goddard W III, Merinov B, Van Duin A, Jacob T, Blanco M, Molinero V, Jang SS, Jang YH (2006) Multiparadigm multiscale simulations for fuel cell catalysts and membranes. Mol Simulat 32:251–268
130. Filhol JS, Neurock M (2006) Elucidation of the electrochemical activation of water over Pd by first principles. Angew Chem Int Ed 45:402–406
131. Janik MJ, Neurock M (2007) A first principles analysis of the electro-oxidation of CO over Pt (111). Electrochim Acta 52:5517–5528
132. Janik MJ, Taylor CD, Neurock M (2009) First-principles analysis of the initial electroreduction steps of oxygen over Pt(111). J Electrochem Soc 156:B126–B135
133. Kawakami T, Shigemoto I (2005) Molecular-dynamics studies on the structures of polymer electrolyte membranes and the diffusion mechanism of protons and small molecules. In: Nakahara M, Matubayasi N, Ueno M, Yasuoka K, Watanabe K (eds) Water, steam, and aqueous solutions for electric power. Proceeding of the 14th international conferance properties of water and steam. Maruzen Co., Ltd., Japan, pp 415–420
134. Ureta-Zañartul MS, Berríos C, González T, Fernández F, Báez D, Salazar R, Gutiérrez C (2012) Electrocatalytic oxidation of alcohols at gold electrodes in alkaline media. Int J Electrochem Sci 7:8905–8928
135. Anderson AB, Albu TV (2000) Catalytic effect of platinum on oxygen reduction: an ab initio model including electrode potential dependence. J Electrochem Soc 147:4229–4238
136. Rai V, Aryanpour M, Pitsch H (2008) First principles analysis of oxygen-containing adsorbates formed from the electrochemical discharge of water on Pt(111). J Phys Chem C 112:9760–9768
137. Panchenko A, Koper MTM, Shubina TE, Mitchell SJ, Roduner E (2004) Ab initio calculations of intermediates of oxygen reduction on low-index platinum surfaces. J Electrochem Soc 151:A2016–A2027
138. Vassilev P, Koper MTM (2007) Electrochemical reduction of oxygen on gold surfaces. A density functional theory study of intermediations and reaction paths. J Phys Chem C 111:2607–2613
139. Subbaraman R, Strumcnik D, Stamenkovic V, Markovic NM (2010) Three phase interfaces at electrified metal-solid electrolyte systems 1. Study of the Pt(hkl)-Nafion interface. J Phys Chem C 114:8414–8422
140. Ohma A, Fushinobu K, Okazaki K (2010) Influence of Nafion® film on oxygen reduction reaction and hydrogen peroxide formation on Pt electrode for proton exchange membrane fuel cell. Electrochim Acta 55:8829–8838
141. Sousa R, Gonzalez ER (2005) Mathematical modeling of polymer electrolyte fuel cells. J Power Sources 147:32–45
142. Hsu NY, Yen SC, Jeng KT, Chien CC (2006) Impedance studies and modeling of direct methanol fuel cell anode with interface and porous structure perspectives. J Power Sources 161:232–239
143. Du CY, Zhao TS, Xu C (2007) Simultaneous oxygen reduction and methanol oxidation reactions at the cathode of a DMFC. A model-based electrochemical impedance spectroscopy study. J Power Sources 167:265–271
144. Macdonald DD (2006) Reflections on the history of electrochemical impedance spectroscopy. Electrochim Acta 51:1376–1388
145. Oliveira VB, Falcão DS, Rangel CM, Pinto AMFR (2007) A comparative study of approaches for direct methanol fuel cells modeling. Int J Hydrogen Energy 32:415–424
146. Scott K, Taama W, Cruickshank J (1998) Performance a direct methanol fuel cell. J Appl Electrochem 28:289–297
147. Kulikovsky AA, Divisek J, Kornyshev AA (1999) Modeling the cathode compartment of polymer electrolyte fuel cells: dead and active reaction zones. J Electrochem Soc 146:3981–3991

148. Kauranen PS, Skou E (1996) Mixed methanol oxidation/oxygen reduction currents on a carbon supported Pt catalyst. J Electroanal Chem 408:189–198
149. Simoglou A, Argyropoulos P, Martin EB, Scott K, Morris AJ, Taama WM (2001) Dynamic modeling of the voltage response of direct methanol fuel cells and stacks. Part II: feasibility study of model-based scale-up and scale down. Chem Eng Sci 56:6761–6772
150. Simoglou A, Argyropoulos P, Martin EB, Scott K, Morris AJ, Taama WM (2001) Dynamic modeling of the voltage response of direct methanol fuel cells and stacks. Part I: model development and validation. Chem Eng Sci 56:6773–6779
151. Argyropoulos P, Scott K, Shukla AK, Jackson C (2003) A semi-empirical model of the direct methanol fuel cell performance. Part I. Model development and verification. J Power Sources 123:190–199
152. Dohle H, Wippermann K (2004) Experimental evaluation and semi-empirical modeling of i-V characteristics and methanol permeation of a direct methanol fuel cell. J Power Sources 135:152–164
153. Baxter SF, Battaglia VS, White RE (1999) Methanol fuel cell model: anode. J Electrochem Soc 146:437–447
154. Wang J-T, Savinell RF (1994) In: Srinivasan S, Macdonald DD, Khandkar AC (eds) Electrode materials and processes for energy conversion and storage, PV 94-23. The Electrochemical Society Proceedings Series, Pennington, p 326
155. Kulikovsky A (2003) Analytical model of the anode side of DMFC: the effect of non-Tafel kinetics on cell performance. Electrochem Commun 5:530–538
156. Kulikovsky A, Divisek J, Kornyshev AA (2000) Two-dimensional simulation of a direct methanol fuel cell: a new (embedded) type of current collector. J Electrochem Soc 147:953–959
157. Kulikovsky AA (2000) Two-dimensional numerical modeling of a direct methanol fuel cell. J Appl Electrochem 30:1005–1014
158. Dohle H, Divisek J, Jung R (2000) Process engineering of the direct methanol fuel cell. J Power Sources 86:469–477
159. Scott K, Argyropoulos P, Sundmacher K (1999) A model for the liquid feed direct methanol fuel cell. J Electroanal Chem 477:97–110
160. Sundmacher K, Scott K (1999) Direct methanol polymer electrolyte fuel cell: analysis of charge and mass transfer in the vapour-liquid–solid system. Chem Eng Sci 54:2927–2936
161. Argyropoulos P, Scott K, Taama WM (2000) Hydrodynamic modeling of direct methanol liquid feed fuel cell stacks. J Appl Electrochem 30:899–913
162. Siebke A, Meier F, Eigenberger G, Fischer M (2001) Modeling of liquid direct methanol fuel cells. 3rd European Congress of Chemical Engineering (ECCE) Nuremberg, Germany
163. Xu C, Follmann PM, Biegler LT, Jhon MS (2005) Numerical simulation and optimization of a direct methanol fuel cell. Comp Chem Eng 29:1849–1860
164. Bade Shrestha SO, Mohan S (2011) Performance and modeling of a direct methanol fuel cell. In: Proceedings of the World Congress on Engineering vol III London
165. Nordlund J, Lindbergh G (2002) A model for the porous direct methanol fuel cells anode. J Electrochem Soc 149:A1107–A1113
166. Murgia G, Pisani L, Shukla AK, Scott K (2003) A numerical model of a liquid-feed solid polymer electrolyte DMFC and its experimental validation. J Electrochem Soc 150: A1231–A1245
167. Wang ZH, Wang CY (2003) Mathematical modeling of liquid-feed direct methanol fuel cells. J Electrochem Soc 150:A508–A519
168. García BL, Sethuraman VA, Weidner JW, White RE (2004) Mathematical model of a direct methanol fuel cell. J Fuel Cell Sci Technol 1:43–48
169. Guo H, Ma C (2004) 2D analytical model of a direct methanol fuel cell. Electrochem Commun 6:306–312
170. Scharfer P, Schabel W, Kind M (2008) Modeling of alcohol and water diffusion in fuel cell membranes—experimental validation by means of in situ Raman spectroscopy. Chem Eng Sci 63:4676–4684

171. Kulikovsky AA (2005) Model of the flow with bubbles in the anode channel and performance of a direct methanol fuel cell. Electrochem Commun 7:237
172. Kulikovsky AA (2006) Bubbles in the anode channel and performance of a DMFC: asymptotic solutions. Electrochim Acta 51:2003–2011
173. Argyropoulos P, Scott K, Taama WM (1999) One-dimensional thermal model for direct methanol fuel cell stacks. Part I. model development. J Power Sources 79:169–183
174. Argyropoulos P, Scott K, Taama WM (1999) One-dimensional thermal model for direct methanol fuel cell stacks. Part II. Model based parametric analysis and predicted temperature profiles. J Power Sources 79:184–198
175. Schultz T, Sundmacher K (2005) Rigorous dynamic model of a direct methanol fuel cell based on Maxwell-Stefan mass transport equations and a Flory-Huggins activity model: formulation and experimental validation. J Power Sources 145:435–462
176. Meyers JP, Newman J (2002) Simulation of the direct methanol fuel cell I. Thermodynamic framework for a multicomponent membrane. J Electrochem Soc 149:A710–A717
177. Meyers JP, Newman J (2002) Simulation of the direct methanol fuel cell II. Modeling and data analysis of transport and kinetic phenomena. J Electrochem Soc 149:A718–A728
178. Xing L, Scott K, Sun Y-P (2011) Transient response and steady-state analysis of the anode of direct methanol fuel cells based on dual-site kinetics. Int J Electrochem Article 853261
179. Krewer U, Song Y, Sundmacher K, John V, Lübke R, Matthies G, Tobiska L (2004) Chem Eng Sci 59:119–130
180. Yang WW, Zhao TS, Xu C (2007) Three-dimensional two-phase mass transport model for direct methanol fuel cells. Electrochim Acta 53:853–862
181. Omran MP, Farhadi M, Sedighi K (2011) The effect of cell temperature and channel geometry on the performance of a direct methanol fuel cell. J Fuel Cell Sci Technol 8:061004
182. Lam A, Wetton B, Wilkinson DP (2011) One-dimensional model for a direct methanol fuel cell with a 3D anode structure. J Electrochem Soc 158:B29–B35
183. Basri S, Kamarudin SK, Daud WRW, Yaakub Z, Ahmad MM, Hashim N, Hasran UA (2009) Unsteady-state modeling for a passive liquid-feed DMFC. Int J Hydrogen Energy 34:5759–5769
184. Chen R, Zhao TS (2005) Mathematical modeling of a passive-feed DMFC with heat transfer effect. J Power Sources 152:122–130
185. Cai W, Li S, Feng L, Zhang J, Song D, Xing W, Liu C (2011) Transient behavior analysis of a new designed passive direct methanol fuel cell fed with highly concentrated methanol. J Power Sources 196:3781–3789
186. Xu C, Faghri A (2010) Water transport characteristics in a passive liquid-feed DMFC. Int J Heat Mass Transfer 53:1951–1966
187. Andreadis G, Song S, Tsiakaras P (2006) Direct ethanol fuel cell anode simulation model. J Power Sources 157:657–665
188. Sarris I, Tsiakaras P, Song S, Vlachos S (2006) A three-dimensional CFD model of direct ethanol fuel cells: anode flow bed analysis. Solid State Ion 177:2133–2138
189. Suresh NS, Jayanti S (2009) Modelling of cross-over effects in direct ethanol fuel cells (DEFCs), AIChE proceedings, paper:161991
190. Andreadis G, Tsiakaras P (2006) Ethanol crossover and direct ethanol PEM fuel cell performance modeling and experimental validation. Chem Eng Sci 61:7497–7508
191. Pramanik H, Basu S (2010) Modeling and experimental validation of overpotentials of a direct ethanol fuel cell. Chem Eng Process 49:635–642
192. Meyer M, Melke J, Gerteisen D (2011) Modeling and simulation of a direct ethanol fuel cell considering multistep electrochemical reactions, transport processes and mixed potentials. Electrochim Acta 56:4299–4307
193. Tang D, Chen H, Hou QH, Lv HM, Ni HJ (2011) Two-dimensional numerical simulations of tubular cathode in a direct ethanol fuel cell. Adv Mat Res 311–313:2362–2366
194. Antolini E, Gonzalez ER (2010) The electro-oxidation of carbon monoxide, hydrogen/carbon monoxide and methanol in acid medium on Pt-Sn catalysts for low-temperature fuel cells: a

comparative review of the effect of Pt-Sn structural characteristics. Electrochim Acta 55:6485–6490
195. Kulikovsky AA (2011) A model for carbon and Ru corrosion due to methanol depletion in DMFC. Electrochim Acta 56:9846–9850
196. Franco AA (2012) PEMFC degradation modeling and analysis. In: Hartnig C, Roth C (eds) Polymer electrolyte membrane and direct methanol fuel cell technology (PEMFCs and DMFCs). Volume 1: Fundamentals and performance. Woodhead, Cambridge, UK
197. Franco AA (2013) Toward a bottom-up multiscale modeling framework for the transient analysis of PEM fuel cells operation. In: Franco AA (ed) Polymer electrolyte fuel cells: science, applications and challenges. CRC Press/Taylor & Francis Group, Boca Raton
198. Moçotéguy P, Druart F, Bultel Y, Besse S, Rakotondrainibe A (2007) Monodimensional modeling and experimental study of the dynamic behavior of proton exchange membrane fuel cell stack operating in dead-end mode. J Power Sources 167:349–357
199. Jang SS, Merinov BV, Jacob T, Goddard III WA (2005) In: International conference on solid state ionics-15, Abstract P498, International Solid State Ionics
200. Wang JX, Springer TE, Adzic RR (2006) Dual –pathway kinetic equation for the hydrogen oxidation reaction on Pt electrodes. J Electrochem Soc 153:A1732–A1740
201. Krapf D, Quinn BM, Wu M-Y, Zandbergen HW, Dekker C, Lemay SG (2006) Experimental observation of nonlinear ionic transport at the nanometer scale. Nanoletters 6:2531–2535
202. Greeley J, Norskov JK (2007) Electrochemical dissolution of surface alloys in acids: thermodynamic trends from first-principles calculations. Electrochim Acta 52:5829–5836
203. Choukroun R, de Caro D, Chaudret B, Lecante P, Snoeck E (2001) H_2-induced structural evolution in non-crystalline rhodium nanoparticles. New J Chem 25:525–527
204. Zhu L, Wang R, King TS, DePristo AE (1997) Effects of chemisorption on the surface. Composition of bimetallic catalysts. J Catal 167:408–411
205. Franco AA (2010) A multiscale modeling framework for the transient analysis of electrochemical power generators—From theory to the engineering practice, Habilitation Manuscript (H.D.R.), Université Claude Bernard Lyon 1.
206. Franco AA (2005) A physical multi-scale model of the electrochemical dynamics in a polymer electrolyte fuel cell—An infinite dimensional bond graph approach, Ph.D. thesis, Université Claude Bernard Lyon 1
207. Lopes Oliveira LF, Laref S, Mayousse E, Jallut C, Franco AA (2012) A multiscale physical model for the transient analysis of PEM Water Electrolyzer Anodes. Phys Chem Chem Phys 14:10215–10224
208. Cheah SK, Sycardi O, Guetaz L, Lemaire O, Gelin P, Franco AA (2011) CO impact on the stability properties of PtxCoy nanoparticles in PEMFC anodes: mechanistic insights from a combined experimental and modeling approach. J Electrochem Soc 158:B1358–B1367
209. Malek K, Franco AA (2011) Microstructural resolved modeling of aging mechanisms in PEMFC. J Phys Chem B 115:8088–8101
210. Ferreira de Morais R, Loffreda D, Sautet P, Franco AA (2011) Multi-scale Modeling Methodology to predict electrochemical observables from ab initio data: application to the ORR in a Pt(111)-based PEMFC. Electrochim Acta 56:10842–10856
211. Coulon R, Bessler W, Franco AA (2010) Modeling chemical degradation of a polymer electrolyte membrane and its impact on fuel cell performance. ECS Trans 25:259–273
212. Franco AA, Passot S, Fugier P, Billy E, Guillet N, Guetaz L, De Vito E, Mailley S (2009) Pt_xCo_y catalysts degradation in PEFC environments: mechanistic insights—part I: multiscale modeling. J Electrochem Soc 156:B410–B424
213. Franco AA, Guinard M, Barthe B, Lemaire O (2009) Impact of carbon monoxide on PEFC catalyst carbon support degradation under current-cycled operating conditions. Electrochim Acta 54:5267–5279
214. Franco AA, Tembely M (2007) Transient multi-scale model of aging mechanisms in a PEFC cathode. J Electrochem Soc 154:B712–B723

215. Franco AA, Schott P, Jallut C, Maschke B (2006) A dynamic mechanistic model of an electrochemical interface. J Electrochem Soc 153:A1053–A1061
216. Maschke B, van der Schaft AJ (2001) Canonical interdomain coupling in distributed parameter systems: an extension of the symplectic gyrator. In: Proceedings of the international mechanical engineering congress and exposition. ASME, New York
217. Couenne F, Jallut C, Maschke B, Breedveld P, Tayakout M (2006) Bond graph modeling for chemical reactors. Math Comp Model Dynam Syst 12:159–174
218. Franco AA, Jallut C, Maschke B (2006) Multi-scale bond graph model of the electrochemical dynamics in a fuel cell. In: Troch I, Breitenecker F (eds) Proceeding 5th Mathmod conference, Vienna P103
219. Franco AA, Bessler WG, Coulon R, Ichinose D (2013) Paper in preparation
220. Eikerling M, Malek K, Wang Q (2008) Catalyst layer modeling: structure, properties, and performance. In: Zhang JJ (ed) PEM fuel cells catalysts and catalyst layers—fundamentals and applications. Springer, London
221. www.modeling-electrochemistry.com
222. Franco AA, Xue KH (2013) Carbon-based electrodes for lithium air batteries: scientific and technological challenges from a modeling perspective. ECS J Solid State Sci Technol 2(10): M3084
223. Franco AA (2013) Multiscale modeling and numerical simulation of rechargeable lithium ion batteries: concepts, methods and challenges. RSC Adv 3(32):13027–13058

Chapter 9
Applications and Durability of Direct Methanol Fuel Cells

Esteban A. Franceschini and Horacio R. Corti

Abstract Many companies are making significant efforts in the development of prototypes of DAFC (mainly DMFC) for replace batteries (battery charge and auxiliary power units) in portable devices. Some of the most relevant prototypes are summarized; however, most of these devices are not ready to be commercialized due to the high cost and low power reached. Furthermore, for the massive application of the DAFC technologies is necessary solve some of the drawbacks (as miniaturization, products balance, cost reduction, etc.). The cost of the prototypes is analyzed as well as the degradation of the components that affects the durability of the devices.

9.1 Introduction

Considering all the types of Direct Alcohol Fuel Cells (DAFC) currently in development, that using methanol (DMFC) is closer to massive commercialization. DMFC exhibits higher current and power densities than fuel cells using ethanol, ethylene glycol, etc., mainly due to the difficulty for breaking the C–C bonds of higher alcohols. For this reason this Chapter will be mainly devoted to review the applications of DMFC.

At the end of twentieth century DMFC-driven low emission vehicles were visualized as "a strategy for the future" [1], although it was recognized that the main breakthrough needed for compete with hydrogen-fed fuel cells was the formulation of new membranes able to sustain high temperatures (150–160 °C)

E.A. Franceschini
Grupo Celdas de Combustible, Departamento de Física de la Materia Condensada, Centro Atómico Constituyentes, Comisión Nacional de Energía Atómica (CNEA), Av General Paz 1499 (1650), San Martín, Buenos Aires, Argentina

H.R. Corti (✉)
Departamento de Física de la Materia Condensada, Centro Atómico Constituyentes, CNEA, and INQUIMAE (Universidad de Buenos Aires – CONICET), Buenos Aires, Argentina
e-mail: hrcorti@cnea.gov.ar

H.R. Corti and E.R. Gonzalez (eds.), *Direct Alcohol Fuel Cells: Materials, Performance, Durability and Applications*, DOI 10.1007/978-94-007-7708-8_9,

with low methanol crossover. In Chap. 6 we discussed the efforts to develop such membranes and concluded the in spite of the advances observed during the last years we are still far to reach temperatures high enough for enhancing methanol electro-oxidation at the levels required for vehicular applications.

A decade ago Lipman and Sterling [2] reviewed the fuel cell vehicle (FCV) prototypes and commercialization status of the major automakers. They concluded that while several companies (General Motors, Toyota, Nissan, Renault, Hyundai) favored the use of hydrocarbon fuels with onboard conversion to hydrogen, others like Ford and Honda were focused on direct-hydrogen FCV. Only Daimler-Chrysler supported DMFC in a prototype known as DMFC Go-Cart. Moreover, they estimated the introduction of FCV into limited fleet applications by 2003–2005, with broader introduction in private consumers by 2008–2010, a prophecy that was not fulfilled. On contrary, as mentioned in Chap. 1, the lower power density and higher costs of DMFCs as compared to conventional hydrogen-feed PEM fuel cells, discouraged its development as power source for electric vehicles, despite the advantages of methanol over compressed hydrogen due to the existence of a well-established infrastructure for the distribution and supply of liquid fuel all over the world.

The potential of small direct methanol fuel cells for replacing batteries in portable electronic devices was considered early in twenty-first century, on the basis of their energy density, technological feasibility, safety and cost [3, 4]. Thus, most of the projected applications of DMFC are related to power supply of portable devices in the 1–40 W power range (cell phones, PDAs, Tablets, Notebooks, etc.) and mobile devices with higher power (golf carts/garden, marine, Unmanned Aerial Vehicles -UAVs-, etc.). A few DMFC stacks have reached a power as high as 3 kW at 100 °C (designed for the transportation application in a small one-person vehicle), or even 6 kW (based in the same design stack as the prototype of 3 kW) [4, 5].

The purpose of using DAFC technology (and particularly DMFC) on portable devices, replacing current Li-ion technology and other advanced batteries, is the elimination of some of battery drawbacks such as, low autonomy, long recharge times (in many cases in the order of the autonomy), low lifetime, (usually from 300 to 1,000 charge/discharge cycles) [6], which converts them in consumables.

Although a DMFC operating with dilute aqueous methanol solutions can achieve a good performance for some portable applications, the use of diluted methanol decreases the specific energy of the liquid fuel [7]. Thus, in Fig. 9.1, the theoretical specific energy of a DMFC operating with different methanol concentrations is compared to a state-of-the-art Li-ion battery. The change of the specific energy of the fuel with the methanol concentration of solution, represented by the dashed line, indicates that the specific energy of the methanol solution with a concentration above 2 M will be greater than that of Li-ion batteries (~ 350 Wh · dm^{-3} [7]), while the specific energy of pure methanol is as high as 4,900 Wh · dm^{-3} [8, 9], that is, 14 times higher than that of the battery. However, the data in Fig. 9.1 also show that if the overall energy conversion efficiency, η, of the DMFC system is taken into account, only a small portion of the energy stored in the fuel solution can be converted into net power output and the final specific energy of the DMFC will be significantly reduced. For instance, considering an efficiency of 20 %, the specific energy of a DMFC system can be greater than the Li-ion only if the

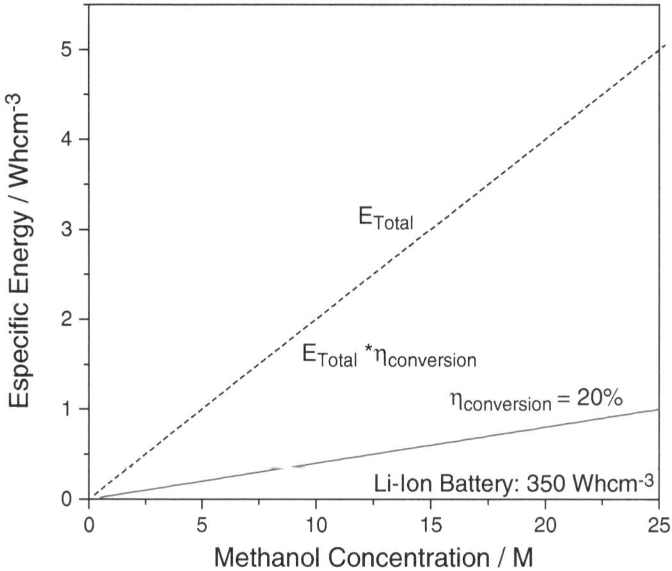

Fig. 9.1 Variation in specific energy of a DMFC system with methanol concentration (Adapted from Ref. [7] with permission)

concentration of the solution of methanol in the fuel cartridge is greater than 9 M. Therefore the specific energy of a DMFC with such efficiency would become higher than a Li-ion battery if the fuel cell operates with concentrated methanol [7].

However, in conventional DMFC designs, an increase in the concentration of methanol feeding the anode will lead to an increase in the rate of methanol crossover, decreasing the cell voltage. For this reason, the strategy for increasing the specific energy of the DMFC system is to use concentrated methanol, but in the meantime the anodic diffusion layer must provide an optimal wetting to the concentrated solutions of methanol, allowing that the methanol be consumed in the catalytic layer as it arrives avoiding its accumulation in the electrode-membrane interface, so that the rate of methanol crossover can be minimized [10]. Although the strategy of using high concentrations of methanol (or pure methanol) is challenging, some progress has been made over the past few years, as it will be shown through the examples of the following sections.

The main problems regarding the replacement of batteries by direct alcohol fuel cells are related to the largest volume required by the fuel cells, as compared to the batteries which have become highly compact (because DAFCs have not reached yet high efficiencies), elimination of residues of the methanol partial oxidation (generally mixtures of water with formic acid, methyl formate, and formaldehyde), and the high temperature which can reach the DAFC (up to around 85 °C for cells using Nafion® membranes) [11, 12].

Portable DMFC power systems have been considered as power suppliers for mobile phones, personal wireless devices, laptop, military communication systems, robotic devices, healthcare systems, multipurpose emergency power sources,

Table 9.1 U. S. DOE 2013 targets for performance, cost and durability of DMFC

Power	Specific power ($W.kg^{-1}$)	Power density ($W.dm^{-3}$)	Energy density ($Wh.dm^{-3}$)	Cost ($U\$S.W^{-1}$)	Durability (hours)
<2 W	10	15	300	100[a]	5,000
2–25 W	50	50	700	6	5,000
25–250 W	50	100	1,000	3	5,000

[a]Cost in U$S/system

marine applications, mountaineering, etc. [13]. Portable/micro PEM fuel cells commercialized has grown from less than 3,000 units in 205 up to almost 10,000 units in 2009 [14], and almost a quarter of them consist of DMFC. The main advantage of DMFC over lithium batteries is that the former has a gravimetric energy density 70–80 times higher than that of Li-ion batteries, significantly reducing the weight to be transported. On the other hand, a third generation multimedia mobile phone which perform several functions such as exchanging information via the Internet and digital mobile broadcasting, will require nearly 3,79 Ah per day, which will need three 5 h charges for 1 day use [15]. DMFC technology is attractive for driving such hungry-power devices replacing advanced batteries.

An additional fact to consider for the use of portable DAFC is the regulation for carrying flammable organic liquids in aircrafts. The International Civil Aviation Organization (ICAO) has approved the transport of certain types of fuel cells, and refill cartridges in aircraft cabins. Particularly, it allows the transport of a device (i.e. mobile phones, laptops, etc.) powered by a fuel cell (using methanol, formic acid or butane) and two refill cartridges, with a maximum volume of 200 cm^3 [16, 17].

The U.S. Department of Energy (DOE) supports the Energy Efficiency and Renewable Energy (EERE) Fuel Cell Technologies Program, with the mission of enabling the widespread commercialization of fuel cells. In this context, in 2009 DOE has proposed 2011 and 2013 targets for the performance, cost, and durability of DMFC with power up to 250 W. The 2013 targets are summarized in Table 9.1 [18], where the specific power and power density are based on the fuel cell system excluding the fuel tank and any hybridization batteries, while the energy density is based on total fuel cell system including the fuel for 5 h of operation at the rated power. The cost is defined for a scale production of 25,000 units per year and durability is the point which the system rated power degrades by more than 20 %.

9.2 Configuration and Operation Modes of DMFC Prototypes

In Chap. 1 (Sects. 1.5.5 and 1.5.6) we described the active and passive mode of DAFC operation and different configurations reported in R&D works. Here we will briefly summarize the configuration and operation modes of DMFC prototypes, which are closely related to the fuel management and the miniaturization of the system.

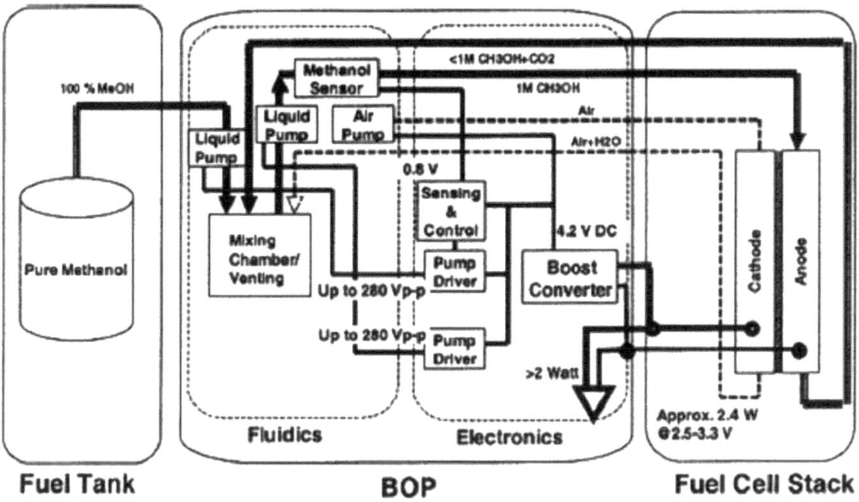

Fig. 9.2 Diagram of the active DMFC showing the balance of plant (BOP) components (Reproduced from Ref. [20] with permission)

Methanol and air/oxygen accessibility to the catalysts layers and the removal of CO_2 and intermediate oxidation products are key factors in the performance and degradation of the overall system. Kamaruddin et al. [19] have recently reviewed the fuel delivery system in DMFC that has to be optimized in order to achieve the maximum performance, measured in terms of fuel or cell efficiencies. These authors discussed the main features of active fuel delivery, starting with the active fuel delivery system, shown in Fig. 9.2, developed by Xie et al. [20] at Motorola. In this active 2 W system pure methanol is mixed with water from the cathode in a mixing chamber that also collect the methanol/CO_2 mix from the anode, in that way methanol is diluted to approximately 4 % v/v in order to minimize crossover through the Nafion membrane. A methanol sensor controls the concentration of methanol at the exit of the mixing chamber, and triggers a liquid pump to add more methanol if necessary. Other piezo-driven liquid pumps recirculate the diluted methanol solution through the fuel cell stack, supplies the required oxygen to the cathode, removes the water from the cathode and pumps both the air and the water back to the mixing chamber.

Power consumption of the pumps is critical for portable applications, and new ideas emerge for extending the performance of active systems using thermo-pneumatic, Lorentz force, and piezoelectric valve-less micro-pumps [19]. MTI Micro has modified the former active system by internally transferring the water produced at the cathode to the anode without using a conventional pump [21], a concept used in the MTI Micro's Mobion charger (see Sect. 9.3).

Zhao et al. [22] reviewed with passive supply of reactants, emphasizing that the mass-transport mechanisms of reactants and products in passive DMFCs that have no liquid pumps and gas blowers/compressors are different from those in active DMFCs. They concluded that the key issues for small (micro) DMFC are:

1. Use of a high methanol concentration but maintaining an adequate concentration in the anode catalyst layer at a given current density to maximize the system specific energy and cell performance. For achieving this aim it is fundamental an optimum design of the fuel supply system (that allows the orientation-independent operation of the fuel cell), the anode current-collector, and the anode diffusion layer.
2. Enhance the oxygen supply to the cathode catalyst layer and the removal of water from the cathode diffusion layer by utilizing passive forces, such as the capillary action and gravity.
3. Supply the water required for the methanol oxidation in the anode by transporting water from the cathode, so that the fuel cell can be operated with pure methanol and with minimum flooding at the cathode. This can be achieved by innovation in the design of the membrane electrode assembly.
4. The materials and the design of the cell structure should minimize the heat dissipation. Therefore, the fuel cell can operate at a sufficiently high temperature to improve the cell performance.
5. Improve the current stack design to make the overall system more compact and achieve an effective management of heat, water and air under the operating condition of feeding a high-fuel concentration.

The power needs for a specific portable application require a fuel cell stack where the unit cells are connected in series or in parallel. High-current or high-voltage output can be achieved by increasing the MEA area or the number of unit cells that are connected, respectively. Most stacks are in-series connections and, according to the arrangement of unit cells, can be classified into two types: bipolar stacks and monopolar stacks [22]. Bipolar systems use graphite or metal to connect anode and cathode and consequently are too heavy for portable applications such as mobile phones [23]. Therefore, passive micro DMFCs adopt the monopolar arrangement, as that shown in Fig. 9.3 (left) [24], where the air-breathing cathode is fully exposed to ambient air in order to maximize oxygen and water transport.

A similar design was adopted by Hahn et al. at Fraunhofer Institute [25] to develop small planar PEM fuel cells with areas between 1 mm^2 and approximately 1 cm^2 using MEMS technology. In this design of DMFC, shown in Fig. 9.3 (right), conventional diffusion layers were replaced by a microstructure fabricated at the inner sides of the current collectors, with a thickness of 200 μm, which serve as flow field and are designed to assure a homogenous gas supply all over the MEA. The current output of this 1 cm × 1 cm cell was 40 mA at 1.5 V and exhibited stable long-term operation at 80 $mW.cm^{-2}$.

Kamarudin and coworkers [23] developed a dual side stack where two MEAs were stacked between a gasket, a current collector and anode and cathode plates, as illustrated in Fig. 9.4 (left).

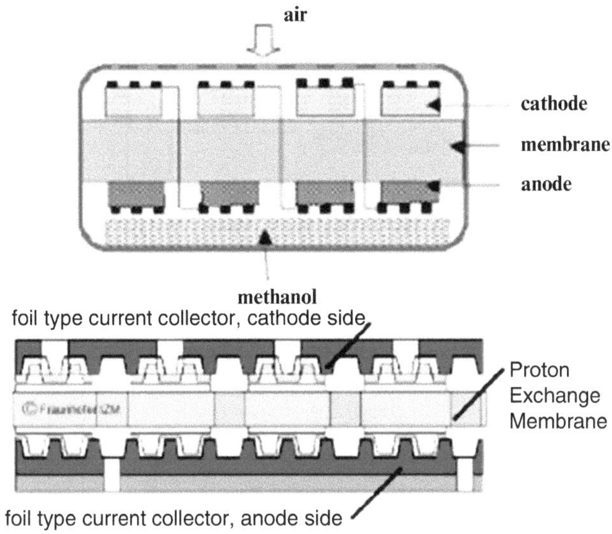

Fig. 9.3 *Top*: scheme of a passive monopolar stack with built-in methanol reservoir [24]. *Bottom*: MEMS-based micro DMFC [25] (Reprinted from Refs. [24] and [25] with permission)

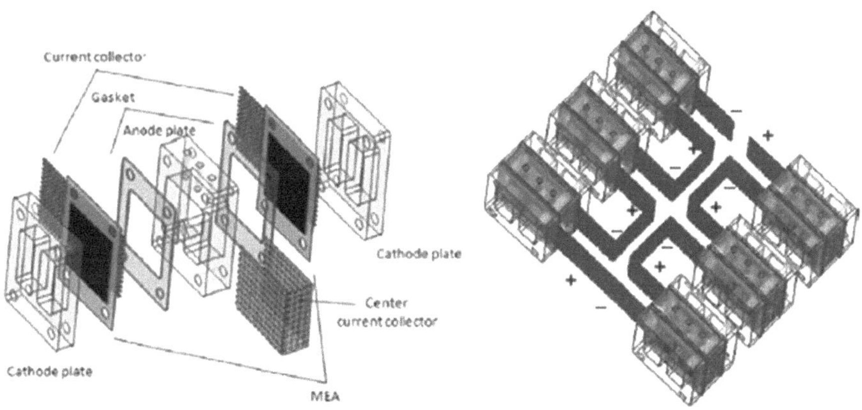

Fig. 9.4 *Left*: dual side stack. *Right*: six dual side stacks connected in series (Reproduced from Ref. [23] with permission)

The anode plate at the center of the stack used as a methanol reservoir permit a more efficient fuel distribution and reduces the stack weight. In Fig. 9.4 (right) six dual side stacks were connected in series to generate a 600 mW mobile phone charger that developed a power density of 25 mW.cm^{-2}.

The first step in a possible total or partial replacement of Li-Ion batteries by DMFC is the construction of integrated DMFC/supercapacitor systems

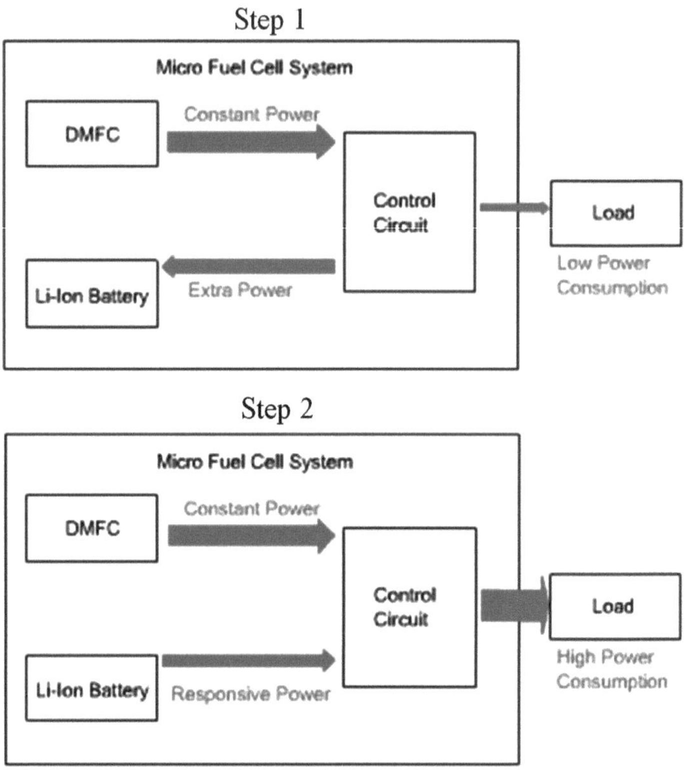

Fig. 9.5 Behavior of the integrated DMFC/Li-Ion in conditions of low power consumption (*Step 1*) and high power consumption (*Step 2*) (Adapted from Ref. [27])

(or DMFC/Li-ion systems), where the fuel cell is used for charging these storage energy systems [26]. This integrated system is able of a quick response to power peaks demands in systems such as mobile communications devices. An example of this concept has been developed by Sony Corporation [27], and it is illustrated in Fig. 9.5. A DMFC delivers constant power to a control circuit that under low power requirement (e.g. standby cell) deliveries the excess power to a Li-Ion battery, which is kept charged waiting for high power requirement (Step 1 in Fig. 9.5).

On power demand (e.g. in call mobile phone) the control circuit supplies the full fuel cell power to the mobile phone circuit, and takes the extra power from the Li-ion battery (Step 2 in Fig. 9.5) [27].

Recently, Zhu et al. [28] tested a 20 W DMFC/supercapacitor hybrid energy system using a maximum power point tracking algorithm, which allow a power flow to and from the supercapacitor bank through a bi-directional converter for close regulation of the DC bus voltage. Thus, the inherently slow DMFC is prevented from having to respond to the varying load demand.

9.3 Market Potential

In the last three decades there has been a large growth in demand for portable electronic devices. A clear example of this growth is that of battery-powered laptops. In 1986 notebooks represented only 2 % of the world market [29] while in 2008 the sale of notebooks reached 145.9 million units (exceeding even the sale of desktops [30–33]), and 177.7 million units in 2009.

Other niche of application for DAFC is mobile communications. Gartner Inc. estimated that, in 2010, approximately 5,600 million the number of mobile phones were in use around the world [34]. According to Frost & Sullivan, in 2009, digital cameras represented the third largest market for portable electronics, following mobile phones and laptops, with approximately 110 million units that year [35].

Moreover it is also promising the netbook market (35 million sold in 2010), and the more recent tablets with 18 million units sold in 2010 and 63.6 million in 2011, while Gartner Inc. estimates that sales of tablets will reach 300 million by 2015 [36].

In addition, the increasing number of supplementary functions present in portable electronic devices (digital photo/video cameras with GPS, laptops with bluetooth readers, Wi-Fi; mobile phones with touch screens, recording/playback of sound files etc.) leads to the need of batteries increasingly large and heavy, in order to maintain the devices autonomy and, in turn, generating the additional problem of increasing the recharge time.

As diverse as the applications proposed for DAFC are the requirements they must satisfy in order to respond appropriately to the demands of each application. Figure 9.6 shows a scheme of the powers required for the most common applications that face a DMFC, as candidates to replace batteries [37].

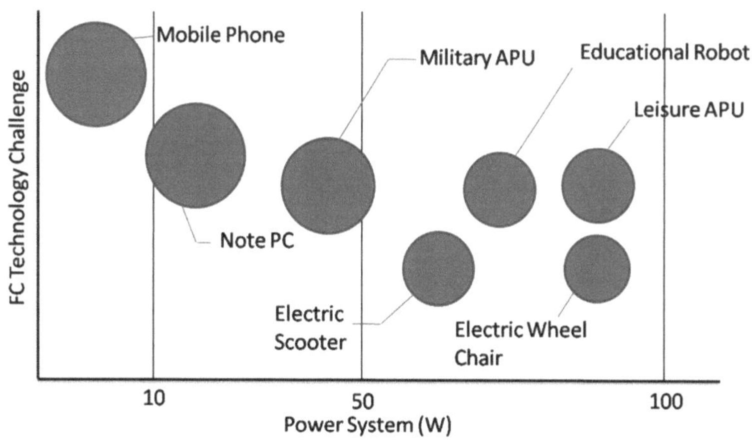

Fig. 9.6 Scheme of the power needed for different applications in contraposition of the technologic challenge (Adapted from Ref. [37])

9.4 DMFC Development

Figure 9.7 shows how evolved the number of patents of devices, materials, and applications related to DMFC and DEFC [38]. It can be seen that the largest number of patents on DMFC, were published in 2007, while patents on DEFC are growing continuously since 2002 and over passed those for DMFC since 2008, which suggests that DEFC technology has not yet reached its maturity, and in the coming years a large development of DEFC is expected.

One of the obstacles to overcome is the largest volume occupied by DMFC (even when the volumetric energy density of methanol is higher than the Li-Ion batteries), due to the low efficiency with the current DMFC technology.

In order to improve that situation two objectives were proposed, the first is to increase the fuel cell efficiency, for enhancing the autonomy of the devices (for example, a DMFC able to power a laptop during 10–20 h without recharge, as compared with a 6-cell lithium battery with an autonomy of around 3 h in a standard notebook [6]).

The second aim is the miniaturization of devices, on which has focused the work of many research groups and companies. Thus, in 2004 Toshiba Inc. presented a direct methanol fuel cell with a size of 22 × 56 × 4.5 mm with a small tank of methanol (2 cm^3 in volume and 8.5 g weight) incorporated, with an output power of 100 mW, which, according to the company is sufficient to maintain a mp3 device operating over 20 h [39].

Both, the feed of alcohol into the device and the output of the reaction products, occur in a passive way and the user carries out the fuel refilling manually through a syringe, without changing cartridges, and reducing the cost of recharging.

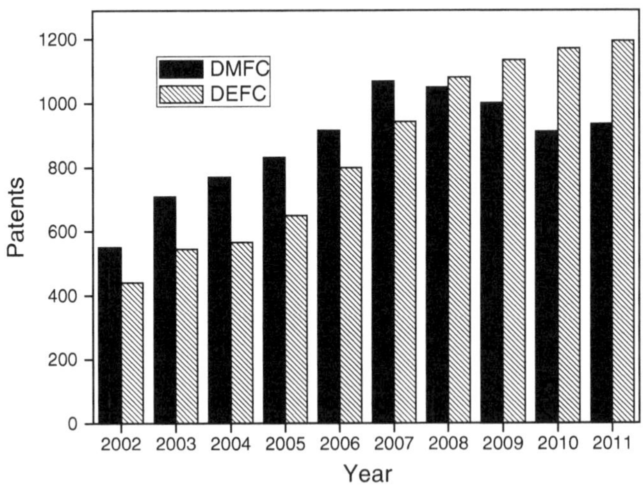

Fig. 9.7 Trend in the number of patents in DMFC and DEFC during the last decade

MTI Micro introduced in 2004 the integrated direct methanol fuel cell technology Mobion®, and in 2007 the Mobion Chip®. This small device, with a volume of only 9 cm^3, is fed with pure methanol to produce 84 $mW.cm^{-2}$. These prototypes are passive DMFC and their design include a fuel conditioning system which allows operate in a temperature range of 0–40 °C, independent of the humidity and cell position, which is particularly useful in portable applications, such as mobile phones [40].

Although the final products of oxidation of alcohols should be CO_2 and H_2O, the reactions are not complete and produce a low amount of semi-oxidized products (which depend on the fuel used) that are generally toxic. Commonly, the residues can be vented without inconvenience or danger to the user.

A novel mobile phone charger which operates with a direct methanol fuel cell feed with a 99 % methanol/water solution was co-developed by Fujitsu Laboratories Ltd. in partnership with NTT DoCoMo, and consists of a passive fuel cell charger with a cartridge (volume of 18 cm^3). It has an average output power of 1 W, but can respond to peaks of up to 9 W, also charging a lithium-ion battery at 5.4 V and 700 mA [41, 42]. In this device the reaction products drain using the lateral slots, so that the products are vented during generation avoiding its accumulation [43].

Not all DAFC systems allow this kind of products removal. Thus, in the case of DEFC using diluted ethanol, the CO_2 production efficiency is much lower than that of DMFC and the amount of liquid semi-oxidized products is even greater [44–46].

Table 9.2 summarizes some of the most prominent DAFC powered prototypes developed for portable applications by different companies around the world in the last decade, as well as some conceptual devices which are not even prototypes, but demonstrate the growing interest of companies in the replacement of lithium-ion batteries by DAFC.

9.5 DMFC Powered Portable Devices

This section summarizes several protoypes and commercial DMFC developed by companies to power different portable electronic devices (Table 9.2) as well as portable multipurpose generators (Table 9.3).

9.5.1 Battery Chargers and Back-Up Devices

There is a relatively wide range of battery chargers are mainly based in DMFC. Different companies have launched in recent years systems that allow battery recharge without power grid, as well as to increase the autonomy of systems.

In 2009, Panasonic Corporation presented a prototype of battery charger based on DMFC with an output power of 20 W for mobile applications (volume: 360 cm^3,

Table 9.2 DMFC power sources for portable applications

Developer	Year	Product	Fuel	Power/power density	Comments
Toshiba Inc	2004	Battery replacement	Methanol	100 mW	mp3 device operating over 20 h [39]
MTI Micro	2007	Mobion chip	Methanol	84 $mW.cm^{-2}$	Portable applications [40]
Fujitsu Labs. Ltd./NTT DoCoMo	2004	Battery charger	Methanol 99 %	1 W (peaks of up to 9 W)	Can charge a Li battery at 5.4 V/0.7 A [41, 42]
Panasonic Corporation	2009	Battery charger	Methanol	20–100 W	Promises about 5,000 h of use [47]
Samsung SDI and SAIT	2006	Battery charger	Concentrated methanol	2.6 W	Promises at least 400 h. Output potential of 4.2 V, an energy density of 55 $Wh \cdot dm^{-3}$[37]
Toshiba Inc	2009	Dynario®	Concentrated methanol	2 W	3,000 units were sold in 2009 at a cost of ~U$S 300 [49]
Samsung SDI	2009	Auxiliary power unit	Pure methanol	25 W (peaks of up to 55 W)	Prototype for powering military portable communication system based on active DMFC technology [50, 51]
Toshiba Inc	2005	Battery replacement	Methanol	300 mW	Enough to maintain a mp3 device turned on up to 60 h [57]
KDDI Corp./Hitachi Ltd.	2005	Smartphone W32H	Methanol 60 %	300 mW	Hybrid DMFC/Li-ion battery with a capacity 2.5 times the conventional value [57]
Samsung SDI	2008	Battery replacement	Methanol		Power to a laptop for about 15 h [60]
NEC	2003	Battery replacement	Methanol 10 % w/w	14 W (peaks of up to 24 W)	A single refueling promises about 5 h of use. [61, 62]
NEC	2004	Battery replacement	Methanol	70 $mW{\cdot}cm^{-2}$	Allows running the laptop continuously for approximately 10 h [63, 64]

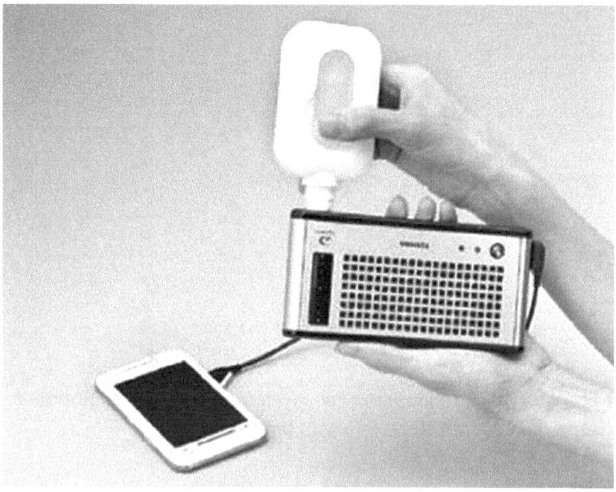

Fig. 9.8 DMFC Dynario charger from Toshiba (Reproduced from Ref. [19] with permission)

weight: 350 g, excluding fuel). Using this technology the company aims to develop a 100 W prototype (volume: 2,000 cm^3, weight: 2 kg, excluding fuel) during 2012 for backup purposes, combining the DMFC with a high capacity lithium battery, which aims to bring to market an integrated generation/storage power system. This system promises about 5,000 h of use (with an intermittent operation of 8 h per day) powering, for example, a notebook. Panasonic solved the problem of low durability of their previous prototypes developing a technology of microporous carbon gas diffusion layer that allows the use of concentrated methanol as fuel, the use of dilute solutions requires higher fuel flows, increasing erosion of the GDE [47].

Samsung SDI and SAIT also developed in 2006 a small portable charger (phone, PDA, cameras, etc.), with a power of 2.6 W and a lifetime greater than 400 h, using replaceable methanol cartridges. This charger has an output potential of 4.2 V, an energy density of 55 $Wh \cdot dm^{-3}$, and operates independently of the cell position, which is an important feature for portable devices, though not exclusive [37]. The passive charger produced by Samsung SDI//SAIT has a thickness of only 5 mm, a volume of only 150 cm^3, a weight of 180 g, and uses a concentrated methanol cartridge of 26.7 cm^3 [48].

Toshiba Corporation introduced in 2009 its DMFC prototype Dynario® [49]. This portable charging device has an output potential of 5 V at a current of 400 mA, which not only allows recharging phones, cameras, and GPS batteries, but also for devices with higher requirements, such as tablets and any system that can be charged via USB. This device has a size of 150 × 21 × 74.5 mm and a weight of 280 g (without fuel), it contains an integrated cartridge, not interchangeable, with a volume of 14 cm^3 developed *ad-hoc* by Toyo Seikan Kaisha, Ltd. which uses a concentrated methanol solution. This cartridge can be refilled by the user with an external fuel tank, as shown in Fig. 9.8 [19].

Dynario is a hybrid DMFC//Li-Ion device comprising a rechargeable Li-Ion battery with 660 $mA \cdot h$ (Sanyo Electric), which works buffering the device response to the requirements, allowing on-fly reloading of methanol without stop

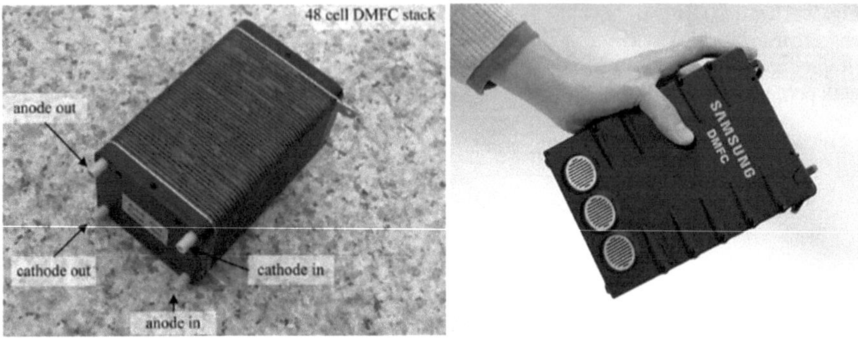

Fig. 9.9 Stack and prototype system for military portable power based on active DMFC (Reproduced from Ref. [54] with permission)

recharging the portable device. This charger was one of the few devices of this type to come out for sale. A unique stock of 3,000 units were sold on line in Japan, in 2009 at a cost of 29,800 yens (~U$S 300) while the pack of 5 cartridges of methanol solution of 50 cm^3 had a cost of 3,150 yen (~U$S 35) [49].

Samsung SDI Company introduced in 2009 a prototype of active DMFC technology for powering military portable communication systems based with an average power of 25 W [50, 51], coupled to an smart Li-Ion battery, that allows to the system respond to peaks up to 55 W, at a potential of 16.4 V. This device has a total volume 2.52 dm^3, a total weight of 1.87 kg, and operates independent of position in the temperature interval −5 °C to 45 °C. The system uses 300 cm^3 cartridges of pure methanol, and was developed for operations from 72 to 96 h. In the case of operation during 72 h the device needs about six DMFC methanol cartridges with a total weight of 3.7 kg and an energy density of 510 $W{\cdot}h{\cdot}kg^{-1}$, which is equivalent to a weight of 8.16 kg of a primary battery [BA-5590] [52] and 10.88 kg of a secondary battery [BB-2590] [53] with the additional advantage of the instant recharge. The use of active direct methanol systems has the disadvantage of the need of small pumps, which consume energy and fuel, and increase noise level. However, Samsung SDI claims that the system emits less than 45 dBA, equivalent to the noise level present in a silence library.

Kwon et al. [54] have analyzed the performance of this prototype, shown in Fig. 9.9, taking into account the stack current and the power spent by the balance of plant (BOP) parts, which include liquid and air pumps, fan and methanol concentration sensor. The use of a smart battery allows the system to provide power during the start-up and improves the dynamic characteristics when the load varies. The average voltage cell efficiency was 43 %, while the output energy is 23.3 % of the total methanol energy (4.8 $Wh.cm^{-3}$). The power losses are distributed as follow: 55 % in the stack, 12.0 % in the BOP and conversion, 9.2 % due to methanol crossover and evaporation. Compared to conventional systems this prototype generates the power at maximum fuel efficiency regardless of the load condition.

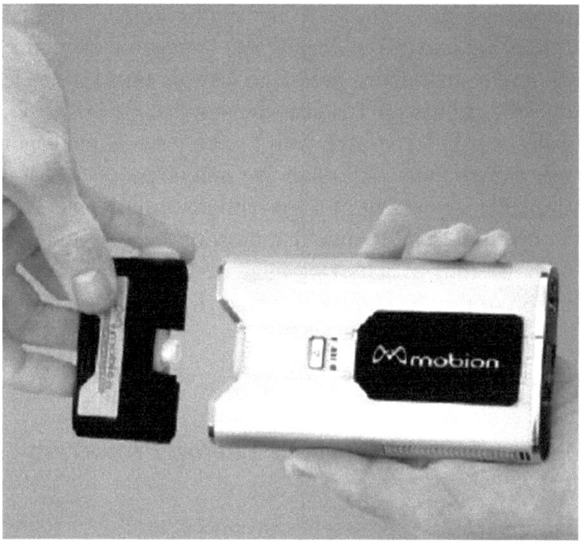

Fig. 9.10 Mobion chip based battery charger (MTI Micro) (Reproduced from Ref. [55] with permission)

Another battery charger based on DMFC is the Mobion prototype introduced by MTI Micro [55]. This portable device, shown in Fig. 9.10, employs a cartridge of pure methanol, with a volume of 25 cm^3, and using a single recharge of methanol could fully recharge the battery of a mobile phone at least 10 times. That charge it is equivalent to 100 h of video playback on a mp4 device or about 6,000 shots with a professional digital camera. This type of DMFC were designed not only as a charger, but also to be used in replace of batteries in professional digital cameras, GPS devices (for more than 60 h of continuous power) [7, 56], and cell phones, where feed is performed by injecting pure methanol directly into a not interchangeable fuel cartridge.

9.5.2 Battery Replacement

Toshiba introduced in 2005 two models of mp3 players powered by DMFC systems. The first is a flash mp3 player with a 3.5 cm^3 reservoir and a power of 100 mW, able to operate for more than 35 h with a single fuel charge. Its dimensions are 35 $\times$ 110 $\times$ 20 mm and only 78 g weight [57].

The second model has a small hard drive similar to those used in other models available in today's market. The higher capacity of the alcohol tank (10 cm^3, compared to the 3.5 cm^3 tank of the original prototype), allows to the 300 mW DMFC to maintain the player turned on up to 60 h. Its size is 65 $\times$ 125 $\times$ 27 mm and its weight is 270 g [57].

In 2005 KDDI Corporation in association with Hitachi Ltd. presented the smartphone prototype W32H. This device uses an inner cartridge of 3 cm^3 that is reloaded with a ~60 w/w % methanol solution from an external pen-shaped cartridge. It uses a passive type DMFC and is capable of producing 300 mW of power [57]. A similar smartphone prototype was presented the same year by KDDI Corporation in association with Toshiba (the smartphone A5509T), having the same power but using a 7 cm^3 cartridge with a 99.5 w/w % methanol solution, which promises a higher autonomy (2.5 times the autonomy of a Li-ion battery), weighing ~160 g, similar to the weight of a conventional smartphone feed with a Li-ion battery.

It is important to note that most of these prototypes are still under development with the aims of achieving increased fuel efficiency (contributing to the miniaturization of the device) and a longer life, and it is difficult to assess a date for their commercialization.

9.5.3 DMFC Powered Laptop Prototypes

Different DMFC prototypes have been developed for notebooks powered by DMFC, with the aim of increasing the low autonomy given by the Li-Ion battery [58, 59].

Samsung SDI has developed a prototype of DMFC for use in laptops which is quoted to have a durability almost twice as compared to other systems being developed. SAIT has reduced the amount of catalyst required by 50 %, by developing a mesoporous carbon material, which supports highly efficient 3 nm nanocatalyst particles. In addition, SAIT has developed a unique concept of nanocomposite membrane to reduce methanol crossover by more than 90 %. This composite uses a 30–100 μm thick proton-conducting membrane with a proton conductivity of 0.1 $S.cm^{-1}$. The DMFC has an energy density of 650 $Wh \cdot dm^{-3}$, and fed with about 200 cm^3 of liquid methanol can supply power to a laptop for about 15 h. The cell measures 23 cm × 8.2 cm × 5.3 cm, and its weight is less than 1 kg [60].

The Japanese company NEC presented two DMFC prototypes clearly showing the rapid progress of this technology. The first prototype of a laptop powered by an integrated direct methanol fuel cell, launched in 2003, was triggered by the application of carbon nanohorns, a form of carbon nanotubes, previously developed by the company in collaboration with the Ministry of Economy, Trade and Industry (METI) and the New Energy and Industrial Technology Development Organization (NEDO). The DMFC used in this device, generates an average power of 14 W, can respond to peak power of up to 24 W at a potential of 12 V, with a weight of 900 g., without considering the 300 g. of fuel storage in a cartridge. This system uses as a fuel aqueous methanol (10 w/w%), which is feed directly into the notebook from a special container, as shown in Fig. 9.11 (left). A single refueling promises a range of about 5 h of continuous use [61, 62].

Although this prototype had already met many of the desired characteristics for replacing Li-ion batteries in portable systems, NEC launched in 2004 a second prototype (Fig. 9.11 (right)) with a higher surface power (70 $mW{\cdot}cm^{-2}$), which allows running the laptop continuously for approximately 10 h (twice the autonomy that the earlier prototype) [63, 64].

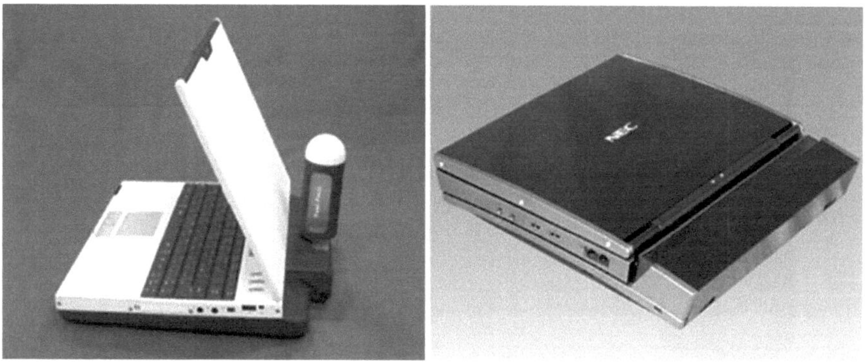

Fig. 9.11 NEC's laptops powered with DMFC. *Left*: DMFC coupled to the refilling tank (2003), *right*: second generation (2004) (Reproduced from Ref. [58] with permission)

9.5.4 Portable Generators

DMFC systems in the power range of 25–65 W for leisure uses, and from 45 to 110 W for industrial users, have been developed by Smart Fuel Cells in Germany., EFOY fuel cell, shown in Fig. 9.12, works like a "mobile power socket" to automatically recharge batteries and are available for use 24 h a day when connected to the methanol tank. Several thousand units are already aboard recreational vehicles, boats, and vacation homes [65]. The industrial version of EFOY DMFC are used in different off-of-grid applications, such as the oil and gas sector (chemical injection pumps, air compressors, and surveillance), traffic management (road weather stations, enforcement vehicles, traffic cameras and counters), and stationary monitoring of critical infrastructures [66]. In these applications portable generators can be connected to fuel cartridges and accumulators for high currents, providing an autonomy of up to a year without any costly servicing.

CPI (UK) and Hitachi (Japan) joined in 2009 to develop a 100 W DMFC with an energy efficiency over 25 % for powering remote surveillance cameras and highways signage [50, 67]. IRD in Denmark developed 500 and 800 W DMFC that works as a backup for emergency batteries and also makes sure that Lismore Lighthouse still shows its light to ships in the Inner Hebrides of Scotland, replacing the old diesel generator [50, 68].

Oorja Protonics commercializes a 1,000 W DMFC designed for use in forklifts, and other vehicles in the materials handling industry. The DMFC is fitted inside the battery tray of the existing machinery and continuously charges the vehicle battery [50, 65]. In Table 9.3 are summarized the performance characteristics of DMFC portable generators.

As it can be observed, the power and energy densities of these systems (including the methanol tank) are still far from reaching the targets reported in Table 9.1. Regarding the cost, the reported prices for the EFOY systems are close to 80 U$S. W^{-1}, meaning that the U.S. DOE targets will not be achieved by 2013. Finally, it should be noted that system durability are not reported for these portable generators.

Fig. 9.12 25 and 65 W portable DMFC generators from Smart Fuel Cells (Reproduced from Ref. [65] with permission)

Table 9.3 Main performance features of DMFC portable generators (Adapted from Ref. [65] with permission)

System	Power (W)	Volume (dm^3)	Weight (kg)	Energy output (Wh)	Run time (h)	Power density ($W.dm^{-3}$)	Energy density ($Wh.dm^{-3}$)
Jenny	25	2.8 (0.6)	1.3 (0.36)	480	19	7.4	141
EFOY1600 –M5	65	24 (7.8)	7.3 (4.3)	4,500	69	2.0	142
EFOY1600 –M10	65	24 (14)	7.3 (8.4)	9,100	140	1.7	239
EFOY1600 –M28	65	24 (41.5)	7.3 (24)	25,200	388	1.0	385
Oorja Pac	1,000	51 (22)	125 (17.6)	20,000	20	13.7	274

9.6 Projections

Market and companies forecasts on the introduction of fuel cells, including DMFC, have been generally very optimistic and, so far, none has managed to achieve the targets. For instance, the worldwide total available markets for micro fuel cells for 2011 estimated by Darnell Group in 2003 [69] were (measured in millions of units): 36 for camcorders, 70 for digital cameras, 608 for mobile phones, 102 for laptops, and 79 for PDAs. However, the penetration forecast of fuel cells also reported by Darnell were used to estimate the percentage of the total available market expected to be taken over by fuel cells [58], resulting in 2 % for camcorders and digital cameras, 5 % for mobile phones, 30 % for laptops, and 51 % for PDAs. These predictions were far to be fulfilled and, even worst, a 2010 review by Fuel Cell Today claimed that of the portable units shipped in 2010, only 4 % were DMFCs, the remainder were largely PEMFC systems [50].

Without doubts the deployment of micro DMFC will be linked to the commercialization strategies of big manufacturers that tend to focus on fuel cells incorporated into consumer products. As quoted by Agnolucci [58], the postponement of market introductions from big companies has been common, considering the impact that a faulty product can have on the brand. Thus, it is reasonable for these companies to introduce the new technologies only after acquiring full confidence of its reliability.

It is then improbably that microfuel cells, and consequently microDMFC, be introduced in the near future. Jollie [70] has concluded in 2004 that in the short term selling to specific industries or the army is the best strategy because it relieves manufacturers from developing a product with a wider appeal and extensive distribution network. This claim will be probably still valid in the next years.

Regarding DMFC portable generators (25–1,000 W), they are principally sold for recreational and industrial users in remote locations (off-of grid), as quoted in Sect. 9.5.4. The clean and silent characteristics of DMFC trigger the use for recreational purposes and DMFC are indeed more reliable than solar cells used in this market [58]. In the industrial sector DMFC will compete with ICE and solar power with the advantage of reliability, and lower operation and maintenance costs.

Concerning micro DEFC, it is worth to note that several companies are working on developing prototypes of portable devices (mainly mobile phones) powered with ethanol. Therefore, it is expected that in the coming years, technological advances may lead to devices operating with direct ethanol fuel cells advance to a prototype stage (as currently for the DMFC). Beyond the technological advances that have an accelerated rate (as shown in Fig. 9.7), there are other barriers to overcome before the devices operating with DEFC can be used in a massive way. One of them is the impossibility to transport ethanol on domestic and international flights, problem that was solved for the DMFC.

Some of the new niches of application of DAFC were introduced by companies such as NEC, Nokia, BlackBerry and others (mainly in smartphones). TryiYeh introduced the concept of a mobile smartphone "Cheers" with all the functionality of a conventional smartphone (touchscreen, wifi, mp3, digital camera, etc.) [71]. The shape of the unit a bit like a bottle or a flask, and it works on the basis of ethanol fuel cell. The entire surface of Cheers is a continuous touch display. It should be noted that the model looks quite stylish: it is like a glass rectangular bottle with rounded corners. Other example is a fuel cell backpack vacuum cleaner concept that operates with ethanol, turning the vacuum cleaner into an extremely portable device [72].

9.7 DMFC Prototype Costs

The fuel cell developers will have to focus on cost reduction if they want their applications to become a mass-market product. In every market, the extent to which the cost of fuel cells will have to fall will depend on competing technologies. It is sometimes argued that the price of fuel cell systems could increase because of the limited availability of precious metals like platinum [58, 69].

In the last 15 years the price of platinum has multiplied by four (growth from 21 U $S per gram in 1997 to 85 U$S per gram in 2012), while the price of ruthenium increased from 0.70 U$S per gram to 4 U$S per gram in the same period [73, 74]. Whereas the normally metal loads used in DMFC is approximately 4 $mg.cm^{-2}$, the price of the catalysts metals would be around 5,000 U$S/$m^2$ for $Pt_{50}Ru_{50}/Pt$ MEAs. At the laboratory scale the price of commercial a $Pt_{50}Ru_{50}/Pt$ catalyst supported on Vulcan XC-72, with a metal load of 40 %, is around 2,000 U$S/$m^2$ [75], that is, much lower than the price of the pure metals used as precursors. The current price of the Nafion 117® membrane is 1,600 U$S/$m^2$ [76].

Precious metals like platinum could be easily recycled through a product returning scheme, thereby reducing the contribution to the cost of fuel cells, therefore, the membrane price it is much more significant in the total cost of the MEA than the metals cost. Despite of that, several studies suggest that the use of membranes on large scales decrease the incidence of the cost of membrane in the total cost of the cell, and the cost of the electrodes would be dominant [77].

Clearly, the cost analysis of DAFC systems is highly dependent on the system power, as well as the production volume, so that, different studies can yield apparently contradictory results on the cost influence of the different components of the cell (or stack) in the total device cost.

Saizensen [78] estimates that a production scale of 150,000 fuel cell vehicles annually would decrease the membrane by a factor of 10, so the influence of the electrode in the cost of the stack would be higher than that of the membrane and other components of the cell [79]. James and Kalinoski [80] concluded that the polymer membrane would contributes to about 7 % to the cost of the stack and to about 4 % to that of the complete system in the case of 80 kW PEMFC manufactured at an scale of 500,000 units per year. However, the massive use of DMFC in transportation is highly unexpected and, on the other hand the use of Nafion in commercial DMFC seems to be improbable in the future.

Bipolar plates work as electric conducting materials and impermeable walls. Some of them have a pattern of flow fields like serpentine and are manufactured by CNC machining of a graphite sheet; this process is time-consuming and very expensive today, but will be substituted by inin the marketjection molding systems or hot pressing molding in the near future [79]. There are tremendous efforts to reduce the cost of bipolar plates, trying carbon composites, injection molding of graphite filled polymer and metals such as stainless steel. Bipolar plates are very important as they have a nearly 80 % dominant share of the total stack weight.

As recognized by Agnolucci [58], it is difficult to get hold of authoritative forecasts of the cost of fuel cells due to the commercial interests surrounding. Wee [81] have estimated the manufacturing cost of a DMFC in Korea by assuming that it would be a factor 10 greater than the equivalent Li-ion battery system. Thus, the cost of a 20 W DMFC for powering a laptop is estimated to be U$S 333 (around 17 U$S.W^{-1}), which is almost a factor 3 more expensive than the U.S. DOE target for 2013 (Table 9.1). Unsurprisingly, such a fuel cell docking station is not yet in the market.

Rashidi et al. [82] performed a feasibility study of powering laptops, camcorders and mobile phones with DMFC in comparison with Li-ion batteries, for an operational period of 4 years. The analysis include methanol crossover level and the cost of fuel during that period, and reveals that for a 20 W laptop operating 5 h per day (146,000 Wh) the consume of methanol for 4 years is 905 kg for a cell with a crossover of 10^{-6} mol.cm^{-2}.s^{-1}, but decrease to 73 kg if the crossover is reduced to 10^{-8} mol.cm^{-2}.s^{-1}. Thus, the cost of fuel was estimated in U$S 453 in the former case and U$S 37 in the last case. The authors also concluded that during the first year of operation the battery would have a lower cost compared to the DMFC, but the battery needs to be replaced after 500 charge/discharge cycles, resulting in higher battery cost after 4 years of operation.

The results follow the same trend for the operation of camcorder and a mobile phone.

The use of fuel cells in the army is apparently cost effective in providing remote power. For instance, providing power to one army retraining site costs about U$S 250 with fuel cells and U$S 800 with batteries, but fuel-cell-based portable generators are still much more expensive than diesel generators available in the market [58].

9.8 Fuel Cell Durability

Durable and reliable operation for several thousand hours are considered crucial for the successfully commercialization of DAFC. The factors that determine a PEM fuel cell's lifetime (as platinum-particle dissolution and sintering, carbon-support corrosion, and membrane thinning) is currently studied by many researchers in order of increase the lifetime without increasing cost or losing performance [83–85].The relative contribution of each component's degradation to the degradation of the entire fuel cell is not completely understood yet.

Within a DAFC the components are exposed to aggressive conditions as strong oxidizing (and reducing) media, high temperature, high electrochemical potentials, high electric current and large potential gradients. Although all these perturbations can be studied separately, the complex interplay among all of them makes very difficult to assess their individual contribution to the shortening of the DAFC durability.

In the following sections we will discuss the most common durability tests and the different factors that lead to the power decrease of the fuel cell. Some of them are general to hydrogen-fed PEM fuel cells, while others are specific of DAFC. Kim and Zelenay [85] quoted that performance degradation of DMFC has received little attention compared with the degradation of H_2/air PEM fuel cells systems, probably because lifetime requirements for DMFC systems are generally perceived as less stringent, and have not been defined as precisely as for H_2/air fuel cells designed to replace ICE in the very mature automotive market.

9.8.1 Durability Tests

Durability tests are usually carried out applying a constant current density for hundreds or thousands of hours, as well as cycling the current several times [86, 87]. U.S. DOE has released a first version of a recommended procedure for acquiring 2,000 h of stack-test data under variable load condition for vehicular applications, and to compare cell and stack performance with U.S. DOE targets [88] for PEM fuel cells feed with hydrogen. In the past few years has greatly increased the emphasis on developing accelerated tests, preferably standardized ones. Such accelerated tests are commonly achieved by performing electrode reactions at high overpotential, elevated temperature, exceeding low or high humidification level, and under voltage-cycling conditions [85].

In most reported fuel cell life tests, single cells or stacks were operated under steady-state conditions, because these conditions were most easily applied but, for many applications, unsteady conditions are more appropriate.

In situ cell resistance measurement, ac impedance, gas permeability, postmortem analysis using optical microscopy, SEM, TEM, NMR, IR, X-ray, neutron techniques, and chemical structural analysis have been used to investigate fuel cell failure mechanisms.

The life test under normal operating conditions provides the most relevant PEM lifetime data, although this method is time-consuming and presents some difficulties in data analysis.

To date, four different accelerated parameters or a combination of these parameters have been employed in accelerated life testing: (1) elevated temperature, (2) reduced humidity, (3) open circuit voltage (OCV), and (4) cycling relative humidity (RH), temperature, potential, freeze/thaw, or start/stop conditions.

Nafion membrane degradation is often monitored by changes in gas crossover rate or fluoride-ion emission rate (FER) during the in situ test. Often, the low humidity conditions were combined with OCV testing in order to accelerate hydrogen peroxide formation in the cathode [89, 90].

Fenton's test, which consists in the immersion of the membrane in a H_2O_2 solution (30 w/w%) containing 20 ppm of Fe^{2+}, monitoring the rate of fluoride ions release into the solution, has become a common ex situ accelerated test for membrane durability [91, 92]. Fenton test usually yields lifetimes much greater than the experimentally observed. This rather poor correlation has been explained by considering that excessive peroxide content is used during an accelerated Fenton's test as compared to normal fuel cell operating conditions.

PEM degradation in fuel cells is a result of a complex combination of different processes which were strongly influenced by membrane composition, fabrication (casting or extrusion), and operating conditions. In addition, direct lifetime comparison among different membrane families is almost impossible when different fuel cell operating conditions are applied. Thus, Fenton's test based on peroxide/radical degradation cannot be considered as unique for predicting membrane lifetime.

Certain performance losses of fuel cells during steady-state operation can be fully or partially recovered by stopping and then restarting the life test. These recoverable losses are associated to reversible phenomena, such as cathode catalyst surface oxidation, cell dehydration or incomplete water removal from the catalyst or diffusion layers [85]. Other changes are irreversible and lead to unrecoverable performance losses, such as the decrease in the ECSA of catalysts, cathode contamination with ruthenium, membrane degradation, and delamination of the catalyst layers.

9.8.2 Operational Effects on Fuel Cell Durability

The activity loss of a operating fuel cell is due to different reasons, including the called operational effects that include exposure to impurities, exposure to and start-up from subfreezing conditions, and other operating conditions as potential cycling, fuel starvation, start/stop cycling, and changes in temperature and/or relative humidity.

Impurities that adsorb onto the electrocatalysts surface inhibit the charge-transfer processes, resulting in performance loss. Common fuel impurities include carbon monoxide, ammonia, hydrogen sulfide, hydrogen cyanide, hydrocarbons, formaldehyde, and formic acid [93]. On the cathode side, ambient air may contain impurities such as sulfur dioxide, nitrogen oxides, and particulate matter (including salts) that can affect fuel cell performance [93].

One of the species that affects the activity of the anodic catalysts in fuel cells is CO, which is strongly adsorbed on the surface of platinum blocking and preventing the adsorption of new molecules of alcohol. CO can be introduced into the cell as an impurity in the fuel, or, more generally, generated in situ because of its incomplete oxidation. Several works focused on studying the rate of poisoning of the surface of catalysts (especially platinum). Poisoning rates were found to be clearly dependent on the presence of alloying metals in the platinum matrix. Thus, typical values for the rate of poisoning in nanoparticulated Pt/Ru catalysts used in DMFC are between 0.0044 s^{-1} for $Pt_{0.5}Ru_{0.5}$ and 0.0228 s^{-1} for $Pt_{0.8}Ru_{0.2}$ [94].

Another species generated in situ during operation of the fuel cell is H_2O_2 produced by incomplete reduction of O_2 at the cathode. The presence of H_2O_2, which is a strong oxidant, generates rapid degradation of the various components of the fuel cell. According to several authors, the of peroxide generated by incomplete O_2 reduction on nano-particulated platinum catalysts reaches values between 2 % and 3.5 % of the total reduced O_2 [95, 96], whereas for catalysts having complex structures, such as platinum meso-structured catalysts, with higher O_2 reduction efficiencies, generates between 0.55 % and 1.4 % of peroxide [97].

Another operational effect corresponds to the subfreezing effects. Recent studies suggest that a significant degradation in the performance of H_2/O_2 PEM fuel cells could occur due to thermal subfreezing cycling. Cho et al. [98] observed a 11 % drop in the current at 0.6 V after four thermal cycles of the cell from -10 °C to 80 °C. This behavior is assigned to the increases in porosity and a 25 % loss in the ECSA due to the expansion during water freezing, which in turns increase the contact resistance between the membrane and the catalyst layer.

Nafion conductivity is highly dependent upon the state of water in the polymer, and it has been shown that the activation energy is higher at low temperatures, indicating a possible change in the proton conduction mechanisms [99]. Water inside the Nafion membrane has also been characterized as nonfreezing water, bound freezing water, and free water [100]. The free water behaves like bulk water and freezes at 0 °C, while the bound freezing water is trapped in the channels of Nafion and has its freezing point suppressed depending on the size of the channel, which in turn depends on the degree of hydration of the polymer. Finally, the chemically bound water does not freeze all the way down to -120 °C and is thought to be the source of the relatively high conductivity of Nafion even in the frozen state.

Corti et al. [101] studied the freezing of water and water–methanol mixtures in Nafion 117 and they found that the amount of freezable water, determined from the area of the ice fusion peak of the DSC scan, increased from 1 % to 23 % as the relative humidity increases from 84 % to 100 %. However, the presence of methanol in the membrane increased the amount of freezable water up to 61 % for methanol 25 w/w% aqueous solutions, reaching an apparent limit of 65 % for more concentrated methanol solutions. This result would indicate that the degradation of the MEA in DMFC under freezing cycles could be worse than that observed for hydrogen feed PEM fuel cells.

Other membranes, such as sulfonated polyimide [102] or cross-linked sulfonated PVA [103] exhibit a major amount, although no quantified, of non-freezable water as compared to Nafion. An interesting behavior was observed in the case of PBI and ABPBI membranes doped with phosphoric acid [104], where the amount of frozen water is 2.2 % for PBI and between 1.1 % and 9.1 % at 100 % relative humidity, much lower than that observed for Nafion under similar conditions. When equilibrated with aqueous methanol, PBI membranes exhibit a maximum of 10 wt% of frozen water, while ABPBI membranes, particularly those prepared by high temperature casting, presented very low percentages of frozen water in all methanol concentrations studied, with a maximum value of 0.6 wt% in methanol 25 w/w%. This result is relevant for the application of ABPBI in DMFC because, the fuel cell start up at low temperatures would not be affected.

Krewer et al. [105] investigate changes in the performance of MEAs of DMFC that are caused by undergoing storage at −10 °C and 60 °C under different experimental conditions. They concluded that the effect of storage at −10 °C on an MEA's performance strongly depended upon the MEA properties. Several MEAs were irreversibly damaged; morphological changes were not visible, but electrochemical characterization indicated a decrease in the cathode performance for some of these MEAs, which was correlated to low cathode loadings. Other MEAs became deactivated after storage at low temperature, but their performance could be restored application of a reverse current led to the instant regaining of the original performance. It was found that storage at high methanol concentrations, without the application of a reverse current, prevented freezing, but did not prevent performance decreases indicating that freezing is not the major cause of the decrease in the performance of the studied MEAs.

Two different procedures: the catalyst-coated membrane (CCM) or the catalyst-coated substrate (CCS) process were used by Krewer et al. [105] to prepare several MEAs using different kinds of hydrocarbon membranes and one MEA using a Nafion® membrane. MEAs have cathode catalyst with different Pt black loads, anode catalyst of supported or unsupported Pt-Ru with a loading higher than 5 mg cm^{-2}, and different kinds of diffusion layers having variable thickness, PTFE content, and type of microporous layer. The authors concluded that MEAs with the best tolerance against storage at −10 °C where those prepared with hydrocarbon membranes and with supported and unsupported anode catalyst with high PtRu loadings, and high Pt cathode loadings, whereas MEAs prepared with Nafion® membranes or a low cathode catalyst loading exhibited a significant performance reduction.

Yang et al. [106] studied performance loss in DMFC induced by freeze/thaw cycling. The MEA was prepared with Nafion 115 and commercial Pt/Ru and Pt catalysts supported on carbon were used as anode and cathode, respectively, along with carbon paper with a microporous layer as GDL. Repetitive cycles between −10 °C and 60 °C were applied with intervals of 8 h. After 100 cycles it was found a recoverable degradation of the fuel performance, attributed to the strong adsorption of MOR intermediates at subzero temperatures, which can be recovered by applying a reverse current. Only slight unrecoverable loss, related to structural changes of the MEA, were observed for DMFC operated with oxygen at low current density, while noticeable irreversible degradation was found when the cell was operated at high current density or fed with air.

MEAs can experience different gas concentration conditions between the fuel (and oxygen) inlet and outlet that can lead to different current distributions. Cells arranged in a stack configuration can experience different flows of fuel, air, and coolant resulting from imperfect manifolding and fuel starving. Several authors have noted that, in the case of gross fuel starvation, cell voltages can become negative, as the anode is elevated to positive potentials and the carbon is consumed instead of the absent fuel resulting in permanent damage to the anode catalyst layer [107, 108].

Finally, under prolonged shutdown conditions with the cell continually provided with fuel, both, the alcohol and oxygen permeate the membrane. These effects generate a transient condition in which fuel exists in both electrodes resulting in a reversal current, cathode poisoning and anode catalyst oxidation [100].

9.8.3 Membrane Degradation

The fuel cell membranes are exposed to different types of wear that reduce their lifetime, which can be separated into two areas, chemical degradation, and physical degradation pathway.

The chemical degradation pathway corresponds primarily to effects of H_2O_2 generated as a byproduct during the reduction of O_2. Direct gas mass spectroscopy of the cathode outlet gas indicated the formation of HF, H_2O_2, CO_2, SO, SO_2, H_2SO_2, and H_2SO_3 under OCV durability tests [109]. The susceptibility to peroxide radical attack has been attributed to a trace amount of polymer end groups with residual H-containing terminal bonds [92]. LaConti et al. also postulated that the formed H_2O_2 could react with minor impurities such as Fe^{2+} and Cu^{2+}, forming hydroxyl (OH) and hydroperoxy (OOH) radicals that could attack the membrane [110, 111].

In regard to physical degradation pathway two processes must be considered, the membrane creep, and the microcrack fractures. During the normal operation of a fuel cell, the MEA is put under compressive force between the bipolar plates. Under this constant compressive stress, polymer electrolyte membranes undergo time-dependent deformation. Borup et al. [112] observed that the membrane thickness compressed at 80 °C between two flow field plates was approximately 46 um on the flow field groove, which was about 16 % thinner than that on the flow field land, probably due to elongation of the membrane in the open flow field area. Furthermore, this local stress on the membrane can be a determining factor in the membrane physical degradation, which contributes to the membrane chemical degradation.

The fact that the ionomer swells with water uptake suggests that increases in water uptake as the membrane is exposed to high RH conditions can lead to compressive stresses in the membrane that turns into a tensile residual stresses during drying [113]. These stresses are suggested as a significant contributor to mechanical failures of the membrane. Microcrack fracture resulting from these stresses has been observed under accelerated RH cycling tests as well as under normal operating conditions.

Most of the commonly used membranes used in DMFC exhibit acceptable performance stability after thousands of hours under steady-state operation, including perfluorinated sulfonic acid, PTFE-co-HFP radiation-grafted poly- styrenes, and sPAE [85]. However, membrane durability results under unsteady-state operation are still scarce. Siroma et al. [114] have shown that a significant fraction of Nafion dissolved after 1 week of DMFC operation with methanol solution.

Shin et al. [115] observed that the ICE of a Nafion 115 membrane was reduced by 17 % after operating in a DMFC at 50 °C during 1,000 h, leading to a reduction of 29 % in power density.

9.8.4 Electrocatalyst Stability

DMFC performance loss due to catalyst degradation has been attributed to several factors: a decrement of the electrochemically active surface area (ECSA) of the platinum electrocatalyst supported on a high-surface-area carbon, a loss of cathode activity towards the ORR by surface oxide formation, and ruthenium crossover [83, 85, 116, 117].

There is growing evidence that platinum dissolution plays a major role in the ECSA loss, especially of the cathode catalyst, where high potentials are encountered [118]. Oxidized platinum can then either deposit on existing platinum particles to form larger particles [117] or diffuse into electrochemically inaccessible portions of the MEA (i.e., sites not fulfilling the requirements of gas, electron, and proton access in the three phases region).

Platinum is slowly oxidized at voltages relevant to fuel cell operation (i.e., 0.85 and 0.95 V vs. RHE) [119]. Surface oxygen can be formed with water as the only source of oxygen species; however, the presence of gas-phase oxygen accelerates the oxidation, with the charge equivalent of one monolayer of oxide forming after only 30 min at 0.85 V [119].

A typical recoverable performance loss in DMFC is shown in Fig. 9.13 (top) [120]. The degradation of voltage in the six-cell DMFC stack occurs within the first 30 h of the life test due to the inhibition of the ORR on the cathode due to Pt surface oxidation. The potential is recovered after 100 h following a break of the test and reduction of the cathode catalyst by applying and adequate potential. Eickes et al. [120] concluded that the process of oxide formation is complete after approximately 2 h and it impacts the ORR by change in the electronic properties of the catalyst rather than blocking the active sites. The potential of the cathode can be reduced in an operating DMFC, without an external power supply, by air break, that is, by interruption of the air delivery to the cathode. The methanol crossover, which is quite rapid, reduces the cathode potential to 0.4 V, needed to reduce the surface platinum oxide. The effect of air break on an operating DMFC is shown in Fig. 9.13 (bottom).

Three sintering mechanisms have been proposed to explain ECSA loss of fuel cell catalysts: catalyst dissolution/reprecipitation, migration of Pt particles, and carbon corrosion [85].

Platinum nanoparticles were found to have a higher solubility than bulk platinum [121]. The possible mechanisms for nanoparticle growth include local coalescence of agglomerated particles, agglomeration of nonadjacent crystallites via Pt particle migration and subsequent ripening, and dissolution of the catalyst and subsequent reprecipitation of platinum. The particle growth rates and mechanisms may change

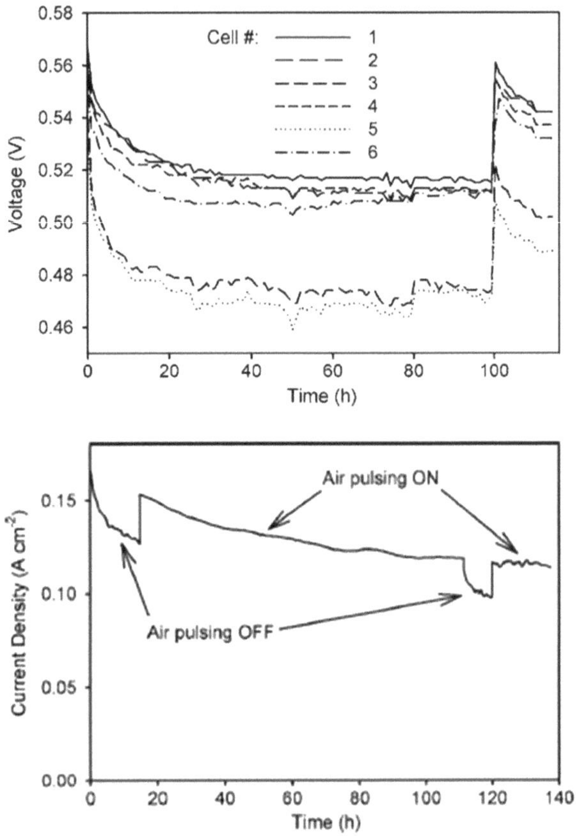

Fig. 9.13 *Top*: voltage degradation in a DMFC during a constant current (80 mA.cm^{-2}) test. *Bottom*: effect of air break (4 s every 100 s) on the long-term performance of air-pulsed DMFC (Reproduced from Ref. [120] with permission)

as a function of electrode potential, cell voltage cycling conditions, current density, state of hydration of the membrane, and operating temperature.

In the case of DMFC the effect of ECSA loss may be limited by the relatively high catalyst loading and the redeposition of dissolved Pt or Ru on the surface of the catalyst nanoparticles [85].

Xin and coworkers have carried out DMFC [122] and DEFC [123] long life-tests. After the DMFC test at 75 °C during 500 h the authors found that the catalyst particle size increases from 2.87 to 3.05 nm in the anode (Pt/Ru), and from 3.71 to 4.31 nm in the cathode (Pt), resulting in a loss of 5.9 % and 13.9 % in the anodic and cathodic specific surface area, respectively. These values are lower than the loss of ECSA, that is 11.4 in the anode and >16.4 in the cathode, meaning that nanoparticle growth is not the only factor affecting ECSA. The authors attributed the excess of degradation to loss of catalytic active components from catalyst layers due to the migration of noble metals, the poisoning of electrocatalysts by intermediates and impurities, and the destruction of proton-conductive passages by the degradation of the polymer electrolyte [122].

A 60 h life-time test on a DEFC with Pt/Sn anodic catalyst showed that the particle size of the anode catalyst increased from 2.3 to 3.3 nm and the cathode from 3.0 to 4.6 nm [123], leading top a loss of 15 % of the maximum power density. The conclusions regarding the causes of the degradation are similar to that to that proposed for DMFC.

Ruthenium crossover is a DMFC specific mechanism of degradation, due to the use of Ru/Pt alloys as catalyst in the anode, which is considered one of the most active catalyst for the MOR. However, the limited stability of the ruthenium in the alloy (or in the nonalloyed fraction of the catalyst) lead to crossover of ruthenium from the anode to the cathode. The presence of Ru in the cathode of DMFC was determined by different authors resorting to CO stripping scans of the cathode [124], X-ray fluorescence [124], EDXS [122, 125] and XRAS [126].

Although the performance of the anode is not modified after 3,000 h of cell operation [127], the leaching of ruthenium from the anode and its accumulation in the cathode leads to deterioration of the DMFC performance with a loss of efficiency for the ORR and a faster oxidation of the catalyst.

The mechanism of ruthenium crossover has two routes. One is currentless and proceed by pure diffusion during the MEA fabrication process [124]. The second is current-assisted and is enhanced by higher anode operating potential. In order to mitigate the ruthenium crossover in DMFC, several procedures have been proposed, including high-temperature cure of the anode catalyst [124], preleach of the "loose" ruthenium phase using acid solutions [124], use of low-permeability membranes [129], and electrochemical removal of mobile ruthenium [128]. Although these approaches do not eliminate completely the ruthenium contamination of the cathode, contribute to reduce the currentless contamination [85].

9.8.5 Corrosion of Catalyst Support and GDL

In addition to loss of the platinum, the carbon support that anchors the platinum crystallites and provides electrical connectivity to the gas-diffusion media and bipolar plates is also subject to degradation. In phosphoric acid fuel cell, graphitized carbons are the standard because of the need for corrosion resistance in high-temperature acid environments [129], but PEM fuel cells have not employed fully graphitized carbons in the catalyst layers, due in large part to the belief that the extra cost could be avoided. Electrochemical corrosion of carbon materials as catalyst supports will cause electrical isolation of the catalyst particles as they are separated from the support or lead to aggregation of catalyst particles, both of which result in a decrease in the electrochemical active surface area of the catalyst and an increase in the hydrophilicity of the surface, which can, in turn, result in a decrease in gas permeability as the pores become more likely to be filled with liquid water films that can hinder gas transport.

Many researchers have noted that the way the GDL interacts with water changes during lifetime tests. The changes seem to occur at the microstructural level (micrometer to submicrometer) but can be seen macroscopically. Evidently, the DAFC operating environment gradually changes the GDL from hydrophobic to hydrophilic, which can degrade fuel cell operation. Maintaining the hydrophobic character of the GDL and Mesoporous Layer (MPL) pores, known as a "GDL hydrophobicity gradient", is important to maintaining mass transport in the fuel cell during lifetime tests [130].

Mercury and water porosimetry measurements have shown that the GDL pore structure changes during lifetime tests. Large pore (30–60 um diameter) volume has decreased, while small-pore volume increases. The loss in large-pore volume is likely due to irreversible fuel cell compression [118].

Delamination of the MPL from the GDL substrate has not been widely reported but may occur during freeze-thaw cycles, as occurs with catalyst-layer delamination from the membrane [131, 132]. A different situation occurs in the GDL/MPL, where the pore diameters are on the order of a micron or larger and the water is not hydrating the sulfonic acid of the ionomer. The volume expansion caused by ice formation can produce large isotropic stresses that can damage the structure of the catalyst layer, the MPL, or the GDL.

There are only a few reported studies that examine mechanical degradation and review the effect of compression of gas-diffusion layers on the performance of fuel cells [133–135]. Lee and Mérida [133] characterized some GDL properties after 300 h. of compression at constant temperature in an ex situ test and found that the dry gas-phase permeabilities remained roughly constant.

References

1. Antonucci V (1999) Direct methanol fuel cells for mobile applications: a strategy for the future. Fuel Cells Bull 1999(7):6–8
2. Lipman T, Sperling D (2003) Market concepts, competing technologies and cost challenges for automotive and stationary applications. In: Vielstich W, Lamm A, Gasteiger H (eds) Handbook of fuel cells: fundamentals, technology and applications, vol 4, part 13. Wiley, Chichester, England, pp 1318–1328
3. Dyer CK (2002) Fuel cells for portable applications. J Power Sources 106:31–34
4. Aricò AS, Srinivasan S, Antonucci V (2001) DMFCs: from fundamental aspects to technology development. Fuel Cells 1:133–161
5. Dillon R, Srinivasan S, Aricò AS, Antonucci V (2004) International activities in DMFC R&D: status of technologies and potential applications. J Power Sources 127:112–126
6. Kleiner K (2006) Assault on batteries. Nature 441:1046–1047
7. Zhao TS, Yang WW, Chen R, Wu QX (2010) Towards operating direct methanol fuel cells with highly concentrated fuel. J Power Sources 195:3451–3462
8. Dyer CK (1999) Replacing the battery in portable electronics. Sci Am 281:88–93
9. Vielstich W (2003) Electrochemical energy conversion – methanol fuel cell as example. J Braz Chem Soc 14:503–509
10. Thomas YRJ, Bruno MM, Corti HR (2012) Characterization of a monolithic mesoporous carbon as diffusion layer for micro fuel cells application. Microp Mesop Mat 155:47–55
11. Lu GQ, Wang CY (2004) Electrochemical and flow characterization of a direct methanol fuel cell. J Power Sources 134:33–40

12. Lim SW, Kim SW, Kim J, Ahn JE, Han HS, Shul YG (2006) Effect of operation parameters on performance of micro direct methanol fuel cell fabricated on printed circuit board. J Power Sources 161:27–33
13. Narayanan SR, Valdez TI (2003) Portable direct methanol fuel cell systems. In: Vielstich W, Lamm A, Gasteiger H (eds) Handbook of fuel cells: fundamentals, technology and applications, vol 4. Wiley, Chichester, England, Part 1, pp 1133–1141
14. Wang Y, Chen KS, Mishler J, Cho SC, Adroher XC (2011) A review of polymer electrolyte membrane fuel cells: technology, applications, and needs on fundamental research. Appl Energy 88:981–1007
15. Chang H, Kim H, Choi YS, Lee W (2009) Critical issues in the commercialization of DMFC and role of membranes. In: Zaidi J, Matsuura T (eds) Polymer membranes for fuel cells. Springer, New York, pp 317–339, Chapter 13
16. Leach GA (2004) Dangerous goods panel (DGP), International Civil Aviation Organization, meeting of the working group of the whole, Abu Dhabi, 4–8 Oct 2004, Fuel Cell Cartridges
17. Leach GA (2007) Dangerous goods panel (DGP), international civil aviation organization, twenty first meeting, Montreal, 5–16 Nov 2007, Carriage of fuel cells by passengers
18. Ho DL, Kopasz JP, Benjamin TG, Podolski WF (2011) The U. S. department of energy efforts in fuel cells for portable power applications. ECS Trans 30:337–343
19. Kamaruddin MZF, Kamarudin SK, Daud WRW, Masdar MS (2013) An overview of fuel management in direct methanol fuel cells. Renew Sustain Energy Rev 24:557–565
20. Xie C, Bostaph J, Pavio J (2004) Development of a 2W direct methanol fuel cell power source. J Power Sources 136:55–65
21. http://www.mtimicrofuelcells.com/technology/differentiation.asp. Accessed 15 May 2013
22. Zhao TS, Chen R, Yang WW, Xu C (2009) Small direct methanol fuel cells with passive supply of reactants. J Power Sources 191:185–202
23. Achmad F, Karamudin SK, Daud WRW, Majlan EH (2011) Passive direct methanol fuel cells for portable electronic devices. Appl Energy 88:1681–1689
24. Kim D, Cho EA, Hong SA, Oh IH, Ha HY (2004) Recent progress in passive direct methanol fuel cells at KIST. J Power Sources 131:172–177
25. Hahn R, Wagner S, Schmitz A, Reichl H (2004) Development of a planar micro fuel cell with thin film and micro patterning technologies. J Power Sources 131:73–78
26. Broussely M, Archdale G (2004) Li-ion batteries and portable power source prospects for the next 5–10 years. J Power Sources 136:386–394
27. Goto S (2008) Micro fuel cell system for mobile consumer electronic devices. In: Sony corp. Small fuel cells for commercial and military applications, 9 ed. Knowledge Press, ISBN-10: 1594301360
28. Zhu GR, Loo KH, Lai YM, Tse CK (2012) Quasi-maximum efficiency point tracking for direct methanol fuel cell in DMFC/supercapacitor hybrid energy system. IEEE Trans Energy Conv 27:561–571
29. http://www.eetimes.com/electronics-news/4080861/Analysis-Did-Intel-underestimate-netbook-success. Accessed 5 Apr 2012
30. http://www.edn.com/article/458578_Notebook_PC_shipments_exceeded_desktops_for_first_time_in_Q3_iSuppli_reports.php. Accessed 5 Apr 2012
31. http://www.smartplanet.com/blog/thinking-tech/occupy-smartphone-top-10-of-mobile-users-occupy-90-of-bandwidth/9751. Accessed 5 Apr 2012
32. http://www.eweek.com/c/a/Windows/Netbooks-Are-Destroying-the-Laptop-Market-and-Microsoft-Needs-to-Act-Now-863307/. Accessed 5 Apr 2012
33. http://support.apple.com/kb/TA30635?viewlocale=en_US. Accessed 5 Apr 2012
34. http://www.gartner.com/it/page.jsp?id=1759714 Gartner. 2010-07-09. Accessed 5 Apr 2012
35. http://www.scra.org/docs/InsideTheIndustry_May_18_08.pdf. Accessed 5 Apr 2012
36. http://www.addictware.com.mx/index.php/comunicaciones/1802-creceran-ventas-de-tablets-ipad-rey. Accessed 5 Apr 2012
37. Yoon SK, Na YS, Joung Y, Park J, Kim Y, Hu L, Song I, Cho H (2009) Direct methanol fuel cell systems for portable applications, fuel cell seminar & exposition 18 Nov 2009, Palm Springs

38. http://www.wipo.int/patentscope/en/. Accessed 20 Feb 2012
39. http://www.dexigner.com/news/1932. Accessed 5 May 2012
40. MTI Micro pushes leadership with its Mobion chip. Fuel Cells Bull 2007(7):1
41. DoCoMo prototype micro DMFC recharger for cell phone handsets. Fuel Cells Bull 2004(11):8
42. Fujitsu, DoCoMo triple cell phone charger capacity. Fuel Cells Bull 2005(8):1
43. Takei F, Cooray NF, Yoshida K, Yoshida H, Ebisu K, Suzuki S, Sawatari N (2005) Development of prototype micro fuel cells for mobile electronics. Fujitsu Sci Tech J 41:191–200
44. García G, Florez-Montaño J, Hernandez-Creus A, Pastor E, Planes GA (2011) Methanol electrooxidation at mesoporous Pt and Pt-Ru electrodes: a comparative study with carbon supported materials. J Power Sources 196:2979–2986
45. Cantane DA, Ambrosio WF, Chatenet M, Lima FHB (2012) Electro-oxidation of ethanol on Pt/C, Rh/C, and Pt/Rh/C-based electrocatalysts investigated by on-line DEMS. J Electroanal Chem 681:56–65
46. Wang H, Abruña HD (2011) Electrocatalysis of direct alcohol fuel cells: quantitative DEMS studies. In: Bocarsly A, Mingos DMP (eds) Fuel cells and hydrogen storage, vol 141 of structure and bonding. Springer, Berlin Heidelberg, pp 33–83
47. Panasonic unveils high-power, durable DMFC. Fuel Cells Bull 2010(1):6–7
48. Samsung shows smallest fuel cell yet for mobile charger, laptop station. Fuel Cells Bull 2007 (2):3
49. Toshiba launches Dynario power source for mobile devices, but only in Japan. Fuel Cells Bull 2009(12):6
50. Dicks AL (2012) PEM fuel cells: applications. Compr Renew Energy 4:203–245
51. Samsung SDI (2009) The introduction of fuel cell. Samsung SDI, Seoul. http://www.samsungsdi.com/generation/fuel-cell-battery.jsp. Accessed 15 Sept 2013
52. http://www.batterystore.com/Saft/SaftPDF/BA5590.pdf. Accessed 5 Apr 2012
53. http://www.maifl.com/pdfs/BB2590.pdf. Accessed 5 Apr 2012
54. Kwon JM, Kim YJ, Cho HJ (2011) High-efficiency active DMFC system for portable applications. IEEE Trans Power Electr 26:2201–2209
55. McConnell VP (2009) Fuel cells feed power-hungry portable electronics. Fuel Cells Bull 2009(6):12–16
56. MTI Micro debuts embedded DMFC GPS prototype. Fuel Cells Bull 2008(5):1
57. Toshiba, Hitachi DMFCs feature in prototype audio players, cell phones. Fuel Cells Bull 2005(11):8
58. Agnolucci P (2007) Economics and market prospects of portable fuel cells. Int J Hydrogen Energy 32:4319–4328
59. Kukkonen C (2008) Creating a global fuel cartridge manufacturing and distribution infrastructure. In: Small fuel cells for commercial and military applications, 9th edn. Knowledge Press, Brookline, MA, Chapter 11, pp 195–232
60. Song I, Cho H, Choi KH, Chang H (2008) Mobile DMFC: enhancement of stack and system stability. In: Small fuel cells for commercial and military applications, 9th edn. Knowledge Press, Brookline, Chapter 10, pp 179–194
61. NEC unveils fully integrated fuel cell notebook PC. Fuel Cells Bull 2003(8):1
62. NEC improves notebook fuel cell. Fuel Cells Bull 2003(11):4
63. Latest DMFC prototypes from Toshiba, Hitachi. Fuel Cells Bull 2003(12):2
64. Smith T (2004) NEC to show laptop with built-in fuel cell. http://www.theregister.co.uk/2004/10/19/ nec_notebook_fuel-cell. Accessed 15 Sept 2013
65. Narayanan SR, Valdez TI (2008) High-energy portable fuel cells power sources. ECS Interface 17:40–45
66. http://www.efoy-pro.com/sites/default/files/sfc_broschuere_en_online_v2_0.pdf. Accessed 15 Sept 2013
67. Hitachi (2010) Hitachi to take part in direct methanol joint evaluation in the UK with CPI and Hitachi-High technologies. Hitachi, Tokyo. http://www.hitachi.com/New/cnews/090224.pdf. Accessed 15 Sept 2013

68. http://www.ird.dk/solutions/DMCF.aspx. Accessed 20 May 2013
69. Darnell Group Inc. Fuel cells for portable power: markets, manufacture and cost. Revised final report (4) for breakthrough technologies & U.S. Fuel Cell Council Submitted 13 Jan 2003, Corona, California; 2003. Report Number: R4
70. Jollie D (2004) Fuel cell market survey: portable applications. Fuel Cell Today, London
71. http://www.yankodesign.com/2009/06/29/take-a-cell-phone-swig/. Accessed 5 Apr 2012
72. http://www.ecofriend.com/entry/eco-gadgets-bacvac-fuel-cell-powered-vacuum-cleaner-for-cable-free-cleaning/. Accessed 5 Apr 2012
73. Johnson M (1998) Annual report & accounts 1998. http://www.matthey.com/AR98/JMAR98.pdf
74. Johnson M (2012) Annual report & accounts 2012. http://www.worldreginfo.com/wdoc.aspx?file=Johnson_Matthey/6/EE245FA6-D12A-4173-828C-E16949960766/231810_rfa_2012_en_gb0004764071.pdf
75. http://www.fuelcellstore.com/en/pc/viewCategories.asp?idCategory=79. Accessed 27 Oct 2012
76. http://www.ion-power.com/products.html#membrane. Accessed 27 Oct 2012
77. Tsuchiya H, Kobayashi O (2004) Mass production cost of PEM fuel cell by learning curve. Int J Hydrogen Energy 29:985–990
78. Saizensen NK (2001) Development front of fuel cells. A special issue of Nikkei mechanical. Nikkei Business Publications, Japan
79. Marcinkoski J, James BD, Kalinoski JA, Podolski W, Benjamin T, Kopasz J (2011) Manufacturing process assumptions used in fuel cell system cost analyses. J Power Sources 196:5282–5292
80. James BD, Kalinoski JA (2008) Mass production cost estimation for direct H_2 PEM fuel cells systems for automotive applications. DOE Hydrogen program 2008 annual merit review, project ID: FC7, 10 Jun 2008, Arlington. http://www.hydrogen.energy.gov/pdfs/review08/fc_7_james.pdf. Accessed 15 Sept 2013
81. Wee JH (2007) A feasibility study on direct methanol fuel cells for laptop computers based on a cost comparison with lithium-ion batteries. J Power Sources 173:424–436
82. Rashidi R, Dincer I, Naterer GF, Berg P (2009) Performance evaluation of direct methanol fuel cells for portable applications. J Power Sources 187:509–516
83. Chalk SG, Miller JF (2006) Key challenges and recent progress in batteries, fuel cells, and hydrogen storage for clean energy systems. J Power Sources 159:73–80
84. Fowler M, Mann RF, Amphlett JC, Peppley BA, Roberge PR (2002) Incorporation of voltage degradation into a generalized steady state electrochemical model for a PEM fuel cell. J Power Sources 106:274–283
85. Kim YS, Zelenay P (2009) Direct methanol fuel cell durability. In: Büchi FN, Inaba M, Schmidt TJ (eds) Polymer electrolytes fuel cell durability. Springer, New York, pp 223–240
86. Borup RL, Inbody MA, Wood DL, Pacheco SD, Guidry DR, Xie J, Tafoya JI, Blom D (2003) Fuel cell reformer and stack durability: gasoline reformate & hydrogen – PEM fuel cell durability, Fuel Cell Seminar, Nov 2003, Miami
87. Wood DL, Xie J, Pacheco SD, Davey JR, Borup RL, Garzon FH, Atanassov P (2004) Durability issues of the PEMFC GDL and MEA under steady-state and drive-cycle operating conditions. Fuel Cell Seminar, San Antonio
88. Borup R, Davey J, Wood D, Garzon F, Inbody M (2005) PEM fuel cell durability. DOE hydrogen program, FY2005 progress report, pp 1034–1045
89. Protsailo L (2006) Development of high temperature membrane and improved cathode catalysts for PEM fuel cells. U. S. DOE Hydrogen Program Review, Arlington
90. Yan Q, Wu J (2005) Durability studies of PEM fuel cell using specified flow plate, 207th meeting electrochem. Soc.; Quebec City, abstract # 1514
91. Kinumoto T, Inaba M, Nakayama Y, Ogata K, Umebayashi R, Tasaka A, Iriyama Y, Abe T, Ogumi Z (2006) Durability of perfluorinated ionomer membrane against hydrogen peroxide. J Power Sources 158:1222–1228

92. Curtin DE, Lousenberg RD, Henry TJ, Tangeman PC, Tisack ME (2004) Advanced materials for improved PEMFC performance and life. J Power Sources 131:41–48
93. Garzon F, Brosha E, Pivovar B, Rockward T, Springer T, Uribe F, Urdampilleta I, Valerio J (2006) Freedom car fuel contaminants: effect on PEMFCs, Annual DOE Fuel Cell Program Review
94. Guo JW, Zhao TS, Prabhuram J, Chen R, Wong CW (2005) Preparation and characterization of a PtRu/C nanocatalyst for direct methanol fuel cells. Electrochim Acta 51:754–763
95. Suarez-Alcantara K, Solorza-Feria O (2008) Kinetics and PEMFC performance of Ru_xMo_y Se_z nanoparticles as a cathode catalyst. Electrochim Acta 53:4981–4989
96. Benitez R, Chaparro AM, Daza L (2005) Electrochemical characterisation of Pt/C suspensions for the reduction of oxygen. J Power Sources 151:2–10
97. Franceschini EA, Bruno MM, Viva FA, Williams FJ, Jobbágy M, Corti HR (2012) Mesoporous Pt electrocatalyst for methanol tolerant cathodes of DMFC. Electrochim Acta 71:173–180
98. Cho E, Ko J-J, Ha HY, Hong S-A, Lee K-Y, Lim T-W, Oh I-H (2003) Characteristics of the PEMFC repetitively brought to temperatures below 0°C. J Electrochem Soc 150: A1667–A1670
99. Cappadonia M, Erning JW, Stimming U (1994) Proton conduction of Nafion-117 membrane between 140 K and room temperature. J Electroanal Chem 376:189–193
100. Borup R, Meyers J, Pivovar B, Kim YS, Mukundan R, Garland N, Myers D, Wilson M, Garzon F, Wood D, Zelenay P, More K, Stroh K, Zawodzinski T, Boncella J, McGrath JE, Inaba M, Miyatake K, Hori M, Ota K, Ogumi Z, Miyata S, Nishikata A, Siroma Z, Uchimoto Y, Yasuda K, Kimijima K, Iwashita N (2007) Scientific aspects of polymer electrolyte fuel cell durability and degradation. Chem Rev 107:3904–3951
101. Corti HR, Nores Pondal F, Buera MP (2006) Low temperature thermal properties of Nafion 117 membranes in water and methanol–water mixtures. J Power Sources 161:799–805
102. Mendil-Jakania H, Davies RJ, Dubard E, Guillermo A, Gebel G (2011) Water crystallization inside fuel cell membranes probed by X-ray scattering. J Membr Sci 369:148–154
103. Hwang BJ, Joseph J, Zeng YZ, Lin CW, Cheng MY (2011) Analysis of states of water in poly (vinyl alcohol) based DMFC membranes using FTIR and DSC. J Membr Sci 369:88–95
104. Nores-Pondal FJ, Buera MP, Corti HR (2010) Thermal properties of phosphoric acid-doped polybenzimidazole membranes in water and methanol–water mixtures. J Power Sources 195:6389–6397
105. Krewer U, Park JY, Lee JH, Cho H, Pak C, You DJ, Lee YH (2009) Low and high temperature storage characteristics of membrane electrode assemblies for direct methanol fuel cells. J Power Sources 187:103–111
106. Yang L, Sun H, Wang S, Jiang L, Sun G (2012) Reversible and irreversible loss in performance in direct methanol fuel cells during freeze/thaw cycles. J Power Sources 219:193–198
107. Wilkinson DP (2002) Improving PEM fuel cells robutness and lifetime with respect to reactant starvation. 202nd meeting of electrochemical society, Salt Lake City. Abstract 113
108. Knights SD, Colbow KM, St-Pierre J, Wilkinson DP (2004) Aging mechanisms and lifetime, PEFC and DMFC. J Power Sources 127:127–134
109. Teranishi K, Kawata K, Tsushima S, Hirai S (2006) Degradation mechanism of PEMFC under open circuit operation. Electrochem Solid-State Lett 9:A475–A477
110. LaConti AB, Hamdan M, McDonald RC (2003) Mechanisms of membrane degradation. In: Vielstich W, Gasteiger HA, Lamm A (eds) Handbook of fuel cells – fundamentals, technology and application, vol 3. Wiley, New York, pp 647–662
111. LaConti AB, Fragala AR, Boyack JR (1977) Proceeding of the symposium on electrode materials and process for energy ConVersion and storage. In: McIntyre JDE, Srinivasan S, Will FG (eds) The electrochemical society, inc., Princeton
112. Borup R, Davy J, Wood D, Garzon F, Inbody M, Guidry D (2005) PEM fuel cell durability. DOE hydrogen program review. US DOE hydrogen program 2005 annual merit review and peer evaluation meeting, 23–26 May, Arlinton

113. Kusoglu A, Karlsson AM, Santare MH, Cleghorn S, Johnson WB (2006) Mechanical response of fuel cell membranes subjected to hydro-thermal loading. J Power Sources 161:987–996
114. Siroma Z, Fujiwara N, Ioroi T, Yamazaki S, Yasuda K, Miyasaki Y (2004) Dissolution of Nafion membrane and recast Nafion film in mixtures of methanol and water. J Power Sources 125:41–45
115. Shin SJ, Balabanovich AI, Kim H, Jeong J, Song J, Kim HT (2009) Deterioration of Nafion 115 membrane in direct methanol fuel cells. J Power Sources 191:312–319
116. Patterson T (2002) Fuel cell technology topical conference proceedings. In: Igweand GJ, Mah D (eds) AIChE Spring National Meeting, New York, p 313
117. Xie J, Wood DL, More KL, Atanassov P, Borup RL (2005) Microstructural changes of membrane electrode assemblies during PEFC durability testing at high humidity conditions. J Electrochem Soc 152:A1011–A1020
118. Darling RM, Meyers JP (2003) Kinetic model of platinum dissolution in PEMFCs. J Electrochem Soc 150:A1523–A1527
119. Paik CH, Saloka GS, Graham GW (2007) Influence of cyclic operation on PEM fuel cell catalyst stability. Electrochem Solid-State Lett 10:B39–B42
120. Eickes C, Piela P, Davey J, Zelenay P (2006) Recoverable cathode performance loss in direct methanol fuel cells. J Electrochem Soc 153:A171–A178
121. Azaroul M, Romand B, Freyssinet P, Disnar JR (2001) Solubility of platinum in aqueous solutions at 25 °C and pHs 4–10 under oxidizing conditions. Geochim Cosmochim Acta 65:4453–4466
122. Chen WM, Sun GQ, Guo JS, Zhao XS, Yan SY, Tian J, Tang SH, Zhou ZH, Xin Q (2006) Test on the degradation of direct methanol fuel cell. Electrochim Acta 51:2391–2399
123. Jiang LH, Sun GQ, Wang SL, Wang GX, Xin Q, Zhou ZH, Zhou B (2005) Electrode catalysts behavior during direct ethanol fuel cell life-time test. Electrochem Commun 7:663–668
124. Piela P, Eickes C, Brosha E, Garzon F, Zelenay P (2004) Ruthenium crossover in direct methanol fuel cell with Pt-Ru black anode. J Electrochem Soc 151:A2053–A2059
125. Gancs L, Hult BN, Hakim N, Mukerjee S (2007) The impact of Ru contamination of a Pt/C electrocatalyst on its oxygen-reducing activity. Electrochem Solid-State Lett 10:B150–B154
126. Sarma LS, Chen CH, Wang GR, Hsueh KL, Huang CP, Sheu HS, Liu DG, Lee JF, Hwang BJ (2007) Investigations of direct methanol fuel cells (DMFC) fading mechanisms. J Power Sources 167:358–365
127. Hamon C, Purdy G, Kim YS, Pivovar B, Zelenay P (2006) Novel process for improved long-term stability of DMFC membrane-electrode assemblies. Proc ECS P2004–21:352–356
128. Choi JH, Kim YS, Bashyan R, Zelenay P (2006) Ruthenium crossover in DMFCs operating with different proton conducting membranes. ECS Trans 1:437–445
129. Landsman DA, Luczak FJ (2003) Catalyst studies and coating technologies. In: Vielstich W, Gasteiger HA, Lamm A (eds) Handbook of fuel cells. Fundamentals, technology, and applications, vol 4. Wiley, Chichester, pp 811–831
130. Wood DL, Grot SA, Fly G (2002) Composite gas distribution structure for fuel cell. US Patent 6,350,539
131. Hou J, Yu H, Zhang S, Sun S, Wang H, Yi B, Ming P (2006) Analysis of PEMFC freeze degradation at −20°C after gas purging. J Power Sources 162:513–520
132. Yan QG, Toghiani H, Lee YW, Liang KW, Causey H (2006) Effect of sub-freezing temperatures on a PEM fuel cell performance, startup and fuel cell components. J Power Sources 160:1242–1250
133. Lee C, Merida W (2007) Gas diffusion layer durability under steady-state and freezing conditions. J Power Sources 164:141–153
134. Wilde PM, Mandle M, Murata M, Berg N (2004) Structural and physical properties of GDL and GDL/BPP combinations and their influence on PEMFC performance. Fuel Cells 4:180–184
135. Lee WK, Ho CH, Van Zee JW, Murthy M (1999) The effects of compression and gas diffusion layers on the performance of a PEM fuel cell. J Power Sources 84:45–51

About the Authors

Mariano M. Bruno was born in San Basilio, Argentina, in 1977. He obtained a B.S. in Chemistry (2001) and Ph.D. in Chemistry (2007) at the National University of Río Cuarto. He was a postdoctoral fellow (2008–2010) in the Fuel Cell Group at the National Commission of Atomic Energy (CNEA) and performed several training stays on the synthesis and characterization of structured materials. Currently he is researcher at the National Commission of Atomic Energy, and at the National Council of Scientific and Technological Research (CONICET). He is author of more than 20 scientific journal articles and patents. His research activities involve the design of carbonous materials for electrochemical energy storage and conversion devices, electrocatalysis, and the development of fuel cell membranes.

H.R. Corti and E.R. Gonzalez (eds.), *Direct Alcohol Fuel Cells: Materials, Performance, Durability and Applications*, DOI 10.1007/978-94-007-7708-8,

Horacio R. Corti was born in Buenos Aires in 1949. Ph.D. in Chemistry at the University of Buenos Aires (1979), where he is Full Professor of Physical Chemistry at the Faculty of Sciences, and postdoctoral fellow at the Central Electricity Research Laboratories (Surrey, UK). He is also Principal Researcher of the National Council of Scientific and Technological Research (CONICET) and head of the Fuel Cells Group at the Department of Physics of Condensed Matter, National Commission of Atomic Energy. His research areas include the transport phenomena, thermodynamics, and electro-chemistry of hydrothermal, supercooled, and glassy aqueous systems, and proton exchange membrane fuel cells. He has published more than 100 articles in international journals, authored several text and technical books, and supervised 10 Ph.D. theses.

Esteban A. Franceschini was born in Córdoba, Argentina, in 1985. Ph.D. in Chemistry at the University of Buenos Aires (2012), and postdoctoral fellow at the Nanomaterials group in the National Commission of Atomic Energy. He is also Assistant Researcher of the National Council of Scientific and Technological Research (CONICET). He has published eight articles in international journals and has presented several works in national and international meetings. His research areas include the mechanical, thermodynamics, and electrochemical properties of proton exchange membranes, and alternative catalysts for PEM fuel cells.

Alejandro A. Franco is Full Professor of Chemistry at the Laboratoire de Réactivité et de Chimie des Solides (CNRS and UPJV, Amiens). He headed the Modeling Group of Electrochemical Systems at CEA (Grenoble) in the period 2006–2013. His research activities concern the understanding of physical electrochemical processes through the use of multiscale modeling approaches and numerical simulation, applied to electrochemical power generators such as Li-ion and Li-air batteries, supercaps, PEM fuel cells and water electrolyzers. He is the inventor of the MEMEPhys computational software and of the MS LIBER-T simulation package scaling up ab initio and microstructural data at the electrochemical device level. He is author of more than 30 patents and 4 book chapters, about fuel cells and electrochemical devices, and his work has been published in several electrochemistry journals and conferences (including more than 40 invited lectures). He was invited Guest Editor of *Electrochimica Acta.*

Janaina Fernandes Gomes was born in São Carlos (Brazil) in 1978. Ph.D. in Science at the University of São Paulo (Brazil) and at the University Paris XI (France) in 2007. She was postdoctoral fellow at the Fritz Haber Institute of the Max Planck Society (Germany) in 2008 and at the Institute of Chemistry of São Carlos of the University of São Paulo from 2009 to 2013. Her research areas include catalysis and electrocatalysis. Her research activities are mainly concentrated in the electro-oxidation of small organic molecules. She has published around 20 scientific articles in international journals.

Ernesto R. Gonzalez got his Ph.D. in Physical Chemistry in 1965 from University of Buenos Aires, Argentina. He was employed in various universities but his principal occupation was Full Professor of the University of São Paulo, Brazil, from 1981 to 2008, where he is Senior Professor since 2008. He has published more than 220 articles in international scientific journals. He is author of nine book chapters and two books. He had supervised 29 Ph.D. theses and 37 M.Sc. theses. He has received 25 prizes and nominations and has delivered 135 conferences internationally. The main research areas are fuel cells, electrocatalysis, oxygen reduction, and electrochemistry in general. During the last 40 years he has coordinated more than 50 research projects.

Fabio H. B. Lima was born in Brazil in 1978. He graduated in Chemistry at University of Sao Paulo, São Carlos, in 2001. He obtained his Ph.D. in Physical Chemistry in the same institution, in 2006, with a stage at the Brookhaven National Laboratory. He spent 12 months as a postdoctoral fellow at the Chemistry Institute of São Carlos (IQSC) between 2006 and 2008. He is currently interested in synthesis and electrocatalysis for reactions involved in electrochemical energy conversion devices such as regenerative fuel cells (H_2/O_2 and $HCOOH/CO_2$), rechargeable metal-air batteries, direct hydrazine, borohydride and ethanol fuel cells. Publication of scientific research includes 36 articles in journals, 3 chapters in books and more than 50 scientific summaries in collective books and proceedings.

Andressa Mota-Lima was born in Natal, Brazil, in 1981. She completed undergraduate studies in Materials Engineering (2004) at the Federal University of Rio Grande do Norte (UFRN), Master degree from the Federal University of São Carlos (2007), and Ph.D. in Physical-Chemistry from University of São Paulo (2012) at the Chemistry Institute of São Carlos (IQSC). She worked 18 months in Dr. Eiswirth's group at the Fritz-Haber Institute, Berlin. Currently, she is research fellow in the team headed by Prof. Dr. Ernesto Gonzalez. Her main research interest is self-organized dynamic in electrochemistry.

Patricia Maria Patrizi Pratta was born in São Carlos (Brazil) in 1972. Ph.D. in Science at the Chemical Department of the Federal University of São Carlos (Brazil) in 2007. She was postdoctoral fellow at the Institute of Chemistry of São Carlos of the University of São Paulo (Brazil) from 2010 to 2011 and at the Chemical Engineering Department of the Federal University of São Carlos (Brazil) from 2012 to 2014. Her research areas include catalysis, electrochemistry and electroanalytical. Her recent research activities are concentrated in the bio-fuel and catalysts synthesis.

F. Javier Rodríguez Varela was born in 1969 in Mexico. PhD in Metallurgical Engineering (2004) and Postdoctoral Fellow (2004–2005) at École Polytechnique de Montréal. He joined the Centre for Research and Advanced Studies (CINVESTAV-IPN), in 2006 as Associate Researcher. Since 2009 he is a Full Professor at CINVESTAV-IPN. He has been the leader of national and international projects supported by CONACYT-Mexico. His research has been published in relevant journals such as *Electrochemistry Communications*, *International Journal of Hydrogen Energy*, and the *Journal of the Electrochemical Society*. He is co-editor of book on fuel cells and he has published 9 chapters in technical monographs. His research interests include electrocatalysis, novel energetic materials, fuel cells and renewable energies. He is the current President of the Mexican Hydrogen Society for the period 2012–2014.

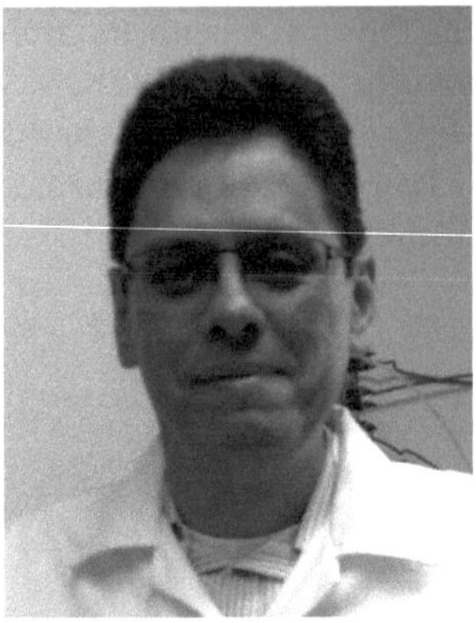

Born in 1954, **O. Solorza-Feria** is Professor in Energetic Electrochemistry at the Chemistry department of the Center of Research and Advances Studies, CINVESTAV-IPN (Mexico). He obtained a B.Sc. in Chemical Engineering (1976) at the National Polytechnic Institute, a Ph.D. in Physical Chemistry (1984) at CINVESTAV-IPN, a post-doctoral fellow at Laboratories of Physique des Liquids et Electrochemie in France (1985), and sabbatical at Solar Energy Materials at Hahn-Meitner-Institut in Germany (1991). His research activities include electrocatalysis of novel materials for PEM and microbial fuel cells. He has published around 115 scientific journal articles and supervised 31 bachelor, 7 masters in science and 11 Ph.D. theses. He is founder of the Mexican Electrochemical Society and the Mexican Hydrogen Society.

Edson A. Ticianelli was born in Brazil in 1952. He is an expert in low temperature fuel cells, particularly the proton exchange membrane and the alkaline systems. Research in these areas includes the development of dispersed catalysts and gas diffusion electrodes for the oxygen reduction and the hydrogen oxidation reactions. He obtained the

B.Sc. degree in 1976, the M.Sc. in 1980 and the Ph.D. in 1985 in Physical Chemistry, all of them at the University of São Paulo. He developed his full academic carrier at the Institute of Chemistry of São Carlos (IQSC), where he is Full Professor since 2003. His publication of scientific research includes 165 articles in journals. He has also published 10 chapters in books and more than 300 summaries/papers in collective books, proceedings and other publications. He has acted as advisor of 19 M.Sc. and 21 Ph.D. theses at IQSC. He was the Director of IQSC (USP) from 2006 to 2010.

Germano Tremiliosi-Filho Ph.D. in Physical Chemistry in 1986 from University of Sao Paulo, at Sao Carlos, Brazil. He is employed at the University of Sao Paulo since 1981 where actually he is Full Professor of Physical Chemistry (2009) at the Institute of Chemistry of Sao Carlos. He made a post-doctorate at the University of Ottawa, Canada (1988–1990) and was a Visiting Professor at the University of Illinois at Urbana-Champaign, USA (1998–1999). He has published more than 130 articles in international scientific journals. He is author of four book chapters and one book. He had supervised 18 Ph.D. thesis and 21 M.Sc. thesis. He has delivered 87 international conferences. The main research areas are electro-catalysis, fuel cells, and alcohol electro-oxidation. He is the Associate Editor of the journal *Electrocatalysis*, published by Springer.

Federico A. Viva was born in Buenos Aires, Argentina, in 1971. He holds a Ph.D. in Chemistry from the University of Southern California and a Licentiate degree in Chemistry from the University of Buenos Aires. In 2001, he moved to Los Angeles (USA) and, after working 2 years in the private sector, he attended graduate studies at the Loker Hydrocarbon Research Institute at USC where he conducted his research on methanol, formic acid, microbial fuel cell and the electrochemical reduction of CO_2. In 2009 he joined the Fuel Cell Group at the National Commission of Atomic Energy (CNEA), where he currently holds a fellow research position. Since 2010 he is also an Associated Researcher at National Council of Scientific and Technology Research (CONICET). Some of his research interests are the electrochemistry of energy conversion and storage devices, including batteries and electrolyzer, and the electrochemical reduction of CO_2.

Index

H.R. Corti and E.R. Gonzalez (eds.), *Direct Alcohol Fuel Cells: Materials, Performance, Durability and Applications*, DOI 10.1007/978-94-007-7708-8,

MIX
Papier aus verantwortungsvollen Quellen
Paper from responsible sources
FSC® C105338

If you have any concerns about our products,
you can contact us on
ProductSafety@springernature.com

In case Publisher is established outside the EU,
the EU authorized representative is:
Springer Nature Customer Service Center GmbH
Europaplatz 3, 69115 Heidelberg, Germany

Printed by Libri Plureos GmbH
in Hamburg, Germany